ETHICS IN INFORMATION TECHNOLOGY

Fourth Edition

ETHICS IN INFORMATION TECHNOLOGY

Fourth Edition

George W. Reynolds
Strayer University

Australia • Brazil • Japan • Korea • Mexico • Singapore • Spain • United Kingdom • United States

COURSE TECHNOLOGY
CENGAGE Learning

Ethics in Information Technology, Fourth Edition
George W. Reynolds

Publisher: Joe Sabatino

Senior Acquisitions Editor: Charles McCormick, Jr.

Senior Product Manager: Kate Mason

Development Editor: Mary Pat Shaffer

Editorial Assistant: Courtney Bavaro

Marketing Director: Keri Witman

Marketing Manager: Adam Marsh

Senior Marketing Communications Manager: Libby Shipp

Marketing Coordinator: Suellen Ruttkay

Production Management and Composition: PreMediaGlobal

Rights Acquisition Specialist: John Hill

Media Editor: Chris Valentine

Senior Art Director: Stacy Jenkins Shirley

Manufacturing Planner: Julio Esperas

Cover Designer: Lou Ann Thesing

Cover Credit: ©iStock Photo

Library of Congress Control Number: 2011938705

ISBN-13: 978-1-111-53412-7

ISBN-10: 1-111-53412-8

Instructor Edition:

ISBN-13: 978-1-111-53413-4

ISBN-10: 1-111-53413-6

Course Technology
20 Channel Center Street
Boston, MA 02210
USA

Cengage Learning is a leading provider of customized learning solutions with office locations around the globe, including Singapore, the United Kingdom, Australia, Mexico, Brazil, and Japan. Locate your local office at: **www.cengage.com/global**

Printed in the United States of America
1 2 3 4 5 6 7 15 14 13 12 11

BRIEF CONTENTS

TABLE OF CONTENTS

PREFACE

We are excited to publish the fourth edition of *Ethics in Information Technology.* This new edition builds on the success of the previous editions and meets the need for a resource that helps readers understand many of the legal, ethical, and societal issues associated with information technology. We have responded to the feedback from our previous edition adopters, students, and other reviewers to create an improved text. We think you will be pleased with the results.

Ethics in Information Technology, Fourth Edition, fills a void of practical business information for business managers and IT professionals. The typical introductory information systems book devotes one chapter to ethics and IT, which cannot possibly cover the full scope of ethical issues related to IT. Such limited coverage does not meet the needs of business managers and IT professionals—the people primarily responsible for addressing ethical issues in the workplace. What is missing is an examination of the different ethical situations that arise in IT as well as practical advice for addressing these issues.

Ethics in Information Technology, Fourth Edition, has enough substance for an instructor to use it in a full-semester course in computer ethics. Instructors can also use the book as a reading supplement for such courses as Introduction to Management Information Systems, Principles of Information Technology, Managerial Perspective of Information Technology, Computer Security, E-Commerce, and so on.

WHAT'S NEW

Ethics in Information Technology, Fourth Edition, has been updated and revised to incorporate the many new developments and ethical issues that have arisen since the last edition. There is new or expanded coverage of the following topics: computer crime and cyberterrorism, the security risks of cloud computing, the use of computers to improve healthcare delivery, implementation of trustworthy computing, assessing the risks of IT, the pros and cons of protecting government data and documents, the role of the Department of Homeland Security in securing cyberspace, computer forensics, and green computing.

All opening vignettes and two-thirds of the end-of-chapter cases are new. Dozens of new real-world examples are presented in each chapter. At least 50 percent of the "Self-Assessment Questions," "Discussion Questions," and "What Would You Do?" exercises are new. Based on reviewer feedback, we have also increased the number of "Discussion Questions" and "What Would You Do?" exercises. Instructors of online courses frequently use these as the basis for discussion forums that allow online students to share a variety of perspectives and experiences and to create a learning community. Such discussions provide students the opportunity to more deeply understand the material while challenging their critical thinking skills. We think you will like these changes and additions.

ORGANIZATION

Each of the 10 chapters in this book addresses a different aspect of ethics in information technology:

- Chapter 1, "An Overview of Ethics," provides an introduction to ethics, ethics in business, and the relevance of discussing ethics in IT. The chapter defines the distinction between morals, ethics, and laws. It presents five reasons why practicing good business ethics is important in business, and it examines the role of the chief ethics officer and board of directors in establishing a strong organizational ethics program. The chapter also outlines the need for an organizational code of ethics and describes key steps in establishing a sound ethics program. It suggests a model for ethical decision making and also discusses four commonly used philosophical approaches to ethical decision making. The chapter ends with a discussion of the role of ethics in IT.
- Chapter 2, "Ethics for IT Workers and IT Users," explains the importance of ethics in the business relationships of IT professionals, including those between IT workers and employers, clients, suppliers, other professionals, IT users, and society. The chapter also emphasizes the significance of IT professional organizations and their codes of ethics, and it discusses the roles that certification and licensing can play in legitimizing professional standards. The chapter also points out the difficulties in licensing IT workers. The chapter touches on some ethical issues faced by IT users—including software piracy, inappropriate use of computing resources, and inappropriate sharing of information—and outlines actions that can be taken to support the ethical practices of IT users. The chapter introduces the topic of compliance and the role the audit committee and members of the internal audit team have in ensuring that both the IT organization and IT users are in compliance with organizational guidelines and policies, as well as various legal and regulatory practices.
- Chapter 3, "Computer and Internet Crime," describes the types of ethical decisions that IT professionals must make, as well as the business needs they must balance when dealing with security issues. The chapter identifies security issues associated with the use of cloud computing and virtualization software. It describes some of the more common hacker attacks, including viruses, worms, Trojan horses, distributed denial-of-service, rootkits, spam, phishing, spear-phishing, smishing, and vishing. In addition to providing a useful classification of computer crimes and their perpetrators, the chapter summarizes the major federal laws that address computer crime. The chapter outlines both how to implement trustworthy computing to manage security vulnerabilities and how to respond to specific security incidents to quickly resolve problems and improve ongoing security measures. A process for performing an assessment of an organization's computers and network from both internal and external threats is presented. The chapter discusses the need for a corporate security policy and offers a number of security-related policy

templates that can help an organization to quickly develop effective security policies.

- Chapter 4, "Privacy," explains how the use of IT affects privacy rights and discusses several key pieces of legislation that have addressed privacy rights over the years. The Fourth Amendment is explained, and laws designed to protect personal financial and health data—as well as the privacy of children—are discussed. Electronic surveillance is covered, along with many laws associated with this activity, including the Foreign Intelligence Surveillance Act and the USA Patriot Act. The chapter also covers the various regulations affecting the export of personal data from one country to another. The chapter explains how the personal information businesses gather using IT can be used to obtain or keep customers (or to monitor employees). It also discusses the concerns of privacy advocates regarding how much information can be gathered, with whom it can be shared, how the information is gathered in the first place, and how it is used. These concerns also extend to the information-gathering practices of law enforcement and government. Identity theft and data breaches are covered along with various tactics used by identity thieves; the chapter also presents some safeguards that can thwart identity theft. The expanding use of electronic discovery and consumer profiling is discussed, and guidelines and principles for treating consumer data responsibly are offered.
- Chapter 5, "Freedom of Expression," addresses issues raised by the growing use of the Internet as a means for freedom of expression, while also examining the types of speech protected by the First Amendment of the U.S. Constitution. The chapter opens with a discussion of WikiLeaks, and it goes on to cover the ways in which the ease and anonymity with which Internet users can communicate may pose problems for people who might be adversely affected by such communication. It describes attempts at using legislation (such as the Communications Decency Act, the Children Online Protection Act, and the Children's Internet Protection Act) and technology, such as Internet filtering, to control access to Internet content that is unsuitable for children or unnecessary in a business environment. The use of John Doe lawsuits to reveal the identities of anonymous posters is discussed. Defamation and hate speech, pornography on the Internet, corporate blogging, and spam are also covered.
- Chapter 6, "Intellectual Property," defines intellectual property and explains the varying degrees of ownership protection offered by copyright, patent, and trade secret laws. Copyright, patent, and trademark infringement are examined, using many examples. Key U.S. and international rules aimed at protecting intellectual property are discussed, including The Prioritizing Resources and Organization for Intellectual Property Act, the General Agreement on Tariffs and Trade, the World Trade Organization Agreement on Trade-Related Aspects of Intellectual Property Rights, the World Intellectual Property Organization Copyright Treaty, and the Digital Millennium Copyright Act. The chapter explains software patents and the use of cross-licensing agreements, as well as defensive publishing and patent trolls.

The use of nondisclosure agreements and noncompete clauses in work contracts is also discussed. Finally, the chapter covers several key issues relevant to ethics in IT, including plagiarism, reverse engineering of software, open source code, competitive intelligence gathering, and cybersquatting.

- Chapter 7, "Software Development," provides a thorough discussion of the software development process and the importance of software quality. The chapter covers issues software manufacturers must consider when deciding "how good is good enough?" with regard to their software products—particularly when the software is safety-critical and its failure can cause loss of human life. Topics include software product liability, risk analysis, and different approaches to quality assurance testing. The chapter also examines Capability Maturity Model Integration (CMMI), the ISO 9000 family of standards, and the Failure mode and effects analysis (FMEA) technique.
- Chapter 8, "The Impact of Information Technology on Productivity and the Quality of Life," examines the effect that IT has had on the standard of living and worker productivity in developed countries. The increase in the use of telework (also known as telecommuting) is discussed, as are the pros and cons of this work arrangement. The chapter also discusses the digital divide and profiles some programs designed to close that gap. The chapter closes with a look at IT's impact on the delivery of health care and healthcare costs.
- Chapter 9, "Social Networking," discusses how people use social networks, identifies common business uses of social networks, and examines many of the ethical issues associated with the use of social networks. The chapter also touches on virtual life communities and the ethical issues associated with virtual worlds.
- Chapter 10, "Ethics of IT Organizations," covers a range of ethical issues facing IT organizations, including those associated with the use of nontraditional workers, such as temporary workers, contractors, consulting firms, H-1B workers, and the use of outsourcing and offshore outsourcing. The chapter also discusses the risks, protections, and ethical decisions related to whistle-blowing, and it presents a process for safely and effectively handling a whistle-blowing situation. In addition to introducing the concept of green computing, the chapter discusses the ethical issues that both IT manufacturers and IT users face when a company is considering how to transition to green computing—and at what cost. It discusses the use of the Electronic Product Environment Assessment Tool to evaluate, compare, and select electronic products based on a set of 51 environmental criteria. Finally, the chapter examines a code of conduct for the electronics and information and communications technology (ICT) industries designed to address ethical issues in the areas of worker safety and fairness, environmental responsibility, and business efficiency.
- Appendix A provides an in-depth discussion of how ethics and moral codes developed over time. Appendices B through D consist of the codes of ethics for several important IT professional organizations. Appendix E provides answers to the end-of-chapter Self-Assessment Questions.

PEDAGOGY

Ethics in Information Technology, Fourth Edition, employs a variety of pedagogical features to enrich the learning experience and provide interest for the instructor and student:

- **Opening Quotation**. Each chapter begins with a quotation to stimulate interest in the chapter material.
- **Vignette**. At the beginning of each chapter, a brief real-world example illustrates the issues to be discussed and piques the reader's interest.
- **Questions to Consider**. Carefully crafted focus questions follow the vignette to further highlight topics that are covered in the chapter.
- **Learning Objectives**. Learning objectives appear at the start of each chapter. They are presented in the form of questions for students to consider while reading the chapter.
- **Key Terms**. Key terms appear in bold in the text and are listed at the end of the chapter. They are also defined in the glossary at the end of the book.
- **Manager's Checklist**. Each checklist provides a practical and useful list of questions to consider when making a business decision.

End-of-Chapter Material

To help students retain key concepts and expand their understanding of important IT concepts and relationships, the following sections are included at the end of every chapter:

- **Summary**. Each chapter includes a summary of the key issues raised. These items relate to the Learning Objectives for each chapter.
- **Self-Assessment Questions**. These questions help students review and test their understanding of key chapter concepts. The answers to the Self-Assessment Questions are included in Appendix E.
- **Discussion Questions**. These more open-ended questions help instructors generate class discussion to move students deeper into the concepts and help them explore the numerous aspects of ethics in IT.
- **What Would You Do?** These exercises present realistic dilemmas that encourage students to think critically about the ethical principles presented in the text.
- **Cases.** In each chapter, three real-world cases reinforce important ethical principles and IT concepts, and show how real companies have addressed ethical issues associated with IT. Questions after each case focus students on its key issues and ask them to apply the concepts presented in the chapter. A set of additional case studies from previous editions will be available at the Cengage Web site to provide the instructor with a wide range of cases from which to select.

ABOUT THE AUTHOR

George W. Reynolds brings a wealth of computer and industrial experience to this project, with more than 30 years of experience in government, institutional, and commercial IT organizations. He has authored over two dozen texts and has taught at the University of Cincinnati, Xavier University (Ohio), Miami University (Ohio), and the College of Mount St. Joseph. He is currently teaching at Strayer University.

Teaching Tools

The following supplemental materials are available when this book is used in a classroom setting. All of these tools are provided to the instructor on a single CD-ROM. You can also find some of these materials on the Cengage Learning Web site at www.cengage.com/sso.

- **Electronic Instructor's Manual.** The Instructor's Manual that accompanies this textbook includes additional instructional material to assist in class preparation, including suggestions for lecture topics. It also includes solutions to all end-of-chapter exercises.
- **ExamView®**. This textbook is accompanied by ExamView, a powerful testing software package that allows instructors to create and administer printed, computer (LAN-based), and Internet exams. ExamView includes hundreds of questions that correspond to the topics covered in this text, enabling students to generate detailed study guides that include page references for further review. The computer-based and Internet testing components allow students to take exams at their computers, and they save the instructor time by grading each exam automatically.
- **PowerPoint Presentations**. This book comes with Microsoft PowerPoint slides for each chapter. The slides can be included as a teaching aid for classroom presentation, made available to students on the network for chapter review, or printed for classroom distribution. The slides are fully customizable. Instructors can either add their own slides for additional topics they introduce to the class or delete slides they won't be covering.
- **Figure Files**. Figure files allow instructors to create their own presentations using figures taken directly from the text.
- **Blackboard® and WebCT™ Level 1 Online Content**. If you use Blackboard or WebCT, the test bank for this textbook is available at no cost in a simple, ready-to-use format.

ACKNOWLEDGMENTS

I wish to express my appreciation to a number of people who helped greatly in the creation of this book: Charles McCormick, Jr., Senior Acquisitions Editor, for his belief in and encouragement of this project; Jennifer Feltri and Divya Divakaran, Content Product Managers, for guiding the book through the production process; Kate Mason, Senior Product Manager, for overseeing and directing this effort; Mary Pat Shaffer, Development Editor, for her tremendous support, many useful suggestions, and helpful edits; Naomi

Friedman, for writing many of the vignettes and cases; Abigail Reip, our photo researcher; and my many students who provided excellent ideas and constructive feedback on the text. I also wish to thank Clancy Martin for writing Appendix A.

In addition, I want to thank an excellent set of reviewers who offered many useful suggestions:

Rochelle Brooks, Viterbo University
Barbara Klein, University of Michigan–Dearborn
David Koppy, Baker College of Auburn Hills
Michael Martel, Ohio University
Roosevelt Martin, Chicago State University
Frances Rampey, Wilson Community College
Carol Segura, Strayer University
Toni Somers, Wayne State University
Richard Smith, The University of Findlay

Last of all, thanks to my family for all their support, and for giving me the time to write this text.

—George W. Reynolds

ETHICS IN INFORMATION TECHNOLOGY

Fourth Edition

CHAPTER 1

AN OVERVIEW OF ETHICS

QUOTE

Ethics are so annoying. I avoid them on principle.
—Darby Conley, cartoonist, *Get Fuzzy* comic strip[1]

VIGNETTE

HP CEO Forced Out; Lands on Feet at Oracle

In early 2005, Mark Hurd was hired as the CEO of Hewlett-Packard Company (HP). Under Hurd's leadership from 2005 to 2009, the HP share price more than doubled, while its revenue grew over 40 percent to $115 billion.[2] This was accomplished by extreme cost-cutting measures, including the reduction of 50,000 jobs and the acquisition of several major technology firms, including Palm, 3Com, and Electronic Data Systems.

In June 2010, an HP contractor accused Hurd of sexual harassment. The contractor's work involved planning various HP-hosted CEO forums over the course of two years, and she often dined alone with Hurd following these events. She claimed that her work for HP stopped after she refused Hurd's advances.[3]

During the ensuing investigation of the sexual harassment charge, HP found evidence of false expense reports covering payments made to the woman. While HP executives said that the sexual harassment charge could not be substantiated, they did find that Hurd had violated HP's standards of business conduct. Michael Holston, HP executive vice president and general counsel, stated that Hurd's actions "demonstrated a profound lack of judgment that seriously undermined his creditability and damaged his effectiveness in leading HP."[4] The board urged Hurd to resign from the company.

Hurd, married with two children, denied making any advances toward the contractor and stated that he never prepared his own expense reports. He refused to resign and instead offered to reimburse the company the disputed expense payments.[5] After numerous discussions with members of the board, Hurd finally agreed to resign and to repay HP the disputed expense payments. He also settled the sexual harassment charges out of court, agreeing to pay the woman an undisclosed amount of his own money.[6]

Within a month of his resignation, computer technology giant Oracle announced that it had hired Mark Hurd as its new co-president. The hiring raised several issues as Hurd's severance package of $40 million included a confidential nondisclosure agreement restricting what Hurd could tell future employers about HP plans and operations.[7] HP filed a lawsuit asking the court to prevent Hurd from taking the job with Oracle, saying "HP is threatened with losing customers, technology, its competitive advantage, its trade secrets and goodwill in amounts which may be impossible to determine."[8]

Hiring disputes are common among technology companies. However, in California (where both Oracle and HP are headquartered), courts have encouraged employee mobility and allowed people who change jobs to continue working in their area of expertise.

Within a month, HP and Oracle settled the dispute and "reaffirmed their long-term strategic partnership." Hurd agreed to "adhere to his obligations to protect HP's confidential information while fulfilling his responsibilities at Oracle." In addition, Hurd agreed to waive his right to over 345,000 restricted shares of HP stock valued at $13.6 million.[9]

Questions to Consider

1. Does the hiring of Mark Hurd raise any ethical concerns about Oracle's business practices?
2. Should the fact that Mark Hurd left HP under disconcerting circumstances affect his future employment at other firms?

LEARNING OBJECTIVES

As you read this chapter, consider the following questions:

1. What is ethics, and why is it important to act according to a code of ethics?
2. Why is business ethics becoming increasingly important?
3. What are organizations doing to improve their business ethics?
4. Why are organizations interested in fostering good business ethics?
5. What approach can you take to ensure ethical decision making?
6. What trends have increased the risk of using information technology in an unethical manner?

WHAT IS ETHICS?

Every society forms a set of rules that establishes the boundaries of generally accepted behavior. These rules are often expressed in statements about how people should behave, and the individual rules fit together to form the **moral code** by which a society lives. Unfortunately, the different rules often have contradictions, and people are sometimes uncertain about which rule to follow. For instance, if you witness a friend copy someone else's answers while taking an exam, you might be caught in a conflict between loyalty to your friend and the value of telling the truth. Sometimes the rules do not seem to cover new situations, and an individual must determine how to apply existing rules or develop new ones. You may strongly support personal privacy, but do you think an organization should be prohibited from monitoring employees' use of its email and Internet services?

The term **morality** refers to social conventions about right and wrong that are so widely shared they become the basis for an established consensus. However, individual views of what is moral may vary by age, cultural group, ethnic background, religion, life experiences, education, and gender. There is widespread agreement on the immorality of murder, theft, and arson, but other behaviors that are accepted in one culture might be unacceptable in another. Even within the same society, people can have strong disagreements over important moral issues. In the United States, for example, issues such as abortion, stem cell research, the death penalty, and gun control are continuously debated, and people on both sides of these debates feel that their arguments are on solid moral ground.

Definition of Ethics

Ethics is a set of beliefs about right and wrong behavior within a society. Ethical behavior conforms to generally accepted norms—many of which are almost universal. However, although nearly everyone would agree that lying and cheating are unethical, opinions about what constitutes ethical behavior can vary dramatically. For example, attitudes toward **software piracy**—the practice of illegally making copies of software or enabling others to access software to which they are not entitled—range from strong opposition to acceptance of the practice as a standard approach to conducting business. In 2009,

43 percent of all software in circulation worldwide was pirated—at a cost of over $51 billion (USD).[10] The countries with the highest piracy rate were Georgia (95%), Zimbabwe (92%), Bangladesh (91%), Moldova (91%), and Armenia (90%). The lowest piracy rates were in the United States (20%), Japan (21%), Luxembourg (21%), and New Zealand (22%).[11]

As children grow, they learn complicated tasks—such as walking, talking, swimming, riding a bike, and writing the alphabet—that they perform out of habit for the rest of their lives. People also develop habits that make it easier to choose between what society considers good or bad. A **virtue** is a habit that inclines people to do what is acceptable, and a **vice** is a habit of unacceptable behavior. Fairness, generosity, and loyalty are examples of virtues, while vanity, greed, envy, and anger are considered vices. People's virtues and vices help define their personal value system—the complex scheme of moral values by which they live.

The Importance of Integrity

Your moral principles are statements of what you believe to be rules of right conduct. As a child, you may have been taught not to lie, cheat, or steal. As an adult facing more complex decisions, you often reflect on your principles when you consider what to do in different situations: Is it okay to lie to protect someone's feelings? Should you intervene with a coworker who seems to have a chemical dependency problem? Is it acceptable to exaggerate your work experience on a résumé? Can you cut corners on a project to meet a tight deadline?

A person who acts with **integrity** acts in accordance with a personal code of principles. One approach to acting with integrity—one of the cornerstones of ethical behavior—is to extend to all people the same respect and consideration that you expect to receive from others. Unfortunately, consistency can be difficult to achieve, particularly when you are in a situation that conflicts with your moral standards. For example, you might believe it is important to do as your employer requests while also believing that you should be fairly compensated for your work. Thus, if your employer insists that you do not report the overtime hours that you have worked due to budget constraints, a moral conflict arises. You can do as your employer requests or you can insist on being fairly compensated, but you cannot do both. In this situation, you may be forced to compromise one of your principles and act with an apparent lack of integrity.

Another form of inconsistency emerges if you apply moral standards differently according to the situation or people involved. To be consistent and act with integrity, you must apply the same moral standards in all situations. For example, you might consider it morally acceptable to tell a little white lie to spare a friend some pain or embarrassment, but would you lie to a work colleague or customer about a business issue to avoid unpleasantness? Clearly, many ethical dilemmas are not as simple as right versus wrong but involve choices between right versus right. As an example, for some people it is "right" to protect the Alaskan wildlife from being spoiled and also "right" to find new sources of oil to maintain U.S. oil reserves, but how do they balance these two concerns?

The Difference Between Morals, Ethics, and Laws

Morals are one's personal beliefs about right and wrong; the term *ethics* describes standards or codes of behavior expected of an individual by a group (nation, organization, profession) to which an individual belongs. For example, the ethics of the law profession demand that defense attorneys defend an accused client to the best of their ability, even if they know that the client is guilty of the most heinous and morally objectionable crime one could imagine.

Law is a system of rules that tells us what we can and cannot do. Laws are enforced by a set of institutions (the police, courts, law-making bodies). Legal acts are acts that conform to the law. Moral acts conform to what an individual believes to be the right thing to do. Laws can proclaim an act as legal, although many people may consider the act immoral—for example, abortion.

The remainder of this chapter provides an introduction to ethics in the business world. It discusses the importance of ethics in business, outlines what businesses can do to improve their ethics, provides advice on creating an ethical work environment, and suggests a model for ethical decision making. The chapter concludes with a discussion of ethics as it relates to information technology (IT).

ETHICS IN THE BUSINESS WORLD

Ethics has risen to the top of the business agenda because the risks associated with inappropriate behavior have increased, both in their likelihood and in their potential negative impact. In the past decade, we have watched the collapse and/or bailout of financial institutions such as Countrywide Financial, Lehman Brothers, and American International Group (AIG) due to unwise and unethical decision making regarding the approval of mortgages and lines of credit to unqualified individuals and organizations. We have also witnessed numerous corporate officers and senior managers sentenced to prison terms for their unethical behavior, including former investment broker Bernard Madoff, who bilked his clients out of an estimated $65 billion. Clearly, unethical behavior has led to serious negative consequences that have had a major global impact.

Several trends have increased the likelihood of unethical behavior. First, for many organizations, greater globalization has created a much more complex work environment that spans diverse cultures and societies, making it more difficult to apply principles and codes of ethics consistently. For example, numerous U.S. companies have moved operations to developing countries, where employees work in conditions that would not be acceptable in most developed parts of the world.

Second, in today's difficult and uncertain economic climate, organizations are extremely challenged to maintain revenue and profits. Some organizations are sorely tempted to resort to unethical behavior to maintain profits. For example, in January 2009, the chairman of the India-based outsourcing firm Satyam Computer Services admitted he had overstated the company's assets by more than $1 billion. The revelation represented India's largest ever corporate scandal and caused the government to step in to protect the jobs of the company's 53,000 employees.[12]

Employees, shareholders, and regulatory agencies are increasingly sensitive to violations of accounting standards, failures to disclose substantial changes in business conditions, nonconformance with required health and safety practices, and production of unsafe or substandard products. Such heightened vigilance raises the risk of financial loss for businesses that do not foster ethical practices or that run afoul of required standards. There is also a risk of criminal and civil lawsuits resulting in fines and/or incarceration for individuals.

A classic example of the many risks of unethical decision making can be found in the Enron accounting scandal. In 2000, Enron employed over 22,000 people and had annual revenue of $101 billion. During 2001, it was revealed that much of Enron's revenue was the result of deals with limited partnerships, which it controlled. In addition, as a result of faulty accounting, many of Enron's debts and losses were not reported in its financial statements. As the accounting scandal unfolded, Enron shares dropped from $90 per share to less than $1 per share, and the company was forced to file for bankruptcy in December 2001. The Enron case was notorious, but many other corporate scandals have recently occurred in spite of safeguards enacted as a result of the Enron debacle.[13] Here are just a few examples of lapses in business ethics by employees in IT organizations:

- In 2005, HP chairperson Patricia Dunn authorized an internal investigation of HP board members suspected of leaking information about HP's long-term strategy to the news media. Such information could have a significant impact on the competitiveness of the company and affect its share price. Three private detectives involved in the investigation allegedly engaged in pretexting—the use of false pretenses—to gain access to the telephone records of HP directors, certain employees, and nine journalists. Allegedly, they obtained and used the targeted individuals' Social Security numbers to impersonate those individuals in calls to the phone company, with the goal of obtaining private phone records. Eventually, Dunn and other members of the board resigned. Employees and shareholders were upset that a board member leaked confidential information about the firm and that unethical means were used to investigate the leaks.[14,15]
- CEO Sanjay Kumar and several other executives at Computer Associates International (CA) were implicated in charges of falsely reporting hundreds of millions of dollars in revenue from software licensing agreements before the deals were finalized. This was done to give the impression that CA was successful in meeting optimistic revenue and earnings growth targets so the stock would be attractive to investors. Kumar was forced to resign from the board, charged with securities fraud, convicted, and sentenced to 12 years in jail. He was also ordered to repay $798 million in restitution.[16]
- In 2010, former IBM executive Robert Moffat confessed to providing confidential tech industry information to an alleged insider trader with whom he was having an affair. The woman passed on this information to traders at multiple hedge funds; ultimately, numerous other individuals in the tech and financial industries were also charged with insider trading. Moffat was fined $50,000 and sentenced to six months of prison with two years of post-release supervision. One tidbit that Moffat shared related to plans by Advanced Micro Devices (AMD) to set up a separate company for its chip manufacturing and to partially finance the venture with funds from Abu Dhabi investors. Moffat was aware of the plan because IBM had a licensing agreement with AMD.[17]

Even lower-level employees can find themselves in the middle of ethical dilemmas, as these examples illustrate:

- A low-level government employee was the first customer at a new Dunkin' Donuts. He was asked for his name and address, and he provided the store manager a state business card. Subsequently, the employee received a letter from the president of the company that included a $100 gift card. The employee checked with one member of the legal department and was told that he must return the card. Later, another member of the same department told him he could keep the card. This employee was definitely caught in an ethical dilemma due to conflicting advice from his company's legal department.[18]

- A staff member on the U.S. House Ethics Committee accidently leaked a sensitive document containing the names of over 30 lawmakers under investigation by the Ethics Committee and Office of Congressional Ethics. The staffer was working at home on a computer with file-sharing software that was not secure. The staff member's innocent mistake gave others the opportunity to access the document without authorization. The list eventually found its way to the *Washington Post*.[19]

This is just a small sample of the scandals that have led to an increased focus on business ethics within many IT organizations.

Why Fostering Good Business Ethics Is Important

Organizations have at least five good reasons for promoting a work environment in which employees are encouraged to act ethically when making business decisions:

- Gaining the goodwill of the community
- Creating an organization that operates consistently
- Fostering good business practices
- Protecting the organization and its employees from legal action
- Avoiding unfavorable publicity

Gaining the Goodwill of the Community

Although organizations exist primarily to earn profits or provide services to customers, they also have some fundamental responsibilities to society. Often they declare these responsibilities in a formal statement of their company's principles or beliefs. See Figure 1-1 for an example.

Our Values

As a company, and as individuals, we value integrity, honesty, openness, personal excellence, constructive self-criticism, continual self-improvement, and mutual respect. We are committed to our customers and partners and have a passion for technology. We take on big challenges, and pride ourselves on seeing them through. We hold ourselves accountable to our customers, shareholders, partners, and employees by honoring our commitments, providing results, and striving for the highest quality.

FIGURE 1-1 Microsoft's statement of values

Credit: Microsoft Statement of Values, "Our Values," from www.microsoft.com. Reprinted by permission.

All successful organizations, including technology firms, recognize that they must attract and maintain loyal customers. Philanthropy is one way in which an organization can demonstrate its values in action and to make a positive connection with its stakeholders. (A **stakeholder** is someone who stands to gain or lose, depending on how a situation is resolved.) As a result, many organizations initiate or support socially responsible activities, which may include making contributions to charitable organizations and nonprofit institutions, providing benefits for employees in excess of any legal requirements, and devoting organizational resources to initiatives that are more socially desirable than profitable. Table 1-1 provides a few examples of some of the many socially responsible activities supported by major IT organizations.

TABLE 1-1 Examples of IT organizations' socially responsible activities

Organization	Examples of socially responsible activities
Cisco Systems, Inc.	One of the core values at Cisco is empowering people to help themselves; Cisco supports that value by donating networking equipment to various nonprofit organizations. The firm also encourages employee volunteerism, resulting in employees working over 148,000 volunteer hours.[20,21]
Comcast Corporation	The company made charitable contributions of more than $406 million in 2009, representing 10% of 2008 profits.[22]
Dell Inc.	Committed to a goal of giving 1% of pretax profits to charitable causes; during 2010, Dell team members volunteered over 201,000 hours. [23,24]
Google	The company's Corporate Giving Council oversees millions of dollars in scholarships, and the firm supports employees' volunteerism by sponsoring an annual service event called GoogleServe.[25]
IBM	IBM's Employee Charitable Contributions Campaign is a long-standing tradition that in recent years has exceeded $166 million in giving.[26]
Oracle	Oracle's charitable contributions for 2009 amount to over $2 billion—or 9.3% of its 2008 profit.[27]
SAP, North America	The company supports several major corporate responsibility initiatives aimed at improving education, matches employee gifts to nonprofit agencies and schools, and encourages and supports employee volunteerism.[28]

Source Line: Copyright © Cengage Learning. Adapted from multiple sources. See End Notes 20, 21, 22, 23, 24, 25, 26, 27, 28.

The goodwill that socially responsible activities create can make it easier for corporations to conduct their business. For example, a company known for treating its employees well will find it easier to compete for the best job candidates. On the other hand, companies viewed as harmful to their community may suffer a disadvantage. For example, a corporation that pollutes the environment (see Figure 1-2) may find that adverse publicity reduces sales, impedes relationships with some business partners, and attracts unwanted government attention.

FIGURE 1-2 Corporations that pollute the environment place themselves at a disadvantage
Credit: © Antonio V. Oquias/Shutterstock.com

Creating an Organization That Operates Consistently

Organizations develop and abide by values to create an organizational culture and to define a consistent approach for dealing with the needs of their stakeholders—shareholders, employees, customers, suppliers, and the community. Such consistency ensures that employees know what is expected of them and can employ the organization's values to help them in their decision making. Consistency also means that shareholders, customers, suppliers, and the community know what they can expect of the organization—that it will behave in the future much as it has in the past. It is especially important for multinational or global organizations to present a consistent face to their shareholders, customers, and suppliers no matter where those stakeholders live or operate their business. Although each company's value system is different, many share the following values:

- Operate with honesty and integrity, staying true to organizational principles
- Operate according to standards of ethical conduct, in words and action
- Treat colleagues, customers, and consumers with respect
- Strive to be the best at what matters most to the organization
- Value diversity
- Make decisions based on facts and principles

Fostering Good Business Practices

In many cases, good ethics can mean good business and improved profits. Companies that produce safe and effective products avoid costly recalls and lawsuits. Companies that provide excellent service retain their customers instead of losing them to competitors. Companies that develop and maintain strong employee relations enjoy lower turnover rates and better employee morale. Suppliers and other business partners often place a priority on working with companies that operate in a fair and ethical manner. All these factors tend to increase revenue and profits while decreasing expenses. As a result, ethical companies should have a favorable stock price performance over unethical companies.

Figure 1-3 compares the stock price index for the 2010 world's 100 most ethical companies (as determined by the research-based Ethisphere Institute, an international think tank dedicated to the creation, advancement, and sharing of best practices in business ethics, corporate social responsibility, anticorruption, and sustainability) versus two other broadly used stock price indexes—the Standard and Poor 500 stocks and the Financial Times and Stock Exchange 100 stocks. Although this graph seems to confirm the superior stock price performance of ethical companies (shown as the WME Index), the time period involved is relatively short.

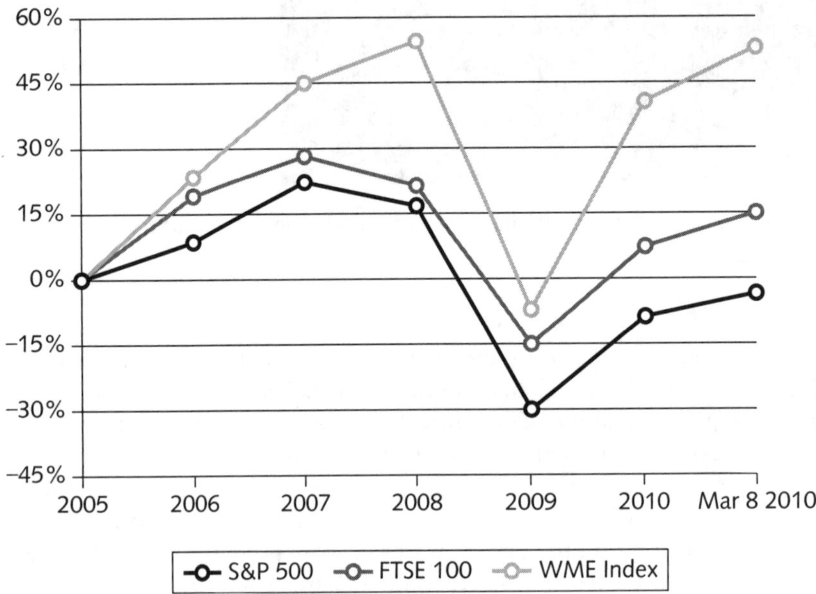

FIGURE 1-3 Stock performance of the world's 100 most ethical companies

Credit: Ethisphere Magazine, Quarter 1, 2010.

In coming up with its fourth annual list, Ethisphere reviewed nominations from companies in more than 100 countries and 36 industries. "The methodology for the World's Most Ethical Companies includes reviewing codes of ethics, litigation, and regulatory infraction histories; evaluating the investment in innovation and sustainable business practices; looking at activities designed to improve corporate citizenship; and studying nominations from senior executives, industry peers, suppliers and customers."[29]

On the other hand, bad ethics can lead to bad business results. Bad ethics can have a negative impact on employees, many of whom may develop negative attitudes if they perceive a difference between their own values and those stated or implied by an organization's actions. In such an environment, employees may suppress their tendency to act in a manner that seems ethical to them and instead act in a manner that will protect them against anticipated punishment. When such a discrepancy between employee and organizational ethics occurs, it destroys employee commitment to organizational goals and objectives, creates low morale, fosters poor performance, erodes employee involvement in organizational improvement initiatives, and builds indifference to the organization's needs.

Protecting the Organization and Its Employees from Legal Action

In a 1909 ruling (*United States v. New York Central & Hudson River Railroad Co.*), the U.S. Supreme Court established that an employer can be held responsible for the acts of its employees even if the employees act in a manner contrary to corporate policy and their employer's directions.[30] The principle established is called *respondeat superior*, or "let the master answer."

In October 2010, the CEO and the general counsel of IT solutions and services provider GTSI Corporation were forced by the Small Business Administration (SBA) to resign, while three other top GTSI executives were suspended over allegations that employees of GTSI were involved in a scheme with its contracting partners that resulted in the firm receiving money set aside for small businesses. GTSI, which has over 500 employees and revenue over $760 million, was providing services to the Department of Homeland Security in partnership with contractors who qualified as small businesses, but GTSI—as a subcontractor—was actually performing most of the services and being paid most of the fees.[31] In this case, top executives were punished for the acts of several unidentified employees.

A coalition of several legal organizations, including the Association of Corporate Counsel, the U.S. Chamber of Commerce, the National Association of Manufacturers, the National Association of Criminal Defense Lawyers, and the New York State Association of Criminal Defense Lawyers, argues that organizations should "be able to escape criminal liability if they have acted as responsible corporate citizens, making strong efforts to prevent and detect misconduct in the workplace."[32] One way to do this is to establish effective ethics and compliance programs. However, some people argue that officers of companies should not be given light sentences if their ethics programs fail to deter criminal activity within their firms.

Avoiding Unfavorable Publicity

The public reputation of a company strongly influences the value of its stock, how consumers regard its products and services, the degree of oversight it receives from government agencies, and the amount of support and cooperation it receives from its business partners. Thus, many organizations are motivated to build a strong ethics program to avoid negative publicity. If an organization is perceived as operating ethically, customers, business partners, shareholders, consumer advocates, financial institutions, and regulatory bodies will usually regard it more favorably.

Improving Corporate Ethics

Research by the Ethics Resource Center (ERC) found that only one in four organizations has a well-implemented ethics and compliance program. The ERC has defined the following characteristics of a successful ethics program:

- Employees are willing to seek advice about ethics issues.
- Employees feel prepared to handle situations that could lead to misconduct.
- Employees are rewarded for ethical behavior.
- The organization does not reward success obtained through questionable means.
- Employees feel positively about their company.[33]

The risk of unethical behavior is increasing, so the improvement of business ethics is becoming more important. The following sections explain some of the actions corporations can take to improve business ethics.

In its 2009 National Business Ethics Survey, based on responses from over 3,000 individuals, the ERC found evidence of some improvement in ethics in the workplace as summarized in Table 1-2.[34]

TABLE 1-2 Conclusions from the 2009 National Business Ethics Survey

Finding	2007 survey results	2009 survey results
Employees who said they witnessed misconduct on the job	56%	49%
Employees who said they reported misconduct when they saw it	58%	63%
Employees who felt pressure to commit an ethics violation	10%	8%
ERC's measure of strength of ethical culture in the workplace	53%	62%

Source Line: Ethics Resource Center, "2009 National Business Ethics Survey, Ethics in Recession," © 2009, www.ethics.com/research/nbes.com.

Appointing a Corporate Ethics Officer

A **corporate ethics officer** (also called a **corporate compliance officer**) provides an organization with vision and leadership in the area of business conduct. This individual "aligns the practices of a workplace with the stated ethics and beliefs of that workplace, holding people accountable to ethical standards."[35]

Organizations send a clear message to employees about the importance of ethics and compliance in their decision about who will be in charge of the effort and to whom that individual will report. Ideally, the corporate ethics officer should be a well-respected, senior-level manager who reports directly to the CEO. Ethics officers come from diverse backgrounds, such as legal staff, human resources, finance, auditing, security, or line operations.

Not surprisingly, a rapid increase in the appointment of corporate ethics officers typically follows the revelation of a major business scandal. The first flurry of appointments began following a series of defense-contracting scandals during the administration of Ronald Reagan in the late 1980s—when firms used bribes to gain inside information that they could use to improve their contract bids. A second spike in appointments came in the early 1990s, following the new federal sentencing guidelines that stated that "companies with effective compliance and ethics programs could receive preferential treatment during prosecutions for white-collar crimes."[36] A third surge followed the myriad accounting scandals of the early 2000s. Another increase in appointments can be expected in the aftermath of the mortgage loan scandals uncovered beginning in 2008.

The ethics officer position has its critics. Many are concerned that if one person is appointed head of ethics, others in the organization may think they have no responsibility in this area. On the other hand, Odell Guyton—who has been the director of compliance at Microsoft for a decade—feels a point person for ethics is necessary, otherwise "how are you going to make sure it's being done, when people have other core responsibilities? That doesn't mean it's on the shoulders of the compliance person alone."[37]

Typically the ethics officer tries to establish an environment that encourages ethical decision making through the actions described in this chapter. Specific responsibilities include the following:

- Responsibility for compliance—that is, ensuring that ethical procedures are put into place and consistently adhered to throughout the organization
- Responsibility for creating and maintaining the ethics culture that the highest level of corporate authority wishes to have
- Responsibility for being a key knowledge and contact person on issues relating to corporate ethics and principles[38]

Unfortunately, simply naming a corporate ethics officer does not automatically improve an organization's ethics; hard work and effort are required to establish and provide ongoing support for an organizational ethics program.

Ethical Standards Set by Board of Directors

The board of directors is responsible for the careful and responsible management of an organization. In a for-profit organization, the board's primary objective is to oversee the organization's business activities and management for the benefit of all stakeholders, including shareholders, employees, customers, suppliers, and the community. In a nonprofit organization, the board reports to a different set of stakeholders—in particular, the local community that the nonprofit serves.

A board of directors fulfills some of its responsibilities directly and assigns others to various committees. The board is not normally responsible for day-to-day management and operations; these responsibilities are delegated to the organization's management team. However, the board is responsible for supervising the management team.

Board members are expected to conduct themselves according to the highest standards for personal and professional integrity, while setting the standard for company-wide ethical conduct and ensuring compliance with laws and regulations. Employees will "get the message" if board members set an example of high-level ethical behavior. If they don't set a good example, employees will get that message as well. Importantly, board members must create an environment in which employees feel they can seek advice about appropriate business conduct, raise issues, and report misconduct through appropriate channels.

Establishing a Corporate Code of Ethics

A **code of ethics** is a statement that highlights an organization's key ethical issues and identifies the overarching values and principles that are important to the organization and its decision making. Codes frequently include a set of formal, written statements about the purpose of an organization, its values, and the principles that should guide its employees' actions. An organization's code of ethics applies to its directors, officers, and employees, and it should focus employees on areas of ethical risk relating to their role in the organization, offer guidance to help them recognize and deal with ethical issues, and provide mechanisms for reporting unethical conduct and fostering a culture of honesty and accountability within the organization. An effective code of ethics helps ensure that employees abide by the law, follow necessary regulations, and behave in an ethical manner.

The **Sarbanes–Oxley Act of 2002** was passed in response to public outrage over several major accounting scandals, including those at Enron, WorldCom, Tyco, Adelphia,

Global Crossing, and Qwest—plus numerous restatements of financial reports by other companies, which clearly demonstrated a lack of oversight within corporate America. The goal of the bill was to renew investors' trust in corporate executives and their firm's financial reports. The act led to significant reforms in the content and preparation of disclosure documents by public companies. However, the Lehman Brothers accounting fiasco and resulting collapse as well as other similar examples raise questions about the effectiveness of Sarbanes–Oxley in preventing accounting scandals.[39]

Section 404 of the act states that annual reports must contain a statement signed by the CEO and CFO attesting that the information contained in all of the firm's Securities and Exchange Commission (SEC) filings is accurate. The company must also submit to an audit to prove that it has controls in place to ensure accurate information. The penalties for false attestation can include up to 20 years in prison and significant monetary fines for senior executives. Section 406 of the act also requires public companies to disclose whether they have a code of ethics and to disclose any waiver of the code for certain members of senior management. The SEC also approved significant reforms by the NYSE and NASDAQ that, among other things, require companies listed with them to have codes of ethics that apply to all employees, senior management, and directors.

A code of ethics cannot gain company-wide acceptance unless it is developed with employee participation and fully endorsed by the organization's leadership. It must also be easily accessible by employees, shareholders, business partners, and the public. The code of ethics must continually be applied to a company's decision making and emphasized as an important part of its culture. Breaches in the code of ethics must be identified and dealt with appropriately so the code's relevance is not undermined.

Each year, *Corporate Responsibility* magazine rates U.S. publicly held companies using a statistical analysis of corporate ethical performance in several categories. (For 2010, the categories were environment, climate change, human rights, employee relations, governance, philanthropy, and financial.) Intel Corporation, the world's largest chip maker, has been rated in the top 25 every year since the list began in 2000, and was rated #2 in 2010.[40] As such, Intel is recognized as one of the most ethical companies in the IT industry. A summary of Intel's code of ethics is shown in Figure 1-4. A more detailed version is spelled out in a 22-page document (Intel Code of Conduct, November 2009, found at *www.intel.com/assets/PDF/Policy/code-of-conduct.pdf*), which offers employees guidelines designed to deter wrongdoing, promote honest and ethical conduct, and comply with applicable laws and regulations. Intel's Code of Conduct also expresses its policies regarding the environment, health and safety, intellectual property, diversity, nondiscrimination, supplier expectations, privacy, and business continuity.

1. Intel conducts business with honesty and integrity
2. Intel follows the letter and spirit of the law
3. Intel employees treat each other fairly
4. Intel employees act in the best interests of Intel and avoid conflicts of interest
5. Intel employees protect the company's assets and reputation

FIGURE 1-4 Intel's five principles of conduct

Credit: Five Principals of Conduct. © Intel Corporation. Reprinted by permission.

Conducting Social Audits

An increasing number of organizations conduct social audits of their policies and practices. In a **social audit**, an organization reviews how well it is meeting its ethical and social responsibility goals and communicates its new goals for the upcoming year. This information is shared with employees, shareholders, investors, market analysts, customers, suppliers, government agencies, and the communities in which the organization operates. For example, each year Intel prepares its Corporate Responsibility Report, which summarizes the firm's progress toward meeting its ethical and social responsibility goals. Goals are set for improvement in environment, workplace, supply chain, community, and education. In 2009, Intel set specific five-year goals in the areas of environment related to global-warming emissions, energy consumption, water use, chemical and solid waste reduction, and product energy efficiency.[41]

Requiring Employees to Take Ethics Training

The ancient Greek philosophers believed that personal convictions about right and wrong behavior could be improved through education. Today, most psychologists agree with them. Lawrence Kohlberg, the late Harvard psychologist, found that many factors stimulate a person's moral development, but one of the most crucial is education. Other researchers have repeatedly supported the idea that people can continue their moral development through further education, such as working through case studies and examining contemporary issues.

Thus, an organization's code of ethics must be promoted and continually communicated within the organization, from top to bottom. Organizations can do this by showing employees examples of how to apply the code of ethics in real life. One approach is through a comprehensive ethics education program that encourages employees to act responsibly and ethically. Such programs are often presented in small workshop formats in which employees apply the organization's code of ethics to hypothetical but realistic case studies. Employees may also be given examples of recent company decisions based on principles from the code of ethics. It is critical that such training increase the percentage of employees who report incidents of misconduct; thus, employees must be shown effective ways of reporting such incidents. In addition, they must be reassured that such feedback will be acted on and that they will not be subjected to retaliation.

In its 2009 National Business Ethics Survey, the Ethics Resource Center reported that 46 percent of all complaints are reported to an employee's direct supervisor. "Because they are the 'eyes and ears' of the company, these supervisors need adequate resources, support, and training to address the stress created by and the additional misconduct related to the implementation of company tactics" according to the ERC.[42]

Motorola is committed to a strong corporate ethics training program to ensure that its employees conduct its business with integrity. The focus of the training is to clarify corporate values and policies and to encourage employees to report ethical concerns via numerous reporting channels. Motorola investigates all allegations of ethical misconduct. It will take appropriate disciplinary actions if a claim is proven, up to and including dismissal of all involved employees. All salaried employees must complete an online introduction to the ethics program every three years. All managers in newly acquired businesses or high-risk locations must take further classroom ethics training. In 2009, virtual dialogue sessions focusing on trade compliance, insider trading, fraud, gifts and entertainment, and third-party due diligence were delivered to 775 employees via 10 webinars in the United States and Asia. Ethics Quiz Competition events were held in Korea, India, and Thailand. The goal of these events was to raise employee awareness of Motorola's commitment to ethics standards. These training requirements resulted in over 43,000 hours of ethics training in 2009.[43]

Formal ethics training not only makes employees more aware of a company's code of ethics and how to apply it, but also demonstrates that the company intends to operate in an ethical manner. The existence of formal training programs can also reduce a company's liability in the event of legal action.

Including Ethical Criteria in Employee Appraisals

Managers can ensure that employees are meeting performance expectations by monitoring employee behavior and providing feedback; however, a recent survey of HR professionals revealed that only 43 percent of organizations include ethical conduct as part of an employee's performance appraisal.[44] Those that do so base a portion of their employees' performance evaluations on treating others fairly and with respect; operating effectively in a multicultural environment; accepting personal accountability for meeting business needs; continually developing others and themselves; and operating openly and honestly with suppliers, customers, and other employees. These factors are considered along with the more traditional criteria used in performance appraisals, such as an employee's overall contribution to moving the business ahead, successful completion of projects and tasks, and maintenance of good customer relations.

Creating an Ethical Work Environment

Most employees want to perform their jobs successfully and ethically, but good employees sometimes make bad ethical choices. Employees in highly competitive workplaces often feel pressure from aggressive competitors, cutthroat suppliers, unrealistic budgets, unforgiving quotas, tight deadlines, and bonus incentives. Employees may also be encouraged to do "whatever it takes" to get the job done. In such environments, some employees may feel pressure to engage in unethical conduct to meet management's expectations, especially if the organization has no corporate code of ethics and no strong examples of senior management practicing ethical behavior.

Here are a few examples of how managerial behavior can encourage unethical employee behavior.

- A manager sets and holds people accountable to meet "stretch" goals, quotas, and budgets, causing employees to think, "My boss wants results, not excuses, so I have to cut corners to meet the goals my boss has set."
- A manager fails to provide a corporate code of ethics and operating principles to make decisions so employees think, "Because the company has not established any guidelines, I don't think my conduct is really wrong or illegal."
- A manager fails to act in an ethical manner and instead sets a poor example for others to follow so employees think, "I have seen other successful people take unethical actions and not suffer negative repercussions."
- Managers fail to hold people accountable for unethical actions so employees think, "No one will ever know the difference, and if they do, so what?"

- Managers put a 3-inch-thick binder entitled "Corporate Business Ethics, Policies, and Procedures" on the desks of new employees and tell them to "read it when you have time and sign the attached form that says you read and understand the corporate policy." Employees think, "This is overwhelming. Can't they just give me the essentials? I can never absorb all this."

Table 1-3 provides a manager's checklist for establishing an ethical workplace. The preferred answer to each question is *yes*.

TABLE 1-3 Manager's checklist for establishing an ethical work environment

Question	Yes	No
Does your organization have a code of ethics?		
Do employees know how and to whom to report any infractions of the code of ethics?		
Do employees feel that they can report violations of the code of ethics safely and without fear of retaliation?		
Do employees feel that action will be taken against those who violate the code of ethics?		
Do senior managers set an example by communicating the code of ethics and using it in their own decision making?		
Do managers evaluate and provide feedback to employees on how they operate with respect to the values and principles in the code of ethics?		
Are employees aware of sanctions for breaching the code of ethics?		
Do employees use the code of ethics in their decision making?		

Source Line: Course Technology/Cengage Learning.

Employees must have a knowledgeable resource with whom they can discuss perceived unethical practices. For example, Intel expects employees to report suspected violations of its code of conduct to a manager, the Legal or Internal Audit Departments, or a business unit's legal counsel. Employees can also report violations anonymously through an internal Web site dedicated to ethics. Senior management at Intel has made it clear that any employee can report suspected violations of corporate business principles without fear of reprisal or retaliation.

Including Ethical Considerations in Decision Making

We are all faced with difficult decisions in our work and in our personal life. Most of us have developed a decision-making process that we execute automatically, without thinking about the steps we go through. For many of us, the process generally follows the steps outlined in Figure 1-5.

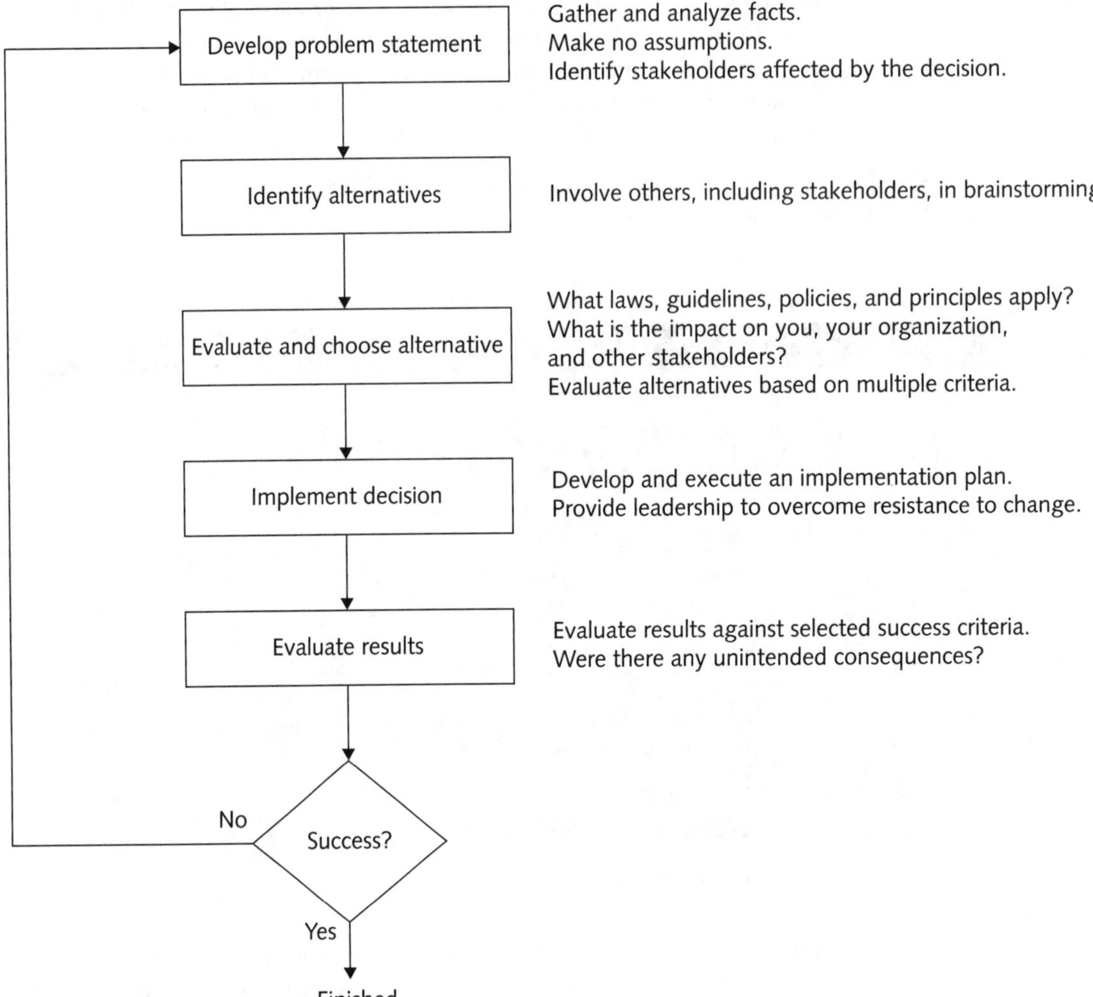

FIGURE 1-5 Decision-making process

Source Line: Course Technology/Cengage Learning.

The following sections discuss this decision-making process further and point out where and how ethical considerations need to be brought into the process.

Develop a Problem Statement

A **problem statement** is a clear, concise description of the issue that needs to be addressed. A good problem statement answers the following questions: What do people observe that causes them to think there is a problem? Who is directly affected by the problem? Is there anyone else affected? How often does the problem occur? What is the impact of the problem? How serious is the problem? Development of a problem statement is the most critical step in the decision-making process. Without a clear statement of the problem or the decision to be made, it is useless to proceed. Obviously, if the problem is stated incorrectly, the decision will not solve the problem.

You must gather and analyze facts to develop a good problem statement. Seek information and opinions from a variety of people to broaden your frame of reference. During this process, you must be extremely careful not to make assumptions about the situation. Simple situations can sometimes turn into complex controversies because no one takes the time to gather the facts. For example, you might see your boss receive what appears to be an employment application from a job applicant and then throw the application into the trash after the applicant leaves. This would violate your organization's policy to treat each applicant with respect and to maintain a record of all applications for one year. You could report your boss for failure to follow the policy, or you could take a moment to speak directly to your boss. You might be pleasantly surprised to find out that the situation was not as it appeared. Perhaps the "applicant" was actually a salesperson promoting a product for which your company had no use, and the "application" was marketing literature.

Part of developing a good problem statement involves identifying the stakeholders and their positions on the issue. Stakeholders often include others beyond those directly involved in an issue. Identifying the stakeholders helps you understand the impact of your decision and could help you make a better decision. Unfortunately, it may also cause you to lose sleep from wondering how you might affect the lives of others. However, by involving stakeholders in the decision, you can work to gain their support for the recommended course of action. What is at stake for each stakeholder? What does each stakeholder value, and what outcome does each stakeholder want? Do some stakeholders have a greater stake because they have special needs or because the organization has special obligations to them? To what degree should they be involved in the decision?

The following list includes one example of a good problem statement as well as two examples of poor problem statements:

- Good problem statement: Our product supply organization is continually running out of stock of finished products, creating an out-of-stock situation on over 15 percent of our customer orders, resulting in over $300,000 in lost sales per month.
- Poor problem statement: We need to implement a new inventory control system. (This is a possible solution, not a problem statement.)
- Poor problem statement: We have a problem with finished product inventory. (This is not specific enough.)

Identify Alternatives

During this stage of decision making, it is ideal to enlist the help of others, including stakeholders, to identify several alternative solutions to the problem. Brainstorming with others will increase your chances of identifying a broad range of alternatives and determining the best solution. On the other hand, there may be times when it is inappropriate to involve others in solving a problem that you are not at liberty to discuss. In providing participants information about the problem to be solved, offer just the facts, without your opinion, so you don't influence others to accept your solution.

During any brainstorming process, try not to be critical of ideas, as any negative criticism will tend to shut down the discussion, and the flow of ideas will dry up. Simply write down the ideas as they are suggested.

Evaluate and Choose an Alternative

Once a set of alternatives has been identified, the group must evaluate them based on numerous criteria, such as effectiveness at addressing the issue, the extent of risk associated with each alternative, cost, and time to implement. An alternative that sounds attractive but that is not feasible will not help solve the problem.

As part of the evaluation process, weigh various laws, guidelines, and principles that may apply. You certainly do not want to violate a law that can lead to a fine or imprisonment for yourself or others. Are there any corporate policies or guidelines that apply? Does the organizational code of ethics offer guidance? Do any of your own personal principles apply?

Also consider the likely consequences of each alternative from several perspectives: What is the impact on you, your organization, other stakeholders (including your suppliers and customers), and the environment?

The alternative selected should be ethically and legally defensible; be consistent with the organization's policies and code of ethics; take into account the impact on others; and, of course, provide a good solution to the problem.

Philosophers have developed many approaches to aid in ethical decision making. Four of the most common approaches, summarized in Table 1-4 and discussed in the following sections, provide a framework for decision makers to reflect on the acceptability of their actions and evaluate their moral judgments. People must find the appropriate balance among all applicable laws, corporate principles, and moral guidelines to help them make decisions. (For a more in-depth discussion of ethics and moral codes, see Appendix A.)

TABLE 1-4 Four common approaches to ethical decision making

Approach to dealing with moral issues	Principle
Virtue ethics approach	The ethical choice best reflects moral virtues in yourself and your community.
Utilitarian approach	The ethical choice produces the greatest excess of benefits over harm.
Fairness approach	The ethical choice treats everyone the same and shows no favoritism or discrimination.
Common good approach	The ethical choice advances the common good.

Source Line: Course Technology/Cengage Learning.

Virtue Ethics Approach

The **virtue ethics approach** to decision making focuses on how you should behave and think about relationships if you are concerned with your daily life in a community. It does not define a formula for ethical decision making, but suggests that when faced with a complex ethical dilemma, people do either what they are most comfortable doing or what they think a person they admire would do. The assumption is that people are guided by their virtues to reach the "right" decision. A proponent of virtue ethics believes that a disposition to do the right thing is more effective than following a set of principles and rules, and that people should perform moral acts out of habit, not introspection.

Virtue ethics can be applied to the business world by equating the virtues of a good businessperson with those of a good person. However, businesspeople face situations that are peculiar to a business setting, so they may need to tailor their ethics accordingly. For example, honesty and openness when dealing with others are generally considered virtues; however, a corporate purchasing manager who is negotiating a multimillion dollar deal might need to be vague in discussions with potential suppliers.

A problem with the virtue ethics approach is that it doesn't provide much of a guide for action. The definition of *virtue* cannot be worked out objectively; it depends on the circumstances—you work it out as you go. For example, bravery is a great virtue in many circumstances, but in others it may be foolish. The right thing to do in a situation also depends on which culture you're in and what the cultural norm dictates.

Utilitarian Approach

The **utilitarian approach** to ethical decision making states that you should choose the action or policy that has the best overall consequences for all people who are directly or indirectly affected. The goal is to find the single greatest good by balancing the interests of all affected parties.

Utilitarianism fits easily with the concept of value in economics and the use of cost-benefit analysis in business. Business managers, legislators, and scientists weigh the benefits and harm of policies when deciding whether to invest resources in building a new plant in a foreign country, to enact a new law, or to approve a new prescription drug.

A complication of this approach is that measuring and comparing the values of certain benefits and costs is often difficult, if not impossible. How do you assign a value to human life or to a pristine wildlife environment? It can also be difficult to predict the full benefits and harm that result from a decision.

Fairness Approach

The **fairness approach** focuses on how fairly actions and policies distribute benefits and burdens among people affected by the decision. The guiding principle of this approach is to treat all people the same. However, decisions made with this approach can be influenced by personal bias, without the decision makers even being aware of their bias. If the intended goal of an action or a policy is to provide benefits to a target group, other affected groups may consider the decision unfair.

Common Good Approach

The **common good approach** to decision making is based on a vision of society as a community whose members work together to achieve a common set of values and goals. Decisions and policies that use this approach attempt to implement social systems, institutions, and environments that everyone depends on and that benefit all people. Examples include an effective education system, a safe and efficient transportation system, and accessible and affordable health care.

As with the other approaches to ethical decision making, there are complications with the common good approach. People clearly have different ideas about what constitutes the common good, which makes consensus difficult. In addition, maintaining the common good often requires some groups to bear greater costs than others—for instance, homeowners pay property taxes to support public schools, but apartment dwellers do not.

Implement the Decision

Once an alternative is selected, it should be implemented in an efficient, effective, and timely manner. This is often much easier said than done, because people tend to resist change. In fact, the bigger the change, the greater is the resistance to it. Communication is the key to helping people accept a change. It is imperative that someone whom the stakeholders trust and respect answer the following questions: Why are we doing this? What is wrong with the current way we do things? and What are the benefits of the new way for you? A transition plan must be defined to explain to people how they will move from the old way of doing things to the new way. It is essential that the transition be seen as relatively easy and pain free.

Evaluate the Results

After the solution to the problem has been implemented, monitor the results to see if the desired effect was achieved, and observe its impact on the organization and the various stakeholders. Were the success criteria fully met? Were there any unintended consequences? This evaluation may indicate that further refinements are needed. If so, return to the problem development step, refine the problem statement as necessary, and work through the process again.

ETHICS IN INFORMATION TECHNOLOGY

The growth of the Internet, the ability to capture and store vast amounts of personal data, and greater reliance on information systems in all aspects of life have increased the risk that information technology will be used unethically. In the midst of the many IT breakthroughs in recent years, the importance of ethics and human values has been underemphasized—with a range of consequences. Here are some examples that raise public concern about the ethical use of information technology:

- Many employees have their email and Internet access monitored while at work, as employers struggle to balance their need to manage important company assets and work time with employees' desire for privacy and self-direction.
- Millions of people have downloaded music and movies at no charge and in apparent violation of copyright laws at tremendous expense to the owners of those copyrights.
- Organizations contact millions of people worldwide through unsolicited email (spam) as an extremely low-cost marketing approach.
- Hackers break into databases of financial and retail institutions to steal customer information, then use it to commit identity theft—opening new accounts and charging purchases to unsuspecting victims.
- Students around the world have been caught downloading material from the Web and plagiarizing content for their term papers.

- Web sites plant cookies or spyware on visitors' hard drives to track their online purchases and activities.

This book is based on two fundamental tenets. First, the general public needs to develop a better understanding of the critical importance of ethics as it applies to IT; currently, too much emphasis is placed on technical issues. Unlike most conventional tools, IT has a profound effect on society. IT professionals and users need to recognize this fact when they formulate policies that will have legal ramifications and affect the well-being of millions of consumers.

The second tenet on which this book is based is that in the business world, important decisions are too often left to the technical experts. General business managers must assume greater responsibility for these decisions, but to do so they must be able to make broad-minded, objective decisions based on technical savvy, business know-how, and a sense of ethics. They must also try to create a working environment in which ethical dilemmas can be discussed openly, objectively, and constructively.

Thus, the goals of this text are to educate people about the tremendous impact of ethical issues in the successful and secure use of information technology; to motivate people to recognize these issues when making business decisions; and to provide tools, approaches, and useful insights for making ethical decisions.

Summary

- Even within the same society, people can have strong disagreements over moral issues.
- Ethics has risen in importance because the risks associated with inappropriate behavior have increased, both in their likelihood and in their potential negative impact.
- Organizations have five good reasons for creating an environment that encourages employees to act ethically: (1) to gain the goodwill of the community, (2) to create an organization that operates consistently, (3) to foster good business practices, (4) to protect the organization and its employees from legal action, and (5) to avoid unfavorable publicity.
- An organization with a successful ethics program is one in which employees are willing to seek advice about ethical issues that arise; employees feel prepared to handle situations that could lead to misconduct; employees are rewarded for ethical behavior; employees are not rewarded for success gained through questionable means; and employees feel positively about their company.
- The corporate ethics officer (or corporate compliance officer) ensures that ethical procedures are put into place and are consistently adhered to throughout the organization, creates and maintains the ethics culture, and serves as a key resource on issues relating to corporate principles and ethics.
- Managers' behavior and expectations can strongly influence employees' ethical behavior.
- Most of us have developed a simple decision-making model that includes these steps: (1) develop a problem statement, (2) identify alternatives, (3) evaluate and choose an alternative, (4) implement the decision, and (5) evaluate the results.
- You can incorporate ethical considerations into decision making by identifying and involving the stakeholders; weighing various laws, guidelines, and principles—including the organization's code of ethics—that may apply; and considering the impact of the decision on you, your organization, stakeholders, your customers and suppliers, and the environment.
- Philosophers have developed many approaches to ethical decision making. Four common philosophies are the virtue ethics approach, the utilitarian approach, the fairness approach, and the common good approach.

Key Terms

code of ethics
common good approach
corporate compliance officer
corporate ethics officer
ethics
fairness approach
integrity
law
moral code
morality
morals
problem statement
Sarbanes–Oxley Act
social audit
software piracy
stakeholder
utilitarian approach
vice
virtue
virtue ethics approach

Self-Assessment Questions

The answers to the Self-Assessment Questions can be found in Appendix E.

Choose the word(s) that best complete the following sentences.

1. The term ____________ refers to social conventions about right and wrong that are so widely shared they become the basis for an established consensus.
2. ____________ is a set of beliefs about right and wrong behavior within a society.
3. ____________ are habits of unacceptable behavior.
4. A person who acts with integrity acts in accordance with a personal ____________.
5. ____________ are one's personal beliefs about right and wrong.
6. The principle called ____________ states that an employer can be held responsible for the acts of its employee even if the employee acts in a manner contrary to corporate policy and the employer's direction.
7. The public ____________ of an organization strongly influences the value of its stock, how consumers regard its products and services, the degree of oversight it receives from government agencies, and the amount of support and cooperation it receives from its business partners.
8. The corporate ethics officer provides the organization with ____________ and ____________ in the area of business conduct.
9. The ____________ is responsible for the careful and responsible management of an organization.
10. ____________ requires public companies to disclose whether they have codes of ethics and disclose any waiver to their code of ethics for certain members of senior management.
11. The goal of the Sarbanes–Oxley Act was to ____________.
12. ____________ highlights an organization's key ethical issues and identifies the overarching values and principles that are important to the organization and its decision-making process.
13. A(n) ____________ enables an organization to review how well it is meeting its ethical and social responsibility goals and communicate new goals for the upcoming year.
14. ____________ makes employees more aware of a company's code of ethics and how to apply it, as well as demonstrates that the company intends to operate in an ethical manner.
15. The most important part of the decision-making process is ____________.
16. The ____________ approach to an ethical decision focuses on how fairly actions and policies distribute benefits and burdens among people affected by the decision.
17. ____________ is a process for generating a number of alternative solutions to a problem.

Discussion Questions

1. There are many ethical issues about which people hold very strong opinions—abortion, gun control, and the death penalty, to name a few. If you were a team member on a project with someone whom you knew held an opinion different from yours on one of these issues, would it affect your ability to work effectively with this person? Why or why not?
2. Identify two important life experiences that helped you define your own personal code of ethics.
3. Do you think that the importance of ethical behavior in business is increasing or decreasing? Defend your position.
4. Do you believe that an organization should be able to escape criminal liability for the acts of its employees if it has acted as a responsible corporate citizen, making strong efforts to prevent and detect misconduct in the workplace? Why or why not?
5. The Ethics Resource Center identified five characteristics of a successful ethics program. Suggest a sixth characteristic, and defend your choice.
6. Which incident has a greater negative impact on an organization: an unethical act performed by an hourly worker or the same act performed by a senior manager of the organization? Explain your answer. Should the hourly worker be treated differently than the senior manager who committed the unethical act? Why or why not?
7. It is a common and acceptable practice for managers to hold people accountable to meet "stretch" goals, quotas, and budgets. How can this be done in a way that does not encourage unethical behavior on the part of employees?
8. Describe a hypothetical situation in which the action you would take is not legal, but it is ethical. Describe a hypothetical situation where the action you would take is legal, but not ethical.
9. It is easier to establish an ethical work environment in a nonprofit organization than in a for profit organization. State three facts or opinions that support this hypothesis. State three facts or opinions that refute the hypothesis.
10. This chapter discusses four approaches to dealing with moral issues. Which approach is closest to your way of analyzing moral issues? Now that you are aware of different approaches, do you think you might modify your approach to include other perspectives? Explain why or why not.
11. Is it possible for an employee to be successful in the workplace without acting ethically?
12. The Hewlett-Packard Board of Directors has forced two high-level executives to resign over internal scandals—first Patricia Dunn (see David A. Kaplan, "Scandal at HP: The Boss Who Spied on Her Board," *Newsweek*, September 10, 2006) and now Mark Hurd. Can you find evidence that turnover at the top has affected the company and/or its stockholders?
13. Should software piracy within the boundaries of third-world countries be tolerated to allow those countries an opportunity to move into the information age?
14. Without revealing the name of your employer, comment on the efforts of your employer to promote a work environment in which employees are encouraged to act ethically.
15. Is ethics training really just a waste of time that will not change the behavior of employees?

What Would You Do?

Use the five-step decision-making process discussed in the chapter to analyze the following situations and recommend a course of action.

1. You are currently being considered for a major promotion within your company to vice president of marketing. In your current position as manager of advertising, you supervise 15 managers and 10 hourly workers. As part of the annual salary review process, you have been given the flexibility to grant your employees an average 4 percent annual salary increase; however, you are strongly considering a lower amount. This would ensure that your department's expenses stay under budget and would send the message that you are able to control costs. How would you proceed?
2. You are the customer service manager for a small software manufacturer. The newest addition to your 10-person team is Aubrey, a recent college graduate. She is a little overwhelmed by the volume of calls, but is learning quickly and doing her best to keep up. Today, as you performed your monthly review of employee email, you were surprised to see that Aubrey has received several messages from employment agencies. One message says, "Aubrey, I'm sorry you don't like your new job. We have lots of opportunities that I think would much better match your interests. Please call me, and let's talk further." You're shocked and alarmed. You had no idea she was unhappy, and your team desperately needs her help to handle the onslaught of calls generated by the newest release of software. If you're going to lose her, you'll need to find a replacement quickly. You know that Aubrey did not intend for you to see the email, but you can't ignore what you saw. Should you confront Aubrey and demand to know her intentions? Should you avoid any confrontation and simply begin seeking her replacement? Could you be misinterpreting the email? What should you do?
3. As part of your company's annual performance review process, each employee must identify three coworkers to be interviewed by his manager to get a perspective on the employee's overall work performance. Your friend has offered to give you a glowing performance review if you agree to do the same for him. Truth be told, your friend is not a very dependable worker, and his work is often below minimum standards. However, he is a good friend, and you would hate to upset him. What would you do?
4. While mingling with friends at a party, you mention a recent promotion that has put you in charge of evaluating bids for a large computer hardware contract. A few days later, you receive a dinner invitation at the home of an acquaintance who also attended the party. Over cocktails, the conversation turns to the contract you are managing. Your host seems remarkably well-informed about the bidding process and likely bidders. You volunteer information about the potential value of the contract and briefly outline the criteria your firm will use to select the winner. At the end of the evening, the host surprises you by revealing that he is a consultant for several companies in the computer hardware market. Later that night your mind is racing. Did you reveal information that could provide a supplier with a competitive advantage? What are the potential business risks and ethical issues in this situation? Should you report the conversation to someone? If so, whom should you talk to, and what would you say?

5. You are a recent graduate of a well-respected business school, but you are having trouble getting a job. You worked with a professional résumé service to develop a well-written résumé and placed it on several Web sites; you also sent it directly to contacts at a dozen companies. So far you have not even had an invitation for an interview. You know that one of your shortcomings is that you have no real job experience to speak of. You are considering beefing up your résumé by exaggerating the extent of the class project you worked on for a few weeks under the supervision of your brother-in-law. You could reword the résumé to make it sound as if you were actually employed and that your responsibilities were greater than they actually were. What should you do?
6. You have just completed a grueling 10-day business trip calling on two dozen accounts up and down the East Coast. There were even business meetings combined with social events late into the night and on the weekends. On the flight back home at the end of this marathon, you are tired and feeling as if you have not seen your family for a month. As you work on completing your expense report, you say to yourself, "The company does not pay me enough for the work that I do." For more than a few moments, you think about padding your expense report to make up for all the extra hours and time away from your family. Would it be okay to add "extra expenses" to compensate for the hardship of the trip?

Cases

1. Dell Inc. Settles Customer Hardware Dispute Out of Court

Michael Dell founded Dell in 1984 in Austin, Texas, with what was at the time a novel idea—sell computer systems directly to customers and thereby deliver the most effective computing solutions to meet their needs.[45] This approach was so successful that the "Dell model" was taught in business schools as the way to sell custom products directly to customers.

Dell currently segments its customers into four groups: large enterprises, public institutions (health care, education, and government), small and medium businesses, and consumers. Although HP has surpassed Dell in worldwide PC sales, Dell is still the number one provider of PCs to large enterprises around the world and the number one provider to public sector customers in the United States.[46]

Between 2003 and 2004, Advanced Internet Technologies (AIT), an Internet services company based in North Carolina, purchased 2,000 OptiPlex™ computers from Dell. The OptiPlex desktop computer was Dell's primary offering to business and government customers. In 2007, the firm filed suit against Dell alleging that Dell knew long before AIT purchased the computers that there were significant problems with the Dell OptiPlex computers, including, but not limited to, problems with the motherboard and the power supply as well as CPU fan failures that caused overheating, crashes, and lost data.[47] AIT alleged that Dell had refused to take responsibility for the problem and that AIT had lost millions of dollars in business as a result.[48] AIT sought $75,000 and punitive damages for breach of contract, fraud, and deceptive business practices.[49]

A primary cause of the computer problems was bad capacitors provided by Nichicon, a Japanese PC component supplier. A capacitor is a small electrical component on a computer's motherboard that conditions direct current voltage to the components and thus provides a steady power supply. If the capacitor is faulty, it can lose its power and shut down the PC or freeze the screen. Millions of Dell OptiPlex PC computers with the faulty capacitor were shipped to customers such as Wells Fargo, WalMart, the Mayo Clinic, and AIT.[50]

In June 2010, the *New York Times* published an article noting that the failed computers and the manner in which the problem was handled were part of "the decline of one of America's most celebrated and admired companies." In addition, the article cited Dell emails and internal documents that indicated Dell employees had prior knowledge of problems with the computers. In one document, a Dell worker stated, "We need to avoid all language indicating the boards were bad or had 'issues' per our discussion this morning."[51] In offering advice on how to handle complaints about the faulty computers, Dell salespeople were told, "Don't bring this to customer's attention proactively" and "Emphasize uncertainty."[52] Other documents showed that after AIT complained, Dell service people examined the failed computers and informed AIT that the computers had been driven too hard in a hot, confined space. The servers were stacked on steel pantry racks, and Dell suggested a change to the stacking method that would result in cooler operating conditions. Dell sales reps then tried to sell AIT more expensive computers to replace the failed computers.[53]

Dell spokesman, David Frink, wrote in an email that "Dell worked with customers to address their issues, and Dell extended the warranties of all OptiPlex motherboards to January 2008 in order to address the Nichicon capacitor problem. The AIT lawsuit does not involve any current Dell products. Dell is responsive to customer issues and we continue to remain focused on our customers, their needs, and our growing record of superior customer service."[54]

In October 2010, AIT agreed to dismiss the lawsuit against Dell, thus avoiding a trial scheduled to start in February 2011. Financial terms of the agreement were not released.

Discussion Questions

1. How might the situation with AIT been handled more smoothly by Dell to avoid creating such ill will with a key customer?
2. If problems with the capacitors were so widespread, why didn't other customers file suit against Dell?
3. Do you believe that the manner in which the problems with the OptiPlex computers were handled represents an example of unethical behavior on the part of Dell? Why or why not?
4. Compare the manner in which Dell addressed the problems with its computers to the way in which Toyota managed the recall of over 4 million autos due to faulty floor mats and sticking gas pedals between late 2009 and mid-2010 (see *Motor Trend*'s January 2010 article, "The Toyota Recall Crisis," at *www.motortrend.com/features/auto_news/2010/112_1001_toyota_recall_crisis/index.html*.). Which company was more ethical in the way it handled its problems? Why do you believe this?

2. Automatic Renewal of Software Subscriptions

Symantec Corporation and McAfee are two of the more widely known providers of antivirus software. They are competitors, and both offer a wide range of products and services. Symantec was founded in 1982 and has grown into a *Fortune* 500 software firm with recent annual sales of nearly $6 billion. The company provides security, storage, and systems management solutions to meet the needs of individuals, small businesses, and large global organizations. Symantec's Norton brand of consumer products offers a range of security, backup, recovery, and maintenance services to individuals and small businesses.[55] McAfee Associates was founded in 1987 and had recent annual sales of $2 billion. The company is focused exclusively on security and provides solutions and services that protect systems, networks, and mobile devices around the world. Its customers include individual consumers, SMEs (small and medium-sized enterprises), large organizations, and governments. Its headquarters is in Santa Clara, California.[56]

In November 2005, Symantec began enrolling its North American customers in the Norton Automatic Renewal Service and eventually expanded the practice worldwide. Under this program, once a customer is registered, the service automatically bills the credit or debit card on file in the customer's account before the product subscription expires. Shortly before a subscription expires, an email is sent to the customer notifying them that their product subscription is about to be renewed for another year. Norton customers can check to see if they are signed up for this service by logging into their Norton account. Once in their Norton account, the customer may elect to cancel the Automatic Renewal Service.[57]

All McAfee software purchased as a subscription (e.g., VirusScan, Personal Firewall, SpamKiller, and PC Protection Plus) is automatically enrolled in the McAfee Auto-Renewal program. The company states that the Auto-Renewal program is designed to ensure that a subscriber's virus protection continues uninterrupted. Customers who wish to opt out of the program can do so by logging into their account online. Information about the Auto-Renewal program is displayed on the McAfee Web site at the time of purchase; it is also included in the End User License, which must be accepted by the customer at the time of installation. McAfee claims it implemented the Auto-Renewal program because a majority of its customers prefer to have their license renewed automatically rather than having to manually repurchase their subscription each year.[58] The McAfee Auto-Renewal program began in 2001.

As least as early as 2006, purchasers of Norton and McAfee software were complaining about the automatic renewal of their software subscriptions from these companies. These included claims that the customers did not know they had accepted automatic renewal, that they did not receive the promised email prior to automatic billing, that it was difficult to cancel the subscription, and that there were complications uninstalling the software following cancellation.[59,60]

Following extensive investigation of the many complaints, the New York Attorney General found that Symantec and McAfee had not adequately disclosed to consumers that their subscriptions would automatically be renewed and that they would be charged."[61]

In June 2009, McAfee and Symantec each agreed to pay the New York Attorney General's office $375,000 in fines to settle charges they automatically charged customers a software subscription renewal fee without their permission. In addition, the two firms were ordered to make detailed disclosures to consumers about subscription terms and renewal policies at the time of purchase. The companies were ordered to refund fees to consumers who requested a refund within 60 days of being charged. The two companies also stated they would send an email to consumers before and after renewal of the subscription.[62] The Attorney General did not order the firms to stop the practice of automatic subscription renewal.

Seven months after this agreement, an individual from New York alleged that Symantec had automatically renewed his software license and charged his credit card $76.03 with no prior notification. He brought a lawsuit against Symantec claiming that the firm had failed to abide with the settlement. The consumer also charged the company with deceptive business practices and unjust enrichment and sought to have the firm refund all fees generated by automatic renewals. Furthermore, he asked the court to grant the lawsuit class-action status, which would open the case to a pool of tens of thousands of New York consumers.[63] Two months later, the same individual brought charges against McAfee claiming he was charged $78.85 in April 2009 for renewal of the firm's software without his consent.[64]

Discussion Questions

1. Automatic software renewal is a common practice among software providers; however, many feel the process followed by Symantec and McAfee is inherently flawed and destined to create customer ill will. Visit the Web site of either of these companies and follow their software purchase process up to the point of entering method of payment. Based on this experience and the discussion above, identify several weaknesses with the current software purchase and renewal process.
2. Outline an improved software renewal process that would require minimal effort for the consumer but also minimize the likelihood that the license would expire, leaving the consumer's computer open to attacks.
3. Identify two other industries that have implemented an automatic renewal process. In general, how has the public reacted to automatic renewal processes? Identify three positives and three negatives associated with automatic renewal.
4. Do you think automatic renewal of software licenses is an ethical practice or simply a way to ensure an ongoing stream of revenue for the software provider? Explain your answer.

3. Is There a Place for Ethics in IT?

On March 15, 2005, Michael Schrage published an article in *CIO* magazine entitled "Ethics, Schmethics," which stirred up a great deal of controversy in the IT community. In the article, Schrage proposed that CIOs (chief information officers) "should stop trying to do the 'right thing' when implementing IT and focus instead on getting their implementations right." Schrage argued that *ethics* had become a buzzword, just like *quality* in the 1980s; he asserted that the demand for ethical behavior interferes with business efficiency.

In the article, Schrage provided a few scenarios to back up his opinion. In one such example, a company is developing a customer relationship management (CRM) system, and the staff is working very hard to meet the deadline. The company plans to outsource the maintenance and support of the CRM system once it is developed, meaning there is a good chance that two-thirds of the IT staff will be laid off. Would you disclose this information? Schrage answered, "I don't think so."

In another scenario, Schrage asked readers if they would consider deliberately withholding important information from their boss if they knew that its disclosure would provoke his or her immediate counterproductive intervention in an important project. Schrage said he would withhold it. Business involves competing values, he argued, and trade-offs must be made to keep business operations from becoming paralyzed.[65]

Schrage was hit with a barrage of responses accusing him of being dishonorable, short-sighted, and lazy. Other feedback provided new perspectives on his scenarios that Schrage had not considered in his article. For example, an IT manager at Boise State University argued that doing the right thing is good for business. Not disclosing layoffs, she argued, is a trick that only works once. Remaining employees will no longer trust the company and may pursue jobs where they can feel more secure. New job applicants will think twice before joining a company with a reputation for exploiting employees. Other readers responded to that scenario by suggesting that the company could try to maintain loyalty by offering incentives for those who stayed or by providing job-placement services to departing employees.

Addressing the second scenario, another reader, Dewey, suggested that not giving the boss important information could backfire on the employee: "What if your boss finds out the truth? What if you were wrong and the boss could have helped? Once your boss knows that you lied once, will he believe you the next time?"

Another reader had actually worked under an unproductive, reactive, meddling boss. Based on his experience, he suggested speaking to the boss about the problem at an appropriate time and place. In addition, the reader explained that as situations arose that required him to convey important information that might elicit interference, he developed action plans and made firm presentations to his boss. The boss, the reader assured Schrage, will adapt.

Some readers argued that CIOs must consider the company's long-term needs rather than just the current needs of a specific project. Others argued that engaging in unethical behavior, even for the best of purposes, crosses a line that eventually leads to more serious transgressions. Some readers suspected that Schrage had published the article to provoke outrage. Another reader agreed with Schrage, arguing that ethics has to "take a back seat to budgets and schedules" in a large organization. This reader explained, "At the end of the day, IT is business."

Discussion Questions

1. Discuss how a CIO might handle Schrage's scenarios using the suggested process for ethical decision making presented in this chapter.
2. Discuss the possible short-term losses and long-term gains in implementing ethical solutions for each of Schrage's scenarios.
3. Must businesses choose between good ethics and financial benefits? Explain your answer using Schrage's scenarios as examples.
4. What do you think Schrage means when he says that CIO's "should stop trying to do the 'right thing' when implementing IT and focus instead on getting their implementations right"? Do you agree?

End Notes

[1] Darby Conley, *Get Fuzzy*, August 8, 2007.

[2] Ashlee Vance, "H.P. Ousts Chief for Hiding Payments to Friend," *New York Times*, August 6, 2010, www.nytimes.com/2010/08/07/business/07hewlett.html.

[3] Jordan Robertson, Associated Press, "Oracle Names Ex-HP CEO Mark Hurd Co-President," *Washington Times*, www.washingtontimes.com/news/2010/sep/7/oracle-names-ex-hp-ceo-mark-hurd-co-president.

[4] Aaron Ricadela, "HP Contractor Didn't Have Sex with Hurd, Lawyer Says," *BusinessWeek*, August 6, 2010, www.businessweek.com/news/2010-08-06/hp-contractor-didn-t-have-sex-with-hurd-lawyer-says.html.

[5] Ashlee Vance, "H.P. Ousts Chief for Hiding Payments to Friend," *New York Times*, August 6, 2010, www.nytimes.com/2010/08/07/business/07hewlett.html.

[6] "Hurd Settles Sexual Harassment Allegation: Report," *Market Watch*, August 8, 2010, www.marketwatch.com/story/hurd-settles-sexual-harassment-allegations-ap-2010-08-08.

[7] Jordan Robertson, Associated Press, "Oracle Names Ex-HP CEO Mark Hurd co-President," *Washington Times*, www.washingtontimes.com/news/2010/sep/7/oracle-names-ex-hp-ceo-mark-hurd-co-president.

[8] Tom Gilesand David Rovella, "HP's Bid to Bar Hurd's Oracle Move May Be Long Shot," *BusinessWeek*, September 8, 2010, www.businessweek.com/news/2010-09-08/hp-s-bid-to-bar-hurd-s-oracle-move-may-be-long-shot.html.

[9] Chris Kanaracus, "Update: HP, Oracle Settle Hurd Litigation, Affirm Partnership," *Computerworld*, September 20, 2010, www.computerworld.com/s/article/9186861/Update_HP_Oracle_settle_Hurd_litigation_affirm_partnership.

[10] Business Software Alliance, "Seventh Annual BSA/IDC Global Software Piracy Study," May 2010, http://portal.bsa.org/globalpiracy2009/studies/09_Piracy_Study_Report_A4_final_111010.pdf.

[11] Business Software Alliance, "Seventh Annual BSA/IDC Global Software Piracy Study," May 2010, http://portal.bsa.org/globalpiracy2009/studies/09_Piracy_Study_Report_A4_final_111010.pdf.

[12] Ketaki Gokhale, "Satyam Tumbles in Mumbai Trading After Posting Loss Amid Financial Probe," Bloomberg.com, September 30, 2010, www.bloomberg.com/news/2010-09-30/satyam-tumbles-in-mumbai-trading-after-company-reports-full-year-loss.html.

[13] Prof. Leigh Tesfatsion, "The Enron Scandal and Moral Hazard," Iowa State University, November 10, 2010, www2.econ.iastate.edu/classes/econ353/tesfatsion/enron.pdf.

[14] Patrick Thibodeau, "Case Against Dunn in HP Scandal Dismissed," *Computerworld*, March 14, 2007.

[15] David A. Kaplan, "Scandal at HP: The Boss Who Spied on Her Board," *Newsweek*, September 20, 2006.

[16] Bob Brown, "Tech's Most Notorious CEO Scandals," *Network World*, August 9, 2010.

[17] Paul McDougall, "Fallen IBM Exec Sentenced to Prison," *InformationWeek*, September 14, 2010, www.informationweek.com/news/hardware/unix_linux/showArticle.jhtml?articleID=227400389.

[18] "Government Ethics Rules are Poorly Implemented," *Columbus Underground* (discussion board), www.columbusunderground.com/forums/topic/government-ethics-rules-are-poorly-implemented.

[19] Kim Zetter, "Ethics Committee Staffer Leaks Secrets on File-Sharing Network, *Wired*, November 2, 2009.

[20] Cisco Systems, Inc., "Employee Volunteerism/Giving," www.cisco.com (accessed October 20, 2010).

[21] Cisco Systems, Inc., "CSR Report Highlights," www.cisco.com/web/about/ac227/csr2010/assets/pdfs/CSR2010_Highlights.pdf (accessed January 7, 2011).

[22] Jacquelyn Smith, "America's Most Generous Companies," *Forbes*, October 28, 2010, www.forbes.com/2010/10/28/most-generous-companies-leadership-corporate-citizenship-philanthropy.html.

[23] Dell Inc., "Dell Giving is Serious Business," Direct2Dell, August 2009, http://en.community.dell.com/dell-blogs/Direct2Dell/b/direct2dell/archive/2009/08/27/dell-giving-is-serious-business.aspx.

[24] Dell Inc., "Dell Corporate Responsibility Report 2010," http://i.dell.com/sites/content/corporate/corp-comm/en/Documents/dell-fy10-cr-report.pdf.

[25] Google, "Philanthropy @ Google," www.google.org/googlers.html (accessed October 24, 2010).

[26] IBM, "ECCC Contributions at Record Levels; Over $36 Million Raised in 2008," w3.ibm.com/chq/ecccinfo (accessed October 22, 2010).

[27] Jacquelyn Smith, "America's Most Generous Companies," *Forbes*, October 28, 2010, www.forbes.com/2010/10/28/most-generous-companies-leadership-corporate-citizenship-philanthropy.html.

[28] SAP, "Corporate Social Responsibility," www.sap.com/usa/about/csr/index.epx (accessed October 24, 2010).

[29] Ethisphere Institute, "Ethisphere Announces 2010 World's Most Ethical Companies," *BusinessWire*, March 22, 2010, www.businesswire.com/news/home/20100322005479/en/Ethisphere-Announces-2010-World%E2%80%99s-Ethical-Companies.

[30] *United States v New York Central & Hudson River R. Co*, 212 U.S. 509 (1909), http;//supreme.justia.com/us/212/509/case.html.

[31] John Foley, "Amid Contract Scandal, A Shakeup and Lingering Questions," *InformationWeek*, October 22, 2010, www.informationweek.com/news/government/policy/showArticle.jhtml?articleID=227900598.

[32] Paula, J.Desio, "Ethics and Compliance Programs May Get Their Day in Court," Ethics Resource Center, www.ethics.org/ethics-today/1208/policy-report.html.

[33] Ethics Resource Center, "2007 Ethics Resource Center's National Business Ethics Survey," www.ethics.org/research/nbes.com.

[34] Ethics Resource Center, "2009 National Business Ethics Survey, Ethics in Recession," www.ethics.org/nbes/download.html.

[35] "What is an Ethics Officer?", *WiseGEEK*, www.wisegeek.com/what-is-an-ethics-officer.htm.

[36] Hannah Clark, "Chief Ethics Officers: Who Needs Them?" *Forbes*, October 23, 2006, www.forbes.com/2006/10/23/leadership-ethics-hp-lead-govern-cx_hc_1023ethics.html.

[37] Hannah Clark, "Chief Ethics Officers: Who Needs Them?" *Forbes*, October 23, 2006, www.forbes.com/2006/10/23/leadership-ethics-hp-lead-govern-cx_hc_1023ethics.html.

38 "Corporate-Ethics US, "Three Main Responsibilities of an Ethics Officer," www.corporate-ethics.us/EO.htm, (accessed October 30, 2010).

39 Kate Benner, "Is Sarbanes Oxley a Failure?", *Fortune*, March 24, 2010.

40 "CR's 100 Best Corporate Citizens 2010," *Corporate Responsibility*, www.thecro.com/files/CR100Best.pdf.

41 Intel Corporation, "Intel 2009 Corporate Responsibility Report," www.intel.com/about/corporateresponsibility/report/index.htm (accessed November 1, 2010).

42 Audra Bianca, "Ethics Awareness Training," *eHow*, www.ehow.com/about_6574961_ethics-awareness-training.html.

43 Motorola, "Ethics Awareness and Training," http://responsibility.motorola.com/index.php/ourapproach/busconduct/eat (accessed on November 22, 2010).

44 "Performance Reviews Often Skip Ethics, HR Professionals Say," Ethics Resource Center, June 12, 2008, www.ethics.org/about-erc/press-releases.asp?aid=1150.

45 Dell, Inc., "Our Story," http://content.dell.com/us/en/corp/d/corp-comm/our-story-facts-about-dell.aspx (accessed October 17, 2010).

46 Dell, Inc., "Our Story," http://content.dell.com/us/en/corp/d/corp-comm/our-story-facts-about-dell.aspx (accessed October 17, 2010).

47 Chloe Albanesius, "Dell Accused of Shipping Desktops Known to be Faulty," *PC Magazine*, June 30, 2010, www.pcmag.com/article2/0,2817,2365875,00.asp.

48 Ashlee Vance, "Suit Over Faulty Computers Highlights Dell's Decline," *New York Times*, June 28, 2010, www.nytimes.com/2010/06/29/technology/29dell.html.

49 Agam Shah, "Dell Refutes Withholding Evidence in Faulty-PC Case," *CIO*, August 13, 2010, www.cio.com/article/603320/Dell_Refutes_Withholding_Evidence_in_Faulty_PC_Case.

50 Ashlee Vance, "Suit Over Faulty Computers Highlights Dell's Decline," *New York Times*, June 28, 2010, www.nytimes.com/2010/06/29/technology/29dell.html.

51 Ashlee Vance, "Suit Over Faulty Computers Highlights Dell's Decline," *New York Times*, June 28, 2010, www.nytimes.com/2010/06/29/technology/29dell.html.

52 Ashlee Vance, "Suit Over Faulty Computers Highlights Dell's Decline," *New York Times*, June 28, 2010, www.nytimes.com/2010/06/29/technology/29dell.html.

53 Ashlee Vance, "Suit Over Faulty Computers Highlights Dell's Decline," *New York Times*, June 28, 2010, www.nytimes.com/2010/06/29/technology/29dell.html.

54 Agam Shah, "Dell Knew Some PCs Were Faulty, Court Documents Reveal," *CIO*, June 20, 2010.

55 Symantec Corporation, "Business Overview," www.symantec.com/about/profile/business.jsp (accessed November 20, 2010).

56 McAfee Associates, "McAfee Facts," www.mcafee.com/us/local_content/brochures/mcafee_factsheet.pdf (accessed November 20, 2010).

57 Symantec Corporation, "Norton Automatic Renewal Service," www.symantec.com/support/autorenew.html (accessed on November 6, 2010.

58 "What is the McAfee Auto-Renewal Program?", http://service.mcafee.com/FAQDocument.aspx?lc=1033&id=CS40083 (accessed November 6, 2010).

59 "Anyone Having Trouble Cancelling Symantec Auto-Renewal?", http://forums.pcworld.com/index.php?/topic/31391-anyone-having-trouble-cancelling-symantec-auto-renewal (accessed November 6, 2010).

60 "McAfee Unauthorized Auto Renewal," *BusinessReporter.net*, April 27, 2008, www.businessreporter.net/article/mcafee-unauthorized-auto-renewal-458-1.html (accessed November 6, 2010).

61 Robert McMillan, "Symantec, McAfee To Pay Fines Over Auto-Renewals," *Computerworld*, June 10, 2009, www.computerworld.com/s/article/9134221/Symantec_McAfee_to_pay_fines_over_auto_renewals.

62 Gregg Keizer, "Symantec Hit with Class-Action Lawsuit over Auto-Renewals," *Computerworld*, February 7, 2010, www.computerworld.com/s/article/9153118/Symantec_hit_with_class_action_lawsuit_over_auto_renewals.

63 Gregg Keizer, "Symantec Hit with Class-Action Lawsuit over Auto-Renewals," *Computerworld*, February 7, 2010, www.computerworld.com/s/article/9153118/Symantec_hit_with_class_action_lawsuit_over_auto_renewals.

64 Gregg Keizer, "N.Y. Man Sues McAfee Over Antivirus Auto-Renewal Fees," *Computerworld*, March 29, 2010, www.computerworld.com/s/article/9174329/N.Y._man_sues_McAfee_over_antivirus_auto_renewal_fees.

65 Michael Schrage, "Ethics, Shmethics," *CIO*, April 5, 2005, www.cio.com.au/article/185611/ethics_shmethics.

CHAPTER 2

ETHICS FOR IT WORKERS AND IT USERS

QUOTE

To become an able and successful man in any profession, three things are necessary: nature, study, and practice.

—Henry Ward Beecher, American minister and social reformer

VIGNETTE

Texas Health Resources Works to Ensure Compliance with HIPAA in Its Electronic Health Record System

A major emphasis of corporate ethics is ensuring compliance to a set of legal standards or to a set of corporate values that are fundamental to how the organization conducts business. A key goal of compliance is to promote the right behaviors and discourage undesirable ones. Due to the life-and-death nature of many medical decisions and the extremely private nature of patient data, corporate compliance is especially crucial in the design of successful IT systems that serve healthcare facilities.

Texas Health Resources operates a large nonprofit healthcare delivery system that provides services to patients in 16 counties in north-central Texas through 24 acute-care and short-stay hospitals and 18 outpatient facilities.[1] Texas Health employs more than 20,500 workers, and its recent operating revenue was $2.9 billion.[2]

Texas Health has a long track record of implementing IT initiatives to improve patient care. It has repeatedly been selected as a member of the *InformationWeek 500*, a list that tracks the nation's most innovative IT organizations. It has also been recognized as a winner of the American Hospital Association's "Most Wired" award for its extensive and innovative use of IT.[3]

One of Texas Health's most challenging and impactful IT projects was the investment of more than $200 million over five years to implement improved information technology, including an organization-wide electronic health record (EHR) system.[4] The EHR captures and stores patient data, including medical history, prescribed medications, and lab results. The EHR can be quickly and securely accessed by authorized users involved in treatment of the patient. The goal is to improve the overall treatment of the patient, increase efficiency, and reduce the chance of medical errors.

Not only was the EHR system designed to meet stringent technical and business requirements, it also had to comply with federal law and the organization's own high ethical standards. These two sets of requirements were sometimes in conflict with each other. Business needs required that care-givers and insurance companies gain instant access to patient records. However, the federal Health Insurance Portability and Accountability Act (HIPAA) mandates that health agencies keep the medical records of patients private and carefully restrict access to this information. So the Texas Health EHR system was designed to enable healthcare providers to quickly access their patients' medical records, but only after they provide proper authentication. The system requires additional authentication for healthcare providers to access the records of patients who are not under their immediate care. To ensure that there is no inappropriate access, the EHR system records who is accessing patients' records, which enables administrators to audit and review individual patient cases.

The successful implementation of the EHR system required the effort and cooperation of dozens of people, including hospital administrators, physicians, staff, and IT workers. Assigning the right people to lead the project helped Texas Health ensure that the project was run efficiently and that it met its technical and compliance goals. One of the key players on the project was Debbie Jowers, the administrative director for applications at Texas Health. In this role, she manages a group of 125 employees who watch over the more than 300 IT applications used at Texas Health. She also managed the budget for the EHR project. Jowers has always had a passion for health care, although she knew she would never be a nurse or physician. She has fulfilled her passion by working in the IT field for over 25 years delivering innovations to improve patient care. Debbie was named one of the 2009 *CIO* magazine's Ones to Watch award winners. The award recognizes the next generation of IT leaders who have never stopped investing in their talents and who excel in leading innovation, business strategy, project execution, team building, and organization change. Edward Marxim, CIO at Texas Health, speaks about Debbie: "She seeks to continuously improve and doesn't take no for an answer. She is passionate about her work and Texas Health and it shows with the results she drives from herself and her team."[5] Debbie received a B.A. from the University of Missouri–St. Louis with an emphasis in both management information systems and accounting; she is also a Certified Public Accountant. In addition, Debbie is the vice president of the Dallas–Fort Worth chapter of the Society for Information Management.

Questions to Consider

1. What can senior management do to help ensure that dedicated and competent workers have been assigned to key information technology projects?
2. What does it take to provide IT services in a way that is both effective and ethical?

LEARNING OBJECTIVES

As you read this chapter, consider the following questions:

1. What key characteristics distinguish a professional from other kinds of workers, and is an IT worker considered a professional?
2. What factors are transforming the professional services industry?
3. What relationships must an IT worker manage, and what key ethical issues can arise in each?
4. How do codes of ethics, professional organizations, certification, and licensing affect the ethical behavior of IT professionals?
5. What is meant by compliance, and how does it help promote the right behaviors and discourage undesirable ones?

IT PROFESSIONALS

A **profession** is a calling that requires specialized knowledge and often long and intensive academic preparation. Over the years, the United States government adopted labor laws and regulations that required a more precise definition of what is meant by a *professional* employee. The United States Code of federal regulations defines a "professional employee" as one who is engaged in the performance of work:

> "(i) requiring knowledge of an advanced type in a field of science or learning customarily acquired by a prolonged course of specialized intellectual instruction and study in an institution of higher learning or a hospital (as distinguished from knowledge acquired by a general academic education, or from an apprenticeship, or from training in the performance of routine mental, manual, mechanical, or physical activities);
> (ii) requiring the consistent exercise of discretion and judgment in its performance;
> (iii) which is predominantly intellectual and varied in character (as distinguished from routine mental, manual, mechanical, or physical work); and
> (iv) which is of such character that the output produced or the result accomplished by such work cannot be standardized in relation to a given period of time."[6]

In other words, professionals such as doctors, lawyers, and accountants require advanced training and experience; they must exercise discretion and judgment in the course of their work; and their work cannot be standardized. Many people would also expect professionals to contribute to society, to participate in a lifelong training program (both formal and informal), to keep abreast of developments in their field, and to assist other professionals in their development. In addition, many professional roles carry special rights and responsibilities. Doctors, for example, prescribe drugs, perform surgery, and request confidential patient information while maintaining doctor–patient confidentiality.

Are IT Workers Professionals?

Many business workers have duties, backgrounds, and training that qualify them to be classified as professionals, including marketing analysts, financial consultants, and IT specialists. A partial list of IT specialists includes programmers, systems analysts, software engineers, database administrators, local area network (LAN) administrators, and chief information officers (CIOs). One could argue, however, that not every IT role requires "knowledge of an advanced type in a field of science or learning customarily acquired by a prolonged course of specialized intellectual instruction and study," to quote again from the United States Code. From a *legal* perspective, IT workers are not recognized as professionals because they are not licensed by the state or federal government. This distinction is important, for example, in malpractice lawsuits, as many courts have ruled that IT workers are not liable for malpractice because they do not meet the legal definition of a professional.

The Changing Professional Services Industry

Although not legally classified as professionals, IT workers are considered part of the professional services industry, which is experiencing immense changes that impact how members of this industry must think and behave to be successful. Ross Dawson, author and CEO of the consulting firm Advanced Human Technology, identifies seven forces that are changing the nature of professional services: client sophistication, governance, connectivity, transparency, modularization, globalization, and commoditization.[7]

Client Sophistication

Clients are more aware of what they need from service providers, are more willing to look outside their own organization to get the best possible services, and are better able to drive a hard bargain to get the best possible services at the lowest possible cost. These forces are pushing professional services organizations to offer higher-quality services tailored to meet the specific needs of a wide variety of clients.

Governance

Major scandals and tougher laws enacted to attempt to avoid future scandals (e.g., Sarbanes–Oxley) have created an environment in which there is less trust and more oversight in client–service provider relationships. For example, today it is commonly accepted that information technology service organizations that host or process the data of their customers must show that they have adequate controls and safeguards in place. The Statement on Auditing Standards (SAS) No. 70 is a widely recognized standard (developed by the American Institute of Certified Public Accountants) for performing an audit of such a service organization. Completion of an SAS No. 70 audit means that a service organization has been through an in-depth audit of its control objectives and activities and that it has adequate controls and safeguards in place in order to host or process data belonging to its customers.[8]

Connectivity

Clients and service providers have built their working relationships on the expectation that they can communicate easily and instantly around the globe through videoconferences and teleconferences, email, text messaging, instant messaging, and a variety of Web applications—through an array of wired and wireless devices.

Transparency

Transparency involves any attempt to reveal and clarify any information or processes that were previously hidden or unclear. Increased transparency leads to increased trust and cooperation between clients and service providers. Clients expect to be able to see work-in-progress in real time, and they expect to be able to influence that work. No longer are clients willing to wait until the end product is complete before they weigh in with comments and feedback.

Modularization

Modularization is the act of breaking down a production or business process into smaller components. In the fabrication of an object, modularization can result in the use of interchangeable parts on a production line. In the professional services industry, modularization enables more options to be created in shorter time spans, at lower costs and with less complexity. Clients are able to break down their business processes into the fundamental steps and decide which they will perform themselves and which they will outsource to service providers. This often leads to highly effective, low-cost solutions.

Globalization

Globalization is the "process of interaction and integration among the people, companies, and governments of different nations."[9] Globalization has been going on for thousands of years but has speeded up dramatically in the last few decades as advances in telecommunications infrastructure, including the Internet, have made it easier for people to communicate and do business internationally.

As a result of globalization, an organization's supply, production, logistics, and distribution networks are extended globally—leading to worldwide purchasing, integrated customer service, and global brands. Clients are able to evaluate and choose among service providers around the globe, making the service provider industry extremely competitive. Outsourcing and offshoring of business processes has become commonplace.

Commoditization

Commoditization is the transformation of goods or services into commodities that offer nothing to differentiate themselves from those offered by competitors. Commoditized goods and services are sold strictly on the basis of price. Clients look at the delivery of low-end services (e.g., staff augmentation to complete a project) as a commodity service for which price is the primary criterion for choosing a service provider. For the delivery of high-end services (e.g., development of an IT strategic plan), clients may still seek to form a partnership with their service providers.

Professional Relationships That Must Be Managed

IT workers typically become involved in many different relationships, including those with employers, clients, suppliers, other professionals, IT users, and society at large as illustrated in Figure 2-1. In each relationship, an ethical IT worker must act honestly and appropriately. These various relationships are discussed in the following sections.

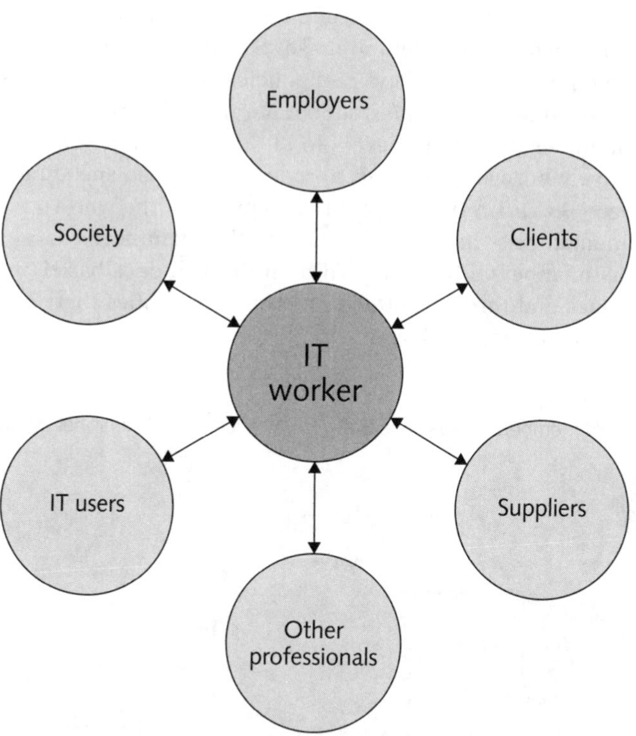

FIGURE 2-1 Professional relationships IT workers must manage
Source Line: Course Technology/Cengage Learning.

Relationships Between IT Workers and Employers

IT workers and employers have a critical, multifaceted relationship that requires ongoing effort by both parties to keep it strong. An IT worker and an employer typically agree on fundamental aspects of this relationship before the worker accepts an employment offer. These issues may include job title, general performance expectations, specific work responsibilities, drug-testing requirements, dress code, location of employment, salary, work hours, and company benefits. Many other aspects of this relationship may be addressed in a company's policy and procedures manual or in the company's code of conduct, if one exists. These issues may include protection of company secrets; vacation policy; time off for a funeral or an illness in the family; tuition reimbursement; and use of company resources, including computers and networks.

Other aspects of this relationship develop over time as the need arises (for example, whether the employee can leave early one day if the time is made up another day). Some aspects are addressed by law—for example, an employee cannot be required to do anything illegal, such as falsify the results of a quality assurance test. Some aspects are specific to the role of the IT worker and are established based on the nature of the work or project—for example, the programming language to be used, the type and amount of documentation to be produced, and the extent of testing to be conducted.

As the stewards of an organization's IT resources, IT workers must set an example and enforce policies regarding the ethical use of IT. IT workers often have the skills and knowledge to abuse systems and data or to allow others to do so. Software piracy is an area in which IT workers may be tempted to violate laws and policies. Although end users often get the blame when it comes to using illegal copies of commercial software, software piracy in a corporate setting is sometimes directly traceable to IT staff members—either they allow it to happen or they actively engage in it, often to reduce IT-related spending.

The **Business Software Alliance (BSA)** is a trade group that represents the world's largest software and hardware manufacturers. Its mission is to stop the unauthorized copying of software produced by its members. BSA is funded both through dues based on member companies' software revenues and through settlements from companies that commit piracy. For a list of BSA members, see Table 2-1.

TABLE 2-1 Worldwide and policy council members of Business Software Alliance (as of November 2010)

Adobe	Altium	Apple	Autodesk
AVEVA	AVG	Bentley Systems	CA
Cadence Design Systems	Cisco Systems	CNC Software–Mastercam	Corel
Dassault Systèmes Solid-Works Corporation	Dell	HP	IBM
Intel	Intuit	Kaspersky	McAfee
Microsoft	Mindjet	Progress Software	PTC
Quark	Quest	Rockwell Automation	Siemens PLM Software, Inc.
Stone Bond Technologies	Sybase	Symantec	Synopsys

Source Line: Business Software Alliance, "BSA Members," © 2011, www.bsa.org/country/BSA%20and%20Members/Our%20Members.aspx.

More than 100 BSA lawyers and investigators prosecute thousands of cases of software piracy each year. BSA investigations are usually triggered by calls to the BSA hotline (1-888-NO-PIRACY), reports sent to the BSA Web site, and referrals from member companies. Many of these cases are reported by disgruntled employees or former employees. In 2009, BSA initiated more than 16,000 end-user enforcement actions and shut down over 36,000 online auctions of unlicensed software.[10] When BSA finds cases of software piracy, it assesses heavy monetary penalties.

Failure to cooperate with the BSA can be extremely expensive. The cost of criminal or civil penalties to a corporation and the people involved can easily be many times more expensive than the cost of "getting legal" by acquiring the correct number of software licenses. Software manufacturers can file a civil suit against software pirates with penalties of up to $100,000 per copyrighted work. Furthermore, the government can criminally prosecute violators and fine them up to $250,000, incarcerate them for up to five years, or both.

In August 2010, AutoTrader.com, the automotive Web site for buying and selling used and new cars, agreed to pay $400,000 to BSA to settle claims of pirated software from Adobe, Autodesk, Corel, and Quest that was installed on its computers. In addition, the firm was required to delete all copies of the unlicensed software, purchase licenses of whatever software was required to become compliant, and implement improved software asset management processes to avoid future problems.[11]

In September 2010, an unidentified worker provided information on his company's use of pirated software to the BSA and received a bounty of £10,000 (approximately $15,500 USD) while his company is facing fines and license fees in excess of £100,000 (approximately $155,000 USD).[12]

Trade secrecy is another area that can present challenges for IT workers and their employers. A **trade secret** is information, generally unknown to the public, that a company has taken strong measures to keep confidential. It represents something of economic value that has required effort or cost to develop and that has some degree of uniqueness or novelty. Trade secrets can include the design of new software code, hardware designs, business plans, the design of a user interface to a computer program, and manufacturing processes. Examples include the Colonel's secret recipe of 11 herbs and spices used to make the original KFC chicken, the formula for Coke, and Intel's manufacturing process for the i7 quad core processing chip. Employers worry that employees may reveal these secrets to competitors, especially if they leave the company. As a result, companies often require employees to sign confidentiality agreements and promise not to reveal the company's trade secrets.

Prior to its acquisition by Cambridge Silicon Radio, SiRF Technology, Inc. was a maker of Global Positioning System (GPS) chipsets used by wireless location-aware programs for auto navigation systems and mobile phones. In 2009, the firm's director of software development resigned and set up a company "in order to develop and sell location-based services utilizing trade secrets stolen from SiRF." This individual was later arrested on charges that he stole valuable trade secrets from his former employer and faces a possible 10-year jail sentence.[13]

Another issue that can create friction between employers and IT workers is whistle-blowing. **Whistle-blowing** is an effort by an employee to attract attention to a negligent, illegal, unethical, abusive, or dangerous act by a company that threatens the public interest. Whistle-blowers often have special information based on their expertise or position within the offending organization. For example, an employee of a chip manufacturing company may know that the chemical process used to make the chips is dangerous to employees and the general public. A conscientious employee would call the problem to management's attention and try to correct it by working with appropriate resources within the company. But what if the employee's attempt to correct the problem through internal channels was thwarted or ignored? The employee might then consider becoming a whistle-blower and reporting the problem to people outside the company, including state or federal agencies that have jurisdiction. Obviously, such actions could have negative consequences on the employee's job, perhaps resulting in retaliation and firing.

In 2007, a whistle-blower lawsuit was filed by a former Oracle contract specialist alleging that Oracle failed to offer federal government customers the same discounts that it offered to its other large customers. As a result, Oracle allegedly overcharged the federal government on a software contract that ran from 1998 to 2006.[14] As a result of this whistle-blower lawsuit, the U.S. Department of Justice sued Oracle in 2010 claiming that it had overcharged the government hundreds of millions of dollars.[15]

Relationships Between IT Workers and Clients

IT workers provide services to clients; sometimes those "clients" are coworkers who are part of the same organization as the IT worker. In other cases, the client is part of a different organization. In relationships between IT workers and clients, each party agrees to provide something of value to the other. Generally speaking, the IT worker provides hardware, software, or services at a certain cost and within a given time frame. For example, an IT worker might agree to implement a new accounts payable software package that meets the client's requirements. The client provides compensation, access to key contacts, and perhaps a work space. This relationship is usually documented in contractual terms—who does what, when the work begins, how long it will take, how much the client pays, and so on. Although there is often a vast disparity in technical expertise between IT workers and their clients, the two parties must work together to be successful.

Typically, the client makes decisions about a project on the basis of information, alternatives, and recommendations provided by the IT worker. The client trusts the IT worker to use his or her expertise and to act in the client's best interests. The IT worker must trust that the client will provide relevant information, listen to and understand what the IT worker says, ask questions to understand the impact of key decisions, and use the information to make wise choices among various alternatives. Thus, the responsibility for decision making is shared between client and IT worker.

One potential ethical problem that can interfere with the relationship between IT workers and their clients involves IT consultants or auditors who recommend their own products and services or those of an affiliated vendor to remedy a problem they have detected. Such a situation has the potential to undermine the objectivity of an IT worker due to a **conflict of interest**—a conflict between the IT worker's (or the IT firm's) self-interest and the interests of the client. For example, an IT consulting firm might be hired to assess a firm's IT strategic plan. After a few weeks of analysis, the consulting firm might provide a poor rating for the existing strategy and insist that its proprietary products and services are required to develop a new strategic plan. Such findings would raise questions about the vendor's objectivity and whether its recommendations can be trusted.

EMC (a data storage company), Hewlett-Packard, IBM, and Price Waterhouse Coopers each paid millions of dollars to the federal government to settle claims that their firms gave illegal kickbacks to consulting firms that recommended their products to government agencies. "Influencer fees," frequently amounting to hundreds of thousands of dollars, were paid to consultants who helped federal agencies purchase and implement its computer systems. While some of the companies defended these payments as legitimate rebates and discounts, the Justice Department argued that the payments created a conflict of interest for the consultants whose recommendations may have resulted in federal agencies paying too much for some products or making unnecessary purchases.[16]

Problems can also arise during a project if IT workers find themselves unable to provide full and accurate reporting of the project's status due to a lack of information, tools, or experience needed to perform an accurate assessment. The project manager may want to keep resources flowing into the project and hope that problems can be corrected before anyone notices. The project manager may also be reluctant to share status information because of contractual penalties for failure to meet the schedule or to develop certain system functions. In this situation, the client may not be informed about a problem until it has become a crisis. After the truth comes out, finger-pointing and heated discussions about cost overruns, missed schedules, and technical incompetence can lead to charges of fraud, misrepresentation, and breach of contract.

Fraud is the crime of obtaining goods, services, or property through deception or trickery. Fraudulent misrepresentation occurs when a person consciously decides to induce another person to rely and act on a misrepresentation. To prove fraud in a court of law, prosecutors must demonstrate the following elements:

- The wrongdoer made a false representation of material fact.
- The wrongdoer intended to deceive the innocent party.
- The innocent party justifiably relied on the misrepresentation.
- The innocent party was injured.

The filing of false medical claims has increased in recent years and is likely to continue to increase as changes in medical care coverage evolve and are debated in the courts. As more patient information is being implemented in electronic medical health records, computer fraud involving that information is increasingly likely. In response to this threat, the Health Care Fraud Prevention and Enforcement Action Team (HEAT) was formed as a joint initiative of the Department of Justice, the Department of Health and Human Services Office of Inspector General, and the Centers for Medicare and Medicaid Services. The goal is to cut down on the amount of healthcare fraud by coordinating the efforts of multiple agencies. In time, the Department of Justice and the Department of Health and Human Services hope to develop a system for sharing data and monitoring fraud attempts in real time. Within its first year of operation, HEAT sought $500 million in restitution to the Medicare program from some 300 criminal healthcare fraud cases involving over 560 defendants. During that time, HEAT also recovered over $2 billion in civil penalties.[17]

Misrepresentation is the misstatement or incomplete statement of a material fact. If the misrepresentation causes the other party to enter into a contract, that party may have the legal right to cancel the contract or seek reimbursement for damages.

As an example of misrepresentation, consider the experience of BSkyB, which operates a multichannel TV and radio service that offers access to over 500 TV and radio channels to some 10 million households in the United Kingdom and Ireland.[18] In 2000, BSkyB contracted with Electronic Data Systems (EDS)—which was acquired by Hewlett-Packard in 2008 and is now known as HP Services—to implement a customer relationship management computer system. However, EDS failed to provide adequate resources to the project and missed key project milestones. BSkyB terminated the contract with EDS after eight months of little to no progress. BSkyB then sued EDS, claiming it misrepresented the time and cost required to deliver a customer relationship management system. Due to extensive discussions between the two firms, the lawsuit did not reach court until 2007. Finally, in June 2010, the lawsuit was settled, and HP Services was required to pay BSkyB a total amount of $461 million. The judge determined that in order to win the BSkyB deal, EDS had deliberately lied about the time needed to complete the project. The judge also ruled "that EDS failed to exercise reasonable skill and care or to conform to good industry practice."[19]

Breach of contract occurs when one party fails to meet the terms of a contract. Further, a **material breach of contract** occurs when a party fails to perform certain express or implied obligations, which impairs or destroys the essence of the contract. Because there is no clear line between a minor breach and a material breach, determination is made on a case-by-case basis. "When there has been a material breach of contract, the non-breaching party can either: (1) rescind the contract, seek restitution of any compensation paid under the contract to the breaching party, and be discharged from any further performance under the contract; or (2) treat the contract as being in effect and sue the breaching party to recover damages."[20]

When IT projects go wrong because of cost overruns, schedule slippage, lack of system functionality, and so on, aggrieved parties might charge fraud, fraudulent misrepresentation, and/or breach of contract. Trials can take years to settle, generate substantial legal fees, and create bad publicity for both parties. As a result, the vast majority of such disputes are settled out of court, and the proceedings and outcomes are concealed from the public. In addition, IT vendors have become more careful about protecting themselves from major legal losses by requiring that contracts place a limit on potential damages.

Most IT projects are joint efforts in which vendors and customers work together to develop a system. Assigning fault when such projects go wrong can be difficult; one side might be partially at fault while the other side is mostly at fault. Consider the following frequent causes of problems in IT projects:

- The customer changes the scope of the project or the system requirements.
- Poor communication between customer and vendor leads to performance that does not meet expectations.
- The vendor delivers a system that meets customer requirements, but a competitor comes out with a system that offers more advanced and useful features.
- The customer fails to reveal information about legacy systems or databases that make the new system extremely difficult to implement.

Clients and vendors often disagree about who is to blame in such circumstances. For example, EMFinders is a technology company that makes a product to help in the rapid location of missing children and adults. Its EmSeeQ product is an inexpensive, wearable device and activation service that can locate an individual and report that location directly to a 911 operator. Project Lifesaver International (PLI) is a nonprofit organization formed in 1999 that provides services and products to public safety agencies across the United States, Canada, and Australia involved in the search and rescue of individuals with Alzheimer's, autism, dementia, and other cognitive impairments. EmFinders accused PLI of breach of contract for failing to adopt its EmSeeQ system as its primary technology, refusing to engage in marketing activities for which it was paid, and for going so far as to publicly disparage both EmFinders and the EmSeeQ system. All this after what EmSeeQ judged to be a successful test and evaluation of its product by PLI.[21] Project Lifesaver responded to that lawsuit by countersuing EmFinders for fraud, breach of contract, and defamation—arguing that the EmSeeQ system failed independent testing in multiple geographic areas.[22]

Relationships Between IT Workers and Suppliers

IT workers deal with many different hardware, software, and service providers. Most IT workers understand that building a good working relationship with suppliers encourages the flow of useful communication as well as the sharing of ideas. Such information can lead to innovative and cost-effective ways of using the supplier's products and services that the IT worker may never have considered.

IT workers can develop good relationships with suppliers by dealing fairly with them and not making unreasonable demands. Threatening to replace a supplier who can't deliver needed equipment tomorrow, when the normal industry lead time is one week, is aggressive behavior that does not help a working relationship.

Suppliers strive to maintain positive relationships with their customers in order to make and increase sales. To achieve this goal, they may sometimes engage in unethical actions—for example, offering an IT worker a gift that is actually intended as a bribe. Clearly, IT workers should not accept a bribe from a vendor, and they must be careful in considering what constitutes a bribe. For example, accepting invitations to expensive dinners or payment of entry fees for a golf tournament may seem innocent to the recipient, but it may be perceived as bribery by an auditor.

Bribery is that act of providing money, property, or favors to someone in business or government in order to obtain a business advantage. An obvious example is a software supplier sales representative who offers money to another company's employee to get its business. This type of bribe is often referred to as a kickback or a payoff. The person who offers a bribe commits a crime when the offer is made, and the recipient is guilty of bribery upon accepting the offer. Various states have enacted bribery laws, which have sometimes been used to invalidate contracts involving bribes but have seldom been used to make criminal convictions.

The **Foreign Corrupt Practices Act (FCPA)** makes it a crime to bribe a foreign official, a foreign political party official, or a candidate for foreign political office. The act applies to any U.S. citizen or company and to any company with shares listed on any U.S. stock exchange. However, a bribe is not a crime if the payment was lawful under the laws of the foreign country in which it was paid. Penalties for violating the FCPA are severe—corporations face a fine of up to $2 million per violation, and individual violators may be fined up to $100,000 and imprisoned for up to five years.

The FCPA also requires corporations whose securities are listed in the United States to meet U.S. accounting standards by having an adequate system of internal controls, including maintaining books and records that accurately and fairly reflect their transactions. The goal of these standards is to prevent companies from using slush funds or other means to disguise payments to foreign officials. A firm's business practices and its accounting information systems must be frequently audited by both internal and outside auditors to ensure that they meet these standards.

The FCPA permits facilitating payments that are made for "routine government actions," such as obtaining permits or licenses; processing visas; providing police protection; providing phone services, power, or water supplies; or facilitating actions of a similar nature. Thus, it is permissible under the FCPA to pay an official to perform some official function faster (for example, to speed customs clearance) but not to make a different substantive decision (for example, to award business to one's firm).[23]

In 2006, U.S.-based Lucent merged with French-based Alcatel to form Alcatel-Lucent. The company provides solutions to deliver voice, data, and video communications services to end users in more than 130 countries.[24] Prior to this acquisition, Lucent was accused by the Department of Justice of bribery. It is alleged that Lucent spent millions of dollars arranging tours for Chinese officials to the United States, Europe, Australia, and other countries in exchange for winning some telecommunications equipment deals. Lucent agreed to pay a $2.5 million fine in 2008.[25] Also in 2008, a former executive of Alcatel was sentenced to 2 ½ years in prison for conspiring with a Costa Rican consultant to channel $2.5 million in bribes to win contracts with a government-controlled mobile phone operator. It is estimated that bribes amounting to $15 million may have been paid to Costa Rican officials by agents working on the firm's behalf. Shortly afterward, an Alcatel spokesman announced that "after careful review, the company has decided to reduce the number of outside consultants and agents." Some Alcatel-Lucent managers complained privately that the move would make it more difficult for the firm to compete for business in some countries.[26]

There is growing global recognition of the need to prevent corruption. The United Nations Convention Against Corruption is a legally binding global treaty designed to fight bribery and corruption. During its November 2010 meeting, Finance Ministers and Central Bank Ministers of members of the Group of 20 (G20), which includes Argentina, China, India, Japan, Russia, the United Kingdom, the United States, and 13 other countries, pledged to implement this treaty effectively. In particular, the countries pledged to put in place mechanisms for the recovery of property from corrupt officials through international cooperation in tracing, freezing, and confiscating assets. Members also pledged to adopt and enforce laws against international bribery and put in place rules to protect whistle-blowers.[27]

In some countries, gifts are an essential part of doing business. In fact, in some countries, it would be considered rude not to bring a present to an initial business meeting. In the United States, a gift might take the form of free tickets to a sporting event from a personnel agency that wants to get on your company's list of preferred suppliers. At what point does a gift become a bribe, and who decides?

The key distinguishing factor is that no gift should be hidden. A gift may be considered a bribe if it is not declared. As a result, most companies require that all gifts be declared and that everything but token gifts be declined. Some companies have a policy of pooling the gifts received by their employees, auctioning them off, and giving the proceeds to charity.

When it comes to distinguishing between bribes and gifts, the perceptions of the donor and the recipient can differ. The recipient may believe he received a gift that in no way obligates him to the donor, particularly if the gift was not cash. The donor's intentions, however, might be very different. Table 2-2 shows some distinctions between bribes and gifts.

TABLE 2-2 Distinguishing between bribes and gifts

Bribes	Gifts
Are made in secret, as they are neither legally nor morally acceptable	Are made openly and publicly, as a gesture of friendship or goodwill
Are often made indirectly through a third party	Are made directly from donor to recipient
Encourage an obligation for the recipient to act favorably toward the donor	Come with no expectation of a future favor for the donor

Source Line: Course Technology/Cengage Learning.

Relationships Between IT Workers and Other Professionals

Professionals often feel a degree of loyalty to the other members of their profession. As a result, they are often quick to help each other obtain new positions but slow to criticize each other in public. Professionals also have an interest in their profession as a whole, because how it is perceived affects how individual members are viewed and treated. (For example, politicians are not generally thought to be very trustworthy, but teachers are.) Hence, professionals owe each other an adherence to the profession's code of conduct. Experienced professionals can also serve as mentors and help develop new members of the profession.

A number of ethical problems can arise among members of the IT profession. One of the most common is **résumé inflation**, which involves lying on a résumé and claiming competence in an IT skill that is in high demand. Even though an IT worker might benefit in the short term from exaggerating his or her qualifications, such an action can hurt the profession and the individual in the long run. Customers—and society in general—might become much more skeptical of IT workers as a result.

Some studies have shown that around 30 percent of all U.S. job applicants exaggerate their accomplishments, while roughly 10 percent "seriously misrepresent" their backgrounds.[28] Résumé exaggeration is an even bigger problem in Asia. According to a recent survey conducted by the University of Hong Kong and a Hong Kong–based company specializing in preemployment screening, over 62 percent of respondents confessed to exaggerating their years of experience, previous positions held, and job responsibilities with 33 percent confessing to having exaggerated even more.[29] Table 2-3 lists the areas of a résumé that are most prone to exaggeration.

TABLE 2-3 Most frequent areas of résumé falsehood or exaggeration

Area of exaggeration	How to uncover the truth
Dates of employment	Thorough reference check
Job title	Thorough reference check
Criminal record	Criminal background check
Inflated salary	Thorough reference check
Education	Verification of education claims with universities and other training organizations
Professional licenses	Verification of license with accrediting agency
Working for fictitious company	Thorough background check

Source Line: Lisa Vaas, "Most Common Resume Lies," The Ladders, © July 17, 2009, www.theladders.com/career-advice/most-common-resume-lies.

Another ethical issue that can arise in relationships between IT workers and other professionals is the inappropriate sharing of corporate information. Because of their roles, IT workers may have access to corporate databases of private and confidential information about employees, customers, suppliers, new product plans, promotions, budgets, and so on.

As discussed in Chapter 1, this information is sometimes shared inappropriately. It might be sold to other organizations or shared informally during work conversations with others who have no need to know.

In 2010, an Apple employee was arrested on charges of accepting kickbacks for more than $1 million over several years from several of the firm's Asian suppliers. The money allegedly was paid in exchange for inside information about product plans for the Apple iPhone® and iPod®. The suppliers allegedly used the insider information to tailor their products in anticipation of Apple's future needs, thus gaining a competitive advantage. The Apple employee and an accomplice are charged with wire fraud, money laundering, receiving kickbacks, and other charges.[30]

Relationships Between IT Workers and IT Users

The term **IT user** refers to a person who uses a hardware or software product; the term distinguishes end users from the IT workers who develop, install, service, and support the product. IT users need the product to deliver organizational benefits or to increase their productivity.

IT workers have a duty to understand a user's needs and capabilities and to deliver products and services that best meet those needs—subject, of course, to budget and time constraints. IT workers also have a key responsibility to establish an environment that supports ethical behavior by users. Such an environment discourages software piracy, minimizes the inappropriate use of corporate computing resources, and avoids the inappropriate sharing of information.

Relationships Between IT Workers and Society

Regulatory laws establish safety standards for products and services to protect the public. However, these laws are less than perfect, and they cannot safeguard against all negative side effects of a product or process. Often, professionals can clearly see what effect their work will have and can take action to eliminate potential public risks. Thus, society expects members of a profession to provide significant benefits and to not cause harm through their actions. One approach to meeting this expectation is to establish and maintain professional standards that protect the public.

Clearly, the actions of an IT worker can affect society. For example, a systems analyst may design a computer-based control system to monitor a chemical manufacturing process. A failure or an error in the system may put workers or residents near the plant at risk. As a result, IT workers have a relationship with society members who may be affected by their actions. There is currently no single, formal organization of IT workers that takes responsibility for establishing and maintaining standards that protect the public. However, as discussed in the following sections, there are a number of professional organizations that provide useful professional codes of ethics to guide actions that support the ethical behavior of IT workers.

Professional Codes of Ethics

A **professional code of ethics** states the principles and core values that are essential to the work of a particular occupational group. Practitioners in many professions subscribe to a code of ethics that governs their behavior. For example, doctors adhere to varying versions of the 2,000-year-old Hippocratic oath, which medical schools offer as an affirmation to their graduating classes. Most codes of ethics created by professional organizations have two main parts: the first outlines what the organization aspires to become, and the second typically lists rules and principles by which members of the organization are expected to abide. Many codes also include a commitment to continuing education for those who practice the profession. (For examples of professional codes of ethics, see Appendices B through D.)

Laws do not provide a complete guide to ethical behavior. Just because an activity is not defined as illegal does not mean it is ethical. However, a professional code of ethics cannot be expected to provide an answer to every ethical dilemma—no code can be a definitive collection of behavioral standards. However, following a professional code of ethics can produce many benefits for the individual, the profession, and society as a whole:

- *Ethical decision making*—Adherence to a professional code of ethics means that practitioners use a common set of core values and beliefs as a guideline for ethical decision making.
- *High standards of practice and ethical behavior*—Adherence to a code of ethics reminds professionals of the responsibilities and duties that they may be tempted to compromise to meet the pressures of day-to-day business. The code also defines acceptable and unacceptable behaviors to guide professionals in their interactions with others. Strong codes of ethics have procedures for censuring professionals for serious violations, with penalties that can include the loss of the right to practice. Such codes are the exception, however, and few exist in the IT arena.
- *Trust and respect from the general public*—Public trust is built on the expectation that a professional will behave ethically. People must often depend on the integrity and good judgment of a professional to tell the truth, abstain from giving self-serving advice, and offer warnings about the potential negative side effects of their actions. Thus, adherence to a code of ethics enhances trust and respect for professionals and their profession.
- *Evaluation benchmark*—A code of ethics provides an evaluation benchmark that a professional can use as a means of self-assessment. Peers of the professional can also use the code for recognition or censure.

Professional Organizations

No one IT professional organization has emerged as preeminent, so there is no universal code of ethics for IT workers. However, the existence of such organizations is useful in a field that is rapidly growing and changing. In order to stay on top of the many new developments in their field, IT workers need to network with others, seek out new ideas, and continually build on their personal skills and expertise. Whether you are a freelance programmer or the CIO of a *Fortune* 500 company, membership in an organization of IT workers enables you to associate with others of similar work experience, develop working relationships, and exchange ideas. These organizations disseminate information through email, periodicals, Web sites, meetings, and conferences. Furthermore, in recognition of the need for professional standards of competency and conduct, many of these organizations have developed codes of ethics. Five of the most prominent IT-related professional organizations are highlighted in the following sections.

Association for Computing Machinery (ACM)

The Association for Computing Machinery (ACM) is a computing society founded in 1947 with over 97,000 student and professional members in more than 100 countries. It is international in scope with an ACM Europe, ACM India, and ACM China organization. ACM currently publishes over 50 journals and magazines and 30 newsletters—including *Communications of the ACM* (ACM's primary publication), *Tech News* (coverage of timely topics for IT professionals), and *XRDS* (for both graduate and undergraduate students considering computing careers), *RISKS Forum* (a moderated dialogue on risks to the public from computers and related systems), and *eLearn* (an online magazine about online education and training). The organization also offers a substantial digital library of bibliographic information, citations, articles, and journals. The ACM sponsors 34 special-interest groups (SIGs) representing major areas of computing. Each group provides publications, workshops, and conferences for information exchange.[31]

Institute of Electrical and Electronics Engineers Computer Society (IEEE-CS)

The Institute of Electrical and Electronics Engineers (IEEE) covers the broad fields of electrical, electronic, and information technologies and sciences. The IEEE-CS is one of the oldest and largest IT professional associations, with about 85,000 members. Founded in 1946, the IEEE-CS is the largest of the 38 societies of the IEEE. The IEEE-CS helps meet the information and career development needs of computing researchers and practitioners with technical journals, magazines, books, conferences, conference publications, and online courses. It also offers a Certified Software Development Professional (CSDP) program for experienced professionals and a Certified Software Development Associate (CSDA) credential for recent college graduates. The society sponsors many conferences, applications-related and research-oriented journals, local and student chapters, technical committees, and standards working groups.[32]

In 1993, the ACM and IEEE-CS formed a Joint Steering Committee for the Establishment of Software Engineering as a Profession. The initial recommendations of the committee were to define ethical standards, to define the required body of knowledge and recommended practices in software engineering, and to define appropriate curricula to acquire knowledge. The Software Engineering Code of Ethics and Professional Practice documents the ethical and professional responsibilities and obligations of software engineers. After a thorough review process, version 5.2 of the Software Engineering Code of Ethics was adopted by both the ACM and IEEE-CS in 1999.[33] This code of ethics is provided in Appendix B.

Association of Information Technology Professionals (AITP)

The Association of Information Technology Professionals (AITP) started in Chicago in 1951, when a group of machine accountants got together and decided that the future was bright for the IBM punched-card tabulating machines they were operating—a precursor of the modern electronic computer. They were members of a local group called the Machine

Accountants Association (MAA), which first evolved into the Data Processing Management Association in 1962 and finally the AITP in 1996.[34]

The AITP provides IT-related seminars and conferences, information on IT issues, and forums for networking with other IT workers. Its mission is to provide superior leadership and education in information technology, and one of its goals is to help members make themselves more marketable within their industry. The AITP also has a code of ethics and standards of conduct, which are presented in Appendix C. The standards of conduct are considered to be rules that no true IT professional should violate.

SysAdmin, Audit, Network, Security (SANS) Institute

The SysAdmin, Audit, Network, Security (SANS) Institute provides information security training and certification for a wide range of individuals, such as auditors, network administrators, and security managers. Each year, its programs train some 12,000 people, and a total of more than 165,000 security professionals around the world have taken one or more of its courses. SANS publishes a weekly news digest (*NewsBites*), a weekly security vulnerability digest (*@Risk*), and flash security alerts.[35] The SANS IT Code of Ethics is provided in Appendix D.

At no cost, SANS makes available a collection of some 1,200 research documents about various topics of information security. SANS also operates Internet Storm Center—a program that monitors malicious Internet activity and provides a free early warning service to Internet users—and works with Internet service providers to thwart malicious attackers.

Certification

Certification indicates that a professional possesses a particular set of skills, knowledge, or abilities, in the opinion of the certifying organization. Unlike licensing, which applies only to people and is required by law, certification can also apply to products (e.g., the Wi-Fi CERTIFIED logo assures that the product has met rigorous interoperability testing to ensure that it will work with other Wi-Fi-certified products) and is generally voluntary. IT-related certifications may or may not include a requirement to adhere to a code of ethics, whereas such a requirement is standard with licensing.

Numerous companies and professional organizations offer certifications, and opinions are divided on their value. Many employers view them as a benchmark that indicates mastery of a defined set of basic knowledge. On the other hand, because certification is no substitute for experience and doesn't guarantee that a person will perform well on the job, some hiring managers are rather cynical about the value of certifications. Most IT employees are motivated to learn new skills, and certification provides a structured way of doing so. For such people, completing a certification provides clear recognition and correlates with a plan to help them continue to grow and advance in their careers. Others view certification as just another means for product vendors to generate additional revenue with little merit attached.

Deciding on the best IT certification—and even whether to seek a certification—depends on the individual's career aspirations, existing skill level, and accessibility to training. Is certification relevant to your current job or the one to which you aspire? Does the company offering the certification have a good reputation? What is the current and potential future demand for skills in this area of certification?

Vendor Certifications

Many IT vendors—such as Cisco, IBM, Microsoft, SAP, and Oracle—offer certification programs for their products. Workers who successfully complete a program can represent themselves as certified users of a manufacturer's product. Depending on the job market and the demand for skilled workers, some certifications might substantially improve an IT worker's salary and career prospects. Certifications that are tied to a vendor's product are relevant for job roles with very specific requirements or certain aspects of broader roles. Sometimes, however, vendor certifications are too focused on technical details of the vendor's technology and do not address more general concepts.

To become certified, one must pass a written exam. Because of legal concerns about whether other types of exams can be graded objectively, most exams are presented in a multiple-choice format. A few certifications, such as the Cisco Certified Internetwork Expert (CCIE) certification, also require a hands-on lab exam that demonstrates skills and knowledge. It can take years to obtain the necessary experience required for some certifications. Courses and training material are available to help speed up the preparation process, but some training costs can be expensive. Depending on the certification, study materials can cost $1,000 or more, and in-class formal training courses often cost more than $10,000.

Industry Association Certifications

There are many available industry certifications in a variety of IT-related subject areas. Their value varies greatly depending on where people are in their career path, what other certifications they possess, and the nature of the IT job market. Listed below are ten certifications that can be instrumental in landing a new position or help boost one's salary according to a survey of 17,000 IT professionals by Dice Learning, a technical training organization.[36]

- Project Management Professional
- Microsoft Certified Systems Engineer
- Microsoft Certified Professionals
- Microsoft Certified Systems Administrators
- Cisco Certified Network Associate
- Network+
- A+
- Security+
- Certified Information Systems Security Professional
- Information Technology Infrastructure Library

The requirements for certification generally require that the individual has the prerequisite education and experience and that he or she sits for and passes an exam. In order to remain certified, the individual must typically pay an annual certification fee, earn continuing education credits, and—in some cases—pass a periodic renewal test.

Certifications from industry associations generally require a higher level of experience and a broader perspective than vendor certifications; however, industry associations often lag in developing tests that cover new technologies. The trend in IT certification is to move from purely technical content to a broader mix of technical, business, and behavioral competencies, which are required in today's demanding IT roles. This trend is evident in industry association certifications that address broader roles, such as project management and network security.

Due to the ongoing need for strong project managers, some of the most widely recognized and most sought-after certifications come from the Project Management Institute, which offers certification at several different levels, as summarized in Table 2-4.

TABLE 2-4 Available PMI certifications

Certificate	Intended for	Primary eligibility requirement
Project Management Professional (PMP)	Individuals who lead and direct projects	3–5 years of project management experience, depending on level of college education
Certified Associate of Project Management (CAPM)	Individuals who contribute to project teams	1,500 hours of experience or 23 hours of project management education
Program Management Professional (PgMP)	Individuals who achieve organizational objectives through defining and overseeing projects and resources	4–7 years of both project management and program management experience, depending on level of college education
PMI Scheduling Professional (PMI-SP)	Individuals who develop and maintain project schedules	3,500–5,000 hours of project scheduling experience and 30–40 hours of project scheduling education, depending on level of college education
PMI Risk Management Professional (PMI-RMP)	Individuals who assess and identify project risks, mitigate threats, and capitalize on unique project opportunities	3,000–4,500 hours of project risk management experience and 30–40 hours of project risk management education, depending on level of college education

Source Line: Course Technology/Cengage Learning.

There are over 400,000 people worldwide who have earned some form of PMI certification.[37] PMI certification requires that a person meet specific education and experience requirements, agree to follow the PMI Code of Ethics, and pass the necessary PMI exam for the desired certification. College business majors are well advised to obtain PMI's Certified Associate of Project Management prior to their graduation as a means of opening up future employment opportunities.

Government Licensing

In the United States, a **government license** is government-issued permission to engage in an activity or to operate a business. It is generally administered at the state level and often requires that the recipient pass a test of some kind. Some professionals must be licensed, including certified public accountants (CPAs), lawyers, doctors, various types of medical- and day-care providers, and some engineers.

States have enacted legislation to establish licensing requirements and protect public safety in a variety of fields. For example, Texas passed the Engineering Registration Act after a tragic school explosion at New London, Texas, in 1937. Under the act and subsequent revisions, only duly licensed people may legally perform engineering services for the public, and public works must be designed and constructed under the direct supervision of a licensed professional engineer. People cannot call themselves engineers or professional engineers unless they are licensed, and violators are subject to legal penalties. Most states have similar laws.

The Case for Licensing IT Workers

The days of simple, stand-alone information systems are over. Modern systems are highly complex, interconnected, and critically dependent on one another. Highly integrated enterprise resource planning (ERP) systems help multibillion-dollar companies control all of their business functions, including forecasting, production planning, purchasing, inventory control, manufacturing, and distribution. Complex computers and information systems manage and control the nuclear reactors of power plants that generate electricity. Medical information systems monitor the vital statistics of hospital patients on critical life support. Every year, local, state, and federal government information systems are entrusted with generating and distributing millions of checks worth billions of dollars to the public.

As a result of the increasing importance of IT in our everyday lives, the development of reliable, effective information systems has become an area of mounting public concern. This concern has led to a debate about whether the licensing of IT workers would improve information systems. Proponents argue that licensing would strongly encourage IT workers to follow the highest standards of the profession and practice a code of ethics. Licensing would also allow for violators to be punished. Without licensing, there are no requirements for heightened care and no concept of professional malpractice.

Issues Associated with Government Licensing of IT Workers

Australia, Great Britain, and the Canadian provinces of Ontario and British Columbia have adopted licensing for software engineers. The National Council of Examiners for Engineering and Surveying (NCEES) has developed a professional exam for electrical engineers and computer engineers. However, there are many reasons why there are few international or national licensing programs for IT professionals:

- *There is no universally accepted core body of knowledge*. The core **body of knowledge** for any profession outlines agreed-upon sets of skills and abilities that all licensed professionals must possess. At present, however, there are no universally accepted standards for licensing programmers, software engineers, and other IT workers. Instead, various professional societies, state agencies, and federal governments have developed their own standards.
- *It is unclear who should manage the content and administration of licensing exams*. How would licensing exams be constructed, and who would be responsible for designing and administering them? Would someone who passes a license exam in one state or country be accepted in another state or country? In a field as rapidly changing as IT, workers must commit to

ongoing, continuous education. If an IT worker's license were to expire every few years (like a driver's license), how often would practitioners be required to prove competence in new practices in order to renew their license? Such questions would normally be answered by the state agency that licenses other professionals.

- *There is no administrative body to accredit professional education programs*. Unlike the American Medical Association for medical schools or the American Bar Association for law schools, no single body accredits professional education programs for IT. Furthermore, there is no well-defined, step-by-step process to train IT workers, even for specific jobs such as programming. There is not even broad agreement on what skills a good programmer must possess; it is highly situational, depending on the computing environment.
- *There is no administrative body to assess and ensure competence of individual workers*. Lawyers, doctors, and other licensed professionals are held accountable to high ethical standards and can lose their license for failing to meet those standards or for demonstrating incompetence. The AITP standards of conduct state that professionals should "take appropriate action in regard to any illegal or unethical practices that come to [their] attention. However, [they should] bring charges against any person only when [they] have reasonable basis for believing in the truth of the allegations and without any regard to personal interest." The AITP code addresses the censure issue much more forcefully than other IT codes of ethics, although it has seldom, if ever, been used to censure practicing IT workers.

IT Professional Malpractice

Negligence has been defined as not doing something that a reasonable person would do, or doing something that a reasonable person would not do. **Duty of care** refers to the obligation to protect people against any unreasonable harm or risk. For example, people have a duty to keep their pets from attacking others and to operate their cars safely. Similarly, businesses must keep dangerous pollutants out of the air and water, make safe products, and maintain safe operating conditions for employees.

The courts decide whether parties owe a duty of care by applying a **reasonable person standard** to evaluate how an objective, careful, and conscientious person would have acted in the same circumstances. Likewise, defendants who have particular expertise or competence are measured against a **reasonable professional standard**. For example, in a medical malpractice suit based on improper treatment of a broken bone, the standard of measure would be higher if the plaintiff were an orthopedic surgeon rather than a general practitioner. In the IT arena, consider a hypothetical negligence case in which an employee inadvertently destroyed millions of customer records in an Oracle database. The standard of measure would be higher if the plaintiff were a licensed, Oracle-certified database administrator (DBA) with 10 years of experience rather than an unlicensed systems analyst with no DBA experience or specific knowledge of the Oracle software.

If a court finds that a defendant actually owed a duty of care, it must then determine whether the duty was breached. A **breach of the duty of care** is the failure to act as a reasonable person would act. A breach of duty might consist of an action, such as throwing a lit cigarette into a fireworks factory and causing an explosion, or a failure to act when there is a duty to do so—for example, a police officer not protecting a citizen from an attacker.

Professionals who breach the duty of care are liable for injuries that their negligence causes. This liability is commonly referred to as **professional malpractice**. For example, a CPA who fails to use reasonable care, knowledge, skill, and judgment when auditing a client's books is liable for accounting malpractice. Professionals who breach this duty are liable to their patients or clients, and possibly to some third parties.

Courts have consistently rejected attempts to sue individual parties for computer-related malpractice. Professional negligence can only occur when people fail to perform within the standards of their profession, and software engineering is not a uniformly licensed profession in the United States. Because there are no uniform standards against which to compare a software engineer's professional behavior, he or she cannot be subject to malpractice lawsuits.

IT USERS

Chapter 1 outlined the general topic of how corporations are addressing the increasing risks of unethical behavior. This section focuses on improving employees' ethical use of IT, which is an area of growing concern as more companies provide employees with PCs, access to corporate information systems and data, and the Internet.

Common Ethical Issues for IT Users

This section discusses a few common ethical issues for IT users. Additional ethical issues will be discussed in future chapters.

Software Piracy

As mentioned earlier in this chapter, software piracy in a corporate setting can sometimes be directly traceable to IT professionals—they might allow it to happen, or they might actively engage in it. Corporate IT usage policies and management should encourage users to report instances of piracy and to challenge its practice.

The software piracy rate in China exceeds 80 percent, so it is clear that the IT professionals in many Chinese organizations do not take a strong stand against the practice. Even so, Microsoft was able to win a $320,000 court award in China against Shanghai-based Dazhong Insurance for running 450 illegal copies of nine different Microsoft programs.[38]

Sometimes IT users are the ones who commit software piracy. A common violation occurs when employees copy software from their work computers for use at home. When confronted, the IT user's argument might be: "I bought a home computer partly so I could take work home and be more productive; therefore, I need the same software on my home computer as I have at work." However, if no one has paid for an additional license to use the software on the home computer, this is still piracy.

The increasing popularity of the Android smartphone operating system has created a serious software piracy problem. Some IT end users have figured out how to download applications from the Android Market Web site without paying for them and then use the software or sell it to others. One legitimate Android application developer complained that his first application was pirated within a month and that the number of downloads from the pirate's site were greater than his own. Professional developers become discouraged as they watch their sales sink while pirates' sales rocket.[39]

Inappropriate Use of Computing Resources

Some employees use their computers to surf popular Web sites that have nothing to do with their jobs, participate in chat rooms, view pornographic sites, and play computer games. These activities eat away at worker productivity and waste time. Furthermore, activities such as viewing sexually explicit material, sharing lewd jokes, and sending hate email could lead to lawsuits and allegations that a company allowed a work environment conducive to racial or sexual harassment. A survey by the Fawcett Society found that one in five men admit to viewing porn at work, while a separate study found that 30 percent of mobile workers are viewing porn on their Web-enabled phones.[40,41] Organizations typically fire frequent pornography offenders and take disciplinary action against less egregious offenders. A two-and-a-half-year probe dating back to 2007 found 31 workers at the SEC who violated rules regarding the use of government computers to view pornographic material. Each of the workers was disciplined according to the extent of his or her viewing, and several workers were fired.[42]

Inappropriate Sharing of Information

Every organization stores vast amounts of information that can be classified as either private or confidential. Private data describes individual employees—for example, their salary information, attendance data, health records, and performance ratings. Private data also includes information about customers—credit card information, telephone number, home address, and so on. Confidential information describes a company and its operations, including sales and promotion plans, staffing projections, manufacturing processes, product formulas, tactical and strategic plans, and research and development. An IT user who shares this information with an unauthorized party, even inadvertently, has violated someone's privacy or created the potential that company information could fall into the hands of competitors. For example, if an employee accessed a coworker's payroll records via a human resources computer system and then discussed them with a friend, it would be a clear violation of the coworker's privacy.

Hundreds of thousands of leaked State Department documents were posted on the Web site *http://cablegate.wikileaks.org* in late 2010. As of this writing, it appears that the source of the leaks was a low-level IT user (an Army private) with access to confidential documents. The documents revealed details of behind-the-scenes international diplomacy, often divulging candid comments from world leaders and providing particulars of U.S. tactics in Afghanistan, Iran, and North Korea.[43] The leaked documents strained relations between the United States and some of its allies. It also is likely that the incident will lead to less sharing of sensitive information with the United States because of concerns over further disclosures.

Supporting the Ethical Practices of IT Users

The growing use of IT has increased the potential for new ethical issues and problems; thus, many organizations have recognized the need to develop policies that protect against abuses. Although no policy can stop wrongdoers, it can set forth the general rights and responsibilities of all IT users, establish boundaries of acceptable and unacceptable behavior, and enable management to punish violators. Adherence to a policy can improve services to users, increase productivity, and reduce costs. Companies can take several of the following actions when creating an IT usage policy.

Establishing Guidelines for Use of Company Software

Company IT managers must provide clear rules that govern the use of home computers and associated software. Some companies negotiate contracts with software manufacturers and provide PCs and software so that IT users can work at home. Other companies help employees buy hardware and software at corporate discount rates. The goal should be to ensure that employees have legal copies of all the software they need to be effective, regardless of whether they work in an office, on the road, or at home.

Defining the Appropriate Use of IT Resources

Companies must develop, communicate, and enforce written guidelines that encourage employees to respect corporate IT resources and use them to enhance their job performance. Effective guidelines allow some level of personal use while prohibiting employees from visiting objectionable Internet sites or using company email to send offensive or harassing messages.

Structuring Information Systems to Protect Data and Information

Organizations must implement systems and procedures that limit data access to just those employees who need it. For example, sales managers may have total access to sales and promotion databases through a company network, but their access should be limited to products for which they are responsible. Furthermore, they should be prohibited from accessing data about research and development results, product formulas, and staffing projections if they don't need it to do their jobs.

Installing and Maintaining a Corporate Firewall

A **firewall** is a hardware or software device that serves as a barrier between an organization's network and the Internet; a firewall also limits access to the company's network based on the organization's Internet usage policy. A firewall can be configured to serve as an effective deterrent to unauthorized Web surfing by blocking access to specific objectionable Web sites. (Unfortunately, the number of such sites is continually growing, so it is difficult to block them all.) A firewall can also serve as an effective barrier to incoming email from certain Web sites, companies, or users. It can even be programmed to block email with certain kinds of attachments (e.g., Microsoft Word® documents), which reduces the risk of harmful computer viruses.

Table 2-5 provides a manager's checklist for establishing an IT usage policy. The preferred answer to each questions is *yes*.

TABLE 2-5 Manager's checklist for establishing an IT usage policy

Question	Yes	No
Is there a statement that explains the need for an IT usage policy?		
Does the policy provide a clear set of guiding principles for ethical decision making?		
Is it clear how the policy applies to the following types of workers? • Employees • Part-time workers • Temps • Contractors		
Does the policy address the following issues? • Protection of the data privacy rights of employees, customers, suppliers, and others • Control of access to proprietary company data and information • Use of unauthorized or pirated software • Employee monitoring, including email, wiretapping and eavesdropping on phone conversations, computer monitoring, and surveillance by video • Respect of the intellectual rights of others, including trade secrets, copyrights, patents, and trademarks • Inappropriate use of IT resources, such as Web surfing, personal emailing, and other use of computers for purposes other than business • The need to protect the security of IT resources through adherence to good security practices, such as not sharing user IDs and passwords, using "hard-to-guess" passwords, and frequently changing passwords • The use of the computer to intimidate, harass, or insult others through abusive language in emails and by other means		
Are disciplinary actions defined for IT-related abuses?		
Is there a process for communicating the policy to employees?		
Is there a plan to provide effective, ongoing training relative to the policy?		
Has a corporate firewall been implemented?		
Is the corporate firewall maintained?		

Source Line: Course Technology/Cengage Learning.

Compliance

Compliance means to be in accordance with established policies, guidelines, specifications, or legislation. Records management software, for example, may be developed in compliance with the U.S. Department of Defense's Design Criteria Standard for Electronic Management Software applications (known as *DoD 5015*) that defines mandatory functional requirements for records management software used within the Department of Defense. Commercial software used within an organization should be distributed in compliance with the vendor's licensing agreement.

In the legal system, compliance usually refers to behavior in accordance with legislation—such the Sarbanes–Oxley Act of 2002, which established requirements for internal controls to govern the creation and documentation of accurate and complete financial statements, or HIPAA, which requires employers to ensure the security and privacy of employee healthcare data. Failure to be in conformance to specific pieces of legislation can lead to criminal or civil penalties specified in that legislation.

Failure to be in compliance with legislation can also lead to lawsuits from stakeholders, such as shareholders. For instance, a number of shareholder lawsuits were filed against Pfizer, the world's largest pharmaceutical company, and the firm was ordered by the courts to pay $2.3 billion to settle federal marketing investigations in 2009 that it had marketed numerous drugs for unapproved purposes. (This was the firm's fourth settlement in seven years over illegal marketing.) In response, Pfizer established a $75 million fund to pay the shareholders' attorneys' fee and to finance a newly created Regulatory and Compliance Committee to monitor legal, regulatory, and marketing activities for the company.[44]

Demonstrating compliance with multiple government and industry regulations, many with similar but sometimes conflicting requirements, has become a major challenge. As a result, many organizations have implemented specialized software to track and record compliance actions, hired management consultants to provide advice and training, and even created a new position, the Chief Compliance Officer (CCO), to deal with the issues.

In 1972, the SEC recommended that publicly held organizations establish audit committees.[45] The **audit committee** of a board of directors provides assistance to the board in fulfilling its responsibilities with respect to the oversight of the following areas of activity:

- The quality and integrity of the organization's accounting and reporting practices and controls, including the financial statements and reports
- The organization's compliance with legal and regulatory requirements
- The qualifications, independence, and performance of the company's independent auditor (a certified public accountant who provides a company with an accountant's opinion but who is not otherwise associated with the company)
- The performance of the company's internal audit team

In some cases, audit committees have uncovered violations of law and reported their findings to appropriate law enforcement agencies. For example, the audit committee of Sensata Technology (which designs, manufactures, and distributes electronic sensors and controls) conducted an investigation into whether certain company officials had violated foreign bribery laws in connection with a business deal in China. As a result of that investigation, the audit committee reported possible Foreign Corrupt Practices Act violations to the SEC and the Justice Department.[46]

In addition to an audit committee, most organizations also have an internal audit department whose primary responsibilities are to

- Determine that internal systems and controls are adequate and effective
- Verify the existence of company assets and maintain proper safeguards over their protection

- Measure the organization's compliance with its own policies and procedures
- Insure that institutional policies and procedures, appropriate laws, and good practices are followed
- Evaluate the adequacy and reliability of information available for management decision making

Although the members of the internal audit team are not typically experts in detecting and investigating financial statement fraud, they can offer advice on how to develop and test policies and procedures that result in transactions being recorded in accordance with generally accepted accounting principles (GAAP). This can go a long way toward deterring financial statement fraud. Quite often in cases of financial statement fraud, senior management (including members of the audit committee) ignored or tried to suppress the recommendations of the internal audit team, especially when red flags were raised.

The audit committee and members of the internal audit team have a major role in ensuring that both the IT organization and IT users are in compliance with the various organizational guidelines and policies as well as various legal and regulatory practices.

Summary

- The key characteristics that distinguish professionals from other kinds of workers are as follows: (1) they require advanced training and experience, (2) they must exercise discretion and judgment in the course of their work, and (3) their work cannot be standardized.
- A professional is expected to contribute to society, to participate in a lifelong training program, to keep abreast of developments in the field, and to help develop other professionals.
- From a legal standpoint, a professional has passed the state licensing requirements (if they exist) and earned the right to practice there.
- From a legal perspective, IT workers are not recognized as professionals because they are not licensed by the state or federal government. As a result, IT workers are not liable for malpractice.
- As members of the professional services industry, IT workers must be cognizant of seven major factors that are transforming the professional services industry: (1) increased client sophistication, (2) greater governance requirements, (3) increased connectivity, (4) more transparency, (5) increased need for modularization, (6) growing globalization, and (7) greater commoditization.
- IT professionals typically become involved in many different relationships, each with its own set of ethical issues and potential problems.
- In relationships between IT professionals and employers, important issues include setting and enforcing policies regarding the ethical use of IT, the potential for whistle-blowing, and the safeguarding of trade secrets.
- In relationships between IT professionals and clients, key issues revolve around defining, sharing, and fulfilling each party's responsibilities for successfully completing an IT project.
- A major goal for IT professionals and suppliers is to develop good working relationships in which no action can be perceived as unethical.
- In relationships between IT workers, the priority is to improve the profession through activities such as mentoring inexperienced colleagues and demonstrating professional loyalty.
- Résumé inflation and the inappropriate sharing of corporate information are relevant problems.
- In relationships between IT professionals and IT users, important issues include software piracy, inappropriate use of IT resources, and inappropriate sharing of information.
- When it comes to the relationship between IT workers and society at large, the main challenge for IT workers is to practice the profession in ways that cause no harm to society and provide significant benefits.
- A professional code of ethics states the principles and core values that are essential to the work of an occupational group.
- A code of ethics serves as a guideline for ethical decision making, promotes high standards of practice and ethical behavior, enhances trust and respect from the general public, and provides an evaluation benchmark.
- Several IT-related professional organizations have developed a code of ethics, including ACM, AITP, IEEE-CS, and SANS.
- These codes usually have two main parts—the first outlines what the organization aspires to become, and the second typically lists rules and principles that members are expected to

live by. The codes also include a commitment to continuing education for those who practice the profession.

- Many people believe that the licensing and certification of IT workers would increase the reliability and effectiveness of information systems.
- Licensing and certification raise many issues, including the following: (1) there is no universally accepted core body of knowledge on which to test people; (2) it is unclear who should manage the content and administration of licensing exams; (3) there is no administrative body to accredit professional education programs; and (4) there is no administrative body to assess and ensure competence of individual professionals.
- The audit committee and members of the internal audit team have a major role in ensuring that both the IT organization and IT users are in compliance with organizational guidelines and policies as well as various legal and regulatory practices.

Key Terms

audit committee
body of knowledge
breach of contract
breach of duty of care
bribery
Business Software Alliance (BSA)
certification
commoditization
compliance
conflict of interest
duty of care
firewall
Foreign Corrupt Practices Act (FCPA)
fraud
globalization
government license
IT user
material breach of contract
misrepresentation
modularization
negligence
profession
professional code of ethics
professional malpractice
reasonable person standard
reasonable professional standard
résumé inflation
trade secret
transparency
whistle-blowing

Self-Assessment Questions

The answers to the Self-Assessment Questions can be found in Appendix E.

1. A professional is someone who:
    a. requires advanced training and experience
    b. must exercise discretion and judgment in the course of his or her work
    c. does work that cannot be standardized
    d. all of the above

2. Although end users often get the blame when it comes to using illegal copies of commercial software, software piracy in a corporate setting is sometimes directly traceable to ____________.
3. The mission of the Business Software Alliance is to ____________.
4. Whistle-blowing is an effort by an employee to attract attention to a negligent, illegal, unethical, abusive, or dangerous act by a company that threatens the public interest. True or False?
5. ____________ is the crime of obtaining goods, services, or property through deception or trickery.
6. ____________means to be in accordance with established policies, guidelines, specifications, or legislation.
7. Society expects professionals to act in a way that:
   a. causes no harm to society
   b. provides significant benefits
   c. establishes and maintains professional standards that protect the public
   d. all of the above
8. Most organizations have a(n) ____________ team with primary responsibilities to determine that internal systems and controls are adequate and effective.
9. ____________ is a process that one undertakes voluntarily to prove competency in a set of skills.
   a. Licensing
   b. Certification
   c. Registering
   d. all of the above
10. Senior management (including members of the audit committee) has the option of ignoring or suppressing recommendations of the internal audit committee. True or False?
11. ____________ has been defined as not doing something that a reasonable person would do or doing something that a reasonable person would not do.
12. A ____________ states the principles and core values that are essential to the work of a particular occupational group.

Discussion Questions

1. How do you distinguish between misrepresentation and embellishment of one's professional accomplishments in order to win a contract to complete a major project? Provide an example of an embellishment that would not be considered misrepresentation.
2. Do laws provide a complete guide to ethical behavior? Can an activity be legal but not ethical?

3. What is professional malpractice? Can an IT worker be sued for professional malpractice? Why or why not?
4. Does charging by the hour encourage unethical behavior on the part of contract workers and consultants?
5. What must IT professionals do to ensure that the projects they lead meet the client's expectations and do not lead to charges of fraud, fraudulent misrepresentation, or breach of contract?
6. Should all IT professionals be either licensed or certified? Why or why not?
7. What commonalities do you find among the IT professional codes of ethics discussed in this chapter? What differences are there? Do you think there are any important issues not addressed by these codes of ethics?
8. What issues could arise if you were to report a failure to follow good accounting principles and practices to a member of your firm's audit committee?

What Would You Do?

1. Your old roommate from college was recently let go from his firm during a wave of employee terminations to reduce costs. You two have kept in touch over the six years since school, and he has asked you to help him get a position in the IT organization where you work. You offered to review his résumé, make sure that it gets to the "right person," and even put in a good word for him. However, as you read the résumé, it is obvious that your friend has greatly exaggerated his accomplishments at his former place of work and even added some IT-related certifications that you are sure he never earned. What would you do?
2. You are a U.S. citizen currently on a three-year assignment as the IT manager for your company's Armenian manufacturing plant. The company is a U.S.-based *Fortune* 1000 company, and the Armenian plant employs 1,500 workers. The plant budgeted for 250 new computers and associated software to replace all computers over six years old. The cost of the software licenses is $125,000 (USD). Unfortunately, the plant has been hit hard by the worldwide financial crises, and you have been told that your department must reduce its budget by $100,000 (USD). You have been focusing on two options. The first is not to purchase all the software licenses needed and instead make illegal copies of existing software. (The software piracy rate in Armenia is 90 percent.) The second is to terminate the employment of two IT employees. You have identified who would be terminated based on their recent job performance and their relatively low level of anticipated future work activity. What would you do?
3. You are the human resources contact for your firm's IT organization. The daughter of the firm's CIO is scheduled to participate in a job interview for an entry-level position in the IT organization next week. It will be your responsibility to meet with the three people who will interview her to form an assessment and make a group decision about whether or not she will be offered the position and, if so, at what salary. How do you handle this situation?
4. You are in charge of awarding all PC service contracts for your employer. In recent emails with the company's PC service contractor, you casually exchanged ideas about home

landscaping, your favorite pastime. You also mentioned that you would like to have a few Bradford pear trees in your yard. Upon returning from a vacation, you discover three mature trees in your yard, along with a thank-you note in your mailbox from the PC service contractor. You really want the trees, but you didn't mean for the contractor to buy them for you. You suspect that the contractor interpreted your email comment as a hint that you wanted him to buy the trees. You also worry that the contractor still has the email. If the contractor sent your boss a copy, it might look as if you were trying to solicit a bribe. Can the trees be considered a bribe? What would you do?

5. Your organization is preparing to submit a bid for a multimillion-dollar contract in South America. The contract is extremely important to your firm and would represent its first contract in South America. While meeting with your local contacts, you are introduced to a consultant who offers to help your firm prepare and submit its bid as well as to negotiate with the prospective customer company. The consultant is quite impressive in his knowledge of local government officials and managers and executives at the customer's company. The fee requested is only 1 percent of the potential value of the contract, but it is unclear exactly what the consultant will do. Later that day, your local contacts tell you that the use of such consultants is common. They say that they are familiar with this particular consultant and that he has a good reputation for getting results. Your company has never worked with such consultants in the past, and you are uncertain how to proceed.
6. You are an experienced, mid-level manager in your firm's IT organization. One of your responsibilities is to screen résumés for job openings in the organization. You are in the process of reviewing more than 100 résumés you received for a position as an Oracle database administrator. Your goal is to trim the group down to the top five candidates to invite to an in-house interview. About half the résumés are from IT workers with less than three years of experience who claim to have one or more Oracle certifications. There are also a few candidates with over five years of impressive experience but no Oracle certifications listed on their résumés. You were instructed to only include candidates with an Oracle certification in the list of finalists. However, you are concerned about possible résumé inflation and the heavy emphasis on certification versus experience. What would you do?

Cases

1. IBM and the State of Indiana Involved in a Breach of Contract Dispute

In December 2006, IBM and the Indiana Family and Social Services Administration (FSSA) entered into a ten-year, $1.16 billion contract to modernize the state's processes and systems for determining welfare eligibility. The state expected to generate $500 million in administrative costs savings over the life of the contract.[47]

FSSA claims it began to notice problems in the new system as early as the project's initial rollout to 10 northern Indiana counties in October 2007. As a result, further expansion was delayed. The state's lawyers wrote: "IBM assured FSSA that if the Region 2 rollout was implemented, IBM would recognize some efficiencies and economies of scale that would improve performance." Accordingly, FSSA agreed to roll out the system to the next region.[48]

By May 2008, the system had expanded into 59 of Indiana's 92 counties. In January 2009, new FSSA secretary Anne Murphy took over and halted any further expansion until IBM submitted a corrective action plan. She set a deadline of July 2009, and her request included the stipulation that the contract be cancelled if IBM failed to improve the situation by September 2009.[49] IBM estimated that addressing the issues would cost $180 million. In October 2009, the state announced it had cancelled the deal because IBM failed to make the proposed improvements to the satisfaction of the state.[50]

In May 2010, the state of Indiana sued IBM for $1.3 billion, claiming breach of contract. The Indiana FSSA claimed that system processing errors resulted in incorrect denials of benefits and delays in processing claims bringing harm to in-need citizens. The claims mishandling rate had climbed from 4 percent to 18 percent under the new system.[51] FSSA spokesman Marcus Barlow stated that "there was more staff working on eligibility during IBM's tenure than before IBM came, yet the results show that once IBM put their system in place, timeliness got worse, error rates went higher. Backlogs got larger."[52]

When the FSSA defined the project in 2006, they told IBM that, for staffing flexibility and efficiency, they wanted a system that would not assign one citizen to a single caseworker. Thus, IBM designed a task-based process that involved outsourcing 1,500 former FSSA employees to IBM. These workers interacted with welfare applicants to gather the necessary data to apply for welfare. Once these workers completed their tasks, the application was turned over to some 700 FSSA state workers who used the accumulated data to determine benefits eligibility.[53]

An IBM spokesman asserted that while there were delays in the system, it was because there were an insufficient number of workers to handle the number of claims. In addition, IBM pointed out that during contract negotiations with IBM, FSSA specified that the system be able to handle up to 4,200 applications per month. However, during the severe recession of 2008–2010, the number of applications frequently exceeded 10,000 per month.[54] The IBM spokesman made it clear that changing from the assigned caseworker approach was Indiana's idea and was not proposed by IBM.[55] FSSA has since implemented a hybrid system that incorporates the "successful elements of the old welfare delivery system" and a "modernized system." This system assigns caseworkers to welfare recipients and allows for more face-to-face contact.

In its lawsuit, Indiana is demanding that IBM refund the $437 million the state already paid to IBM. Indiana also wants reimbursement of all overtime pay state employees earned working longer hours due to problems with the system. In addition, Indiana insists that IBM be liable for any federal penalties or damages from any lawsuits filed by others because of delays in payments to citizens. IBM is countersuing Indiana to keep the $400 million it was already paid and for an additional $53 million for the equipment it left in place, which FSSA workers are now using.[56]

In a press release issued at the time the lawsuit was filed, IBM claimed that Indiana had acknowledged that the new system had reduced fraud that was estimated to cost over $100 million per year, led to creation of 1,000 new jobs, and reduced Indiana's operating expenses by $40 million per year for 2008 and 2009 with projected saving of hundreds of millions in upcoming years.[57]

Discussion Questions

1. Experienced observers point out that the development of a state social services system is always exceedingly difficult. Multiagency interaction and interdependence often leads to delays and complications in getting requirements finalized and agreed upon. And even if that is accomplished, changes in welfare policies by the state or federal government can

render those requirements invalid and require considerable rework. Given the problems that IBM encountered on this contract, should it decline the future opportunities it may have to propose a new solution for a state social services system?

2. Present a strong argument that the state of Indiana is entitled to reimbursement of all funds paid to IBM as well as reimbursement of all overtime employees were paid due to fix problems associated with the new system. Now present a strong argument that IBM should be allowed to keep all funds it has received so far for this new system.
3. Imagine that you were a judge trying this case. Form three questions whose answers would help you decide what to do in this dispute.

2. Accounting Problems at Dell Inc.

In August 2006, Dell Inc. revealed that it was being investigated by the SEC over matters related to revenue recognition and financial reporting. At the same time, Dell launched an internal accounting investigation led by the firm's audit committee. During the year-long investigation, which employed an outside law firm, Willkie Farr & Gallagher of New York, and the accounting firm KPMG LLP, over 5 million documents were reviewed. As a result of the investigation, Dell "identified a number of accounting errors, evidence of misconduct, and deficiencies in the financial control environment."[58,59] Dell eventually restated its financial statements relating to fiscal years 2003, 2004, 2005, and 2006 (including the quarterly reports within those years) as well as the first quarter of fiscal 2007.[60]

Even though the amount of earnings that was restated amounted to just $92 million (a relatively small amount for a company that earned $3 billion on sales of $49 billion in 2005), the timing of the accounting manipulations occurred during critical quarterly reporting periods when Dell earnings was just barely meeting Wall Street expectations.[61,62]

In addition to its accounting problems, Dell was struggling during this time to deal with personnel changes and the loss of U.S. market share to Hewlett-Packard. In January 2007, founder and chairman of the board Michael Dell ousted his successor Kevin Rollins, who had been CEO since July 2004, and returned to day-to-day management of the firm.[63] Michael Dell promised to cooperate with the SEC investigation and work to regain Dell's market leadership position and profitability.

In one of his first moves upon returning to the helm, Michael Dell won a reprieve from the NASDAQ stock exchange, which had threatened to de-list the company (an action that would be disastrous to its shareholders and the price of the stock) because it had missed deadlines for filing three quarterly earnings reports (Form 10-Q) and its 2007 annual report (Form 10-K) with the SEC.[64] The 2007 10-K report for the fiscal year ending February 2, 2007 normally due in March was not filed until October 30, 2007.[65] (A publicly traded company is required to file a 10-Q report on its financial performance within 35 days after the end of each of its first three quarters. Dell's second quarter ended on August 4; however, Dell could not meet this deadline due to the financial irregularities uncovered.)

Upon his return as CEO, Dell also implemented a number of changes—forcing out many senior managers and eliminating salary bonuses. In a departure from the direct sales model that once provided the firm a competitive advantage, Dell began selling personal computers in WalMart and other retailers. Dell shifted from its strategy of no acquisitions and began buying other companies in an effort to further improve profits and growth. In May 2007, Dell announced that it would reduce its workforce by some 8,800 employees or roughly 10 percent to improve profit margins.[66]

In a conference call to a number of market analysts in August 2007, Donald Carty, longtime Dell board member and chairman of the audit committee, acknowledged that Dell would be tightening its financial controls and that some employees had been fired, reassigned, or fined as a result of the investigation. However, Carty refused to say whether all Dell executives involved in the scandal had been terminated from the company. He also refused to identify any individual executive that was implicated. Carty did admit that Dell "did not maintain an effective control environment. Accounting adjustments came to be viewed as an acceptable device to compensate for operational shortfalls."[67] At a Citigroup Technology Conference in September 2007, Michael Dell stated, "I was not involved in or aware of any of the accounting irregularities. And certainly I'm not proud of what occurred at our company, but I'm proud of the company overall."[68]

Dell's problems continued into 2010, when the SEC announced in July that its five-year investigation had found that "Dell failed to properly disclose payments the company received from chipmaker Intel for using its products, rather than chips from rival Advanced Micro Devices." In fiscal year 2003, those payments represented 10 percent of Dell's operating income. In 2006, that number grew to 38 percent, and for the first quarter of 2007, the Intel payments represented 76 percent of Dell's operating income. When Dell announced it had decided to use AMD chips in its Dimension desktop computers, Intel cut its payments to Dell. The SEC alleged that Dell failed to disclose this as the cause for firm's sharp drop in earnings in the second quarter of 2007. "Dell manipulated its accounting over an extended period to project financial results that the company wished it had achieved, but could not. Dell was only able to meet Wall Street targets consistently during this period by breaking the rules" according to Christopher Conte, associate director of the SEC's Division of Enforcement in a July 2010 statement.[69]

The SEC's allegations in regards to Michael Dell were limited to alleged failure to provide adequate disclosures of the firm's commercial relationship with Intel. The allegations did not include any of the separate accounting fraud charges directed at the firm. Nor did the SEC prohibit Michael Dell from leading the firm.[70]

The company and its executives settled without admitting or denying the SEC allegations. Dell Chairman and CEO Michael Dell, former CEO Kevin Rollins, and former CFO James Schneider were charged with disclosure violations. Schneider, Vice President of Finance Nicholas Dunning, and former Assistant Controller Leslie Jackson were charged with improper accounting. Dell Inc. agreed to pay a $100 million civil penalty for failing to disclose material information to investors and for fraudulent accounting. Michael Dell was assessed $4 million in fines, former CEO Kevin Rollins accepted a $4 million penalty, and former CFO James Schneider agreed to pay $3 million.[71]

In addition to problems with the SEC, a lawsuit filed by shareholders alleged that a few executives (including Michael Dell and Kevin Rollins) made misleading statements about the company's financial results and prospects that artificially boosted the stock price. The lawsuit further alleged that the executives then sold off large portions of their own stock before the bad news about the company's financial status became public knowledge. (The stock hit a peak of $42 in January 2005 but had dropped dramatically to under $15 per share within two years.) The lawsuit was initially dismissed, but was appealed to the Fifth Circuit Court of Appeals. Prior to a ruling by this court, the parties agreed to a $40 million settlement payment to the shareholders.[72]

Discussion Questions

1. Is it possible to design accounting systems to detect suspicious entries that have a substantial impact on an organization's profit-and-loss statement? What additional processes would need to be put in place to keep senior management or others from overriding such safeguards?
2. Should the company's own internal audit organization, its audit committee, or its external audit firm have detected the accounting shenanigans that were occurring? Why did members of one of these groups not report the improprieties?
3. What actions could the Dell audit committee take to ensure that such accounting irregularities do not happen again?

3. When Certification Is Justified

On June 13, 2005, Don Tennant, editor-in-chief of *Computerworld*, published an editorial in favor of IT certification and was promptly hit with a barrage of angry responses from IT workers.[73] They argued that testable IT knowledge does not necessarily translate into quality IT work. A worker needs good communication and problem-solving skills as well as perseverance to get the job done well. Respondents explained that hardworking IT workers focus on skills and knowledge that are related to their current projects and don't have time for certifications that will quickly become obsolete. They suspected vendors of offering certification as a marketing ploy and a source of revenue. They accused managers without technical backgrounds of using certification as "a crutch, a poor but politically defensible substitute for knowing what and how well one's subordinates are doing."[74]

Any manager would certainly do well to review these insightful points, yet they beg the question: What useful purposes *can* certification serve within an organization?

Some CIOs and vice presidents of technology assert that many employers use certification as a means of training employees and increasing skill levels within the company. Some companies are even using certification as a perk to attract and keep good employees. Such companies may also enhance their employee training programs by offering a job-rotation program through which workers can acquire certification and experience as well.

Employers are also making good use of certification as a hiring gate both for entry-level positions and for jobs that require specific core knowledge. For example, a company with a Windows Server network might run an ad for a systems integration engineer and require a Microsoft Certified Systems Engineer (MCSE) certification. A company that uses Siebel customer relationship management software might require a new hire to have a certification in the latest version of Siebel.

In addition, specific IT fields such as project management and security have a greater need for certification. As the speed and complexity of production increase within the global marketplace, workers in a variety of industries are showing an increasing interest in project management certification. With mottos like "Do It, Do It Right, Do It Right Now," the Project Management Institute has already certified more than 400,000 people. IT industry employers are beginning to encourage and sometimes require project management certification.

Calls for training in the field of security management go beyond certification. The demand for security workers is expected to continue to grow rapidly in the next few years in the face of growing threats. Spam, computer viruses, spyware, botnets, and identity theft have businesses and government organizations worried. They want to make sure that their security managers can protect their data, systems, and resources.

One of the best recognized security certifications is the CISSP, awarded by the International Information Systems Security Certification Consortium. Yet the CISSP examination, like so many other IT certification examinations, is multiple choice. Employers and IT workers alike have begun to recognize the limitations of these types of examinations. They want to ensure that examinees not only have core knowledge but also know how to use that knowledge—and a multiple-choice exam, even a six-hour, 250-question exam like the CISSP, cannot provide that assurance.

Other organizations are catching on. Sun Microsystems requires the completion of programming or design assignments for some of its certifications. So, although there is no universal call for certification or a uniform examination procedure that answers all needs within the IT profession, certifying bodies are beginning to adapt their programs to better fulfill the evolving needs for certification in IT.

Discussion Questions

1. How can organizations and vendors change their certification programs to test for skills as well as core knowledge? What issues might this introduce?
2. What are the primary arguments against certification, and how can certifying bodies change their programs to overcome these shortcomings?
3. What are the benefits of certification? How might certification programs need to change in the future to better serve the needs of the IT community?

End Notes

[1] Texas Health Resources, "About Us," www.texashealth.org/body.cfm?id=125 (accessed November 7, 2010).

[2] Texas Health Resources, "Facts about Texas Health Resources," www.texashealth.org/body.cfm?id=2818 (accessed December 9, 2010).

[3] Texas Health Resources, "Advanced Technology," www.texashealth.org/landing_subsite_nt.cfm?id=3346 (accessed November 7, 2010).

[4] Texas Health Resources, "The Business of Healthcare Report," www.texashealth.org (accessed on November 7, 2010).

[5] Texas Health Resources, "News Release: Texas Health Information Technology Services Director Receives Award from CIO Magazine, Executive Council," www.texashealth.org/body.cfm?id=1629&action=detail&ref=665 (accessed November 7, 2010).

[6] U.S. Code, Title 5, Part III, Subpart F, Chapter 71, Subchapter 1, Section 7103, http://law.justia.com/us/codes/title5/5usc7103.html.

[7] Ross Dawson, "The Seven Mega Trends of Professional Services," www.rossdawsonblog.com/SevenMegaTrendsofPS.pdf (accessed January 27, 2011).

8 SAS70.com, "SAS No 70 Overview," http://sas70.com/sas70_overview.html (accessed January 11, 2011).

9 The Levin Institute/The State University of New York, "Globalization 101: What is Globalization?", www.globalization101.org/What_is_Globalization.html (accessed January 27, 2011).

10 Business Software Alliance, "Progress Through the Economic Storm: 2009 Year in Review," www.bsa.org/country/~/media/Files/General/YIR2009.ashx (accessed January 28, 2011).

11 "AutoTrader Pays $400,000 to Settle Claims of Unlicensed Software," *SecurityWeek*, August 24, 2010, www.securityweek.com/autotrader-pays-400000-settle-claims-unlicensed-software-use#.

12 Dean Wilson, "BSA Hands Out £10K Bounty to Unlicensed Software Whistleblower," *TechEYE.net*, September 6, 2010, www.techeye.net/software/bsa-hands-out-10k-bounty-to-unlicensed-software-whistleblower.

13 Robert McMillan, "Man Charged with Stealing Secrets from Wireless Company Sirf," *Computerworld*, November 16, 2010, www.computerworld.com/s/article/9196878/Man_charged_with_stealing_secrets_from_wireless_company_Sirf.

14 Patrick Thibodeau, "DOJ Involvement in Whistleblower Case Turns UP Heat on Oracle," *Computerworld*, June 21, 2010, www.computerworld.com/s/article/9178231/DOJ_involvement_in_whistleblower_case_turns_up_heat_on_Oracle_.

15 Justin Blum and David Voreacos, "Oracle Sued by U.S. in Case Claiming Overcharges," *BusinessWeek*, July 29, 2010, www.businessweek.com/news/2010-07-29/oracle-sued-by-u-s-in-case-claiming-overcharges.html.

16 Brandon Bailey, "HP Pays $55 Million in Illegal Kickback Settlement," *MercuryNews.com*, August 30, 2010.

17 Nicole Lewis, "Healthcare Fraud Losses Narrowing," *InformationWeek*, May 17, 2010, www.informationweek.com/news/healthcare/policy/showArticle.jhtml?articleID=224900055.

18 BSkyB, "Sky Our Products and Services," http://corporate.sky.com/about_sky/what_we_do/our_products_and_services.htm (accessed January 28, 2011).

19 Leo King, "HP Pays BSkyB £318m in Final Settlement of CRM Lawsuit," *Computerworld-UK*, June 7, 2010, www.computerworlduk.com/news/it-business/20593/hp-pays-bskyb-318m-in-final-settlement-of-crm-lawsuit.

20 Henry R. Cheeseman, "*Contemporary Business Law*, 3rd ed. (Upper Saddle River, NJ: Prentice Hall, 2000, 292).

21 EMFinders, "Press Release: EMFinders Terminates Contract with Project Lifesaver International and Files Breach of Contract Suit," *The Street*, November 1, 2010.

22 Project Lifesaver International, "Press Release: Project Lifesaver International Countersues EmFinders for Fraud, Defamation & Breach of Contract After EmSeeQ Tracking Device Fails Independent Testing," Yahoo! News, November 29, 2010.

[23] United States Department of Justice, "Foreign Corrupt Practices Act: Antibribery Provisions," www.justice.gov/criminal/fraud/fcpa/docs/lay-persons-guide.pdf (accessed January 28, 2011).

[24] Alcatel-Lucent, "Company Overview," www.alcatel-lucent.com/wps/portal/AboutUs/Overview.

[25] "Alcatel-Lucent Fined US$2.5 Million for Bribery in China," *China CSR*, January 2, 2008, www.chinaacsr.com/en/eoo8/01/02/1990-alcatel-lucent-fined-us25-million-for-bribery-in-china.

[26] Carol Matlack, "Alcatel-Lucent Says 'No Thanks' to Middlemen," *Europe Insight, Bloomberg Businessweek* (blog), January 7, 2009, www.businessweek.com/blogs/europeinsight/archives/2009/01/alcatel-lucent.html.

[27] "G20 Throws Weight Behind Global Anti-Corruption Treaty," *TrustLaw*, November 12, 2010, www.trust.org/trustlaw/news/g20-throws-weight-behind-global-anti-corruption-treaty.

[28] Ropella, "Hiring Smart: How to Avoid the Top Ten Mistakes," www.ropella.com/index.php/knowledge/recruitingProcessArticles/hiring_smart (accessed January 28, 2011).

[29] Leo Ma, "Resume Exaggeration in Asia Pacific," *Ezine Articles*, http://ezinearticles.com/?Resume-Exaggeration-in-Asia-Pacific&id=4788569 (accessed January 28, 2011).

[30] Paul McDougall, "Apple Worker Pleads Not Guilty to Kickbacks," *InformationWeek*, August 17, 2010, www.informationweek.com/news/software/operating_systems/showArticle.jhtml?articleID=226700356.

[31] Association for Computing Machinery, www.acm.org (accessed January 26, 2011).

[32] IEEE Computer Society, "About Us—About the Computer Society," www.computer.org/portal/web/about (accessed January 28, 2011).

[33] IEEE Computer Society, "Computer Society and ACM Approve Software Engineering Code of Ethics," The IEEE Computer Society, www.computer.org/cms/Computer.org/Publications/code-of-ethics.pdf.

[34] Association of Information Technology Professionals, "About AITP: History," www.aitp.org/organization/about/history/history.jsp.

[35] SysAdmin, Audit, Network, Security (SANS) Institute, "Information Security Training, Certification & Research," www.sans.org/about/sans.php.

[36] Denise Dubie, "Job Prospects Boosted by Microsoft, Cisco, and CompTIA Training", *ComputerworkUK*, March 11, 2010, www.computerworlduk.com/news/careers/19322/job-prospects-boosted-by-microsoft-cisco-and-comptia-training.

[37] Project Management Institute, "What are PMI Certifications," www.pmi.org/Certification/What-are-PMI-Certifications.aspx (accessed January 27, 2011).

[38] Owen Fletcher, "Microsoft Wins Piracy Case Against Chinese Company," *Computerworld*, April 22, 2010, www.pcworld.com/article/194851/microsoft_wins_piracy_case_against_chinese_company.html.

[39] John Cox, "Android Software Piracy Rampant Despite Google's Efforts to Curb," *Network World*, September 29, 1010, www.networkworld.com/news/2010/092910-google-android-piracy.html.

[40] Andres Millington, "Porn in the Workplace is Now a Major Board-Level Concern for Business," *Business Computing World*, April 23, 2010, www.businesscomputingworld.co.uk/porn-in-the-workplace-is-now-a-major-board-level-concern-for-business.

[41] Dean Wilson, "Third of Mobile Workers Distracted by Porn, Report Finds," *TechEYE.net*, June 14, 2010, www.techeye.net/mobile/third-of-mobile-workers-distracted-by-porn-report-finds.

[42] Jaikumar Vijayan, "SEC Workers Spent Hours at Work Watching Online Porn," *Computerworld*, April 23, 2010, www.computerworld.com/s/article/9175948/SEC_workers_spent_hours_at_work_watching_online_porn.

[43] Associated Press, "WikiLeaks Reveals Sensitive Diplomacy," *Cincinnati Enquirer*, November 28, 2010.

[44] Duff Wilson, "Pfizer Plans $75 Million Fund to Address Shareholder Suits," *New York Times*, December 3, 2010, www.nytimes.com/2010/12/04/business/04drug.html.

[45] Annemarie K. Keinath and Judith C. Walo, "Audit Committees Responsibilities," *The CPA Journal Online*, www.nysscpa.org/cpajournal/2004/1104/essentials/p22.htm.

[46] Shareholders Foundation, Inc. "Press Release: Sensata Technologies Holding N.V. Under Investor Investigation Over Possible Foreign Bribery," *PRLog*, October 26, 2010, www.prlog.org/11024869-sensata-technologies-holding-nv-under-investor-investigation-over-possible-foreign-bribery.html.

[47] "IBM Closes in on $1.16bn Indiana Deal," *Computer Business Review*, November 29, 2006, www.cbronline.com/news/ibm_closes_in_on_116bn_indiana_deal (accessed November 12, 2010).

[48] Associated Press, "Indiana: IBM Welfare Intake Work Flawed from Start," *Indianapolis Business Journal*, July 21, 2010, www.ibj.com/indiana-ibm-welfare-intake-work-flawed-from-start/PARAMS/article/21227.

[49] Ken Kusmer, Associated Press, "IBM on Notice over Indiana Welfare Deal, *FortWayne.com*, www.newssentinel.com/apps/pbcs.dll/article?AID=/20090708/NEWS/907080335 (accessed December 19. 2010).

[50] Audrey B., "IBM vs. Indiana: Big Blue Makes Indiana See Red," *Seeking Alpha* (blog), May 18, 2010, http://seekingalpha.com/article/205668-ibm-vs-indiana-big-blue-makes-indiana-see-red.

[51] Robert Charette, "Indiana and IBM Sue Each Other Over Failed Outsourcing Contract," *IEEE Spectrum Risk Factor* (blog), May 14, 2010, http://spectrum.ieee.org/riskfactor/computing/it/indiana-and-ibm-sue-each-other-over-failed-outsourcing-contract.

[52] Andy Opsahl, "IBM and Indiana Suing Each Other Over Canceled Outsourcing Deal," *Government Technology*, May 13, 2010, www.govtech.com/health/IBM-and-Indiana-Suing-Each-Other.html.

[53] Andy Opsahl, "IBM and Indiana Suing Each Other Over Canceled Outsourcing Deal," *Government Technology*, May 13, 2010, www.govtech.com/health/IBM-and-Indiana-Suing-Each-Other.html.

[54] Andy Opsahl, "IBM and Indiana Suing Each Other Over Canceled Outsourcing Deal," *Government Technology*, May 13, 2010, www.govtech.com/health/IBM-and-Indiana-Suing-Each-Other.html.

[55] Andy Opsahl, "IBM and Indiana Suing Each Other Over Canceled Outsourcing Deal," *Government Technology*, May 13, 2010, www.govtech.com/health/IBM-and-Indiana-Suing-Each-Other.html.

[56] Andy Opsahl, "IBM and Indiana Suing Each Other Over Cancelled Outsourcing Deal," *Government Technology*, May 13, 2010, www.govtech.com/health/IBM-and-Indiana-Suing-Each-Other.html.

[57] IBM, "Press Release: IBM Seeks Enforcement of Indiana Welfare Contract," May 13, 2010, www-03.ibm.com/press/us/en/pressrelease/31641.wss.

[58] Edward F. Moltzen, "Dell Accounting Scandal 'Not A Happy Story'—CFO," *CRN*, August 16, 2007, www.crn.com/201800702/printablearticle.htm.

[59] Ben Ames, "Dell to Delay Filings, Investigation Finds Misconduct," *Computerworld*, March 29, 2007, www.computerworld.com/s/article/print/9015089/Dell_to_delay_filings_investiga tion_finds_misconduct?taxonomyName=Security&taxonomyId=17.

[60] Agam Shah, "Dell Restates Earnings After Internal Probe," *Computerworld*, October 31, 2007, www.computerworld.com/s/article/9044858/Dell_restates_earnings_after_internal_probe.

[61] Edward F. Moltzen, "Dell Accounting Scandal 'Not A Happy Story'—CFO," *CRN*, August 16, 2007, www.crn.com/201800702/printablearticle.htm.

[62] Agam Shah, "Dell Restates Earnings After Internal Probe," *Computerworld*, October 31, 2007, www.computerworld.com/s/article/9044858/Dell_restates_earnings_after_internal_probe.

[63] Agam Shah, "Dell Restates Earnings After Internal Probe," *Computerworld*, October 31, 2007, www.computerworld.com/s/article/9044858/Dell_restates_earnings_after_internal_probe.

[64] Agam Shah, "Dell Restates Earnings After Internal Probe," *Computerworld*, October 31, 2007, www.computerworld.com/s/article/9044858/Dell_restates_earnings_after_internal_probe.

[65] Dell Inc., "Investors—SEC Filings," http://phx.corporate-ir.net/phoenix.zhtml?c=101133&p=irol-sec&control_selectgroup=Annual%20Filings.

[66] Ben Ames, "Dell to Pay $48.5M to Ousted CEO," *PCWorld*, August 9, 2007, www.pcworld.com/article/135760/dell_to_pay_485m_to_ousted_ceo.html.

[67] Edward F. Moltzen, "Dell Accounting Scandal 'Not A Happy Story'—CFO," *CRN*, August 16, 2007, www.crn.com/201800702/printablearticle.htm.

[68] Erica Ogg, "Michael Dell Speaks Out On Accounting Scandal," *Silicon.com*, September 6, 2007, www.silicon.com/technology/hardware/2007/09/06/michael-dell-speaks-out-on-accounting-scandal-39168362/.

[69] Brian Caulfield, "Dell Pays for Intel Relationship," *Forbes*, July 22, 2010, www.forbes.com/2010/07/22/intel-legal-investigation-technology-dell.html.

[70] Dell Inc., "Press Release: Dell Reaches Settlement with Securities and Exchange Commission," July 22, 2010, http://content.dell.com/us/en/corp/d/secure/2010-7-22-sec-settlement.aspx.

[71] Brian Caulfield, "Dell Pays for Intel Relationship," *Forbes*, July 22, 2010, www.forbes.com/2010/07/22/intel-legal-investigation-technology-dell.html.

[72] Tanya Roth, "Dell Lawsuit Dismissed, Then Dell Settles for $40 Million," *FindLaw* (blog), December 9, 2009, http://blogs.findlaw.com/decided/2009/12/dell-lawsuit-dismissed-then-dell-settles-for-40-million.html.

[73] Don Tennant, "Certifiably Concerned," *Computerworld*, June 13, 2005.

[74] Don Tennant, "Certifiably Mad?", *Computerworld*, June 20, 2005.

CHAPTER 3

COMPUTER AND INTERNET CRIME

QUOTE

The Internet is the first thing that humanity has built that humanity doesn't understand, the largest experiment in anarchy that we have ever had.
—Eric Schmidt, Google CEO

VIGNETTE

The Iranian Cyber Army

In June 2009, the streets of Tehran erupted in violence after the reelection of President Mahmoud Ahmadinejad. Reformers suspected voting fraud after the incumbent president received nearly 63 percent of the vote. Iranian authorities, however, cracked down hard on the protesters, killing at least 20 people.[1] Since Iran is an authoritarian regime that keeps a tight grip on the domestic media, protesters used the Internet and social networking sites, such as Twitter, to spread news of the protests to the outside world. In fact, Twitter played such a vital role that the U.S. State Department asked Twitter to delay a scheduled maintenance shutdown to avoid disrupting the flow of information out of Iran.[2] As police arrested hundreds of protesters on the street, the unrest moved into cyberspace. Activists launched denial-of-service attacks on 12 Iranian government Web sites, including news services, the ministry of foreign affairs, the ministry of justice, and the national police.[3]

Prior to this crisis, the Iranian government had limited its people's use of the Internet by blocking access to Facebook, YouTube, and the BBC.[4] Many people believe that in the wake of this cyberspace battle, the Iranian government began to take a more aggressive stance. On the evening of December 17, 2009, users logging into Twitter were redirected to a site showing a picture of the Iranian flag and a message: "This site has been hacked by the Iranian Cyber Army."[5] An email address was posted under the flag. The attack was followed by assaults on popular Web sites, including Baidu (the Chinese Google) and internal Iranian targets. The message "Stop being agents for those who are safely in the U.S. and are using you" appeared on the Web sites of the two main Iranian opposition groups.[6]

After the initial attack, many people wondered just who ran the Iranian Cyber Army, what their purpose was, and how they were connected to the Iranian regime. A PBS *Frontline* episode suggested that the Iranian government has been actively recruiting hackers inside Iran and importing technology for the operation through Dubai.[7] Although no definitive proof is available to Western journalists, *Frontline* and others have pointed out that the Iranian Cyber Army carries out attacks with impunity and that the attacks are in line with the goals of the current regime. Additionally, in May 2010, a senior officer in Iran's Revolutionary Guard Corps (IRGC) claimed that the IRGC had succeeded in establishing the world's second largest cyberarmy. However, the officer did not specifically name the group.[8]

Then, after a September 2010 assault on *TechCrunch Europe*, a popular European technology blog, the Internet security company Seculert (a company that builds software to detect cyberthreats affecting corporate networks) discovered the Iranian Cyber Army had used an exploit kit (a package of methods designed to infect a computer with malware) such as those sold by cybercriminals. During the attack, visitors to the *TechCrunch* site were redirected to a server that installed malicious software.[9] Seculert was able to trace the exploit kit to the Iranian Cyber Army because the email address on the administrative site of the exploit kit and the one that had been posted under the picture of the Iranian flag were the same. Seculert found that the Iranian Cyber Army was running a botnet, a network of computers often controlled through malware by one source.[10] Typically, botnet owners rent out space to cybercriminals, such as spammers, to generate revenue. But what was the Iranian Cyber Army doing with one? Seculert speculated that the group was jumping from simple defacement tactics to more serious cybercrime. The statistics page on the exploit kit's administrative Web site indicated that the botnet was infecting 14,000 new PCs per hour. By the time Seculert made its discovery and reported it to law enforcement, the system had already infected an estimated 20 million computers.[11]

In November 2010, shortly after the botnet discovery hit the news, the Iranian Cyber Army hacked the Web site of Farsi1, a popular satellite channel that airs Latin American soap operas dubbed in the Persian language. Iranian officials had previously condemned the channel as a tool of the West's cultural invasion and corruption of Iranian morality. The message posted on the hacked site read: "Rupert Murdoch, the Moby company, the Mohseni family [the three co-owners of the channel], and the Zionists partners should know that they will take the wish to destroy the structure of Iranian families with them to the grave."[12] This message, however, may not have simply been one threat in a larger intimidation strategy carried out by an authoritarian regime. It's quite possible that while visitors sat curiously reading the message, their computer was being bombarded with malware so that it could be hijacked and used for more destructive purposes. Yet exactly what these purposes will be and whether they will culminate in a cyberwar with the West is yet to be seen.

Cyberterrorism is certainly not unique to the Middle East. In an April 2010 survey of information technology (IT) security managers at U.S. federal agencies, over 30 percent reported that their systems had been attacked in the previous year by overseas groups or terrorists organization. Many of these agencies and departments deal with national security, including defense, foreign policy, and homeland security.[13] Cyberterrorists believed to be from China, Russia, and other countries have penetrated the U.S. electrical grid and implanted software that could be triggered to disrupt the system, if not detected.[14] And a former computer engineer at Fannie Mae planted malicious software that would have disabled monitoring alerts and all login data, deleted the root passwords to some 4,500 servers, and then erased all data and backup data. If successful, this attack would have wiped out millions of mortgage records. Fortunately, the planned attack was uncovered in time to prevent any loss of data.[15]

Questions to Consider

1. Is it ethical for the government of a country to use cyberterrorism tactics against opposition groups or another country?
2. What measures must private, public, and government-run organizations put in place to protect against future cyberterrorist attacks?

LEARNING OBJECTIVES

As you read this chapter, consider the following questions:

1. What key trade-offs and ethical issues are associated with the safeguarding of data and information systems?
2. Why has there been a dramatic increase in the number of computer-related security incidents in recent years?
3. What are the most common types of computer security attacks?
4. Who are the primary perpetrators of computer crime, and what are their objectives?
5. What are the key elements of a multilayer process for managing security vulnerabilities based on the concept of reasonable assurance?
6. What actions must be taken in response to a security incident?
7. What is computer forensics, and what role does it play in responding to a computer incident?

IT SECURITY INCIDENTS: A MAJOR CONCERN

The security of information technology used in business is of utmost importance. Confidential business data and private customer and employee information must be safeguarded, and systems must be protected against malicious acts of theft or disruption. Although the necessity of security is obvious, it must often be balanced against other business needs. Business managers, IT professionals, and IT users all face a number of ethical decisions regarding IT security, such as:

- If a firm is a victim of a computer crime, should it pursue prosecution of the criminals at all costs, maintain a low profile to avoid the negative publicity, inform its affected customers, or take some other action?
- How much effort and money should be spent to safeguard against computer crime? (In other words, how safe is safe enough?)
- If a company realizes that is has produced software with defects that make it possible for hackers to attack customer data and computers, what actions should it take?
- What should be done if recommended computer security safeguards make life more difficult for customers and employees, resulting in lost sales and increased costs?

Unfortunately, the number of IT-related security incidents is increasing. According to a benchmark study of U.S. companies conducted by the Ponemon Institute in July 2010, cyberattacks have become common occurrences. Each of the 45 companies included in the study reported that it was the victim of at least one successful attack per week.[16] Table 3-1 shows the increase in occurrence of common computer security incidents at 443 U.S-based organizations that responded to the 2009 CSI Computer Crime and Security Survey.

TABLE 3-1 Most common security incidents

Computer-related security incident	2009	2008
Respondents who experienced malware infection	64.3%	50%
Respondents who experienced denial-of-service attacks	29.2%	21%
Respondents who experienced password sniffing	17.3%	9%
Respondents who experienced Web site defacement	13.5%	6%
Respondents who experienced instant messaging abuse	7.6%	21%

Credit: "2009 Computer Security Institute Computer Crime & Security Survey," courtesy of the Computer Security Institute.

Why Computer Incidents Are So Prevalent

In today's computing environment of increasing complexity, higher user expectations, expanding and changing systems, and growing reliance on software with known vulnerabilities, it is no wonder that the number, variety, and impact of security incidents are increasing dramatically. Computer security incidents occur around the world; Table 3-2 shows the ranking of the six countries with the most malicious activity in 2009, as measured by Symantec Corporation.

TABLE 3-2 Country ranking based on level of malicious activity

Country	Rank 2009	Rank 2008
United States	1	1
China	2	2
Brazil	3	5
Germany	4	3
India	5	11
United Kingdom	6	4

Source Line: "Symantec Internet Security Threat Report: Trends for 2009," © April 2010, http://eval.symantec.com/mktginfo/enterprise/white_papers/b-whitepaper_exec_summary_internet_security_threat_report_xv_04-2010.en-us.pdf.

Increasing Complexity Increases Vulnerability

The computing environment has become enormously complex. Networks, computers, operating systems, applications, Web sites, switches, routers, and gateways are interconnected and driven by hundreds of millions of lines of code. This environment continues to increase in complexity every day. The number of possible entry points to a network expands continually as more devices are added, increasing the possibility of security breaches.

To further complicate matters, workers in many organizations operate in a **cloud computing** environment in which software and data storage are services provided via the Internet ("the cloud"); the services are run on another organization's computer hardware and are accessed via a Web browser. This represents a significant change in how data is stored, accessed, and transferred, and it raises many security concerns. The Cloud Security Alliance is a global nonprofit organization created in 2008 to define and promote the use of best practices for providing security assurance within a cloud computing environment.

Virtualization also introduces further complications into today's computer environment. **Virtualization software** emulates computer hardware by enabling multiple operating systems to run on one computer host. This masks the number and identity of individual physical servers and operating systems from end users to increase resource sharing and system utilization.

Higher Computer User Expectations

Today, time means money, and the faster computer users can solve a problem, the sooner they can be productive. As a result, computer help desks are under intense pressure to respond very quickly to users' questions. Under duress, help desk personnel sometimes forget to verify users' identities or to check whether they are authorized to perform a requested action. In addition, even though most have been warned against doing so, some computer users share their login ID and password with other coworkers who have forgotten their own passwords. This can enable workers to gain access to information systems and data for which they are not authorized.

Expanding and Changing Systems Introduce New Risks

Business has moved from an era of stand-alone computers, in which critical data was stored on an isolated mainframe computer in a locked room, to an era in which personal computers connect to networks with millions of other computers, all capable of sharing information. Businesses have moved quickly into e-commerce, mobile computing, collaborative work groups, global business, and interorganizational information systems. Information technology has become ubiquitous and is a necessary tool for organizations to achieve their goals. However, it is increasingly difficult to keep up with the pace of technological change, successfully perform an ongoing assessment of new security risks, and implement approaches for dealing with them.

Increased Reliance on Commercial Software with Known Vulnerabilities

In computing, an **exploit** is an attack on an information system that takes advantage of a particular system vulnerability. Often this attack is due to poor system design or implementation. Once the vulnerability is discovered, software developers create and issue a "fix," or patch, to eliminate the problem. Users of the system or application are responsible for obtaining and installing the patch, which they can usually download from the Web. (These fixes are in addition to other maintenance and project work that software developers perform.) Any delay in installing a patch exposes the user to a potential security breach. The need to install a fix to prevent a hacker from taking advantage of a known system vulnerability can create an ethical dilemma for system support personnel trying to balance a busy work schedule. Should they install a patch that, if left uninstalled, could lead to a security breach or should they complete assigned project work so that the anticipated project savings and benefits from the project can begin to accrue on schedule?

The Open Source Vulnerability Database (OSVDB), the Computer Emergency Response Team (CERT), and the Common Vulnerabilities and Exposures (CVE) are all organizations that capture and report information about software vulnerabilities. Table 3-3 shows that while the number of reported vulnerabilities declined from 2006 to 2008, the number of vulnerabilities reported to these three organizations still ranged from 15 to over 22 per day in 2008.

TABLE 3-3 Total number of reported vulnerabilities

Year	OSVDB	CERT	CVE
2008	8,188	6,058	5,634
2007	8,922	7,235	6,515
2006	10,709	8,064	6,608

Source Line: Jaziar Radianti, Jose J. Gonzalez, and Eliot Rich, "A Quest for a Framework to Improve Software Security: Vulnerability Black Markets Scenario," ISDC 2009 Conference, © 2009, www.systemdynamics.org/conferences/2009/proceed/papers/P1386.pdf.

In recent years, the number of vulnerabilities uncovered in application software has been much greater than those discovered in operating systems or browser software. All these bugs and potential vulnerabilities create a serious work overload for the software developers responsible for providing security fixes and for the IT workers responsible for installing the patches.

In mid-December 2010, Microsoft announced that it was delivering 17 security updates to patch 40 vulnerabilities in Windows, Internet Explorer, Office, SharePoint, and Exchange. For the year, Microsoft issued 266 patches to fix vulnerabilities—more than one per work day.[17] On the same timing, Mozilla announced that it was delivering patches for 13 vulnerabilities in its Firefox browser software.[18] Some observers have commented that the sheer volume of patches raises ethical issues concerning the software development practices of major software development firms and their apparent emphasis on release of new software over the release of quality software.

The timing of all these patches was less than ideal for overworked IT organizations that were forced to add this patch work to their already busy work schedules cut short by the holidays and during an especially busy time of the year, especially for retailers. It is likely that many IT organizations waited until after the start of the new year to make time to install the patches, thus exposing their systems to those vulnerabilities for a few additional weeks.

Clearly, it can be difficult to keep up with all the required patches. Of special concern is a **zero-day attack** that takes place before the security community or software developer knows about the vulnerability or has been able to repair it.

U.S. companies increasingly rely on commercial software with known vulnerabilities. Even when vulnerabilities are exposed, many corporate IT organizations prefer to use already installed software "as is" rather than implement security fixes that will either make the software harder to use or eliminate "nice-to-have" features suggested by current users or potential customers that will help sell the software.

Types of Exploits

There are numerous types of computer attacks, with new varieties being invented all the time. This section will discuss some of the more common attacks, including the virus, worm, Trojan horse, distributed denial-of-service, rootkit, spam, phishing, spear-phishing, smishing, and vishing.

While we usually think of such exploits being aimed at computers, smartphones such as Apple's iPhone, Research In Motion's BlackBerry, and numerous smartphones based on Google's Android operating system are becoming increasingly computer capable. They can store personal identity information, including credit card numbers and bank account numbers. They can be used to surf the Web and transact business electronically. The more people use their smartphones for these purposes, the more attractive these devices become as a target for cyberthieves. Some IT security experts warn that it will not be long before we see exploits directed at these devices to steal users' data or turn them into remote-controlled bots. "There are more phones on the planet than computers. And it's easier to steal money from phones," according to Mikko Hypponen, chief research officer at security firm F-Secure Corp.[19]

Viruses

Computer virus has become an umbrella term for many types of malicious code. Technically, a **virus** is a piece of programming code, usually disguised as something else, that causes a computer to behave in an unexpected and usually undesirable manner. Often a virus is attached to a file, so that when the infected file is opened, the virus executes. Other viruses sit in a computer's memory and infect files as the computer opens, modifies, or creates them. Most viruses deliver a "payload," or malicious software that causes the computer to perform in an unexpected way. For example, the virus may be programmed to display a certain message on the computer's display screen, delete or modify a certain document, or reformat the hard drive.

A true virus does not spread itself from computer to computer. A virus is spread to other machines when a computer user opens an infected email attachment, downloads an infected program, or visits infected Web sites. In other words, viruses spread by the action of the "infected" computer user.

Macro viruses have become a common and easily created form of virus. Attackers use an application macro language (such as Visual Basic or VBScript) to create programs that infect documents and templates. After an infected document is opened, the virus is executed and infects the user's application templates. Macros can insert unwanted words, numbers, or phrases into documents or alter command functions. After a macro virus infects a user's application, it can embed itself in all future documents created with the application.

Worms

Unlike a computer virus, which requires users to spread infected files to other users, a **worm** is a harmful program that resides in the active memory of the computer and duplicates itself. Worms differ from viruses in that they can propagate without human intervention, sending copies of themselves to other computers by email or Internet Relay Chat (IRC).

The negative impact of a worm attack on an organization's computers can be considerable—lost data and programs, lost productivity due to workers being unable to use their computers, additional lost productivity as workers attempt to recover data and programs, and lots of effort for IT workers to clean up the mess and restore everything to as close to normal as possible. The cost to repair the damage done by each of the Code Red, SirCam, and Melissa worms was estimated to exceed $1 billion, with that of the Conficker, Storm, and ILOVEYOU worms totaling well over $5 billion.[20,21]

Trojan Horses

A **Trojan horse** is a program in which malicious code is hidden inside a seemingly harmless program. The program's harmful payload might be designed to enable the hacker to destroy hard drives, corrupt files, control the computer remotely, launch attacks against other computers, steal passwords or Social Security numbers, or spy on users by recording keystrokes and transmitting them to a server operated by a third party.

A Trojan horse can be delivered as an email attachment, downloaded from a Web site, or contracted via a removable media device such as a CD/DVD or USB memory stick. Once an unsuspecting user executes the program that hosts the Trojan horse, the malicious payload is automatically launched as well—with no telltale signs. Common host programs include screen savers, greeting card systems, and games.

Software that accompanied the Energizer DUO—a USB-powered nickel-metal hydride battery recharger—contained a Trojan horse that enables hackers to gain access to a Windows personal computer. The software was designed to show the status of the battery-charging operation. The software was sold in the United States, Latin America, and Asia, but Energizer removed the software from its download site once it learned of the problem. The Energize DUO has since been discontinued.[22]

Another type of Trojan horse is a **logic bomb**, which executes when it is triggered by a specific event. For example, logic bombs can be triggered by a change in a particular file, by typing a specific series of keystrokes, or by a specific time or date.

Distributed Denial-of-Service (DDoS) Attacks

A **distributed denial-of-service attack (DDoS)** is one in which a malicious hacker takes over computers via the Internet and causes them to flood a target site with demands for data and other small tasks. A distributed denial-of-service attack does not involve infiltration of the targeted system. Instead, it keeps the target so busy responding to a stream of automated requests that legitimate users cannot get in—the Internet equivalent of dialing a telephone number repeatedly so that all other callers hear a busy signal. The targeted machine "holds the line open" while waiting for a reply that never comes, and eventually the requests exhaust all resources of the target, as illustrated in Figure 3-1.

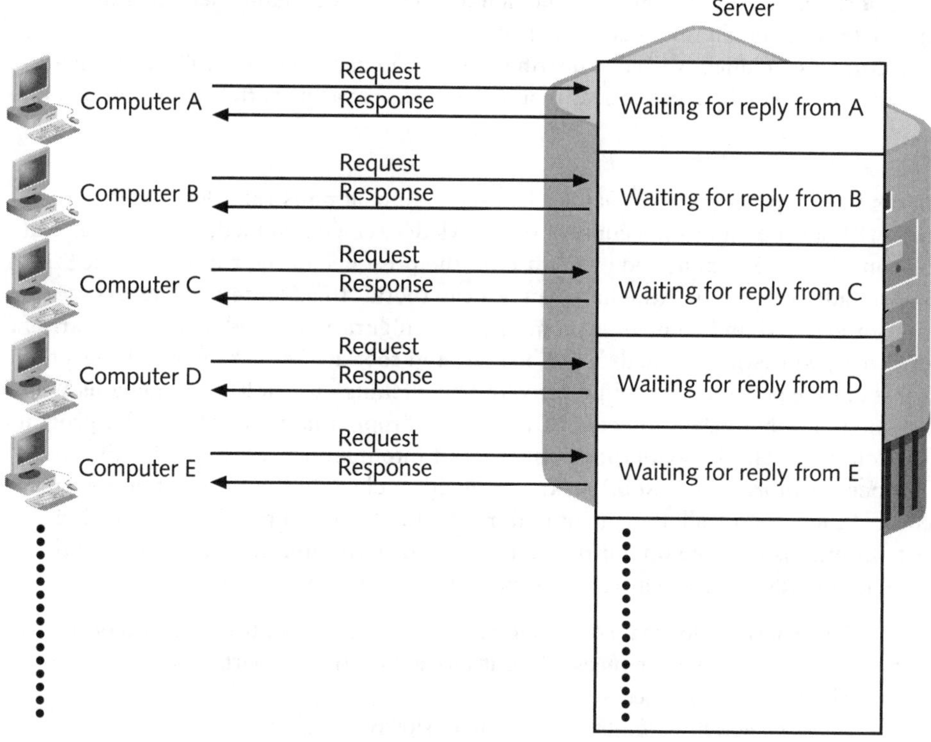

FIGURE 3-1 Distributed denial-of-service attack
Source Line: Course Technology/Cengage Learning.

The software to initiate a denial-of-service attack is simple to use and readily available at hacker sites. A tiny program is downloaded surreptitiously from the attacker's computer to dozens, hundreds, or even thousands of computers all over the world. The term **botnet** is used to describe a large group of such computers, which are controlled from one or more remote locations by hackers, without the knowledge or consent of their owners. Based on a command by the attacker or at a preset time, the botnet computers (called **zombies**) go into action, each sending a simple request for access to the target site again and again—dozens of times per second.

The zombies involved in a denial-of-service attack are often seriously compromised and are left with more enduring problems than their target. As a result, zombie machines need to be inspected to ensure that the attacker software is completely removed from the system. In addition, system software must often be reinstalled from a reliable backup to reestablish the system's integrity, and an upgrade or patch must be implemented to eliminate the vulnerability that allowed the attacker to enter the system.

Operation Payback, a group of online activists, launched repeated attacks on the Web sites of the Motion Picture Association of America and the Recording Industry Association of America because those organizations had stepped up their efforts to eliminate the unlicensed copying of music and movies. Most of the attacks involved the use of a botnet consisting of thousands of computers to launch a distributed denial-of-service attack so that legitimate users could not access the site.[23]

Botnets are also frequently used to distribute spam and malicious code. The collective processing capacity of some botnets exceeds that of the world's most powerful supercomputers.

Rootkits

A **rootkit** is a set of programs that enables its user to gain administrator level access to a computer without the end user's consent or knowledge. Once installed, the attacker can gain full control of the system and even obscure the presence of the rootkit from legitimate system administrators. Attackers can use the rootkit to execute files, access logs, monitor user activity, and change the computer's configuration. Rootkits are one part of a blended threat, consisting of the dropper, loader, and rootkit. The dropper code gets the rootkit installation started and can be activated by clicking on a link to a malicious Web site in an email or opening an infected PDF file. The dropper launches the loader program and then deletes itself. The loader loads the rootkit into memory; at that point the computer has been compromised. Rootkits are designed so cleverly that it is difficult even to discover if they are installed on a computer. The fundamental problem with trying to detect a rootkit is that the operating system currently running cannot be trusted to provide valid test results. Here are some symptoms of rootkit infections:

- The computer locks up or fails to respond to input from the keyboard or mouse.
- The screen saver changes without any action on the part of the user.
- The taskbar disappears.
- Network activities function extremely slowly.

When it is determined that a computer has been infected with a rootkit, there is little to do but reformat the disk; reinstall the operating system and all applications; and reconfigure the user's settings, such as mapped drives. This can take hours, and the user may be left with a basic working machine, but all locally held data and settings may be lost.

The Alureon rootkit (which also goes by the names TDL, TLD3, and Tidserv) is able to infect 32-bit and 64-bit Windows personal computers. This malware changes configurations to enable hackers to access files on the compromised computer. It can also alter the Windows directory, download corrupt files, and monitor user activities to obtain valuable information including logon details. The rootkit was initially successful only at infecting 32-bit systems; however, within a few months of its release, a new version of the rootkit began circulating. This version is able to sidestep anti-rootkit protections that Microsoft built into 64-bit Windows.[24]

Spam

Email spam is the abuse of email systems to send unsolicited email to large numbers of people. Most spam is a form of low-cost commercial advertising, sometimes for questionable products such as pornography, phony get-rich-quick schemes, and worthless stock. Spam is also an extremely inexpensive method of marketing used by many legitimate organizations. For example, a company might send email to a broad cross section of potential customers to announce the release of a new product in an attempt to increase initial sales. Spam is also used to deliver harmful worms and other malware.

The cost of creating an email campaign for a product or service is several hundred to a few thousand dollars, compared to tens of thousands of dollars for direct-mail campaigns. In addition, email campaigns take only a couple of weeks to develop, compared with three months or more for direct-mail campaigns, and the turnaround time for feedback averages 48 hours for email as opposed to weeks for direct mail. However, the benefits of spam to companies can be largely offset by the public's generally negative reaction to receiving unsolicited ads.

Spam forces unwanted and often objectionable material into email boxes, detracts from the ability of recipients to communicate effectively due to full mailboxes and relevant emails being hidden among many unsolicited messages, and costs Internet users and service providers millions of dollars annually. It takes users time to scan and delete spam email, a cost that can add up if they pay for Internet connection charges on an hourly basis. It also costs money for Internet service providers (ISPs) and online services to transmit spam, which is reflected in the rates charged to all subscribers.

The Controlling the Assault of Non-Solicited Pornography and Marketing (CAN-SPAM) Act went into effect in January 2004. The act says that it is legal to spam, provided the messages meet a few basic requirements—a spammer cannot disguise its identity by using a false return address, the email must include a label specifying that it is an ad or a solicitation, and the email must include a way for recipients to indicate that they do not want future mass mailings.

Several companies—such as Google, Microsoft, and Yahoo!—offer free email services. Spammers often seek to use email accounts from such major, free, and reputable Web-based email service providers, as their spam can be sent at no charge and is less likely to be blocked. Spammers can defeat the registration process of the free email services by launching a coordinated bot attack that can sign up for thousands of email accounts. These accounts are then used by the spammers to send thousands of untraceable email messages for free. An amazing 91.9 percent of all email worldwide is spam, with 95 percent of all spam generated by botnets, according Symantec.[25]

A partial solution to this problem is the use of CAPTCHA to ensure that only humans obtain free accounts. **CAPTCHA (Completely Automated Public Turing Test to Tell Computers and Humans Apart)** software generates and grades tests that humans can pass but all but the most sophisticated computer programs cannot. For example, humans can read the distorted text in Figure 3-2, but simple computer programs cannot.

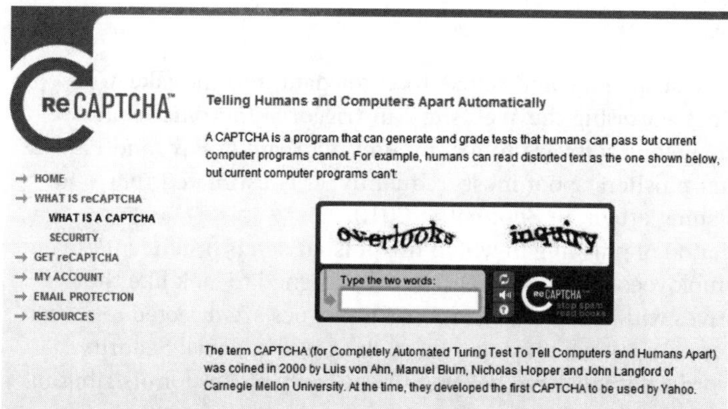

FIGURE 3-2 Example of CAPTCHA

Source Line: CAPTCHA example from www.recaptcha.net. Courtesy of Carnegie Mellon University.

Phishing

Phishing is the act of fraudulently using email to try to get the recipient to reveal personal data. In a phishing scam, con artists send legitimate-looking emails urging the recipient to take action to avoid a negative consequence or to receive a reward. The requested action may involve clicking on a link to a Web site or opening an email attachment. These emails, such as the one shown in Figure 3-3, lead consumers to counterfeit Web sites designed to trick them into divulging personal data.

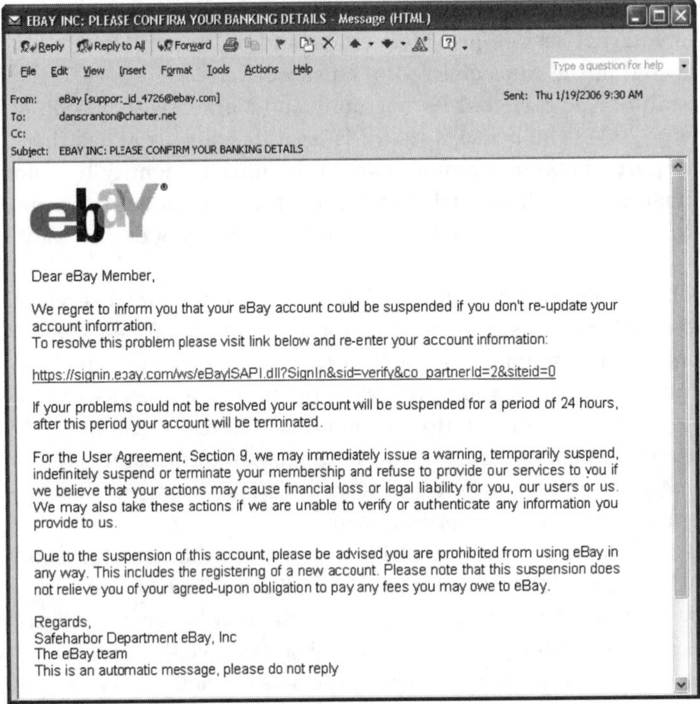

EBAY INC: PLEASE CONFIRM YOUR BANKING DETAILS - Message (HTML)

From: eBay [suppor:_id_4726@ebay.com]
Sent: Thu 1/19/2306 9:30 AM
To: danscranton@charter.net
Cc:
Subject: EBAY INC: PLEASE CONFIRM YOUR BANKING DETAILS

ebY

Dear eBay Member,

We regret to inform you that your eBay account could be suspended if you don't re-update your account information.
To resolve this problem please visit link below and re-enter your account information:

https://signin.e3ay.com/ws/eBayISAPI.dll?SignIn&sid=verify&co_partnerId=2&siteid=0

If your problems could not be resolved your account will be suspended for a period of 24 hours, after this period your account will be terminated.

For the User Agreement, Section 9, we may immediately issue a warning, temporarily suspend, indefinitely suspend or terminate your membership and refuse to provide our services to you if we believe that your actions may cause financial loss or legal liability for you, our users or us. We may also take these actions if we are unable to verify or authenticate any information you provide to us.

Due to the suspension of this account, please be advised you are prohibited from using eBay in any way. This includes the registering of a new account. Please note that this suspension does not relieve you of your agreed-upon obligation to pay any fees you may owe to eBay.

Regards,
Safeharbor Department eBay, Inc
The eBay team
This is an automatic message, please do not reply

FIGURE 3-3 Example of phishing
Source Line: Course Technology/Cengage Learning.

Savvy users often become suspicious and refuse to enter data into the fake Web sites; however, sometimes just accessing the Web site can trigger an automatic and unnoticeable download of malicious software to a computer. Citibank, eBay, and PayPal are among the Web sites that phishers spoof most frequently. It is estimated that 1 in 382 emails comprised a phishing attack in September 2010.[26]

Spear-phishing is a variation of phishing in which the phisher sends fraudulent emails to a certain organization's employees. The phony emails are designed to look like they came from high-level executives within the organization. Employees are directed to a fake Web site and then asked to enter personal information, such as name, Social Security number, and network passwords. Botnets have become the primary means for distributing phishing scams.

Phishers have grown quite sophisticated from the days when the subject lines of their email messages read, "Photo of boss at party!" A single eastern-European crime syndicate called Avalanche initiated nearly two-thirds of all phishing attacks in the second half of 2009. The organization set up counterfeit Web sites of some 40 financial services companies, online services, and job search providers. Avalanche hosted its dummy domains on a massive botnet of compromised PCs and used advanced techniques to hide its host server.[27]

Smishing and Vishing

Smishing is another variation of phishing that involves the use of Short Message Service (SMS) texting; in a smishing scam, people receive a legitimate-looking text message on their phone telling them to call a specific phone number or to log on to a Web site. This is often under done under the guise that there is a problem with their bank account or credit card that requires immediate attention. However, the phone number or Web site is phony and is used to trick unsuspecting victims into providing personal information such as a bank account number, personal identification number, or credit card number. This information can be used to steal money from victims' bank accounts, charge purchases on their credit cards, or open new accounts. In some cases, if victims log onto a Web site, malicious software is downloaded onto their phones that provides criminals with access to information stored on the phones. The number of smishing scams increases around the holidays as people use their cell phones to make online purchases. **Vishing** is similar to smishing except that the victims receive a voice mail telling them to call a phone number or access a Web site.[28] Here are two examples of smishing crimes:

- Account holders at a credit union were sent a text about an account problem and were told to call a phone number provided in the text. If they did so, they were asked to provide personal information that allowed criminals to steal funds from their accounts within 10 minutes of the phone call.
- Bank customers received a text stating that it was necessary to reactivate their automated teller machine (ATM) card. Those who called the phone number in the text were asked to provide their ATM card number, PIN, and the expiration date. Thousands of victims had money stolen from their accounts.[29]

Financial institutions, credit card companies, and other organizations whose customers may be targeted by criminals in this manner need to be on the alert for phishing, vishing, and smishing scams. They must be prepared to act quickly and decisively without alarming their customers if such a scam is detected. Recommended action steps for institutions and organizations include the following:

- Companies should educate their customers about the dangers of phishing, vishing, and smishing through letters, recorded messages for those calling into the company's call center, and articles on the company's Web site.
- Call center service employees should be trained to detect customer complaints that indicate a scam is being perpetrated. They should attempt to capture key pieces of information, such as the callback number the customer was directed to use, details of the phone message or text message, and the type of information requested.

- Customers should be notified immediately if a scam occurs. This can be done via a recorded message for customers phoning the call center, working with local media to place a news article in papers serving the area of the attack, placing a banner on the institution's Web page, and even displaying posters in bank drive-through and lobby areas.
- If it is determined that the calls are originating from within the United States, companies should report the scam to the Federal Bureau of Investigation (FBI).
- Institutions can also try to notify the telecommunications carrier for a particular number to request that they shut down the phone numbers victims are requested to call.[30]

Types of Perpetrators

The people who launch these kinds of computer attacks include thrill seekers wanting a challenge, common criminals looking for financial gain, industrial spies trying to gain a competitive advantage, and terrorists seeking to cause destruction to further their cause. Each type of perpetrator has different objectives and access to varying resources, and each is willing to accept different levels of risk to accomplish his or her objective. Each perpetrator makes a decision to act in an unethical manner to achieve his or her own personal objectives. Knowing the profile of each set of likely attackers, as shown in Table 3-4, is the first step toward establishing effective countermeasures.

TABLE 3-4 Classifying perpetrators of computer crime

Type of perpetrator	Typical motives
Hacker	Test limits of system and/or gain publicity
Cracker	Cause problems, steal data, and corrupt systems
Malicious insider	Gain financially and/or disrupt company's information systems and business operations
Industrial spy	Capture trade secrets and gain competitive advantage
Cybercriminal	Gain financially
Hacktivist	Promote political ideology
Cyberterrorist	Destroy infrastructure components of financial institutions, utilities, and emergency response units

Source Line: Course Technology/Cengage Learning.

Hackers and Crackers

Hackers test the limitations of information systems out of intellectual curiosity—to see whether they can gain access and how far they can go. They have at least a basic understanding of information systems and security features, and much of their motivation comes from a desire to learn even more. The term *hacker* has evolved over the years, leading to its negative connotation today rather than the positive one it used to have. While there is still a vocal minority who believe that hackers perform a service by identifying security weaknesses, most people now believe that a hacker does not have the right to explore public or private networks.

Some hackers are smart and talented, but many are technically inept and are referred to as **lamers** or **script kiddies** by more skilled hackers. Surprisingly, hackers have a wealth of available resources to hone their skills—online chat groups, Web sites, downloadable hacker tools, and even hacker conventions (such as DEFCON, an annual gathering in Las Vegas).

The Web site Gawker hosts a number of irreverent blogs on media, technology, and other issues. A hacker group called Gnois recently claimed responsibility for attacking the Gawker site and compromising the usernames and passwords of people who comment on the Web site. A large number of usernames and passwords were posted to the Pirate Bay (a Swedish Web site that has been involved in numerous lawsuits regarding the downloading of copyrighted material).[31] Potentially, more than a million people could have been affected by the data breach. (A **data breach** is the unintended release of sensitive data or the access of sensitive data by unauthorized individuals.)

Malicious Insiders

A major security concern for companies is the malicious insider—an ever-present and extremely dangerous adversary. Companies are exposed to a wide range of fraud risks, including diversion of company funds, theft of assets, fraud connected with bidding processes, invoice and payment fraud, computer fraud, and credit card fraud. Not surprisingly, fraud that occurs within an organization is usually due to weaknesses in its internal control procedures. As a result, many frauds are discovered by chance and by outsiders—via tips, through resolving payment issues with contractors or suppliers, or during a change of management—rather than through control procedures. Often, frauds involve some form of **collusion**, or cooperation, between an employee and an outsider. For example, an employee in Accounts Payable might engage in collusion with a company supplier. Each time the supplier submits an invoice, the Accounts Payable employee adds $1,000 to the amount approved for payment. The inflated payment is received by the supplier, and the two split the extra money.

Insiders are not necessarily employees; they can also be consultants and contractors. The risk tolerance of insiders depends on whether they are motivated by financial gain, revenge on their employers, or publicity.

Malicious insiders are extremely difficult to detect or stop because they are often authorized to access the very systems they abuse. Although insiders are less likely to attack systems than outside hackers or crackers are, the company's systems are far more vulnerable to them. Most computer security measures are designed to stop external attackers but are nearly powerless against insiders. Insiders have knowledge of individual systems, which often includes the procedures to gain access to login IDs and passwords. Insiders know how the systems work and where the weak points are. Their knowledge of organizational structure and security procedures helps them avoid detection of their actions.

A malicious insider at the Bank of America allegedly wrote a program that ran on the bank's computers and ATMs so that he could make them dispense cash without recording the activity. The insider had been a member of the bank's IT department where he designed and maintained computer systems used by the ATMs. The bank did not disclose how much money was taken, but the scam ran for seven months.[32]

There are several steps organizations can take to reduce the potential for attacks from insiders, including the following:

- Perform a thorough background check as well as psychological and drug testing of candidates for sensitive positions.
- Establish an expectation of regular and ongoing psychological and drug testing as a normal routine for people in sensitive positions.
- Carefully limit the number of people who can perform sensitive operations, and grant only the minimum rights and privileges necessary to perform essential duties.
- Define job roles and procedures so that it is not possible for the same person to both initiate and approve an action.
- Periodically rotate employees in sensitive positions so that any unusual procedures can be detected by the replacement.
- Immediately revoke all rights and privileges required to perform old job responsibilities when someone in a sensitive position moves to a new position.
- Implement an ongoing audit process to review key actions and procedures.

The Defense Advanced Research Projects Agency (DARPA) is exploring new ways to quickly and accurately detect malicious insider activity by monitoring user and network behaviors. According to DARPA, "What sets the insider threat apart from other adversaries is the use of normal tactics to accomplish abnormal and malicious missions."[33]

Organizations must also be concerned about **negligent insiders** who are poorly trained and inadequately managed employees who mean well but have the potential to cause much damage. A survey of U.S. IT managers uncovered that while only one-quarter of them had ever experienced a data breach caused by a malicious insider, a full three-fourths of them had experienced a data breach caused by a negligent insider.[34,35]

Industrial Spies

Industrial spies use illegal means to obtain trade secrets from competitors. Trade secrets are protected by the Economic Espionage Act of 1996, which makes it a federal crime to use a trade secret for one's own benefit or another's benefit. Trade secrets are most often stolen by insiders, such as disgruntled employees and ex-employees.

Competitive intelligence is legally obtained information gathered using sources available to the public. Information is gathered from financial reports, trade journals, public filings, and printed interviews with company officials. Industrial espionage involves using illegal means to obtain information that is not available to the public. Participants might place a wiretap on the phones of key company officials, bug a conference room, or break into a research and development facility to steal confidential test results. An unethical firm may spend a few thousand dollars to hire an industrial spy to steal trade secrets that can be worth a thousand times that amount. The industrial spy avoids taking risks that would expose his employer, as the employer's reputation (an intangible but valuable item) would be considerably damaged if the espionage were discovered. Industrial espionage can involve the theft of new product designs, production data, marketing information, or new software source code. For example, Siemens, a global electronics and electrical engineering firm, sells Simatic WinCC software, which is used in large-scale industrial control systems for manufacturing and utility firms. Siemens recently discovered a highly sophisticated virus designed to infiltrate its software via a vulnerability in the Microsoft Windows operating system on which the firm's software runs. The supervisory control and data acquisition (SCADA) systems that run the Siemens software are seldom connected to the Internet for security reasons; however, this virus spreads when an infected USB memory stick is inserted into a computer. It appears that the virus is being used by industrial spies to glean information from a manufacturer's SCADA software to learn how to counterfeit products.[36]

Cybercriminals

Information technology provides a new and highly profitable venue for cybercriminals, who are attracted to the use of information technology for its ease in reaching millions of potential victims. **Cybercriminals** are motivated by the potential for monetary gain and hack into computers to steal, often by transferring money from one account to another to another—leaving a hopelessly complicated trail for law enforcement officers to follow. Cybercriminals also engage in all forms of computer fraud—stealing and reselling credit card numbers, personal identities, and cell phone IDs. Because the potential for monetary gain is high, they can afford to spend large sums of money to buy the technical expertise and access they need from unethical insiders.

The use of stolen credit card information is a favorite ploy of computer criminals. Fraud rates are highest for merchants who sell downloadable software or expensive items such as electronics and jewelry (because of their high resale value). Credit card companies are so concerned about making consumers feel safe while shopping online that many are marketing new and exclusive zero-liability programs, although the Fair Credit Billing Act limits consumer liability to only $50 of unauthorized charges. When a charge is made fraudulently in a retail store, the bank that issued the credit card must pay the fraudulent charges. For fraudulent credit card transactions over the Internet, the Web merchant absorbs the cost.

A high rate of disputed transactions, known as chargebacks, can greatly reduce a Web merchant's profit margin. However, the permanent loss of revenue caused by lost customer trust has far more impact than the costs of fraudulent purchases and bolstering security. Most companies are afraid to admit publicly that they have been hit by online fraud or hackers because they don't want to hurt their reputations.

Two students at Winona State University in Rochester, Minnesota, engaged in identity theft and ran up over $1 million in fraudulent credit card charges. They created some 180 eBay and 360 PayPal accounts using stolen identities and used them to take orders and payment for advertised items that they sold at significantly reduced prices. The students did not stock the items themselves but instead bought them at full price from various suppliers using the stolen identities. Numerous companies were affected by the scheme because of credit card chargebacks requested by the identity theft victims.[37]

To reduce the potential for online credit card fraud, most e-commerce Web sites use some form of encryption technology to protect information as it comes in from the consumer. Some also verify the address submitted online against the one the issuing bank has on file, although the merchant may inadvertently throw out legitimate orders as a result—for example, a consumer might place a legitimate order but request shipment to a different address because it is a gift. Another security technique is to ask for a card verification value (CVV), the three-digit number above the signature panel on the back of a credit card. This technique makes it impossible to make purchases with a credit card number stolen online. An additional security option is transaction-risk scoring software, which keeps track of a customer's historical shopping patterns and notes deviations from the norm. For example, say that you have never been to a casino and your credit card turns up at Caesar's Palace at 2 a.m. Your transaction-risk score would go up dramatically, so much so that the transaction might be declined.

Some card issuers are issuing debit and credit cards in the form of **smart cards**, which contain a memory chip that is updated with encrypted data every time the card is used. This encrypted data might include the user's account identification and the amount of credit remaining. To use a smart card for online transactions, consumers must purchase a card reader that attaches to their personal computers and enter a personal identification number to gain access to the account. Although smart cards are used widely in Europe, they are not as popular in the United States because of the changeover costs for merchants.

DarkMarket was an infamous cybercriminal forum and online market. It was shut down by the U.S. Secret Service and FBI after it was infiltrated by an FBI agent who worked for two years to gather evidence for indictments of several of its 2,000 members. Members of the DarkMarket bought and sold credit card and personal information along with all forms of malware. Membership was by invitation only, and the DarkMarket code forbade members from scamming each other.[38]

Hacktivists and Cyberterrorists

Hacktivism, a combination of the words *hacking* and *activism*, is hacking to achieve a political or social goal. In late 2010, Anonymous, a group of hacktivists, launched a series of distributed denial-of-service attacks at MasterCard, PayPal, and Visa in retaliation for those companies cutting off their services for the whistle-blowing Web site WikiLeaks. The attacks resulted in the Web sites of these firms being unavailable for up to a few hours to users in the United States, the United Kingdom, France, Germany, and Italy.[39] A hacktivist known as "The Jester" describes himself as a "Hacktivist for good." During 2010, he developed a reputation for targeting Al Qaeda and jihadist Web sites, forcing them to go offline at least temporarily.[40]

A **cyberterrorist** launches computer-based attacks against other computers or networks in an attempt to intimidate or coerce a government in order to advance certain political or social objectives. Cyberterrorists are more extreme in their goals than hacktivists, although there is no clear demarcation line. Because of the Internet, cyberattacks can easily originate from foreign countries, making detection and retaliation much more difficult. Cyberterrorists seek to cause harm rather than gather information, and they use techniques that destroy or disrupt services. They are extremely dangerous, consider themselves to be at war, have a very high acceptance of risk, and seek maximum impact.

Federal Laws for Prosecuting Computer Attacks

Computers came into use in the 1950s. Initially, there were no laws that pertained strictly to computer-related crimes. For example, if a group of criminals entered a bank and stole money at gunpoint, they could be captured and charged with robbery—the crime of seizing property through violence or intimidation. However, by the mid-1970s, it was possible to access a bank's computer remotely using a terminal (a keyboard and monitor), modem, and telephone line. A knowledgeable person could then transfer money (in the form of computer bits) from accounts in that bank to an account in another bank. This act did not fit the definition of robbery, and the traditional laws were no longer adequate to punish criminals who used computer modems.

Over the years, several laws have been enacted to help prosecute those responsible for computer-related crime; these are summarized in Table 3-5. For example, the USA Patriot Act defines cyberterrorism as hacking attempts that cause $5,000 in aggregate damage in one year, damage to medical equipment, or injury to any person. Those convicted of cyberterrorism are subject to a prison term of 5 to 20 years. (The $5,000 threshold is quite easy to exceed, and, as a result, many young people who have been involved in what they consider to be minor computer pranks have found themselves meeting the criteria to be tried as cyberterrorists.)

TABLE 3-5 Federal laws that address computer crime

Federal law	Subject area
USA Patriot Act	Defines cyberterrorism and penalties
Identity Theft and Assumption Deterrence Act (U.S. Code Title 18, Section 1028)	Makes identity theft a Federal crime with penalties up to 15 years imprisonment and a maximum fine of $250,000
Fraud and Related Activity in Connection with Access Devices Statute (U.S. Code Title 18, Section 1029)	False claims regarding unauthorized use of credit cards
Computer Fraud and Abuse Act (U.S. Code Title 18, Section 1030)	Fraud and related activities in association with computers: • Accessing a computer without authorization or exceeding authorized access • Transmitting a program, code, or command that causes harm to a computer • Trafficking of computer passwords • Threatening to cause damage to a protected computer
Stored Wire and Electronic Communications and Transactional Records Access Statutes (U.S. Code Title 18, Chapter 121)	Unlawful access to stored communications to obtain, alter, or prevent authorized access to a wire or electronic communication while it is in electronic storage

Source Line: Course Technology/Cengage Learning.

Now that we have discussed various types of computer exploits, the people who perpetrate these exploits, and the laws under which they can be prosecuted, we will discuss how organizations can take steps to implement a trustworthy computing environment to defend against such attacks.

IMPLEMENTING TRUSTWORTHY COMPUTING

Trustworthy computing is a method of computing that delivers secure, private, and reliable computing experiences based on sound business practices—which is what organizations worldwide are demanding today. Software and hardware manufacturers, consultants, and programmers all understand that this is a priority for their customers. For example, Microsoft has pledged to deliver on a trustworthy computing initiative designed to improve trust in its software products, as summarized in Figure 3-4 and Table 3-6.[41,42]

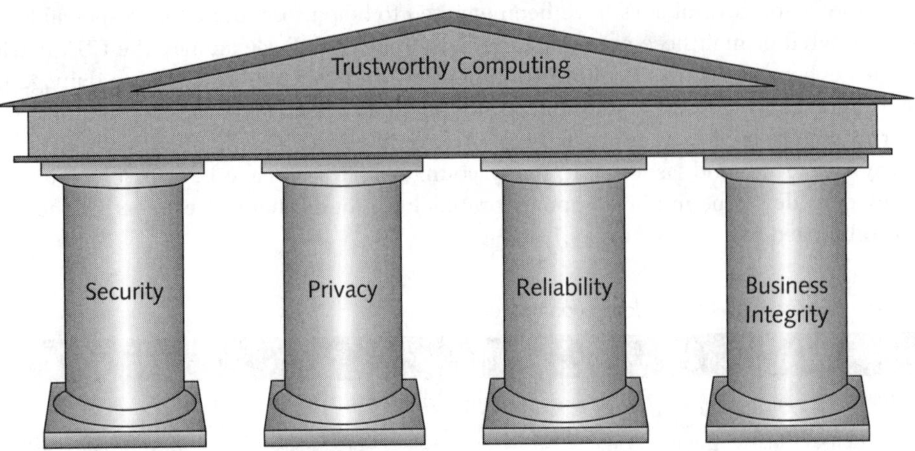

FIGURE 3-4 Microsoft's four pillars of trustworthy computing
Source Line: Course Technology/Cengage Learning.

TABLE 3-6 Actions taken by Microsoft to support trustworthy computing

Pillar	Actions taken by Microsoft
Security	Invest in the expertise and technology required to create a trustworthy environment.
	Work with law enforcement agencies, industry experts, academia, and private sectors to create and enforce secure computing.
	Develop trust by educating consumers on secure computing.
Privacy	Make privacy a priority in the design, development, and testing of products.
	Contribute to standards and policies created by industry organizations and government.
	Provide users with a sense of control over their personal information.
Reliability	Build systems so that (1) they continue to provide service in the face of internal or external disruptions; (2) in the event of a disruption, they can be easily restored to a previously known state with no data loss; (3) they provide accurate and timely service whenever needed; (4) required changes and upgrades do not disrupt them; (5) on release, they contain minimal software bugs; and (6) they work as expected or promised.
Business integrity	Be responsive—take responsibility for problems and take action to correct them. Be transparent—be open in dealings with customers, keep motives clear, keep promises, and make sure customers know where they stand in dealing with the company.

Source Line: Course Technology/Cengage Learning.

The security of any system or network is a combination of technology, policy, and people and requires a wide range of activities to be effective. "Society ultimately expects computer systems to be trustworthy—that is, that they do what is required and expected of them despite environmental disruption, human user and operator errors, and attacks by hostile parties, and that they not do other things."[43] A strong security program begins by assessing threats to the organization's computers and network, identifying actions that address the most serious vulnerabilities, and educating end users about the risks involved and the actions they must take to prevent a security incident. The IT security group must lead the effort to prevent security breaches by implementing security policies and procedures, as well as effectively employing available hardware and software tools. However, no security system is perfect, so systems and procedures must be monitored to detect a possible intrusion. If an intrusion occurs, there must be a clear reaction plan that addresses notification, evidence protection, activity log maintenance, containment, eradication, and recovery. The following sections will discuss these activities.

Risk Assessment

Risk assessment is the process of assessing security-related risks to an organization's computers and networks from both internal and external threats. Such threats can prevent an organization from meeting its key business objectives. The goal of risk assessment is to identify which investments of time and resources will best protect the organization from its most likely and serious threats. In the context of an IT risk assessment, an asset is any hardware, software, information system, network, or database that is used by the organization to achieve its business objectives. A loss event is any occurrence that has a negative impact on an asset, such as a computer contracting a virus or a Web site undergoing a distributed denial-of-service attack. Figure 3-5 illustrates a general security risk assessment process developed by ASIS International.

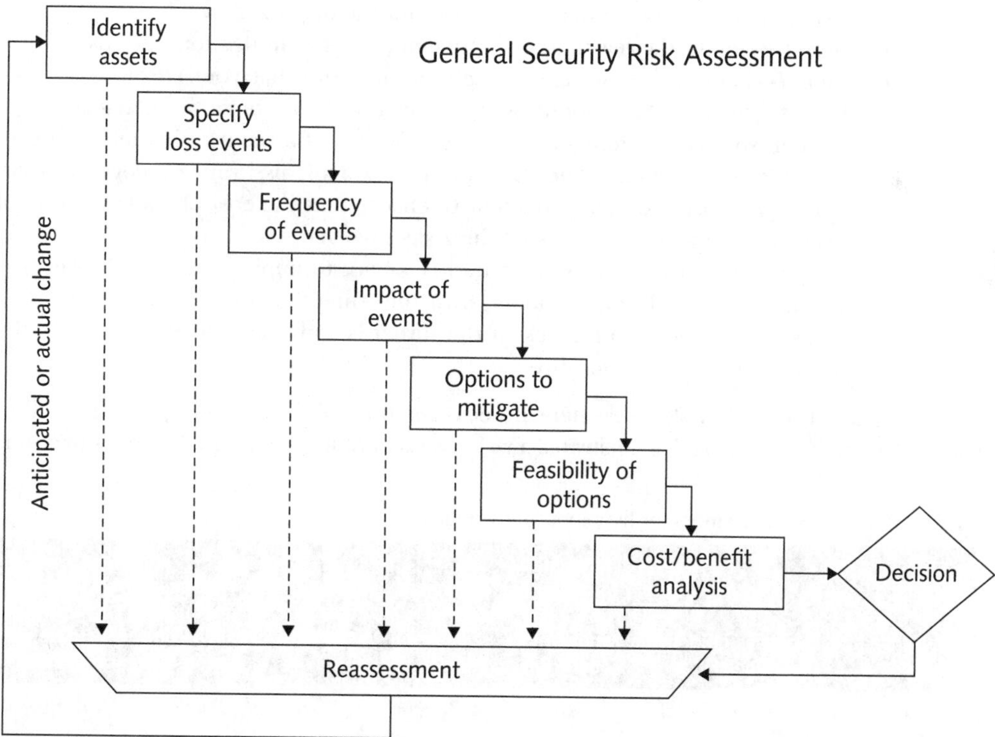

FIGURE 3-5 General security risk assessment

Source Line: General Security Risk Assessment Guidelines, ASIS International (2003). See the Standards and Guidelines page of the ASIS International website (*http://www.asisonline.org*) for revisions and/or updates. Reprinted by permission.

The steps in a general security risk assessment process are as follows:

- **Step 1**—Identify the set of IT assets about which the organization is most concerned. Priority is typically given to those assets that support the organization's mission and the meeting of its primary business goals.
- **Step 2**—Identify the loss events or the risks or threats that could occur, such as a distributed denial-of-service attack or insider fraud.
- **Step 3**—Assess the frequency of events or the likelihood of each potential threat; some threats, such as insider fraud, are more likely to occur than others.
- **Step 4**—Determine the impact of each threat occurring. Would the threat have a minor impact on the organization, or could it keep the organization from carrying out its mission for a lengthy period of time?
- **Step 5**—Determine how each threat can be mitigated so that it becomes much less likely to occur or, if it does occur, has less of an impact on the organization. For example, installing virus protection on all computers makes it much less likely for a computer to contract a virus. Due to time and resource limitations, most organizations choose to focus on those threats that have a high (relative to all other threats) frequency and a high (relative to all other threats) impact. In other words, first address those threats that are likely to occur and that would have a high negative impact on the organization.
- **Step 6**—Assess the feasibility of implementing the mitigation options.
- **Step 7**—Perform a cost-benefit analysis to ensure that your efforts will be cost effective. No amount of resources can guarantee a perfect security system, so organizations must balance the risk of a security breach with the cost of preventing one. The concept of **reasonable assurance** recognizes that managers must use their judgment to ensure that the cost of control does not exceed the system's benefits or the risks involved.
- **Step 8**—Make the decision on whether or not to implement a particular countermeasure. If you decide against implementing a particular countermeasure, you need to reassess if the threat is truly serious and, if so, identify a less costly countermeasure.

The general security risk assessment process—and the results of that process—will vary by organization. Table 3-7 illustrates a risk assessment for a hypothetical organization.

TABLE 3-7 Risk assessment for hypothetical company

Risk	Business objective threatened	Estimated probability of such an event occurring	Estimated cost of a successful attack	Probability × cost = expected cost	Assessment of current level of protection	Relative priority to be fixed
Distributed denial-of-service attack	24/7 operation of a retail Web site	40%	$500,000	$200,000	Poor	1

(Continued)

Risk	Business objective threatened	Estimated probability of such an event occurring	Estimated cost of a successful attack	Probability × cost = expected cost	Assessment of current level of protection	Relative priority to be fixed
Email attachment with harmful worm	Rapid and reliable communications among employees and suppliers	70%	$200,000	$140,000	Poor	2
Harmful virus	Employees' use of personal productivity software	90%	$50,000	$45,000	Good	3
Invoice and payment fraud	Reliable cash flow	10%	$200,000	$20,000	Excellent	4

Source Line: Course Technology/Cengage Learning.

A completed risk assessment identifies the most dangerous threats to a company and helps focus security efforts on the areas of highest payoff.

Establishing a Security Policy

A **security policy** defines an organization's security requirements, as well as the controls and sanctions needed to meet those requirements. A good security policy delineates responsibilities and the behavior expected of members of the organization. A security policy outlines *what* needs to be done but not *how* to do it. The details of *how* to accomplish the goals of the policy are typically provided in separate documents and procedure guidelines.

The SANS (SysAdmin, Audit, Network, Security) Institute's Web site offers a number of security-related policy templates that can help an organization to quickly develop effective security policies. The templates and other security policy information can be found at *www.sans.org/security-resources/policies*. The following is a partial list of the templates available from the SANS Institute:

- Ethics Policy—This template defines the means to establish a culture of openness, trust, and integrity in business practices.
- Information Sensitivity Policy—This sample policy defines the requirements for classifying and securing the organization's information in a manner appropriate to its level of sensitivity.
- Risk Assessment Policy—This template defines the requirements and provides the authority for the information security team to identify, assess, and remediate risks to the organization's information infrastructure associated with conducting business.
- Personal Communication Devices and Voicemail Policy—This sample policy describes Information Security's requirements for Personal Communication Devices and Voicemail.[44]

Whenever possible, automated system rules should mirror an organization's written policies. Automated system rules can often be put into practice using the configuration options in a software program. For example, if a written policy states that passwords must be changed every 30 days, then all systems should be configured to enforce this policy automatically. However, users will often attempt to circumvent security policies. For example, half of the employees polled in a Cisco 2010 survey confessed that they ignored or overrode corporate bans on accessing social networks while at work.[45]

When applying system security restrictions, there are some trade-offs between ease of use and increased security; however, when a decision is made to favor ease of use, security incidents sometimes increase. As security techniques continue to advance in sophistication, they become more transparent to end users.

The use of email attachments is a critical security issue that should be addressed in every organization's security policy. Sophisticated attackers can try to penetrate a network via email attachments, regardless of the existence of a firewall and other security measures. As a result, some companies have chosen to block any incoming mail that has a file attachment, which greatly reduces their vulnerability. Some companies allow employees to receive and open email with attachments, but only if the email is expected and from someone known by the recipient. Such a policy can be risky, however, because worms often use the address book of their victims to generate emails to a target audience.

Another growing area of concern is the use of wireless devices to access corporate email, store confidential data, and run critical applications, such as inventory management and sales force automation. Mobile devices such as smartphones can be susceptible to viruses and worms. However, the primary security threat for mobile devices continues to be loss or theft of the device. Wary companies have begun to include special security requirements for mobile devices as part of their security policies. In some cases, users of laptops and mobile devices must use a virtual private network to gain access to their corporate network.

A **virtual private network (VPN)** works by using the Internet to relay communications; it maintains privacy through security procedures and tunneling protocols, which encrypt data at the sending end and decrypt it at the receiving end. An additional level of security involves encrypting the originating and receiving network addresses. Because of the ease of loss or theft, many organizations encrypt all sensitive corporate data stored on handhelds and laptops. Unfortunately, it is hard to apply a single, simple approach to securing all handheld devices because so many manufacturers and models exist.

Educating Employees and Contract Workers

An ongoing security problem for companies is creating and enhancing user awareness of security policies. Employees and contract workers must be educated about the importance of security so that they will be motivated to understand and follow the security policies. This can often be accomplished by discussing recent security incidents that affected the organization. Users must understand that they are a key part of the security system and that they have certain responsibilities. For example, users must help protect an organization's information systems and data by doing the following:

- Guarding their passwords to protect against unauthorized access to their accounts

- Prohibiting others from using their passwords
- Applying strict access controls (file and directory permissions) to protect data from disclosure or destruction
- Reporting all unusual activity to the organization's IT security group
- Taking care to ensure that portable computing and data storage devices are protected (over 600,000 laptops are lost or stolen per year)[46]

Prevention

No organization can ever be completely secure from attack. The key is to implement a layered security solution to make computer break-ins so difficult that an attacker eventually gives up. In a layered solution, if an attacker breaks through one layer of security, there is another layer to overcome. These layers of protective measures are explained in more detail in the following sections.

Installing a Corporate Firewall

Installation of a corporate firewall is the most common security precaution taken by businesses. A firewall stands guard between an organization's internal network and the Internet, and it limits network access based on the organization's access policy (see Figure 3-6).

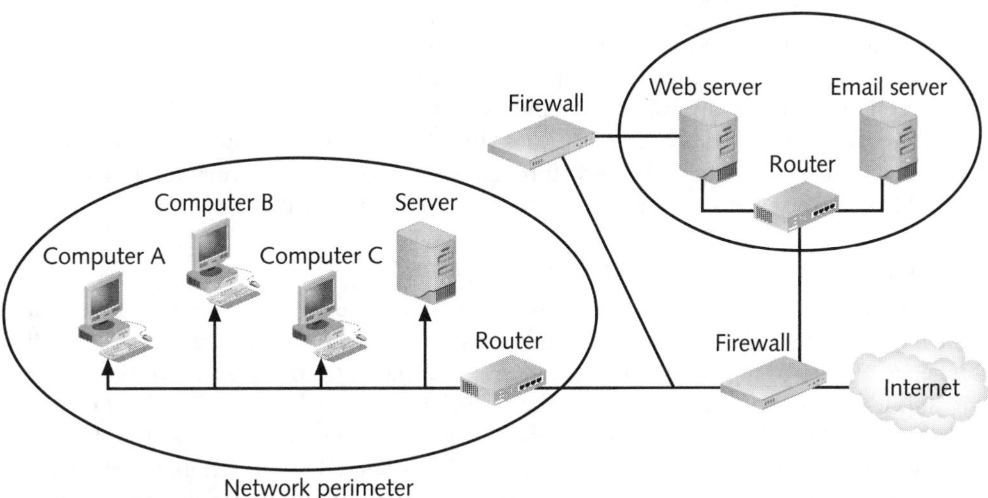

FIGURE 3-6 Firewall
Source Line: Course Technology/Cengage Learning.

Firewalls can be established through the use of software, hardware, or a combination of both. Any Internet traffic that is not explicitly permitted into the internal network is denied entry. Similarly, most firewalls can be configured so that internal network users can be blocked from gaining access to certain Web sites based on such content as sex and violence. Most firewalls can also be configured to block instant messaging, access to newsgroups, and other Internet activities.

Installing a firewall can lead to another serious security issue—complacency. For example, a firewall cannot prevent a worm from entering the network as an email attachment. Most firewalls are configured to allow email and benign-looking attachments to reach their intended recipient.

Table 3-8 lists some of the top-rated firewall software used to protect personal computers. Typically, firewall software sells for $30 to $60 for a single user license.

TABLE 3-8 Popular firewall software for personal computers

Software	Vendor
Zone Alarm Pro	CheckPoint Software Technologies Ltd.
F-Secure Internet Security	F-Secure Corporation
Panda Global Protection	Panda Security
NeT Firewall	NT Kernel Resources
ESET Smart Security 4	ESET

Source Line: "Best Firewall Software–Editor's Choice," All-Internet-Security.com, © January 2011, www.all-internet-security.com/top_10_firewall_software.html.

Intrusion Prevention Systems

An **intrusion prevention system (IPS)** works to prevent an attack by blocking viruses, malformed packets, and other threats from getting into the protected network. The IPS sits directly behind the firewall and examines all the traffic passing through it. A firewall and a network IPS are complementary. Most firewalls can be configured to block everything except what you explicitly allow through; most IPSs can be configured to let through everything except what you explicitly specify should be blocked.

Installing Antivirus Software on Personal Computers

Antivirus software should be installed on each user's personal computer to scan a computer's memory and disk drives regularly for viruses. Antivirus software scans for a specific sequence of bytes, known as a **virus signature**, that indicates the presence of a specific virus. If it finds a virus, the antivirus software informs the user, and it may clean, delete, or quarantine any files, directories, or disks affected by the malicious code. Good antivirus software checks vital system files when the system is booted up, monitors the system continuously for virus-like activity, scans disks, scans memory when a program is run, checks programs when they are downloaded, and scans email attachments before they are opened. Two of the most widely used antivirus software products are Norton AntiVirus from Symantec and Personal Firewall from McAfee.

The United States Computer Emergency Readiness Team (US-CERT) is a partnership between the Department of Homeland Security and the public and private sectors—established in 2003 to protect the nation's Internet infrastructure against cyberattacks. US-CERT serves as a clearinghouse for information on new viruses, worms, and other computer security topics (over 500 new viruses and worms are developed each month[47]). According to US-CERT, most of the virus and worm attacks that the team analyzes use already known programs. Thus, it is crucial that antivirus software be continually updated with the latest virus signatures. In most corporations, the network administrator is responsible for monitoring network security Web sites frequently and downloading updated antivirus software as needed. Many antivirus vendors recommend—and provide for—automatic and frequent updates.

Implementing Safeguards Against Attacks by Malicious Insiders

User accounts that remain active after employees leave a company are potential security risks. To reduce the threat of attack by malicious insiders, IT staff must promptly delete the computer accounts, login IDs, and passwords of departing employees and contractors.

Organizations also need to define employee roles carefully and separate key responsibilities properly, so that a single person is not responsible for accomplishing a task that has high security implications. For example, it would not make sense to allow an employee to initiate as well as approve purchase orders. That would allow an employee to input large invoices on behalf of a "friendly vendor," approve the invoices for payment, and then disappear from the company to split the money with the vendor. In addition to separating duties, many organizations frequently rotate people in sensitive positions to prevent potential insider crimes.

Another important safeguard is to create roles and user accounts so that users have the authority to perform their responsibilities and nothing more. For example, members of the Finance Department should have different authorizations from members of Human Resources. An accountant should not be able to review the pay and attendance records of an employee, and a member of Human Resources should not know how much was spent to modernize a piece of equipment. Even within one department, not all members should be given the same capabilities. Within the Finance Department, for example, some users may be able to approve invoices for payment, but others may only be able to enter them. An effective system administrator will identify the similarities among users and create profiles associated with these groups.

Defending Against Cyberterrorism

In the face of increasing risks of cyberterrorism, organizations need to be aware of the resources available to help them combat this serious threat. The National Cyber Security Division (NCSD) of the Department of Homeland Security (DHS) has as its mission to work "collaboratively with public, private, and international entities to secure cyberspace and America's cyberassets."[48] As such, it has primary responsibility within the federal government for combating cyberterrorism.

The Protected Critical Infrastructure Information Program encourages private industry to share confidential information about the nation's critical infrastructure with DHS under the assurance that the information will be protected from public disclosure.[49] This enables private industry and DHS to work jointly to identify threats and vulnerabilities and to develop countermeasures and defensive strategies.

Critical infrastructures include telecommunications, energy, banking and finance, water, government operations, and emergency services. Specific targets might include telephone-switching systems, an electric power grid that serves major portions of a geographic region, or an air traffic control center that ensures airplanes can take off and land safely. Successful cyberattacks on such targets could cause widespread and massive disruptions to society. Some computer security experts believe that cyberterrorism attacks could be used to create further problems following a major act of terrorism by reducing the ability of fire and emergency teams to respond.

The NCSD's two primary objectives are to build and maintain an effective national security cyberspace response system and to implement a cyber–risk management program for protection of critical infrastructure. NCSD cyberresponse programs and tools include the following:

- National Cyber Alert System—A system that provides current information on cyberthreats to computer users
- US-CERT—A group that analyzes and reduces cyberthreats, disseminates cyberthreat warning information, and coordinates incident response activities
- National Cyber Response Coordination Group—A collection of 13 federal agencies that will help coordinate the federal response in the event of a nationally significant cyber-related incident
- Cyber Cop Portal—An information sharing and collaboration tool used by 5,300 investigators who are involved in electronic crime cases worldwide

The NCSD also works to assess risks through its cyber–risk management programs, including:

- Cyber Storm—An international cybersecurity drill that occurs every two years to evaluate readiness to a major cyber incident
- Software Assurance Program—An effort to reduce software vulnerabilities, minimize exploitation, and improve development of trustworthy software products

Addressing the Most Critical Internet Security Threats

The overwhelming majority of successful computer attacks take advantage of well-known vulnerabilities. Computer attackers know that many organizations are slow to fix problems, which makes scanning the Internet for vulnerable systems an effective attack strategy. The rampant and destructive spread of worms, such as Blaster, Slammer, and Code Red, was made possible by the exploitation of known but unpatched vulnerabilities.

US-CERT regularly updates a summary of the most frequent, high-impact vulnerabilities being reported to them. You can read this summary at *www.us-cert.gov/current*. The actions required to address these issues include installing a known patch to the software and keeping applications and operating systems up to date. Those responsible for computer security must make it a priority to prevent attacks using these vulnerabilities.

Conducting Periodic IT Security Audits

Another important prevention tool is a **security audit** that evaluates whether an organization has a well-considered security policy in place and if it is being followed. For example, if a policy says that all users must change their passwords every 30 days, the audit must check how well that policy is being implemented. The audit should also review who has access to particular systems and data and what level of authority each user has. It is not unusual for an audit to reveal that too many people have access to critical data and that many people have capabilities beyond those needed to perform their jobs. One result of a good audit is a list of items that need to be addressed in order to ensure that the security policy is being met.

A thorough security audit should also test system safeguards to ensure that they are operating as intended. Such tests might include trying the default system passwords that are active when software is first received from the vendor. The goal of such a test is to ensure that all such known passwords have been changed.

Some organizations will also perform a penetration test of their defenses. This entails assigning individuals to try to break through the measures and identify vulnerabilities that still need to be addressed. The individuals used for this test are knowledgeable and are likely to take unique approaches in testing the security measures.

The Information Protection Assessment kit is an assessment tool available from the Computer Security Institute, an organization for information security professionals. The kit can be accessed at *http://gocsi.com/ipak* and is formatted as an Microsoft Excel® spreadsheet that covers 15 categories of security issues (e.g. physical security, business process controls, network security controls, etc.). Each category has approximately 20 statements used to rate the effectiveness of security for that category. Organizations can complete the survey to get a clear measure of the effectiveness of their security program and to define areas that need improvement.

Detection

Even when preventive measures are implemented, no organization is completely secure from a determined attack. Thus, organizations should implement detection systems to catch intruders in the act. Organizations often employ an intrusion detection system to minimize the impact of intruders.

An **intrusion detection system (IDS)** is software and/or hardware that monitors system and network resources and activities, and notifies network security personnel when it identifies possible intrusions from outside the organization or misuse from within the organization. Knowledge-based approaches and behavior-based approaches are two fundamentally different approaches to intrusion detection. Knowledge-based intrusion detection systems contain information about specific attacks and system vulnerabilities and watch for attempts to exploit these vulnerabilities, such as repeated failed login attempts or recurring attempts to download a program to a server. When such an attempt is detected, an alarm is triggered. A behavior-based intrusion detection system models normal behavior of a system and its users from reference information collected by various means. The intrusion detection system compares current activity to this model and generates an alarm if it finds a deviation. Examples include unusual traffic at odd hours or a user in the Human Resources Department who accesses an accounting program that she has never before used.

Response

An organization should be prepared for the worst—a successful attack that defeats all or some of a system's defenses and damages data and information systems. A response plan should be developed well in advance of any incident and be approved by both the organization's legal department and senior management. A well-developed response plan helps keep an incident under technical and emotional control.

In a security incident, the primary goal must be to regain control and limit damage, not to attempt to monitor or catch an intruder. Sometimes system administrators take the discovery of an intruder as a personal challenge and lose valuable time that should be used to restore data and information systems to normal.

A 2010 survey by *CSO* (Chief Security Officer) magazine showed that while 58 percent of the respondents believe they are better prepared to prevent, detect, respond to or recover from a cybercrime than in 2009, only 56 percent said that they have a plan for reporting and responding to a computer crime. Nearly three-fourths (72 percent) of insider incidents are handled internally without legal action or law enforcement involvement.[50]

Incident Notification

A key element of any response plan is to define who to notify and who *not* to notify. Questions to cover include the following: Within the company, who needs to be notified, and what information does each person need to have? Under what conditions should the company contact major customers and suppliers? How does the company inform them of a disruption in business without unnecessarily alarming them? When should local authorities or the FBI be contacted?

Most security experts recommend against giving out specific information about a compromise in public forums, such as news reports, conferences, professional meetings, and online discussion groups. All parties working on the problem need to be kept informed and up to date without using systems connected to the compromised system. The intruder may be monitoring these systems and email to learn what is known about the security breach.

A critical ethical decision that must be made is what to tell customers and others whose personal data may have been compromised by a computer incident. Many organizations are tempted to conceal such information for fear of bad publicity and loss of customers. Because such inaction is perceived to be unethical and harmful, a number of state and federal laws have been passed to force organizations to reveal when customer data has been breached. These laws will be discussed further in the next chapter.

Protection of Evidence and Activity Logs

An organization should document all details of a security incident as it works to resolve the incident. Documentation captures valuable evidence for a future prosecution and provides data to help during the incident eradication and follow-up phases. It is especially important to capture all system events, the specific actions taken (what, when, and who), and all external conversations (what, when, and who) in a logbook. Because this may become court evidence, an organization should establish a set of document handling procedures using the legal department as a resource.

Incident Containment

Often it is necessary to act quickly to contain an attack and to keep a bad situation from becoming even worse. The response plan should clearly define the process for deciding if an attack is dangerous enough to warrant shutting down or disconnecting critical systems from the network. How such decisions are made, how fast they are made, and who makes them are all elements of an effective response plan.

Eradication

Before the IT security group begins the eradication effort, it must collect and log all possible criminal evidence from the system, and then verify that all necessary backups are current, complete, and free of any virus. Creating a forensic disk image of each compromised system on write-only media both for later study and as evidence can be very useful. After virus eradication, the group must create a new backup. Throughout this process, a log should be kept of all actions taken. This will prove helpful during the follow-up phase and ensure that the problem does not recur. It is imperative to back up critical applications and data regularly. Many organizations, however, have implemented inadequate backup processes and found that they could not fully restore original data after a security incident. All backups should be created with enough frequency to enable a full and quick restoration of data if an attack destroys the original. This process should be tested to confirm that it works.

Incident Follow-Up

Of course, an essential part of follow-up is to determine how the organization's security was compromised so that it does not happen again. Often the fix is as simple as getting a software patch from a product vendor. However, it is important to look deeper than the immediate fix to discover why the incident occurred. If a simple software fix could have prevented the incident, then why wasn't the fix installed *before* the incident occurred?

A review should be conducted after an incident to determine exactly what happened and to evaluate how the organization responded. One approach is to write a formal incident report that includes a detailed chronology of events and the impact of the incident. This report should identify any mistakes so that they are not repeated in the future. The experience from this incident should be used to update and revise the security incident response plan.

Creating a detailed chronology of all events will also document the incident for later prosecution. To this end, it is critical to develop an estimate of the monetary damage. Potential costs include loss of revenue, loss in productivity, and the salaries of people working to address the incident, along with the cost to replace data, software, and hardware.

Another important issue is the amount of effort that should be put into capturing the perpetrator. If a Web site was simply defaced, it is easy to fix or restore the site's HTML (Hypertext Markup Language—the code that describes to your browser how a Web page should look). However, what if the intruders inflicted more serious damage, such as erasing proprietary program source code or the contents of key corporate databases? What if they stole company trade secrets? Expert crackers can conceal their identity, and tracking them down can take a long time as well as a tremendous amount of corporate resources.

The potential for negative publicity must also be considered. Discussing security attacks through public trials and the associated publicity not only has enormous potential costs in public relations but real monetary costs as well. For example, a brokerage firm might lose customers who learn of an attack and think their money or records aren't secure. Even if a company decides that the negative publicity risk is worth it and goes after the perpetrator, documents containing proprietary information that must be provided to the court could cause even greater security threats in the future. On the other hand, an organization must decide if it has an ethical or a legal duty to inform customers or clients of a cyberattack that may have put their personal data or financial resources at risk.

Computer Forensics

Computer forensics is a discipline that combines elements of law and computer science to identify, collect, examine, and preserve data from computer systems, networks, and storage devices in a manner that preserves the integrity of the data gathered so that it is admissible as evidence in a court of law. A computer forensics investigation may be opened in response to a criminal investigation or civil litigation. It may also be launched for a variety of other reasons; for example, to retrace steps taken when data has been lost, assess damage following a computer incident, investigate the unauthorized disclosure of personal or corporate confidential data, or to confirm or evaluate the impact of industrial espionage.

Proper handling of a computer forensics investigation is the key to fighting computer crime successfully in a court of law. In addition, extensive training and certification increases the stature of a computer forensics investigator in a court of law. There are numerous certifications related to computer forensics, including the CCE (Certified Computer Examiner), CISSP (Certified Information Systems Security Professional), CSFA (CyberSecurity Forensic Analyst), and GCFA (Global Information Assurance Certification Certified Forensics Analyst). The EnCE Certified Examiner program certifies professionals who have mastered computer investigation methods as well as the use of Guidance Software's EnCase computer forensic software. Numerous universities (both online and traditional) offer degrees specializing in computer forensics. Such degree programs should include training in accounting, particularly auditing, as this is very useful in the investigation of cases involving fraud.

A computer forensics investigator must be knowledgeable about the various laws that apply to the gathering of criminal evidence; see Table 3-9 for a partial list.

TABLE 3-9 Partial list of constitutional amendments and statutes governing the collection of evidence

Law	Subject area
Fourth Amendment	Protects against unreasonable search and seizure
Fifth Amendment	Provides protection from self-incrimination
Wiretap Act (18 U.S.C. 2510-2522)	Regulates the collection of the content of wire and electronic communications
Pen Registers and Trap and Trace Devices Statute (18 U.S.C. 3121-27)	Provides restrictions on the use of pen registers and trap and trace devices. (A pen register is a device that records all numbers dialed from a particular phone. A trap and trace device shows the phone numbers have that made calls to a specific phone.)
Stored Wired and Electronic Communications Act (18 U.S.C 2701-120)	Addresses the disclosure of stored wired and electronic communications and transaction records by Internet service providers.

Source Line: Course Technology/Cengage Learning.

Violation of any one of these laws could result in a case being thrown out of court. It could even result in the investigator being charged with a federal felony, punishable by a fine and/or imprisonment.

Table 3-10 provides a manager's checklist for evaluating an organization's readiness for a security incident. The preferred answer to each question is *yes*.

TABLE 3-10 Manager's checklist for evaluating an organization's readiness for a security incident

Question	Yes	No
Has a risk assessment been performed to identify investments in time and resources that can protect the organization from its most likely and most serious threats?		
Have senior management and employees involved in implementing security measures been educated about the concept of reasonable assurance?		
Has a security policy been formulated and broadly shared throughout the organization?		
Have automated systems policies been implemented that mirror written policies?		
Does the security policy address: • Email with executable file attachments? • Wireless networks and devices? • Use of smartphones deployed as part of corporate rollouts as well as those bought by end users?		
Is there an effective security education program for employees and contract workers?		
Has a layered security solution been implemented to prevent break-ins?		
Has a firewall been installed?		
Is antivirus software installed on all personal computers?		
Is the antivirus software frequently updated?		
Have precautions been taken to limit the impact of malicious insiders?		
Are the accounts, passwords, and login IDs of former employees promptly deleted?		
Is there a well-defined separation of employee responsibilities?		
Are individual roles defined so that users have authority to perform their responsibilities and nothing more?		
Is it a requirement to review at least quarterly the most critical Internet security threats and implement safeguards against them?		
Has it been verified that backup processes for critical software and databases work correctly?		
Has an intrusion detection system been implemented to catch intruders in the act—both in the network and on critical computers on the network?		
Has an intrusion prevention system been implemented to thwart intruders in the act?		
Are periodic IT security audits conducted?		
Has a comprehensive incident response plan been developed?		
Has the security plan been reviewed and approved by legal and senior management?		
Does the plan address all of the following areas: • Incident notification? • Protection of evidence and activity logs? • Incident containment? • Eradication? • Incident follow-up?		

Source Line: Course Technology/Cengage Learning.

Summary

- The security of information technology used in business is of the utmost importance, but it must be balanced against other business needs and issues.
- Increasing complexity, higher computer user expectations, expanding and changing systems, and increased reliance on software with known vulnerabilities have caused a dramatic increase in the number, variety, and impact of security incidents.
- Viruses, worms, Trojan horses, distributed denial-of-service attacks, rootkits, spam, phishing, spear-phishing, smishing, and vishing are among the most common computer exploits.
- A successful computer exploit aimed at several organizations can have a cost impact of more than $1 billion.
- There are many different kinds of people who launch computer attacks, including the hacker, cracker, malicious insider, industrial spy, cybercriminal, hacktivist, and cyberterrorist. Each type has a different motivation.
- Over the years, several laws have been enacted to prosecute those responsible for computer-related crime, including the USA Patriot Act, The Computer Fraud and Abuse Act, The Identity Theft and Assumption Deterrence Act, The Fraud and Related Activities in Connection with Access Devices, and the Stored Wire and Electronic Communications and Transaction Record Access.
- Trustworthy computing is a method of computing that delivers secure, private, and reliable computing experiences based on sound business practices.
- The security of any system is a combination of technology, policy, and people, and it requires a wide range of activities to be effective.
- A strong security program begins by assessing threats to the organization's computers and network, identifying actions that address the most serious vulnerabilities, and educating users about the risks involved and the actions they must take to prevent a security incident.
- The IT security group must lead the effort to implement security policies and procedures, along with hardware and software tools to help prevent security breaches.
- No organization can ever be completely secure from attack. The key to prevention of a computer security incident is to implement a layered security solution to make computer break-ins so difficult that an attacker eventually gives up.
- No security system is perfect, so systems and procedures must be monitored to detect a possible intrusion.
- If an intrusion occurs, there must be a clear reaction plan that addresses notification, evidence protection, activity log maintenance, containment, eradication, and recovery.
- Special measures must be taken to implement safeguards against attacks by malicious insiders and to defend against cyberterrorism.
- Organizations must implement fixes against well-known vulnerabilities.
- Organizations should conduct periodic IT security audits.
- Organizations need to be knowledgeable of and have access to trained experts in computer forensics.

Key Terms

antivirus software
botnet
CAPTCHA
cloud computing
collusion
competitive intelligence
computer forensics
cybercriminal
cyberterrorist
data breach
distributed denial-of-service attack (DDoS)
email spam
exploit
hacker
hacktivism
industrial spy
intrusion detection system (IDS)
intrusion prevention software (IPS)
lamer
logic bomb
negligent insider
phishing
reasonable assurance
risk assessment
rootkit
script kiddie
security audit
security policy
smart cards
smishing
spear-phishing
Trojan horse
trustworthy computing
virtual private network (VPN)
virtualization software
virus
virus signature
vishing
worm
zero-day attack
zombie

Self-Assessment Questions

The answers to the Self-Assessment Questions can be found in Appendix E.

1. According to the 2009 CSI Computer Crime and Security Survey, which of the following was the most common security incident?
   a. malware infection
   b. distributed denial-of-service attacks
   c. password sniffing
   d. instant messaging abuse
2. Computer security incidents occur around the world, with ____________ showing the highest level of malicious activity by country.
3. An attack on an information system that takes advantage of a vulnerability is called a(n) ____________.
4. ____________ is a computing environment in which software and data storage are provided as an Internet service run on another organization's computer hardware and accessed by a Web browser.

5. In recent years, the number of vulnerabilities uncovered in application software has been much greater that those discovered in operating systems or browser software. True or False?

6. A(n) ____________ takes places before the security community or software developer knows about the vulnerability or has been able to repair it.
7. Software that generates and grades tests that humans can pass but that all but the most sophisticated computer programs cannot is called ____________.
8. A person who attacks computers and information systems in order to capture trade secrets and gain a competitive advantage is called a cyberterrorist. True or False?
9. ____________ is a combination of SMS texting and phishing in which people receive a legitimate looking text message on their phone telling them to call a specific phone number or to log on to a Web site.
10. A rootkit is a set of programs that enables its user to gain administrative level access to a computer without the end user's consent or knowledge. True or False?
11. A(n) ____________ is a large group of computers controlled from one or more remote locations by hackers, without the knowledge or consent of their owners.
12. ____________ is a method of computing that delivers secure, private, and reliable computing experiences.
13. The process of assessing security-related risks from both internal and external threats to an organization's computers and networks is called a(n) ____________.
14. The written statement that defines an organization's security requirements as well as the controls and sanctions used to meet those requirements is known as a:
    a. risk assessment
    b. security policy
    c. firewall
    d. none of the above
15. Implementation of a strong firewall provides adequate security for almost any network. True or False?
16. A device that works to prevent an attack by blocking viruses, malformed packets, and other threats from getting into the company network is called a(n):
    a. firewall
    b. honeypot
    c. intrusion prevention system
    d. intrusion detection system

Discussion Questions

1. Do your own research and gather data to show whether computer incidents are increasing or decreasing. Present your results in a summary table with a brief paragraph discussing your findings. Be sure to identify the sources of your data.
2. A successful distributed denial-of-service attack requires downloading software that turns unprotected computers into zombies under the control of the malicious hacker. Should the owners of the zombie computers be fined as a means of encouraging people to better safeguard their computers? Why or why not?
3. Should the use of spam to promote a product or service ever be employed by a legitimate organization? Why or why not?
4. Some IT security personnel believe that their organizations should employ former computer criminals to identify weaknesses in their organizations' security defenses. Do you agree? Why or why not?
5. You have been assigned to be a computer security trainer for your firm's 2,000 employees and contract workers. What are the key topics you would cover in your training program?
6. Why and how do spammers seek to set up email accounts with major, free, and reputable Web-based email service providers?
7. How should a nonprofit charity handle the loss of personal data about its donors? Should law enforcement be involved? Should donors be informed?
8. What measures can an organization take to safeguard against computer incidents initiated by malicious insiders? How might an organization prevent unintended computer incidents initiated by negligent insiders?
9. What is the difference between industrial spying and the gathering of competitive intelligence? Is the use of competitive intelligence ethical or unethical? Why?
10. Should the use of hacktivists by a country against enemy organizations be considered an unethical act of war? Why or why not?

What Would You Do?

1. You are a member of the IT security support group of a large manufacturing company. You have been awakened late at night and informed that someone has defaced your organization's Web site and also attempted to gain access to computer files containing information about a new product currently under development. What are your next steps? How much effort would you spend in tracking down the identity of the hacker?
2. You are a member of the Human Resources Department of a three-year-old software manufacturer that has several products and annual revenue in excess of $500 million. You've just received a request from the manager of software development to hire three notorious crackers to probe your company's software products in an attempt to identify any vulnerabilities. The reasoning is that if anyone can find a vulnerability in your software, they can. This will give your firm a head start on developing patches to fix the problems before anyone can exploit them. You're not sure, and you feel uneasy about

hiring people with criminal records and connections to unsavory members of the hacker/cracker community. What would you do?

3. Imagine that you have decided to consider a career in computer forensics. Do research to determine typical starting positions and salaries for someone with a four-year degree in computer forensics. Do further research to find three universities that offer four-year degrees specializing in computer forensics. Compare the three programs, and choose the best one. Why did you choose this university?
4. You are the CFO of a sporting goods manufacturer and distributor. Your firm has annual sales exceeding $500 million, with roughly 25 percent of your sales coming from online purchases. Today your firm's Web site was not operational for about an hour. The IT group informed you that the site was the target of a distributed denial-of-service attack. You are shocked by an anonymous call later in the day in which a man tells you that your site will be attacked unmercifully unless you pay him $250,000 to stop the attacks. What do you say to the blackmailer?
5. You are the CFO of a midsized manufacturing firm. You have heard nothing but positive comments about the new CIO you hired three months ago. As you watch her outline what needs to be done to improve the firm's computer security, you are impressed with her energy, enthusiasm, and presentation skills. However, your jaw drops when she states that the total cost of the computer security improvements will be $300,000. This seems like a lot of money for security, given that your firm has had no major incident. Several other items in the budget will either have to be dropped or trimmed back to accommodate this project. In addition, the $300,000 is above your spending authorization and will require approval by the CEO. This will force you to defend the expenditure, and you are not sure how to do this. You wonder if this much spending on security is really required. How can you sort out what really needs to be done without appearing to be micromanaging or discouraging the new CIO?
6. It appears that someone is using your firm's corporate directory—which includes job titles and email addresses—to contact senior managers and directors via email. The message requests that they click on a URL that takes them to a Web site that looks as if it were designed by your Human Resources organization. Once at this phony Web site, they are asked to confirm the bank and account number to be used for electronic deposit of their annual bonus check. You are a member of IT security for the firm. What can you do?
7. Do research to capture several opinions on the effectiveness of the Controlling the Assault of Non-Solicited Pornography and Marketing (CAN-SPAM) Act. Would you recommend any changes to this act? If so, what changes would you like to see implemented and why?
8. You are a member of the application development organization for a small but rapidly growing software company that produces patient billing applications for doctors' offices. During work on the next release of your firm's one and only software product, you discover a small programming glitch in the current release that could pose a security risk to users. The probability of the problem being discovered is low, but, if exposed, the potential impact on your firm's 100 or so customers could be substantial: hackers could access private patient data and change billing records. The problem will be corrected in the next

release, scheduled to come out in three months, but you are concerned about what should be done for the users of the current release.

The problem has come at the worst possible time. The firm is seeking approval for a $10 million loan to raise enough cash to continue operations until revenue from the sales of its just-released product offsets expenses. In addition, the effort to develop and distribute the patch, to communicate with users, and to deal with any fallout will place a major drain on your small development staff, delaying the next software release by at least two months. You have your regularly scheduled quarterly meeting with the manager of application development this afternoon; what will you say about this problem?

Cases

1. Computer Forensics

On September 8, 2009, 25-year-old airport limousine driver and former coffee cart vendor Najibullah Zazi rented a car and drove from Denver to New York City.[51] His car was laden with explosives and bomb-building materials. According to the Department of Justice, Zazi's target was the New York City subway system. It is believed Zazi was planning to work with other operatives over the weekend and detonate the bomb the following week. However, after learning he was under investigation, Zazi dumped the evidence and fled back to Denver. On September 19, the FBI arrested him on charges of willfully making false statements to the FBI. Computer forensic investigators with the FBI found bomb-making instructions and Internet searches for hydrochloric acid on Zazi's laptop computer. Investigators also processed video surveillance of Zazi buying large quantities of bomb-making materials at a beauty supply store.[52] Zazi had also emailed himself detailed notes on constructing explosives during an Al Qaeda training session on constructing explosives that he had attended in Afghanistan in 2008. In February 2010, Zazi pled guilty to conspiracy to use weapons of mass destruction against persons or property in the United States, conspiracy to commit murder in a foreign country, and providing material support to Al Qaeda.[53]

In November 2007, a 900-foot-long container ship traveling through dense fog struck the Bay Bridge in San Francisco Bay. Approximately 58,000 gallons of fuel oil seeped through the 100-foot gash in the hull into the water.[54] Over 2,500 birds died during the spill, and wildlife experts estimated that a total of 20,000 perished as a result of the long-term chemical effects of oil exposure.[55] Prosecutors alleged that the captain had failed to use radar and positional fixes or other official navigation aids.[56] However, the crime extended beyond the captain's negligence. Computer forensic investigators found that computer navigational charts had been doctored after the crash, and falsified records, such as passage planning checklists, had been created on ship computers after the crash.[57] The captain was eventually sentenced to 10 months in federal prison after pleading guilty to violating the Clean Water Act and the Migratory Bird Treaty Act.[58] In 2009, the ship's management company, Fleet Management Company Ltd., agreed to pay $10 million in compensation for violating the Oil Pollution Act of 1990.[59]

These two high-profile cases illustrate the central role computer forensic investigators are playing in criminal investigations today. These investigators are at work in both criminal and civil cases exploring everything from murder, kidnapping, and robbery to money laundering and fraud to public corruption, intellectual property theft, and destruction of property by disgruntled employees. Even divorce cases are now making use of computer forensics to uncover evidence of infidelity or locate joint funds that have been hidden by one of the spouses.[60]

Yet perhaps the greatest promise of this fast-developing field of investigation is its potential for preventing crime. On November 18, 2010, police arrested a Florida college student, Daniel Alexander Shana, who had posted on Facebook his plans for carrying out a Columbine High School–type massacre to target people who he felt had bullied him. He boasted that he had purchased a semiautomatic pistol and had registered for a firearms license. Students viewing his Facebook posts reported them to authorities.[61] Computer forensic investigators found that he had viewed videos on Columbine and looked into how to purchase weapons and carry out murder.[62]

As the role of computer forensics has expanded in criminal and civil investigations, the number of jobs available in the fields has grown. The Bureau of Labor Statistics predicts that employment in the field of private detectives and investigators in general will grow by 22 percent between 2008 and 2018.[63] To meet this demand, a number of universities have begun offering undergraduate and graduate degrees in computer forensics. Computer forensic investigators not only analyze, recover, and present data for use as evidence, but they also recover emails, passwords, and encrypted or erased data. They must detect intrusions and probe them. Hence, computer forensic investigators need not only specialized hardware and software, but they also need to have mastered specific methods and techniques. That said, the Bureau of Labor Statistics advises that a degree in computer science or accounting is more helpful than a degree in criminal justice.[64]

Most computer forensic professionals, however, enter the field by getting a job with a law enforcement agency and receiving training while on the job.[65] In addition, universities also offer certificates in computer forensics for those already working in the field, and professional organizations host seminars where people interested in the field can gain expertise. Professionals already working in the field can complete a certificate through an online program.

Once computer forensic professionals gain sufficient on-the-job experience, they frequently branch out into the private sector. Licensing requirements vary from state to state, and certification requirements vary from one professional organization to another. The Bureau of Labor Statistics reported that the median salary for private detectives and investigators in 2008 was $41,760. Although the bureau did not track salary information specifically for a computer forensics investigator, professionals in specialized fields are often able to demand higher compensation.[66]

Most importantly, the Bureau of Labor Statistics reported that job competition in this area is keen. With high-profile cases such as the New York subway bomber and television shows romanticizing the role of computer forensics investigators, it's no wonder people are flocking to the field. Yet even if computer forensics isn't as powerful or glamorous as it appears on TV, the field is becoming more critical to criminal investigation, and increasing expertise will be required as cybercriminals develop more sophisticated means of attack.

Discussion Questions

1. What role did computer forensics play in the high-profile cases of the New York subway bomber and the San Francisco Bay oil spill?

2. Why might computer forensics be more effective at preventing crimes than other forms of criminal investigation?
3. In addition to computer-related training, what other education and background would be ideal for someone who wishes to make a career in computer forensics?

2. Stuxnet and the Future of Cyberwarfare

On February 11, 2010—the 31st anniversary of the Islamic revolution in Iran—President Mahmoud Ahmadinejad declared that Iran was now a nuclear nation. In a televised speech, Ahmadinejad informed the world that Iran was now capable of producing 90 percent enriched uranium, the kind of fuel needed to create a nuclear bomb.[67] The announcement came just months after Iran had tested long-range missiles capable of reaching Israel, U.S. military bases in the Gulf, and parts of Europe.[68] In April 2010, U.S. military officials said that Iran would likely be able to produce enough fuel for a nuclear weapon within a year. The United States chose to avoid a military confrontation, seeking instead tougher economic sanctions on Iran.[69] In June 2010, the United Nations imposed a new set of sanctions against Iran.[70] Yet Iran continued to defy Security Council demands to stop enriching uranium.

Then, on June 17, 2010, in a seemingly unrelated incident, a little-known IT security company in Belarus found malware samples capable of infecting fully patched Windows 7 systems through a USB drive.[71] For a month, nothing happened. Finally, on July 15, security blogger Brian Krebs reported on this new worm, which would soon be known as Stuxnet. Late the following day, Microsoft released a security advisory warning customers that a new worm had been developed that exploited a previously unknown vulnerability—in all Windows operating systems, not just Windows 7. That same day, Siemens notified its customers of a new and highly sophisticated virus that had infiltrated its industrial systems management software via USB drives.[72]

The most startling discovery, however, was that this new worm had capabilities far beyond all other industry-targeting viruses. All previous malware had been used to sabotage or spy on industries. Researchers discovered that Stuxnet could not only spy on industrial systems, but could also take control of and reprogram the specially designed software that controls the operation of the factory or facility, the Supervisory Control and Data Acquisition (SCADA) system. This meant that the hackers could potentially take control over a plant's machinery. Yet by September, Siemens reported that Stuxnet had infected only 14 plants and no malfunctions or damage had occurred as a result of infection.[73] Curious too was the fact that the overwhelming majority of computers targeted by the worm were located in Iran—almost 70 percent.[74]

On November 23, 2010, the International Atomic Energy Agency, the international nuclear regulatory organization, reported a mysterious week-long shut down at the Natanz fuel enrichment plant, Iran's primary nuclear enrichment site and home to a very large number of centrifuges, which are a critical component in the uranium-enrichment process.[75] Was there a connection between the troubles at Natanz and Stuxnet? Iranian officials, including President Mahmoud Ahmadinejad, eventually confirmed that some centrifuges had been damaged through infected software.[76] The media, citing unnamed senior diplomats, speculated that Stuxnet was to blame. Soon afterward, the Institute for Science and International Security (ISIS) in Washington, D.C. reported that the worm had likely decommissioned over 1,000 centrifuges at Natanz.[77]

These suspicions intensified as software engineers began to analyze the Stuxnet code. German industrial systems security expert Ralph Langner ascertained that Stuxnet changes a piece of Siemens code that could easily cause a centrifuge to malfunction.[78] The president of ISIS reported that Iran normally ran at 1,007 cycles per second, but that Stuxnet caused the centrifuges to run at 1,064 cycles per second, causing the motors to break. Langner concluded that Stuxnet had dealt the Iranian nuclear program a major setback—without firing a single missile.[79]

The media wondered just who was behind the Stuxnet worm. Early on, information systems security analysts concluded that Stuxnet could not have been developed by an individual or even a group of cybercriminals. The Stuxnet creators were able to compromise digital certificates of JMicron Technology and Realtek Semiconductor. The code demonstrated a far-reaching knowledge of industrial systems. Analysts concluded that the developers of Stuxnet would have to have access to resources only available to nation-states.[80] Some people suspected Israel. Others pointed to the United States, which, along with China, is thought to have the most advanced cyberwarfare capabilities.[81] Yet Ralph Langer, who is one of the analysts most familiar with the code, believes that at least two countries would have had to work together to develop the worm.[82]

The goal of the most common attacks launched at business and organizations range from financial gain to sabotage.[83] Information systems security analysts have long warned, however, that malware could be developed to take control of SCADA systems. If aimed at critical infrastructure, this malware could be used to trigger major power outages or human-made disasters.[84] Microsoft and Siemen have both developed relatively easy fixes to combat the Stuxnet worm. But Stuxnet proved that malware can be built to control SCADA systems.

In a global survey of over 600 IT professionals in industries that make up critical infrastructure, over three-quarters of those surveyed reported that their SCADA systems were connected to the Internet. Only one-third of the professionals reported that their companies had regulations regarding USB sticks. Of the 14 countries included in the survey, China has implemented the most far-reaching IT systems security regulations, but it was unclear whether government regulations have been effective in preventing attacks since China does not suffer significantly less from high-level attacks. Additionally, certain basic security measures are not implemented in a large percentage of businesses in all 14 countries, such as regularly updating software with patches.[85] The Stuxnet worm has renewed security concerns. Yet whether critical infrastructure organizations will make sufficient changes to protect their SCADA systems remains to be seen.

Discussion Questions

1. How is the Stuxnet worm different from previous malware aimed at industrial systems?
2. Do you think the Stuxnet worm constitutes cyberwarfare? Why or why not?
3. What types of precautions could be taken to protect critical infrastructure from malware that can take control of and reprogram a SCADA system?

3. Whistle-Blower Divides IT Security Community

As a member of the X-Force, Mike Lynn analyzed online security threats for Internet Security Systems (ISS), a company whose clients include businesses and government agencies across the world. In early 2005, Lynn began investigating a flaw in the Internet operating system (IOS) used by Cisco routers. Through reverse engineering, he discovered that it was possible to create a network worm that could propagate itself as it attacked and took control of routers across the Internet. Lynn's discovery was momentous, and he decided that he had to speak out and let IT security professionals and the public know about the danger. "What politicians are talking about when they talk about the Digital Pearl Harbor is a network worm," Lynn said during a presentation. "That's what we could see in the future, if this isn't fixed."[86]

Lynn had informed ISS and Cisco of his intentions to talk at a Black Hat conference—a popular meeting of computer hackers—and all three parties entered discussions with the conference managers to decide what information Lynn would be allowed to convey. Two days before the presentation, Cisco and ISS pulled the plug. Cisco employees tore out 10 pages from the conference booklet, and ISS asked that Lynn speak on a different topic—Voice over Internet Protocol (VoIP) security.

In a dramatic move, Lynn resigned from ISS on the morning of the conference and decided to give the presentation as originally planned. Within a few hours of his presentation, Cisco had filed suit against Lynn, claiming that he had stolen information and violated Cisco's intellectual property rights. "I feel I had to do what's right for the country and the national infrastructure," Lynn explained.[87]

Lynn's words might have sounded more credible had his presentation not been titled "The Holy Grail: Cisco IOS Shellcode and Remote Execution" and had Lynn not chosen a Black Hat annual conference as the venue for his crucial revelation.[88]

Rather than speak to a gathering of Cisco users, who would have responded to the revelation by installing Cisco's patch and putting pressure on Cisco to find additional solutions, Lynn chose an audience that may well have included hackers who viewed the search for the flaw as a holy crusade. Black hats are crackers who break into systems with malicious intent. By contrast, white hats are hackers who reveal vulnerabilities to protect systems. Black Hat is a company that provides IT security consulting, briefings, and training. The CEO of Black Hat, Jeff Moss, also founded DEFCON, an annual meeting of underground hackers who gather to drink, socialize, and talk shop. During the DEFCON conference, which followed the Black Hat conference, hackers worked late into the night trying to find the flaw.

"What Lynn ended up doing was describing how to build a missile without giving all the details. He gave enough details so people could understand how a missile could be built, and they could take their research from there," said one DEFCON hacker.[89]

Once well defined, the line between white hat and black hat has become blurry. Security professionals, law enforcement officials, and other white hats have infiltrated the ranks of the black-garbed renegades at DEFCON annual conferences, and IT companies hire hackers as IT security experts. Microsoft now hosts an annual invitation-only conference called BlueHat. Respectable IT giants such as IBM, Microsoft, and Hewlett-Packard have invited the black hats into the industry, and, in large part, they have accepted the invitation.

Yet Cisco's handling of Lynn had the black hats up in arms. "The whole attempt at security through obscurity is amazing, especially when a big company such as Cisco tries to keep a researcher quiet," exclaimed Marc Maiffret, chief hacker for eEye Digital Security. Maiffret felt that Cisco would have to mend some bridges with the IT security community.[90]

White hats, in the meantime, bombarded the IT media with opinion pieces reminding people that a similar Black Hat disclosure about Microsoft precipitated the creation of the Blaster worm, which tore across the Internet and cost billions of dollars in damage.

Discussion Questions

1. Do you think that Mike Lynn acted in a responsible manner? Why or why not?
2. Do you think that Cisco and ISS were right to pull the plug on Lynn's presentation at the Black Hat conference? Why or why not?
3. Outline a more reasonable approach toward communicating the flaw in the Cisco routers that would have led to the problem being promptly addressed without stirring up animosity among the parties involved.

End Notes

[1] Parisa Hafezi, "Iran Election Violence 'Outrageous,' Says Obama," Reuters, June 26, 2009, www.reuters.com/article/idUSTRE55F54520090626.

[2] "Twitter Hackers Appear to be Shiite Group," *CNN*, December 18, 2009, http://articles.cnn.com/2009-12-18/tech/twitter.hacked_1_hacking-shiite-group-hezbollah?_s=PM:TECH.

[3] Robert McMillan, "With Unrest in Iran, Cyber-attacks Begin," *PC World*, June 15, 2009, www.pcworld.com/article/166714/with_unrest_in_iran_cyberattacks_begin.html.

[4] Robert McMillan, "With Unrest in Iran, Cyber-attacks Begin," *PC World*, June 15, 2009, www.pcworld.com/article/166714/with_unrest_in_iran_cyberattacks_begin.html.

[5] "Twitter hackers Appear to be Shiite Group," *CNN*, December 18, 2009, http://articles.cnn.com/2009-12-18/tech/twitter.hacked_1_hacking-shiite-group-hezbollah?_s=PM:TECH.

[6] Ali Sheikholeslami, "Iran Cyber Army Hacks Major Opposition Web Sites (Update2)," *Bloomberg Businessweek*, February 12, 2010, www.businessweek.com/news/2010-02-12/iran-cyber-army-hacks-major-opposition-web-sites-update2-.html.

[7] Farvartish Rezvaniyeh, "Pulling the Strings of the Net: Iran's Cyber Army," *Frontline*, February 26, 2010, www.pbs.org/wgbh/pages/frontline/tehranbureau/2010/02/pulling-the-strings-of-the-net-irans-cyber-army.html.

[8] Radio Zamaneh, "Iran's Revolutionary Guards Claim 'Iranian Cyber Army'," *Payvand Iran News*, May 21, 2010, www.payvand.com/news/10/may/1225.html.

[9] Mathew J. Schwartz, "Iranian Cyber Army Joins Botnet Business," *InformationWeek*, October 27, 2010, www.informationweek.com/news/security/attacks/showArticle.jhtml?articleID=228000146.

[10] "The 'Iranian Cyber Army' Strikes Back," Seculert Research Lab, October 24, 2010, http://blog.seculert.com/2010/10/iranian-cyber-army-strikes-back.html.

[11] Dana Chivvis, "Iranian Cyber Army' Hacker Group Enters Mercenary Business," *AOL News*, October 25, 2010, www.aolnews.com/surge-desk/article/iranian-cyber-army-hacker-group-enters-the-mercenary-business/19688042.

[12] Golnaz Esfandiari, "Persian Letters, Iranian Cyber Army Hacks Website Of Farsi1," Radio Free Europe/Radio Liberty, November 18, 2010, www.rferl.org/content/Iranian_Cyber_Army_Hacks_Website_Of_Farsi1/2223708.html.

13 Jaikumar Vijayan, "Threat of Cyberattack from Overseas High, Federal IT Execs Say," *Computerworld*, April 7, 2010, www.computerworld.com/s/article/9174906/Threat_of_cyberattacks_from_overseas_high_federal_IT_execs_say.

14 Siobhan Gorman, "Electricity Grid in U.S. Penetrated by Spies," *Wall Street Journal*, April 8, 2009, http://online.wsj.com/article/SB123914805204099085.html.

15 Richi Jennings, "Fannie Mae Sabotaged?" *IT Blogwatch*, *Computerworld* (blog), January 30, 2009, http://blogs.computerworld.com/fannie_mae_sabotaged

16 "First Annual Cost of Cyber Crime Study," Ponemon Institute and ArcSight, July 2010.

17 Gregg Keizer, "Microsoft Slates Another Monster Patch Tuesday," *Computerworld*, December 9, 2010, www.computerworld.com/s/article/9200642/Microsoft_slates_another_monster_Patch_Tuesday.

18 Gregg Keizer, "Mozilla Patches 13 Firefox Security Bugs," *Computerworld*, December 10, 2010, www.computerworld.com/s/article/9200741/Mozilla_patches_13_Firefox_security_bugs.

19 Darlene Storm, "Mobile Malware: You Will Be Billed $90,000 for This Call," *Security is Sexy* (blog), *Computerworld*, Aug 5, 2010, http://blogs.computerworld.com/16661/mobile_malware_you_will_be_billed_90_000_for_this_call.

20 Dancho Danchev, "Conficker's Estimated Economic Cost? $9.1 Billion," Zero Day, *ZDNet* (blog), April 23, 2009, www.zdnet.com/blog/security/confickers-estimated-economic-cost-91-billion/3207.

21 Pelin Aksoy and Laura Denardis, *Information Technology in Theory* (Boston: Cengage Learning, 2007), 299–301.

22 Gregg Keizer, "Energizer Bunny's Software Infects PCs," *Computerworld*, March 7, 2010, www.computerworld.com/s/article/9166978/Energizer_Bunny_s_software_infects_PCs.

23 Robert McMillan, "Groups Used 30,000-Node Botnet in MasterCard, PayPal Attacks," *Computerworld*, December 9, 2010, www.computerworld.com/s/article/9200598/Group_used_30_000_node_botnet_in_MasterCard_PayPal_attacks.

24 Gregg Keizer, "Rootkit with Blue Screen History Now Targets 64-bit Windows," *Computerworld*, August 27, 2010, www.computerworld.com/s/article/9182238/Rootkit_with_Blue_Screen_history_now_targets_64_bit_Windows.

25 David A. Milman, "Spam Wars 2010," *The Computer Repair Blog*, *Computerworld (blog)*, September 28, 2010, http://blogs.computerworld.com/17047/spam_wars_2010.

26 Symantec Corporation, "Press Release: Symantec Announces September 2010 MessageLabs Intelligence Report", September 21, 2010, www.messagelabs.com/resources/press/59940.

27 Dan Kaplan, "'Avalanche' Phishing Slowing, But Was All the 2009 Rage," *SCMagazine*, May 12, 2010, www.scmagazineus.com/avalanche-phishing-slowing-but-was-all-the-2009-rage/article/170052.

28 Elizabeth Montalbano, "FBI Warns of Mobile Cyber Threats," *InformationWeek*, November 29, 2010, www.informationweek.com/news/government/security/showArticle.jhtml?articleID=228400096.

[29] Bill Singer. "The FBI Issues Holiday Warning About Smishing, Vishing and Other Scams by Cyber-Criminals," *Huffington Post*, November 28, 2010, www.huffingtonpost.com/bill-singer/the-fbi-issues-holiday-wa_b_788869.html.

[30] Linda McGlasson, "How to Respond to Vishing Attacks: Bank, State Association Share Tips for Incident Response Plan," *BankInfoSecurity.com*, April 26, 2010, www.bankinfosecurity.com/p_print.php?t=a&id=2457.

[31] Daniel Kennedy, "The Real Lessons of Gawker's Security Mess," *The Firewall*, *Forbes* (blog), December 13, 2010, http://blogs.forbes.com/firewall/2010/12/13/the-lessons-of-gawkers-security-mess.

[32] Robert McMillan, "BofA Insider to Plead Guilty to Hacking ATMs," *Computerworld*, April 7, 2010, www.computerworld.com/s/article/9174991/BofA_insider_to_plead_guilty_to_hacking_ATMs.

[33] Jaikumar Vjayan, "DARPA Launches Insider Threat Detection Effort for Military," *Computerworld*, September 1, 2010, www.computerworld.com/s/article/9183238/DARPA_launches_insider_threat_detection_effort_for_military.

[34] Ponemon Institute, "U.S. Certainty of a Data Breach Study," June 2008.

[35] Lumension Security, Inc., "Anatomy of Insider Risk: Why You Could Be Your Worst Enemy," March 2010, http://i.zdnet.com/whitepapers/Lumension_Whitepaper_AnatomyofInsiderRiskWhyYouCouldBeYourWorstEnemy.pdf.

[36] Robert McMillan, "New Virus Targets Industrial Secrets," *Computerworld*, July 17, 2010, www.computerworld.com/s/article/9179298/New_virus_targets_industrial_secrets.

[37] Lucian Constantin, "Two WSU Students Linked to Major Vietnamese Cybercriminal Ring," Softpedia, January 3, 2011 http://news.softpedia.com/news/Two-WSU-Students-Linked-to-Major-Vietnamese-Cybercriminal-Ring-175737.shtml.

[38] Zeljka Zorz, "Online Cybercriminal DarkMarket Closed, Founder Arrested," *Help Net Security*, January 18, 2010 www.net-security.org/secworld.php?id=8718.

[39] "WikiLeaks: Amazon Taken Offline in Suspected 'Hacktivist' Attack," *The Telegraph*, December 12, 2010, www.telegraph.co.uk/technology/amazon/8198041/Wikileaks-Amazon-taken-offline-in-suspected-hacktivist-attack.html.

[40] Jana Winter, "''Hacktivist' Jester Claims Responsibility for WikiLeaks Attack," *Fox News.com*, December 3, 2010, www.foxnews.com/scitech/2010/12/03/patriotic-hactivist-took-down-wikileaks.

[41] Microsoft Corporation, "The Journey to Trustworthy Computing: Microsoft Execs Report First-Year Progress," January 15, 2003, www.microsoft.com/presspass/features/2003/jan03/01-15twcanniversary.mspx, (accessed January 10, 2011).

[42] Microsoft Corporation, "Trustworthy Computing Continues to Build Momentum," January 13, 2005, www.microsoft.com/presspass/features/2005/jan05/01-13TWCupdate.mspx, (accessed January 10, 2011).

[43] Seymour E. Goodman and Herbert S. Lin, Editors, Toward a Safer and More Secure Cyberspace, page 1, ©2007 Computer Science and Telecommunications Board.

[44] SANS Institute, "Information Security Policy Templates," www.sans.org/security-resources/policies (accessed January 26, 2011).

[45] Allison Diana, "Employees Flout Social Network Security Policies," *InformationWeek*, July 23, 2010, www.informationweek.com/news/security/vulnerabilities/showArticle.jhtml?articleID=226200128.

[46] Ponemon Institute LLC, "Airport Insecurity: The Case of Missing & Lost Laptops," June 30, 2008.

[47] Datasavers, Inc., "Computer and Internet Security," www.datasaversinc.com/computer-and-internet-security (accessed on January 15, 2011).

[48] Department of Homeland Security, "National Cyber Security Division" www.dhs.gov/xabout/structure/editorial_0839.shtm (accessed January 26, 2011).

[49] US-CERT, "About US," www.us-cert.gov/aboutus.html (accessed January 26, 2011).

[50] *CSO*, "2010 Cybersecurity Watch Survey: Cybercrime Increasing Faster Than Some Company Defenses," www.cert.org/archive/pdf/CyberCrimeRelease_final.pdf.

[51] Michael Wilson, "From Smiling Coffee Vendor to Terror Suspect," *New York Times*, September 25, 2009, www.nytimes.com/2009/09/26/nyregion/26profile.html?_r=1.

[52] Regional Computer Forensics Laboratory, "Regional Computer Forensics Laboratory Annual Report for FY 2009, 5.0 Casework/Investigations," www.rcfl.gov/Downloads/Documents/annual_report_web/annual_05-01_casework_09.html (accessed January 26, 2011).

[53] Department of Justice. "Press Release: Najibullah Zazi Pleads Guilty to Conspiracy to Use Explosives Against Persons or Property in U.S., Conspiracy to Murder Abroad, and Providing Material Support to al Qaeda," Federal Bureau of Investigation New York, February 22, 2010, http://newyork.fbi.gov/dojpressrel/pressrel10/nyfo022210.htm.

[54] NOAA's National Ocean Service, "Incident News: M/V Cosco Busan," Office of Response and Restoration, November 7, 2007, www.incidentnews.gov/incident/7708.

[55] International Bird Rescue Research Center "Dark days on San Francisco Bay," www.ibrrc.org/Cosco_Busan_spill_2007.html (accessed January 27, 2011).

[56] "Oil Spill Captain Gets Prison Sentence," *San Francisco Bay Crossings*, www.baycrossings.com/dispnews.asp?id=2211 (accessed January 27, 2011).

[57] Regional Computer Forensics Laboratory, "Fleet Mgt. Ltd. Agrees to Pay $10 Million for Pollution and Obstruction Crimes," August 19, 2009, www.rcfl.gov/index.cfm?fuseAction=Public.N_SV004.

[58] "Oil Spill Captain Gets Prison Sentence," *San Francisco Bay Crossings*, January 27, 2011).

[59] Department of Justice, "Press Release: Cosco Busan Operator Admits Guilt in Causing Oil Spill," Office of Public Affairs, August 13, 2009, www.justice.gov/opa/pr/2009/August/09-enrd-797.html.

[60] Minnesota Lawyers, "Divorce and Computer Forensics," www.nvo.com/beaulier/divorceandforensicevidence (accessed January 27, 2011).

61 "Daniel Shana Threatens To Inflict 'Columbine Take 2' On Lynn University Students," *Huffington Post*, November 19, 2010, www.huffingtonpost.com/2010/11/19/daniel-shana-threatens-to_n_786015.html.

62 "Man Charged with Columbine-Style Plot," MyFoxBoston, November 18, 2010, www.myfoxboston.com/dpp/news/crime_files/crime_watch/man-charged-with-columbine-style-plot-20101118.

63 Bureau of Labor Statistics, "Private Detectives and Investigators," Occupational Outlook Handbook, 2010–11 Edition, www.bls.gov/oco/ocos157.htm (accessed January 27, 2011).

64 Bureau of Labor Statistics, "Private Detectives and Investigators," Occupational Outlook Handbook, 2010–11 Edition," www.bls.gov/oco/ocos157.htm (accessed January 27, 2011).

65 "Computer Forensic Investigator: High-tech Career in Law Enforcement," *Hub Pages*, http://hubpages.com/hub/Computer-Forensic-Investigator-High-tech-Career-in-Law-Enforcement (accessed January 27, 2011).

66 Bureau of Labor Statistics, "Private Detectives and Investigators," Occupational Outlook Handbook, 2010–11 Edition, www.bls.gov/oco/ocos157.htm (accessed January 27, 2011).

67 Nasser Karimi and Ali Akbar Dareini, "Iran Proclaims New Success in Uranium Enrichment," Associated Press, February 11, 2010, http://abcnews.go.com/US/wireStory?id=9803960.

68 "Iran Tests Upgraded Long-Range Missile," *CNN*, December 16, 2009, http://articles.cnn.com/2009-12-16/world/iran.missile_1_surface-to-surface-missile-enrichment-iran?_s=PM:WORLD.

69 "Iran's Nuclear Program," *New York Times*, September 7, 2010, www.nytimes.com/info/iran-nuclear-program.

70 Neil MacFarquhar, "U.N. Approves New Sanctions to Deter Iran," *New York Times*, June 9, 2010, www.nytimes.com/2010/06/10/world/middleeast/10sanctions.html.

71 Krebson Security, "Experts Warn of New Windows Shortcut Flaw," http://krebsonsecurity.com/2010/07/experts-warn-of-new-windows-shortcut-flaw (accessed January 28, 2011).

72 Robert McMillan, "New Spy Rootkit Targets Industrial Secrets," *TechWorld*, July 19, 2010, http://news.techworld.com/security/3232365/new-spy-rootkit-targets-industrial-secrets.

73 Robert McMillan, "Siemens: Stuxnet Worm Hit Industrial Systems," *Computerworld*, September 14, 2010, www.computerworld.com/s/article/print/9185419/Siemens_Stuxnet_worm_hit_industrial_systems?taxonomyName=Network+Security&taxonomyId=142.

74 "Factbox: What is Stuxnet?" September 24, 2010, http://uk.reuters.com/article/idUKTRE68N3PT20100924.

75 David Albright, Andrea Stricker, and Christina Walrond, "IAEA Iran Safeguards Report: Shutdown of Enrichment at Natanz Result of Stuxnet Virus? Downgrade in Role of Fordow Enrichment Site; LEU Production May Have Decreased," Institute for Science and International Security, November 23, 2010, www.isis-online.org/isis-reports/detail/iaea-iran-safeguards-report-shutdown-of-enrichment-at-natanz-result-of-stux.

76 "Ahmadinejad Admits Centrifuges Damaged by Virus," *Jerusalem Post*, November 29, 2010, www.jpost.com/International/Article.aspx?id=197239.

[77] David Albright, Paul Brannan, and Christina Walrond, "Did Stuxnet Take Out 1,000 Centrifuges at the Natanz Enrichment Plant? Preliminary Assessment," Institute for Science and International Security, December 22, 2010, http://isis-online.org/isis-reports/detail/did-stuxnet-take-out-1000-centrifuges-at-the-natanz-enrichment-plant.

[78] Robert McMillan, "Was Stuxnet Built to Attack Iran's Nuclear Program?" *PCWorld*, September 21, 2010, www.pcworld.com/businesscenter/article/205827/was_stuxnet_built_to_attack_irans_nuclear_program.html.

[79] Yaakov Katz, "Stuxnet May Have Destroyed 1,000 Centrifuges at Natanz," *Jerusalem Post*, December 24, 2010, www.jpost.com/Defense/Article.aspx?ID=200843&R=R1.

[80] Robert McMillan, "Was Stuxnet Built to Attack Iran's Nuclear Program?" *PCWorld*, September 21, 2010, www.pcworld.com/businesscenter/article/205827/was_stuxnet_built_to_attack_irans_nuclear_program.html.

[81] Stewart Baker, Shaun Waterman, and George Ivanov, "In the Crossfire: Critical Infrastructure in the Age of Cyber War," Center for Strategic and International Studies, January 28, 2010, http://csis.org/event/crossfire-critical-infrastructure-age-cyber-war.

[82] Yaakov Katz, 'Stuxnet Virus Set Back Iran's Nuclear Program by 2 Years,' *Jerusalem Post,* December 15, 2010, www.jpost.com/IranianThreat/News/Article.aspx?id=199475.

[83] Kim Zetter, "Report: Critical Infrastructures Under Constant Cyberattack Globally" *Wired*, January 28, 2010, www.wired.com/threatlevel/2010/01/csis-report-on-cybersecurity/?intcid=postnav.

[84] Stewart Baker, Shaun Waterman, and George Ivanov, "In the Crossfire: Critical Infrastructure in the Age of Cyber War," Center for Strategic and International Studies, January 28, 2010, http://csis.org/event/crossfire-critical-infrastructure-age-cyber-war.

[85] Stewart Baker, Shaun Waterman, and George Ivanov, "In the Crossfire: Critical Infrastructure in the Age of Cyber War," Center for Strategic and International Studies, January 28, 2010, http://csis.org/event/crossfire-critical-infrastructure-age-cyber-war.

[86] Robert Lemos, "Cisco, ISS File Suit Against Rogue Researcher," *SecurityFocus*, July 27, 2005, www.securityfocus.com/news/11259.

[87] Robert McMillan, "Black Hat: ISS Researcher Quits Job to Detail Cisco Flaws," *InfoWorld*, July 27, 2005, www.infoworld.com/d/security-central/black-hat-iss-researcher-quits-job-detail-cisco-flaws-088.

[88] Bruce Schneier, "Cisco Harrasses Security Researcher," *Schneier on Security* (blog), July 29, 2005, www.schneier.com/blog/archives/2005/07/cisco_harasses.html.

[89] Andy Sullivan, "Hackers Race to Expose Cisco Internet Flaw," *newsgroups.derkeiler.com*, July 31, 2005 http://newsgroups.derkeiler.com/Archive/Comp/comp.dcom.telecom/2005-08/msg00065.html.

[90] Robert Lemos, "Settlement Reached in Cisco Flaw Dispute," *SecurityFocus*, July 29, 2005, www.securityfocus.com/news/11260.

CHAPTER 4

PRIVACY

QUOTE

The makers of our Constitution undertook to secure conditions favorable to the pursuit of happiness.... They conferred, as against the government, the right to be let alone—the most comprehensive of rights and the right most valued by civilized men. To protect that right, every unjustifiable intrusion by the government upon the privacy of the individual, whatever the means employed, must be deemed a violation of the Fourth Amendment.[1]

—Supreme Court Justice Louis Brandeis, 1928

VIGNETTE

Google Collects Unprotected Wireless Network Information

Google's Street View maps allow users to zoom into a location on a map and view actual images of houses, shops, buildings, sidewalks, fields, parked cars, and anything else that can be photographed from the vantage point of a slow-moving vehicle. It's a remarkable tool for those trying to find an auto repair shop, a post office, or a friend's house for the first time. Google launched Street View in a few cities in the United States in May 2007. It gradually expanded to additional U.S. cities and then to other cities around the world. In August 2009, Google began collecting data for Street View in several German cities. Germany, however, has stricter privacy laws than other countries, and prohibits the photographing of private property and people unless they are engaged in a public event, such as a sports match. As a result, Google had to work closely with the country's Data Protection Agency in order to comply with German laws in the hopes of getting its Street View service for Germany online by the end of 2010.[2,3]

In April 2010, a startling admission by Google provoked public outrage in Germany and around the world. It resulted in government probes in numerous countries, as well as several class action lawsuits in the United States. In response to queries by Germany's Data Protection Agency, Google acknowledged on April 27 that, in addition to taking snapshots, its cars were also sniffing out unprotected wireless network information. Google reported, however, that it was only collecting service set identifier (SSID) data—such as the network name—and the media access control (MAC) address—the unique number given to wireless network devices. Google's geo-location services could use this data to more accurately pinpoint the location of a person utilizing a mobile device, such as a smartphone. The company insisted that it was not collecting or storing payload data (the actual data sent over the network).[4]

The German Federal Commissioner for the Data Protection Agency was horrified and requested that Google stop collecting data immediately.[5] Additionally, the German authorities asked to audit the data Google had collected. Google agreed to hand over its code to a third party, the security consulting firm Stroz Friedberg. Nine days later there came another admission: Google had in fact been collecting and storing payload data. But Google insisted that it had only collected fragmented data and made no use of this data.[6] A few days later, Germany announced that it was launching a criminal investigation. Other European nations quickly opened investigations of their own.[7]

By early June, six class action lawsuits claiming that Google had violated federal wiretapping laws had been filed in the United States.[8] In its defense, Google argued that collecting unencrypted payload data is not a violation of federal laws.[9] Google explained that in order to locate wireless hotspots, it used a passive scanning technique, which had picked up payload data by mistake. The company used open source Kismet wireless scanning software that was customized by a Google engineer in 2006.[10] Google insisted that the project's managers were unaware that the software had been programmed to collect payload data when they launched the project. Finally, Google argued that the data it collected was fragmented—not only was the car moving, but it was changing channels five times per second.[11]

However, a civil lawsuit claimed that Google filed a patent for its wireless network scanning system in November 2008 that revealed that Google's system could more accurately locate a router's location—giving Google the ability to identify the street address of the router. The more data collected by the scanning system, the lawsuit contended, the higher the confidence level Google would have in its calculated location of the wireless hotspot.[12]

In the fall of 2010, the U.S. Federal Trade Commission (FTC) ended its investigation, deciding not to take action or impose fines. The FTC recognized that Google had taken steps to amend the situation by ceasing to collect the payload data and hiring a new director of privacy.[13] But by that time, 30 states had opened investigations into the matter.[14] During the course of these and other investigations, Google turned over the data it had collected to external regulators. On October 22, the company announced that not all of the payload data it had collected was fragmentary. It had in fact collected entire email messages, URLs, and passwords.[15] In November, the U.S. Federal Communications Commission announced that it was looking into whether Google had violated the federal Communications Act.[16]

Some analysts believe that Google's behavior follows a trend in the Internet industry: Push the boundaries of privacy issues; apologize, and then push again once the scandal dies down.[17] If this is the case, Google will have to decide, as the possible fines and other penalties accrue, whether this strategy pays off.

Questions to Consider

1. What information about you is being held, who is holding it, and what is this information being used for?
2. What measures are being taken to safeguard this information, and what happens if it is inadvertently disclosed or deliberately stolen?

LEARNING OBJECTIVES

As you read this chapter, consider the following questions:

1. What is the right of privacy, and what is the basis for protecting personal privacy under the law?
2. What are some of the laws that provide protection for the privacy of personal data, and what are some of the associated ethical issues?
3. What is identity theft, and what techniques do identity thieves use?
4. What are the various strategies for consumer profiling, and what are the associated ethical issues?
5. What must organizations do to treat consumer data responsibly?
6. Why and how are employers increasingly using workplace monitoring?
7. What are the capabilities of advanced surveillance technologies, and what ethical issues do they raise?

PRIVACY PROTECTION AND THE LAW

The use of information technology in business requires balancing the needs of those who use the information that is collected against the rights and desires of the people whose information is being used.

Information about people is gathered, stored, analyzed, and reported because organizations can use it to make better decisions. Some of these decisions, including whether or not to hire a job candidate, approve a loan, or offer a scholarship, can profoundly affect people's lives. In addition, the global marketplace and intensified competition have increased the importance of knowing consumers' purchasing habits and financial condition. Companies use this information to target marketing efforts to consumers who are most likely to buy their products and services. Organizations also need basic information about customers to serve them better. It is hard to imagine an organization having productive relationships with its customers without having data about them. Thus, organizations want systems that collect and store key data from every interaction they have with a customer.

However, many people object to the data collection policies of governments and businesses on the grounds that they strip individuals of the power to control their own personal information. For these people, the existing hodgepodge of privacy laws and practices fails to provide adequate protection; rather, it causes confusion that promotes distrust and skepticism, which are further fueled by the disclosure of threats to privacy, such as those detailed in the opening vignette on Google. Indeed, "one of the key factors affecting the growth of e-commerce is the lack of Internet users' confidence in online information privacy."[18]

A combination of approaches—new laws, technical solutions, and privacy policies—is required to balance the scales. Reasonable limits must be set on government and business access to personal information; new information and communication technologies must be designed to protect rather than diminish privacy; and appropriate corporate policies must be developed to set baseline standards for people's privacy. Education and communication are also essential.

This chapter will help you understand the right to privacy, and it presents an overview of developments in information technology that could impact this right. The chapter also addresses a number of ethical issues related to gathering data about people.

First, it is important to gain a historical perspective on the right to privacy. During the debates on the adoption of the United States Constitution, some of the drafters expressed concern that a powerful government would intrude on the privacy of individual citizens. After the Constitution went into effect in 1789, several amendments were proposed that would spell out the rights of individuals. Ten of these proposed amendments were ultimately ratified and became known as the Bill of Rights. So although the Constitution does not contain the word *privacy*, the U.S. Supreme Court has ruled that the concept of privacy is protected by a number of amendments in the Bill of Rights. For example, the Supreme Court has stated that American citizens are protected by the Fourth Amendment when there is a "reasonable expectation of privacy."

The Fourth Amendment is as follows:

> The right of the people to be secure in their persons, houses, papers, and effects, against unreasonable searches and seizures, shall not be violated, and no Warrants shall issue, but upon probable cause, supported by Oath or affirmation, and particularly describing the place to be searched, and the persons or things to be seized.

However, the courts have ruled that *without* a reasonable expectation of privacy, there is no privacy right.

Today, in addition to protection from government intrusion, people want and need privacy protection from private industry. Few laws actually provide such protection, and most people assume that they have greater privacy rights than the law actually provides. Some people believe that only those with something to hide should be concerned about the loss of privacy; however, everyone should be concerned. As the Privacy Protection Study Commission noted in 1977, when the computer age was still in its infancy: "The real danger is the gradual erosion of individual liberties through the automation, integration, and interconnection of many small, separate record-keeping systems, each of which alone may seem innocuous, even benevolent, and wholly justifiable."[19]

Information Privacy

A broad definition of the right of privacy is "the right to be left alone—the most comprehensive of rights, and the right most valued by a free people."[20] Another concept of privacy that is particularly useful in discussing the impact of IT on privacy is the term *information privacy*, first coined by Roger Clarke, director of the Australian Privacy Foundation. **Information privacy** is the combination of communications privacy (the ability to communicate with others without those communications being monitored by other persons or organizations) and data privacy (the ability to limit access to one's personal data by other individuals and organizations in order to exercise a substantial degree of control over that data and its use).[21] The following sections cover concepts and principles related to information privacy, beginning with a summary of the most significant privacy laws, their applications, and related court rulings.

Privacy Laws, Applications, and Court Rulings

This section outlines a number of legislative acts that affect a person's privacy. Note that most of these actions address invasion of privacy by the government. Legislation that protects people from data privacy abuses by corporations is almost nonexistent.

Although a number of independent laws and acts have been implemented over time, no single, overarching national data privacy policy has been developed in the United States. Nor is there an established advisory agency that recommends acceptable privacy practices to businesses. Instead, there are laws that address potential abuses by the government, with little or no restrictions for private industry. As a result, existing legislation is sometimes inconsistent or even conflicting. You can track the status of privacy legislation in the United States at the Electronic Privacy Information Center's Web site (*www.epic.org*).

The discussion is divided into the following topics: financial data, health information, state laws related to security breach notification, children's personal data, electronic surveillance, export of personal data, and access to government records.

Financial Data

Individuals must reveal much of their personal financial data in order to take advantage of the wide range of financial products and services available, including credit cards, checking and savings accounts, loans, payroll direct deposit, and brokerage accounts. To access many of these financial products and services, individuals must use a personal logon name, password, account number, or PIN. The inadvertent loss or disclosure of this personal financial data carries a high risk of loss of privacy and potential financial loss. Individuals should be concerned about how this personal data is protected by businesses and other organizations and whether or not it is shared with other people or companies.

Fair Credit Reporting Act (1970)

The **Fair Credit Reporting Act** of 1970 regulates the operations of credit-reporting bureaus, including how they collect, store, and use credit information. The act, enforced by the U.S. Federal Trade Commission, is designed to ensure the accuracy, fairness, and privacy of information gathered by the credit-reporting companies and to check those systems that gather and sell information about people. The act outlines who may access your credit information, how you can find out what is in your file, how to dispute inaccurate data, and how long data is retained. It also prohibits the credit-reporting bureau from giving out information about you to your employer or potential employer without your written consent.[22]

Fair and Accurate Credit Transactions Act (2003)

The **Fair and Accurate Credit Transactions Act** was passed in 2003 as an amendment to the Fair Credit Reporting Act, and it allows consumers to request and obtain a free credit report once each year from each of the three primary consumer credit reporting companies (Equifax, Experian, and TransUnion). The act also helped establish the National Fraud Alert system to help prevent identity theft. Under this system, consumers who suspect that they have been or may become a victim of identity theft can place an alert on their credit files. The alert places potential creditors on notice that they must proceed with caution when granting credit.[23]

Right to Financial Privacy Act of 1978

The **Right to Financial Privacy Act of 1978** protects the financial records of financial institution customers from unauthorized scrutiny by the federal government. Prior to passage of this act, financial institution customers were not informed if their personal records were being turned over for review by a government authority, nor could customers challenge government access to their records. Under this act, a customer must receive written notice that a federal agency intends to obtain their financial records, along with an explanation of the purpose for which the records are sought. The customer must also be given written procedures to follow if he or she does not wish the records to be made available. In addition, to gain access to a customer's financial records, the government must obtain one of the following:

- an authorization signed by the customer that identifies the records, the reasons the records are requested, and the customer's rights under the act,
- an appropriate administrative or judicial subpoena or summons,
- a qualified search warrant, or
- a formal written request by a government agency (can be used only if no administrative summons or subpoena authority is available)

The financial institution cannot release a customer's financial records until the government authority seeking the records certifies in writing that it has complied with the applicable provision of the act.

The act only governs disclosures to the federal government; it does not cover disclosures to private businesses or state and local governments. The definition of *financial institution* has been expanded to include banks, thrifts, credit unions; money services businesses; money order issuers, sellers, and redeemers; the U.S. Postal Service; securities and futures industries; futures commission merchants; commodity trading advisors; and casinos and card clubs.

Section 358 of the USA PATRIOT Act amended the act to permit disclosure of financial information to any intelligence or counterintelligence agency in any investigation related to international terrorism.[24]

Gramm–Leach–Bliley Act (1999)

The **Gramm–Leach–Bliley Act (GLBA)**, also known as the Financial Services Modernization Act of 1999, was a bank deregulation law that repealed a Depression-era law known as Glass–Steagall. Glass–Steagall prohibited any one institution from offering investment, commercial banking, and insurance services; individual companies were only allowed to offer one of those types of financial service products. GLBA enabled such entities to merge. The emergence of new corporate conglomerates, such as Bank of America, Citigroup, and JPMorgan Chase, soon followed. These one-stop financial supermarkets owned bank branches, sold insurance, bought and sold stocks and bonds, and engaged in mergers and acquisitions. Some people place partial blame for the financial crisis that began in 2008 on the passage of GLBA and the loosening of banking restrictions.[25]

GLBA also included three key rules that affect personal privacy:

- *Financial Privacy Rule*—This rule established mandatory guidelines for the collection and disclosure of personal financial information by financial organizations. Under this provision, financial institutions must provide a privacy notice to each consumer that explains what data about the consumer is gathered, with whom that data is shared, how the data is used, and how the data is protected. The notice must also explain the consumer's right to **opt out**—to refuse to give the institution the right to collect and share personal data with unaffiliated parties. Anytime the privacy policy is changed, the consumer must be contacted again and given the right to opt out. The privacy notice must be provided to the consumer at the time the consumer relationship is formed and once each year thereafter. Customers who take no action automatically **opt in** and give financial institutions the right to share personal data, such as annual earnings, net worth, employers, personal investment information, loan amounts, and Social Security numbers, to other financial institutions.

- *Safeguards Rule*—This rule requires each financial institution to document a data security plan describing its preparation and plans for the ongoing protection of clients' personal data.
- *Pretexting Rule*—This rule addresses attempts by people to access personal information without proper authority by such means as impersonating an account holder or phishing. GLBA encourages financial institutions to implement safeguards against pretexting.

After the law was passed, financial institutions resorted to mass mailings to contact their customers with privacy-disclosure forms. As a result, many people received a dozen or more similar-looking forms—one from each financial institution with which they did business. However, most people did not take the time to read the long forms, which were printed in small type and full of legalese. Rather than making it easy for customers to opt out, the documents required that consumers send one of their own envelopes to a specific address and state in writing that they wanted to opt out—all this rather than sending a simple prepaid postcard that allowed customers to check off their choice. As a result, most customers threw out the forms without grasping their full implications and thus, by default, agreed to opt in to the collection and sharing of their personal data.

Health Information

The use of electronic medical records and the subsequent interlinking and transferring of this electronic information among different organizations has become widespread. Individuals are rightly concerned about the erosion of privacy of data concerning their health. They fear intrusions into their health data by employers, schools, insurance firms, law enforcement agencies, and even marketing firms looking to promote their products and services. The primary law addressing these issues is the Health Insurance Portability and Accountability Act.

Health Insurance Portability and Accountability Act of 1996 (HIPAA)

The **Health Insurance Portability and Accountability Act of 1996 (HIPAA)** was designed to improve the portability and continuity of health insurance coverage; to reduce fraud, waste, and abuse in health insurance and healthcare delivery; and to simplify the administration of health insurance.

To these ends, HIPAA requires healthcare organizations to employ standardized electronic transactions, codes, and identifiers to enable them to fully digitize medical records, thus making it possible to exchange medical data over the Internet. The Department of Health and Human Services developed over 1,500 pages of specific rules governing exchange of such data. At the time of their implementation, these regulations affected more than 1.5 million healthcare providers, 7,000 hospitals, and 2,000 healthcare plans.[26]

Under the HIPAA provisions, healthcare providers must obtain written consent from patients prior to disclosing any information in their medical records. Thus, patients need to sign a HIPAA disclosure form each time they are treated at a hospital, and such a form must be kept on file with their primary care physician. In addition, healthcare providers are required to keep track of everyone who receives information from a patient's medical file.

For their part, healthcare companies must appoint a privacy officer to develop privacy policies and procedures as well as train employees on how to handle sensitive patient data. These actions must address the potential for unauthorized access to data by outside hackers as well as the more likely threat of internal misuse of data. For example, TV media videotaped numerous examples of Rite Aid pharmacies across the country disposing of prescriptions and labeled pill bottles containing patient identification information in trash containers that were readily accessible by the general public. Such careless disposal of patient information violates several HIPAA Privacy Rule requirements and also exposes Rite Aid customers to the risk of identity theft. Following review by the U.S. Department of Health and Human Services, Rite Aid agreed to pay a $1 million fine and to implement a strong program to revise its policies for destroying of protected health information. Rite Aid also agreed to conduct independent assessments of its revised disposal program for the next 20 years.[27]

HIPAA assigns responsibility to healthcare organizations, as the originators of individual medical data, for certifying that their business partners (billing agents, insurers, debt collectors, research firms, government agencies, and charitable organizations) also comply with HIPAA security and privacy rules. This provision of HIPAA has healthcare executives especially concerned, as they do not have direct control over the systems and procedures that their partners implement. Those who misuse data may be fined $250,000 and serve up to 10 years in prison.

Some medical personnel and privacy advocates fear that between the increasing demands for disclosure of patient information and the inevitable complete digitization of medical records, patient confidentiality will be lost. Many think that HIPAA provisions are too complicated and that rather than achieving the original objective of reducing medical industry costs, HIPAA will instead increase costs and paperwork for doctors without improving medical care. All agree that the medical industry must make a substantial investment to achieve compliance.

The American Recovery and Reinvestment Act of 2009, Title XIII, Subtitle D

The **American Recovery and Reinvestment Act of 2009** is a wide-ranging act that authorized $787 billion in spending and tax cuts over a 10-year period. Title XIII, Subtitle D of this act included strong privacy provisions for electronic health records, including banning the sale of health information, promoting the use of audit trails and encryption, and providing rights of access for patients. This section of the act also mandates that within 60 days after discovery of a data breach, each individual whose health information has been exposed must be notified, and if a breach involves 500 or more people, notice must be provided to prominent media outlets.[28]

This act offers protection for victims of data breaches that go unreported for months after discovery, exposing consumers to a high risk of identity theft. For example, a hard drive containing seven years' worth of medical records on 1.5 million people insured by Health Net of Northeast Inc. was reported missing to the Connecticut state attorney general's office a full six months after the drive disappeared. Connecticut Attorney General Richard Blumenthal stated, "Health Net's incomprehensible foot-dragging demonstrates shocking disregard for patients' financial security, as well as loss of their highly sensitive and confidential personal information." He demanded identity theft insurance and reimbursement for at least two years for the nearly half million Connecticut consumers affected.[29]

State Laws Related to Security Breach Notification

Over forty states and U.S. territories have enacted legislation requiring organizations to disclose security breaches involving personal information. Some of the states have extremely stringent laws regarding the reporting a data breach of patient health records. For example, under California law, a data breach involving protected health information must be reported to government agencies and affected individuals within five days of discovery. The Lucile Packard Children's Hospital at Stanford University was fined $250,000 in April 2010 by the California Department of Public Health when it took 17 days to report the theft of a computer with protected health information on 532 patients.[30]

Children's Personal Data

Internet use by children continues to climb; a recent study by the Kaiser Family Foundation found that young Americans ages 8 to 18 spend more than seven and one half hours per day using a smartphone, TV, computer, or other electronic device.[31] Many people feel that there is a need to protect children from being exposed to inappropriate material and online predators; becoming the target of harassment; divulging personal data; and becoming involved in gambling or other inappropriate behavior. To date, only a few laws have been implemented to protect children online, and most of these have been ruled unconstitutional under the First Amendment and its protection of freedom of expression.

Children's Online Privacy Protection Act (1998)

According to the **Children's Online Privacy Protection Act (COPPA)**, any Web site that caters to children must offer comprehensive privacy policies, notify parents or guardians about its data collection practices, and receive parental consent before collecting any personal information from children under 13 years of age. COPPA is meant to give parents control over the collection, use, and disclosure of their children's personal information; it does not cover the dissemination of information to children.

The law has had a major impact and has required many companies to spend hundreds of thousands of dollars to make their sites compliant; other companies eliminated preteens as a target audience.

Echometrix is a New York–based developer of software that gathers information for parents about what their children are doing online. However, parents were unaware that the monitoring program that they used to keep track of their children's online activities was also secretly collecting and analyzing their children's private instant messages for marketing purposes. Echometrix marketed the data to third-party companies as a means to gain insight into what children were saying about certain products and services. The New York State Attorney General reached a settlement with Echometrix under which the firm agreed to stop analyzing and sharing data with third parties and in addition agreed to pay a $100,000 fine to the State of New York.[32]

Although COPPA is a U.S. law, the Federal Trade Commission has stated that the requirements of COPPA will apply to any foreign-operated Web site that is "directed to children in the United States or knowingly collects information from children in the United States."[33] A few other countries, including Canada and Australia, have enacted similar laws to protect preteens.

Family Educational Rights and Privacy Act (1974)

The **Family Educational Rights and Privacy Act (FERPA)** is a federal law that assigns certain rights to parents regarding their children's educational records. These rights transfer to the student once the student reaches the age of 18 or if he or she attends a school beyond the high school level. These rights include

- the right to access educational records maintained by a school;
- the right to demand that educational records be disclosed only with student consent;
- the right to amend educational records; and
- the right to file complaints against a school for disclosing educational records in violation of FERPA[34]

In summary, under FERPA, the presumption is that a student's records are private and not available to the public without the consent of the student. Penalties for violation of FERPA may include a cutoff of federal funding to the educational institution. Educational agencies and institutions *may* disclose education records to the parents of a dependent student, as defined in section 152 of the Internal Revenue Code of 1986, without the student's consent.

Various educational institutions set policies based on their interpretation of FERPA. For example, the University of Washington policy is that notification of grades by email is a violation of FERPA because there is no guarantee of confidentiality on the Internet.[35] In September 2010, an elementary school teacher in Texas was forced to resign after posting photos and personal information of her students in her online blog in violation of FERPA.[36]

Electronic Surveillance

This section covers laws that address government surveillance, including various forms of electronic surveillance. New laws have been added and old laws amended in recent years in reaction to worldwide terrorist activities and the development of new communication technologies.

Communications Act of 1934

The **Communications Act of 1934** established the Federal Communications Commission and gave it responsibility for regulating all non-federal-government use of radio and television broadcasting and all interstate telecommunications—including wire, satellite, and cable—as well as all international communications that originate or terminate in the United States. The act also restricted the government's ability to secretly intercept communications.

Title III of the Omnibus Crime Control and Safe Streets Act (1968, amended 1986)

Title III of the Omnibus Crime Control and Safe Streets Act, also known as the Wiretap Act, regulates the interception of wire (telephone) and oral communications. It allows state and federal law enforcement officials to use wiretapping and electronic eavesdropping, but only under strict limitations. Under this act, a warrant must be obtained from a judge to conduct a wiretap. The judge may approve the warrant only if "there is probable cause [to believe] that an individual is committing, has committed, or is about to commit a particular offense ... [and that] normal investigative procedures have been tried and have failed or reasonably appear to be unlikely if tried or to be too dangerous."[37]

One of the driving forces behind the passage of this act was the case of *Katz v. United States*. Katz was convicted of illegal gambling based on recordings by the FBI of Katz's side of various telephone calls made from a public phone booth; the recordings were made using a device attached to the phone booth. Katz challenged the conviction based on a violation of his Fourth Amendment rights. In 1967, the case finally made it to the Supreme Court, which agreed with Katz. The court ruled that "the Government's activities in electronically listening to and recording the petitioner's words violate the privacy upon which he justifiably relied while using the telephone booth and thus constituted a 'search and seizure' within the meaning of the Fourth Amendment."[38] This ruling helped form the basis for the requirement that there be a reasonable expectation of privacy for the Fourth Amendment to apply.

Title III court orders must describe the duration and scope of the surveillance, the conversations that may be captured, and the efforts to be taken to avoid capture of innocent conversations. In 2009, federal judges authorized 663 wiretaps and state judges authorized 1,713 wiretaps. These led to the arrest of 4,537 people and 678 convictions. Not one request for a wiretap was turned down. These numbers do not include wiretap orders in terrorism investigations, which are authorized by the Foreign Intelligence Surveillance Act Court discussed below.[39]

Since the Wiretap Act was passed, it has been significantly amended by several new laws, including FISA, ECPA, CALEA, and the USA PATRIOT Act. These acts will now be discussed.

Foreign Intelligence Surveillance Act (1978)

The **Foreign Intelligence Surveillance Act (FISA)** of 1978 describes procedures for the electronic surveillance and collection of foreign intelligence information in communications between foreign powers and the agents of foreign powers. **Foreign intelligence** is information relating to the capabilities, intentions, or activities of foreign governments or agents of foreign governments or foreign organizations. The act allows surveillance, without court order, within the United States for up to a year unless the "surveillance will acquire the contents of any communication to which a U.S. person is a party."[40] If a U.S. citizen is involved, judicial authorization is required within 72 hours after surveillance begins.

The act also created the FISA court, which meets in secret to hear applications for orders approving electronic surveillance anywhere within the United States. Each application for a surveillance warrant is made before an individual judge of the court. Such applications are rarely turned down—for the decade 2000 to 2009 there were 16,687 applications for warrants made, and only 11 were denied.[41]

The act also specifies that the U.S. attorney general may request a specific communications common carrier to furnish information, facilities, or technical assistance to accomplish the electronic surveillance.

FISA makes it illegal to intentionally engage in electronic surveillance under appearance of an official act or to disclose or use information obtained by electronic surveillance under appearance of an official act knowing that it was not authorized by statute; this is punishable with a fine of up to $10,000 or up to five years in prison, or both.

Electronic Communications Privacy Act (1986)

The **Electronic Communications Privacy Act of 1986 (ECPA)** deals with three main issues: (1) the protection of communications while in transfer from sender to receiver; (2) the protection of communications held in electronic storage; and (3) the prohibition of devices to record dialing, routing, addressing, and signaling information without a search warrant. ECPA was passed as an amendment to Title III of the Omnibus Crime Control and Safe Streets Act.[42]

Title I of ECPA extends the protections offered under the Wiretap Act to electronic communications, such as email, fax, and messages sent over the Internet. The government is prohibited from intercepting such messages unless it obtains a court order based on probable cause (the same restriction that is in the Wiretap Act relating to telephone calls).

Title II of ECPA (also called the Stored Communications Act) prohibits unauthorized access to stored wire and electronic communications, such as the contents of email inboxes, instant messages, message boards, and social networking sites. However, the law only applies if the stored communications are not readily accessible to the general public. Webmasters who desire protection for their subscribers under this act must take careful measures to limit public access through the use of logon procedures, passwords, and other methods. Under this act, the FBI director or someone acting on his behalf may issue a National Security Letter (NSL) to an Internet service provider to provide various data and records about a service subscriber.

A third part of ECPA establishes a requirement for court-approved law enforcement use of a **pen register**—a device that records electronic impulses to identify the numbers dialed for outgoing calls—or a **trap and trace**—a device that records the originating number of incoming calls for a particular phone number. A recording of every telephone number dialed and the source of every call received can provide an excellent profile of a person's associations, habits, contacts, interests, and activities. A similar type of surveillance has also been applied to email communications to gather email addresses, email header information, and Internet provider addresses. To obtain approval for a pen-register order or a trap-and-trace order, the law enforcement agency only needs to certify that "the information likely to be obtained is relevant to an ongoing criminal investigation." (This requirement is much lower than the probable cause required to obtain a court order to intercept an electronic communication.) A prosecutor does not have to justify the request, and judges are required to approve every request.

Communications Assistance for Law Enforcement Act (1994)

The **Communications Assistance for Law Enforcement Act (CALEA)** was passed by Congress in 1994 and amended both the Wiretap Act and ECPA. CALEA was a hotly debated law because it required the telecommunications industry to build tools into its products that federal investigators could use—after obtaining a court order—to eavesdrop on conversations and intercept electronic communications. The court order can only be obtained if it is shown that a crime is being committed, that communications about the crime will be intercepted, and that the equipment being tapped is being used by the suspect in connection with the crime.[43]

A provision in the act covering radio-based data communication grew from a realization that the Electronic Communications Privacy Act of 1986 failed to cover emerging technologies, such as wireless modems, radio-based electronic mail, and cellular data networks. The ECPA statute outlawed the unauthorized interception of wire-based digital traffic on commercial networks, but the law's drafters did not foresee the growing interest in wireless data networks. Section 203 of CALEA corrected that oversight by effectively covering all publicly available "electronic communication."

With CALEA, the Federal Communications Commission responded to appeals from the Justice Department and other law enforcement officials by requiring providers of Internet phone services and broadband services to ensure that their equipment accommodated the use of law enforcement wiretaps. This equipment includes Voice over Internet Protocol (VoIP) technology, which shifts calls away from the traditional phone network of wires and switches to technology based on converting sounds into data and transmitting them over the Internet. The decision has created a controversy among many who fear that opening VoIP to access by law enforcement agencies will create additional points of attack and security holes that hackers can exploit.

USA PATRIOT Act (2001)

The **USA PATRIOT Act** (Uniting and Strengthening America by Providing Appropriate Tools Required to Intercept and Obstruct Terrorism) of 2001 was passed just after the terrorist attacks of September 11, 2001. It gave sweeping new powers both to domestic law enforcement and U.S. international intelligence agencies, including increasing the ability of law enforcement agencies to search telephone, email, medical, financial, and other records. It also eased restrictions on foreign intelligence gathering in the United States.

Although the act was more than 340 pages long and quite complex (it changed more than 15 existing statutes), it was passed into law just five weeks after being introduced. Legislators rushed to get the act approved in the House and Senate, arguing that law enforcement authorities needed more power to help track down terrorists and prevent future attacks. Critics have argued that the law removed many checks and balances that previously gave courts the opportunity to ensure that law enforcement agencies did not abuse their powers. Critics also argue that many of its provisions have nothing to do with fighting terrorism. Table 4-1 summarizes the key provisions of the PATRIOT Act as they affect the privacy of individuals.

TABLE 4-1 Key provisions of the USA PATRIOT Act

Section	Issue addressed	Summary
201	Wiretapping in terrorism cases	Added several crimes for which federal courts may authorize wiretapping of people's communications
202	Wiretapping in computer fraud and felony abuse cases	Added computer fraud and abuse to the list of crimes for which the FBI may obtain a court order to investigate under Title III of the Wiretap Act
203(b)	Sharing wiretap information	Allows the FBI to disclose evidence obtained under Title III to other federal officials, including "law enforcement, intelligence, protective, immigration, national defense, [and] national security" officials
203(d)	Sharing foreign intelligence information	Provides for disclosure of threat information obtained during criminal investigations to "appropriate" federal, state, local, or foreign government officials for the purpose of responding to the threat
204	FISA pen-register/trap-and-trace exceptions	Exempts foreign intelligence surveillance from statutory prohibitions against the use of pen-register or trap-and-trace devices, which capture "addressing" information about the sender and recipient of a communication; it also exempts the U.S. government from general prohibitions against intercepting electronic communications and allows stored voice-mail communication to be obtained by the government through a search warrant rather than more stringent wiretap orders
206	FISA roving wiretaps	Expands FISA to permit "roving wiretap" authority, which allows the FBI to intercept any communications to or by an intelligence target without specifying the telephone line, computer, or other facility to be monitored
207	Duration of FISA surveillance of non-U.S. agents of a foreign power	Extends the duration of FISA wiretap orders relating to an agent of a foreign power from 90 days to 120 days, and allows an extension in 1-year intervals instead of 90-day increments
209	Seizure of voice-mail messages pursuant to warrants	Enables the government to obtain voice-mail messages under Title III using just a search warrant rather than a wiretap order, which is more difficult to obtain; messages stored on an answering machine, however, remain outside the scope of this section
212	Emergency disclosure of electronic surveillance	Permits providers of communication services (such as telephone companies and Internet service providers) to disclose consumer records to the FBI if they believe immediate danger of serious physical injury is involved; communication providers cannot be sued for such disclosure
214	FISA pen-register/trap-and-trace authority	Allows the government to obtain a pen-register/trap-and-trace device "for any investigation to gather foreign intelligence information"; it prohibits the use of FISA pen-register/trap-and-trace surveillance against a U.S. citizen when the investigation is conducted "solely on the basis of activities protected by the First Amendment"

(Continued)

TABLE 4-1 Key provisions of the USA PATRIOT Act (*Continued*)

Section	Issue addressed	Summary
215	FISA access to tangible items	Permits the FBI to compel production of any record or item without showing probable cause; people served with a search warrant issued under FISA rules may not disclose, under penalty of law, the existence of the warrant or the fact that records were provided to the government. It prohibits investigation of a U.S. citizen when it is conducted solely on the basis of activities protected by the First Amendment.
217	Interception of computer-trespasser communications	Creates a new exception to Title III that permits the government to intercept the "communications of a computer trespasser" if the owner or operator of a "protected computer" authorizes it; it defines a protected computer as any computer "used in interstate or foreign commerce or communication" (because of the Internet, this effectively includes almost every computer)
218	Purpose for FISA orders	Expands the application of FISA to situations in which foreign intelligence gathering is merely a significant purpose rather than the sole purpose
220	Nationwide service of search warrants for electronic evidence	Expands the geographic scope in which the FBI can obtain search warrants or court orders for electronic communications and customer records
223	Civil liability and discipline for privacy violations	Provides that people can sue the government for unauthorized disclosure of information obtained through surveillance
225	Provider immunity for FISA wiretap assistance	Provides immunity from lawsuits for people who disclose information to the government pursuant to a FISA wiretap order, a physical search order, or an emergency wiretap or search
505	Authorizes use of National Security Letters (NSLs) to gain access to personal records	Authorizes the attorney general or a delegate to compel holders of your personal records to turn them over to the government simply by writing an NSL, which is not subject to judicial review or oversight; NSLs can be used against anyone, including U.S. citizens, even if they are not suspected of espionage or criminal activity

Source Line: "Key Provisions of USA Patriot Act," *NewsMax.com*, ©December 22, 2005, http://archive.newsmax.com/archives/articles/2005/12/22/113858.shtml.

One of the more contentious aspects of the USA PATRIOT Act has been the guidelines issued for the use of NSLs. Before the PATRIOT Act was enacted, the FBI could issue an NSL for information about someone only if it had reason to believe the person was a foreign spy. Under the PATRIOT Act, the FBI can issue an NSL to compel banks, Internet service providers, and credit reporting companies to turn over information about their customers without a court order simply on the basis that the information is needed for an ongoing investigation. The FBI issued an average of 50,000 NSLs per year between 2003 and 2006 under these relaxed requirements.[44] The American Civil Liberties Union (ACLU) has challenged the use of NSLs by the FBI in court several times. These lawsuits are in various stages of hearings and appeals. In one lawsuit, *Doe v. Holder*, the Court of the Southern District of New York and, upon appeal, the Second Circuit Court of Appeals ruled that the NSL gag provision—which prohibits NSL recipients from informing anyone, even the person who is the subject of the NSL request, that the government has secretly requested his or her records—violates the First Amendment. The FBI issued an average of about 14,500 NSLs per year between 2004 and 2009.[45]

Recognizing that the PATRIOT Act was both quickly written and very broad in scope, Congress designated several "sunset" provisions that were to terminate on December 31, 2005. A **sunset provision** terminates or repeals a law or portions of it after a specific date unless further legislative action is taken to extend the law. It was March 2006 before Congress decided on the sunset provisions. Only two of the provisions in Table 4-1 were modified:

- Section 215 was extended for four years and altered slightly so that recipients of FISA subpoenas for record searches would have the right to consult with a lawyer to challenge the request in court.
- Section 505 was modified so that the recipient of an NSL could challenge the nondisclosure requirement but no sooner than one year after receiving the NSL. Recipients are not required to disclose the names of any attorneys they consulted, but they are required to tell the FBI if they consulted anyone other than legal counsel.

Export of Personal Data

Various organizations have developed guidelines to ensure that the flow of personal data across national boundaries (transborder data flow) does not result in the unlawful storage of personal data, the storage of inaccurate personal data, or the abuse or unauthorized disclosure of such data. Two sets of these guidelines are discussed in this section.

Organisation for Economic Co-operation and Development Fair Information Practices (1980)

Another agency concerned with privacy is the Organisation for Economic Co-operation and Development (OECD), an international organization consisting of 34 member countries, including Australia, Canada, France, Germany, Italy, Japan, Mexico, New Zealand, the United Kingdom, and the United States. Its goal is to set policy and come to agreement on topics for which multilateral consensus is necessary for individual countries to make progress in a global economy. Dialogue, consensus, and peer pressure are essential to make these policies and agreements stick. The 1980 privacy guidelines set by OECD—also known as the **Fair Information Practices**—are often held up as the model of ethical treatment of consumer data for organizations to adopt. These guidelines are composed of the eight principles summarized in Table 4-2.

TABLE 4-2 Summary of the 1980 OECD privacy guidelines

Principle	Guideline
Collection limitation	The collection of personal data must be limited; all such data must be obtained lawfully and fairly with the subject's consent and knowledge
Data quality	Personal data should be accurate, complete, current, and relevant to the purpose for which it is used
Purpose specification	The purpose for which personal data is collected should be specified and should not be changed
Use limitation	Personal data should not be used beyond the specified purpose without a person's consent or by authority of law
Security safegurds	Personal data should be protected against unauthorized access, modification, or disclosure
Openness principle	Data policies should exist, and a data controller should be identified
Individual participation	People should have the right to review their data, to challenge its correctness, and to have incorrect data changed
Accountability	A data controller should be responsible for ensuring that the above principles are met

Source Line: Organisation for Economic Co-operation and Development, "OECD Guidelines on the Protection of Privacy and Transborder Flows of Personal Data," www.oecd.org/document/18/0,3343,en_2649_34255_1815186_1_1_1_1,00.html.

European Union Data Protection Directive (1998)

The **European Union Data Protection Directive** requires any company doing business within the borders of 15 western European nations to implement a set of privacy directives on the fair and appropriate use of information. Basically, this directive requires member countries to ensure that data transferred to non–European Union (EU) countries is protected. It also bars the export of data to countries that do not have data privacy protection standards comparable to the EU's. The following list summarizes these seven European data privacy principles:

- *Notice*—Tell all customers what is done with their information.
- *Choice*—Give customers a way to opt out of marketing.
- *Onward transfer*—When data is transferred to suppliers or other business partners, companies must observe the notice and choice principles mentioned above and require all recipients of such data to provide at least the same level of protection for such data.
- *Access*—Give customers access to their information.
- *Security*—Protect customer information from unauthorized access.

- *Data integrity*—Ensure that information is accurate and relevant.
- *Enforcement*—Independently enforce the privacy policy.

Initially, EU countries were concerned that the largely voluntary system of data privacy in the United States did not meet the EU directive's stringent standards. Eventually, the U.S. Department of Commerce worked out an agreement with the European Union; only U.S. companies that are certified as meeting safe harbor principles are allowed to process and store data of European consumers and companies. To date, over 2,100 U.S. multinational companies—such as Caterpillar, ExxonMobil, Ford, Gap Inc., Procter & Gamble, Pepsi, and Sony Music Entertainment—that need to exchange employee and consumer data among their subsidiaries to effectively operate their businesses have been certified. In addition, companies such as Facebook, Google, IBM, and Microsoft that provide email, social networking, or cloud computing services involving employee and consumer data also are certified.[46]

However, European countries are concerned because the U.S. government can invoke the USA PATRIOT Act to access the data, thus breaking key principles of the directive. In addition, there is concern because U.S. companies self-certify themselves, with little or no oversight. An independent research company found that 388 of the 2,170 organizations on the Safe Harbor List were not even registered with the Department of Commerce and that 181 companies still on the list had expired certificates. Another 940 companies make no effort to explain how they implement and enforce the safe harbor principles. As a result of these findings, some experts feel that the safe harbor provision has little meaning anymore and caution that Europeans should keep their data in Europe.[47]

BBB Online and TRUSTe

The European philosophy of addressing privacy involves strict government regulation—including enforcement by a set of commissioners—and differs greatly from the U.S. philosophy of having no federal privacy policy. The United States instead relies on self-regulation, which is overseen by the U.S. Department of Commerce, the Better Business Bureau Online (BBB*OnLine*), and TRUSTe.

BBB*OnLine* and TRUSTe are independent initiatives that favor an industry-regulated approach to data privacy. They are both concerned that strict government regulation could have a negative impact on the Internet's use and growth, and that such regulation would be costly to implement and difficult to change or repeal. A Web site operator can apply for a BBB*OnLine* reliability seal or a TRUSTe data privacy seal to demonstrate that his or her site adheres to a high level of data privacy. Having one of the seals can increase consumer confidence in the site operator's desire and ability to manage data responsibly. The seals also help users make more informed decisions about whether to release personal information (such as phone numbers, addresses, and credit card numbers) to a Web site.

An organization must join the Better Business Bureau (BBB) and pay an annual fee ranging from $200 to $7,000, depending on annual sales, before applying for the BBB*OnLine* Reliability Program seal (see Figure 4-1). The BBB*OnLine* seal program identifies online businesses that honor their own stated privacy protection policies; therefore, an accredited Web site must also have adopted and posted an online privacy notice. There are currently 70,311 Web sites that are accredited as meeting the BBB*OnLine* standards.[48]

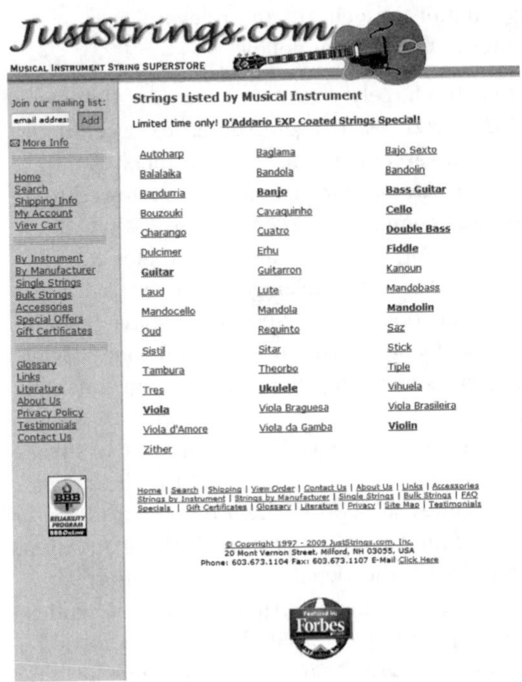

FIGURE 4-1 JustStrings.com displays the BBBOnLine Reliability Program seal
Credit: From www.juststrings.com. Reprinted by permission of JustStrings.com.

For a Web site to receive the TRUSTe seal (see Figure 4-2), its operators must demonstrate that it adheres to established privacy principles. They must also agree to comply with TRUSTe's oversight and consumer resolution process, and pay an annual fee. The privacy principles require the Web site to openly communicate what personal information it gathers, how it will be used, with whom it will be shared, and whether the user has an option to control its dissemination. After operating as a nonprofit organization for 10 years, TRUSTe converted to for-profit status in mid-2008, selling approximately $10 million in newly created stock to the venture capital firm Accel Partners—the same firm that backed eBay and Facebook. Some observers have criticized TRUSTe because it rarely tells users when a Web site has been removed from its program and doesn't require its member Web sites to give users a choice to opt out should the site sell or trade personal data about users to another company. There are currently 1,500 Web sites accredited by TRUSTe.

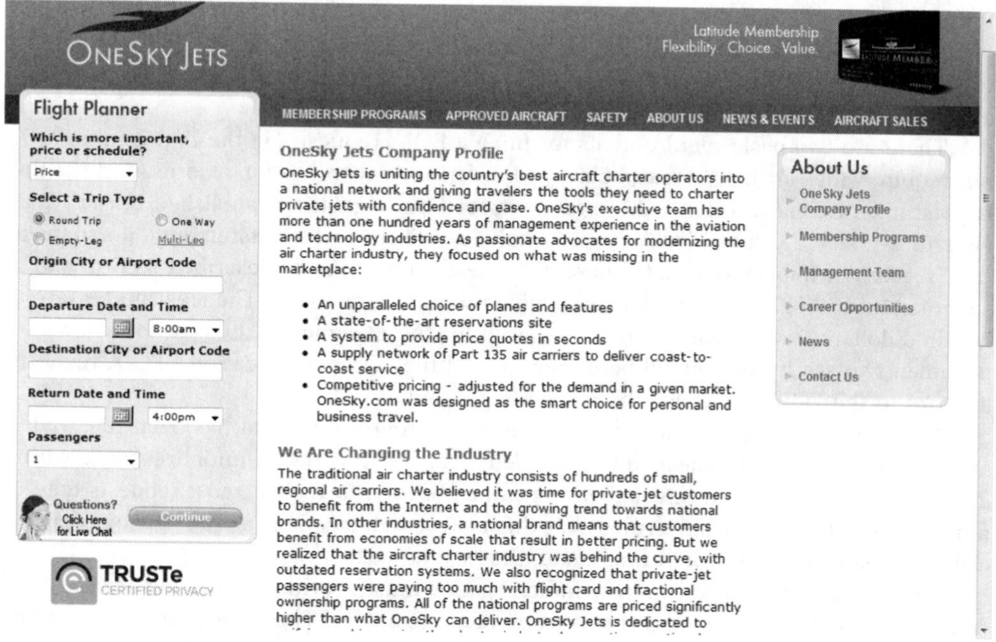

FIGURE 4-2 OneSky Jets displays the TRUSTe seal
Credit: From www.onesky.com. Courtesy of OneSky Jets.

Access to Government Records

The government has a great capacity to store data about each and every one of us and about the proceedings of its various organizations. The Freedom of Information Act enables the public to gain access to certain government records, and the Privacy Act prohibits the government from concealing the existence of any personal data record-keeping systems.

Freedom of Information Act (FOIA) (1966, amended 1974)

The **Freedom of Information Act (FOIA)**, passed in 1966 and amended in 1974, grants citizens the right to access certain information and records of the federal government upon request. FOIA is a powerful tool that enables journalists and the public to acquire information that the government is reluctant to release. The well-defined FOIA procedures have been used to uncover previously unrevealed details about President Kennedy's assassination, determine when and how many times members of Congress or certain lobbyists have visited the White House, obtain budget and spending data about a government agency, and even request information on the "UFO incident" at Roswell in 1947. During 2009, U.S. citizens filed 444,924 requests for government data from the departments of Agriculture, Commerce, Defense, Educations, Energy, Health and Human Services, Homeland Security, Housing and Urban Development, Interior, Justice, Labor, State, Transportation, Treasury and Veterans Affairs, the Environmental

Protection Agency, and the Federal Reserve Board using FOIA procedures.[49] The act is often used by whistle-blowers to obtain records that they would otherwise be unable to get. Citizens have also used FOIA to find out what information the government has about them.

There are two basic requirements for filing a FOIA request: (1) the request must not require wide-ranging, unreasonable, or burdensome searches for records, and (2) the request must be made according to agency procedural regulations published in the *Federal Register*. A typical FOIA request includes the requester's statement: "pursuant to the Freedom of Information Act, I hereby request"; a reasonably described record; and a statement of willingness to pay for reasonable processing charges. (The fees can be several hundred dollars and include the cost to search for the documents, the cost to review documents to see if they should be disclosed, and the cost of duplication.) FOIA requests are sent to the FOIA officer for the responding agency.[50]

Agencies receiving a request must acknowledge that the request has been received and indicate when the request will be fulfilled. The act requires an initial response within 20 working days unless an unusual circumstance occurs. In reality, most requests take much longer. The courts have ruled that this is acceptable as long as the agency treats each request sequentially on a first-come, first-served basis.[51]

If the request is denied, the responding agency must provide the reasons for the denial along with the name and title of each denying officer. The agency must also notify the requester of his or her right to appeal the denial and provide the address to which an appeal should be sent.[52]

Exemptions to the FOIA bar disclosure of information that could compromise national security or interfere with an active law enforcement investigation. Another exemption prevents disclosure of records if it would invade someone's privacy. In this case, a balancing test is applied to evaluate whether the privacy interests at stake are outweighed by competing public interests. Some legislators have called for an additional exemption to protect information shared between the government and private industry regarding attacks against computer and information systems. The sharing of such information, they say, is important for the federal government to protect the nation's critical IT infrastructure from attack; however, private companies are reluctant to share such data. They fear that the FOIA doesn't offer enough assurances that their proprietary data will remain secret.

The Electronic Privacy Information Center (EPIC) filed a lawsuit in 2010 asking the court to intervene with the National Security Administration (NSA) to reveal details of the authority it was granted under National Security Presidential Directive 54, issued during the Bush Administration. The classified directive was used to create the Comprehensive National Security Directive, a multi-billion-dollar cybersecurity program to improve the detection and response to cyberintrusions on federal networks. EPIC previously filed FOIA requests to the NSA asking for information, but the NSA has failed to meet statutory deadlines for providing the requested information. Lawmakers, industry executives, and privacy advocacy groups are also urging the government to release more information on this secret program.[53] Eventually, the NSA released a very limited description of the program that revealed few details.[54]

Privacy Act of 1974

The **Privacy Act of 1974** prohibits U.S. government agencies from concealing the existence of any personal data record-keeping system. Under this law, any agency that maintains such a system must publicly describe both the kinds of information in it and the manner in which the information will be used. The law also outlines 12 requirements that each record-keeping agency must meet, including issues that address openness, individual access, individual participation, collection limitation, use limitation, disclosure limitation, information management, and accountability. The purpose of the act is to provide safeguards for people against invasion of personal privacy by federal agencies. The CIA and law enforcement agencies are excluded from this act; in addition, it does not cover the actions of private industry.

KEY PRIVACY AND ANONYMITY ISSUES

The rest of this chapter discusses a number of current and important privacy issues, including identity theft, electronic discovery, consumer profiling, treating consumer data responsibly, workplace monitoring, and advanced surveillance technology.

Identity Theft

Identity theft occurs when someone steals key pieces of personal information to impersonate a person. This information may include such data as name, address, date of birth, Social Security number, passport number, driver's license number, and mother's maiden name. Using this information, an identity thief may apply for new credit or financial accounts, rent an apartment, set up utility or phone service, and register for college courses—all in someone else's name.

Identity theft is recognized as one of the fastest growing forms of fraud in the United States. Recent research indicates that the annual cost of identity theft in the United States is $221 billion and that some 10 percent of Americans have at some time been a victim. Although half of the victims of identity theft discover their misfortune within three months, some 15 percent of the victims do not find out for four years or more.[55]

Consumers and organizations are becoming more vigilant and proactive in fighting this threat. Many consumers know they can request one free credit report per year from each of the three major credit bureaus and are also aware that they can put fraud alerts on their credit reports and sign up for credit monitoring services. Many consumers are savvy enough to recognize obvious phishing attempts to capture personal data. Organizations and individuals have put into place improved systems and practices to ward off identity thieves, including spyware and antivirus software. Table 4-3 offers additional recommendations for safeguarding your identity data.

TABLE 4-3 Recommendations for safeguarding your identity data

Recommendation	Explanation
Completely destroy digital identity data on used equipment	As it is possible to undelete files and recover data, take necessary actions to ensure that all data is destroyed when you dispose of used computers and data storage devices; consider the use of special software to wipe hard drives completely clean
Shred everything	Identity thieves are not above "dumpster diving" (going through your garbage to find financial statements and bills in order to obtain confidential personal information)
Require retailers to request a photo ID when accepting your credit card	Writing *Request Photo ID* on the back of your credit cards should prompt retailers to request a photo ID before accepting your card
Beware shoulder surfing	Ensure that nobody can look over your shoulder when you enter or write down personal information—at an ATM, filling out forms in public places, and so on
Minimize personal data shown on checks	Do not include a Social Security number or driver's license number on your checks
Minimize time that mail is in your mailbox	Do not leave paid bills in your mailbox for postal pickup; collect mail from your mailbox as soon as possible after it is delivered
Do not use debit cards to pay for online purchases	Victims of credit card fraud are liable for no more than $50 in losses; debit card users can have their entire checking account wiped out
Treat your credit card receipts safely	Always take your credit card receipts from the retailer and keep them for reconciliation purposes; dispose of them by shredding them
Use hard-to-guess passwords and PINs	Do not use names or words in passwords; for a strong password include a mix of capital and small letters with at least one special character ($, #, *)

Source Line: Course Technology/Cengage Learning.

Taking on another's identity can be easy and extremely lucrative, enabling the thief to use a victim's credit cards, siphon off money from bank accounts, and even obtain Social Security benefits. The victim can be left with a credit history in shambles.

Four approaches are frequently used by identity thieves to capture the personal data of their victims: (1) create a data breach to steal hundreds, thousands, or even millions of personal records; (2) purchase personal data from criminals; (3) use phishing to entice users to willingly give up personal data; and (4) install spyware capable of capturing the keystrokes of victims.

Data Breaches

An alarming number of identity theft incidents involve breaches of large databases to gain personal identity information. The breach may be caused by hackers breaking into the database or, more often than one would suspect, by carelessness or failure to follow proper security procedures. For example, a laptop computer containing the unencrypted Social Security numbers of 26.5 million U.S. veterans was stolen from the home of a Veterans Affairs (VA) analyst. The analyst violated existing VA policy by removing the data from his workplace. Table 4-4 identifies the five largest U.S. data breaches.

TABLE 4-4 Largest reported U.S. data breaches

Date incident was reported	Number of records involved	Organization(s) involved
January 20, 2009	130 million	Heartland Payment Systems, Tower Federal Credit Union, Beverly National Bank
January 17, 2007	94 million	The TJX Companies
June 1, 1984	90 million	TRW, Sear Roebuck
October 5, 2009	76 million	National Archives and Records Administration
June 19, 2005	40 million	CardSystems, Visa, Master Card, American Express
May 22, 2006	26.5 million	U.S. Department of Veterans' Affairs

Source Line: Open Security Foundation's DataLossDB, http://datalossdb.org.

The number of data breach incidents is alarming, as is the lack of initiative by some companies in informing the people whose data was stolen. Organizations are reluctant to announce data breaches due to the ensuing bad publicity and potential for lawsuits by angry customers. However, victims whose personal data was compromised during a data breach need to be informed so that they can take protective measures.

The cost to an organization that suffers a data breach can be quite high—by some estimates nearly $200 for each record lost. Nearly half the cost is typically a result of lost business opportunity associated with the customers whose patronage has been lost due to the incident. Other costs include public-relations-related costs to manage the firm's reputation, and increased customer-support costs for information hotlines and credit monitoring services for victims.[56] Hannaford Brothers—a supermarket chain with 167 stores in five northeastern states and Florida—suffered a data breach in which credit and debit card data was captured illegally as cards were swiped at the checkout line. Within days of the breach becoming public knowledge, multiple class actions were filed against the company, alleging that it was negligent for failing to maintain adequate computer security. Hannaford is likely facing years of litigation; tens of millions of dollars in legal fees, settlement costs, and customer credit monitoring services; and reduction of sales revenue due to loss of customer goodwill.[57]

The state of California passed a data security breach notification law in 2002. It was enacted when the state's payroll database was breached and victims were not notified for six weeks. The law requires that "the disclosure shall be in the most expedient time possible and without unreasonable delay, consistent with the legitimate needs of law enforcement.[58] More than half the states have now implemented similar legislation.

Purchase of Personal Data

There is a black market in personal data. Credit card numbers can be purchased in bulk quantity for as little as $.40 each, whereas the logon name and PIN necessary to access a bank account can be had for just $10.[59] A full set of identity information—including date of birth, address, Social Security number, and telephone number—sells for between $1 and $15.[60]

Phishing

As discussed in Chapter 3, phishing is an attempt to steal personal identity data by tricking users into entering information on a counterfeit Web site. Spoofed emails lead consumers to counterfeit Web sites designed to trick them into divulging personal data. Users have learned through sad experience that simply accessing a phishing Web site can trigger an automatic and transparent download of malware known as spyware to a computer.

Spyware

Spyware is keystroke-logging software downloaded to users' computers without the knowledge or consent of the user. It is often marketed as a spouse monitor, child monitor, or surveillance tool. Spyware creates a record of the keystrokes entered on the computer, enabling the capture of account usernames, passwords, credit card numbers, and other sensitive information. The spy can view the Web sites visited as well as transcripts of chat logs. Spyware operates even if the infected computer is not connected to the Internet, continuing to record each keystroke until the next time the user connects to the Internet. Then, the data captured by the spyware is emailed directly to the spy or is posted to a Web site where the spy can view it. Spyware frequently employs sophisticated methods to avoid detection by popular software packages that are specifically designed to combat it. Consumers' fear of spyware has become so widespread that many people now delete email from unknown sources without even opening the messages. This trend is seriously damaging the effectiveness of email as a means for legitimate companies to communicate with customers.

In 2007, the FBI planted spyware on the computer of a 15-year-old student in an attempt to identify him as the person responsible for sending numerous bomb threats to his high school. The FBI first obtained a warrant to allow the agency to install a program called the Computer and Internet Protocol Address Verifier on the student's computer. The software recorded the IP addresses, dates, and times of each communication sent from the student's computer. The student was sentenced to 90 days in juvenile detention and fined $8,852.[61]

Congress passed the Identity Theft and Assumption Deterrence Act of 1998 (full text at *www.ftc.gov/os/statutes/itada/itadact.htm*) to fight identity fraud, making it a federal felony punishable by a prison sentence of 3 to 25 years. The act also assigns the FTC the task of helping victims restore their credit and erase the impact of the imposter. Although people have been convicted under this act, researchers estimate that fewer than 1 in 700 identity crimes leads to a conviction.[62]

Identity Theft Monitoring Services

There are numerous identity theft monitoring services, which offer a wide range of coverage. Basic monitoring services cost about $10 per month and provide protection by monitoring the three major credit reporting agencies (TransUnion, Equifax, and Experian) for anyone using your personal data to apply for a new credit card, cell phone, or loan. More expensive services monitor additional databases (e.g., financial institutions, utilities, and the Department of Motor Vehicles [DMV]). Subscribers to these services receive a phone call or email if suspicious activity is detected.

Electronic Discovery

Discovery is part of the pretrial phase of a lawsuit in which each party can obtain evidence from the other party by various means, including requests for the production of documents. The purpose of discovery is to ensure that all parties will go to trial with as much knowledge as possible. Under the rules of discovery, neither party is able to keep secrets from the other. Should a discovery request be objected to, the requesting party may file a motion to compel discovery with the court.

Electronic discovery (e-discovery) is the collection, preparation, review, and production of electronically stored information for use in criminal and civil actions and proceedings.[63] **Electronically stored information (ESI)** includes any form of digital information, including emails, drawings, graphs, Web pages, photographs, word-processing files, sound recordings, and databases stored on any form of magnetic storage device, including hard drives, CDs, and flash drives. Through the e-discovery process, it is quite likely that various forms of ESI of a private or personal nature (e.g., personal emails) will be disclosed.

The Federal Rules of Procedure define certain processes that must be followed by a party involved in a case in federal court. Under these rules, once a case is filed, the involved parties are required to meet and discuss various e-discovery issues, such as how to preserve discoverable data, how the data will be produced, agreement on the format in which the data will be provided, and whether production of certain electronically stored information will lead to waiver of attorney-client privilege. A key issue is the scope of e-discovery (e.g., how many years of ESI will be requested, what topics and/or individuals need to be included in the e-discovery process, etc.). Often one party will send the other party a litigation hold notice to save relevant data and to suspend data that might be due to be destroyed based on normal data retention rules.

Qualcomm is a U.S. wireless telecommunications research and development company with headquarters in San Diego. One of its competitors is Broadcom Corporation, an Irvine, California–based company that provides products for the delivery of voice, video, data, and multimedia to the home, office, and mobile environment. The case *Qualcomm v. Broadcom* arose over a dispute involving patent rights and royalties, with both companies accusing the other of infringing use of its patents.[64,65] As part of the proceedings, both firms went through an e-discovery process to gather ESI that could be presented as evidence in the trial. During the trial, the judge ruled that Qualcomm had "intentionally withheld tens of thousands of documents" from Broadcom during its e-discovery and, as a result, she levied $8.5 million in sanctions against Qualcomm. The judge then allowed the attorneys for Qualcomm an extra 15 months to search and review an additional 1.6 million documents to be included in the e-discovery process.[66]

It requires extensive time to collect, prepare, and review the tremendous volume of ESI kept by an organization. E-discovery is further complicated because there are often multiple versions of information (such as various drafts) stored in many locations (such as the hard drives of the creator and anyone who reviewed the document, multiple company file servers, and backup tapes). As a result, e-discovery can become so expensive and time consuming that some cases are settled just to avoid the costs.[67]

Traditional software development firms as well as legal organizations have recognized the growing need for improved processes to speed up and reduce the costs associated with the e-discovery. As a result, dozens of companies offer e-discovery software that provides the ability to do the following:

- Analyze large volumes of ESI quickly to perform early case assessments
- Simplify and streamline data collection from across all relevant data sources in multiple data formats
- Cull large amounts of ESI to reduce the number of documents that must be processed and reviewed
- Identify all participants in an investigation to determine who knew what and when

E-discovery raises many ethical issues: Should an organization ever attempt to destroy or conceal incriminating evidence that could otherwise be revealed during discovery? To what degree must an organization be proactive and thorough in providing evidence sought through the discovery process? Should an organization attempt to "bury" incriminating evidence in a mountain of trivial, routine ESI?

Consumer Profiling

Companies openly collect personal information about Internet users when they register at Web sites, complete surveys, fill out forms, or enter contests online. Many companies also obtain information about Web surfers through the use of **cookies**, text files that a Web site can download to visitors' hard drives so that it can identify visitors on subsequent visits. Companies also use tracking software to allow their Web sites to analyze browsing habits and deduce personal interests and preferences. The use of cookies and tracking software is controversial because companies can collect information about consumers without their explicit permission. Outside of the Web environment, marketing firms employ similarly controversial means to collect information about people and their buying habits. Each time a consumer uses a credit card, redeems frequent flyer points, fills out a warranty card, answers a phone survey, buys groceries using a store loyalty card, orders from a mail-order catalog, or registers a car with the DMV, the data is added to a storehouse of personal information about that consumer, which may be sold or shared with third parties. In many of these cases, consumers never explicitly consent to submitting their information to a marketing organization.

Aggregating Consumer Data

Marketing firms aggregate the information they gather about consumers to build databases that contain a huge amount of consumer data. They want to know as much as possible about consumers—who they are, what they like, how they behave, and what motivates them to buy. The marketing firms provide this data to companies so that they can tailor their products and services to individual consumer preferences. Advertisers use the data to more effectively target and attract customers to their messages. Ideally, this means that buyers should be able to shop more efficiently and find products that are well suited for them. Sellers should be better able to tailor their products and services to meet their customers' desires and to increase sales. However, concerns about how this data is used prevent many potential Web shoppers from making online purchases.

Large-scale marketing organizations such as DoubleClick (a subsidiary of Google that develops and provides Internet ad-serving services) employ advertising networks to serve ads to thousands of Web sites. When someone clicks on an ad at a company's Web site, tracking information about the person is gathered and forwarded to DoubleClick, which stores it in a large database. This data includes a record of the ad on which the person clicked and what the person bought. A group of Web sites served by a single advertising network is called a collection of **affiliated Web sites**.

Collecting Data from Web Site Visits

Marketers use cookies to recognize return visitors to their sites and to store useful information about them. The goal is to provide customized service for each consumer. When someone visits a Web site, the site asks that person's computer if it can store a cookie on the hard drive. If the computer agrees, it is assigned a unique identifier, and a cookie with this identification number is placed on its hard drive. Cookies allow marketers to collect **click-stream data**—information gathered by monitoring a consumer's online activity.

During a Web-surfing session, three types of data are gathered. First, as one browses the Web, "GET" data is collected. GET data reveals, for example, that the consumer visited an affiliated book site and requested information about the latest Dean Koontz book. Second, "POST" data is captured. POST data is entered into blank fields on an affiliated Web page when a consumer signs up for a service, such as the Travelocity service that sends an email when fares for flights to favorite destinations change. Third, the marketer monitors the consumer's surfing throughout any affiliated Web sites, keeping track of the information the user sought and viewed. Thus, as a person surfs the Web, a tremendous amount of data is generated for marketers and sellers.

After cookies have been stored on your computer, they make it possible for a Web site to recognize a return visitor and to tailor the ads and promotions presented to that user. The marketer knows what ads have been viewed most recently and makes sure that they aren't shown again, unless the advertiser has decided to market using repetition. The marketer also tracks what sites are visited and uses the data to make educated guesses about the kinds of ads that would be most interesting to you.

Beacons are small pieces of software that run on a Web page and are able to track what a viewer is doing on the page, such as what is being typed or where the mouse is moving. The Dictionary.com Web site places 41 beacons from ad networks on visitors' computers, including several from companies that track visitors' health conditions.[68]

The *Wall Street Journal* conducted a research study in the summer of 2010 that revealed that the tracking of consumers has become highly pervasive and intrusive. One finding of this study was that the nation's 50 most popular Web sites on average installed 64 pieces of tracking software onto the computers of visitors. Microsoft's Web portal, MSN.com, plants a tracking file on each visitor's computer that includes an educated guess of the visitor's income, marital status, number of children, and home ownership. The Encyclopedia Britannica's dictionary Web site Merriam-Webster.com plants a tracking file that scans the Web page the user is viewing and sends ads related to what it sees there. A person getting definitions for eating disorder–related words would see ads for eating disorder medications and treatments on that page and subsequent Web pages.[69]

There are four ways to limit or even stop the deposit of cookies on your hard drive: (1) adjust your browser settings so that your computer will not accept cookies;

(2) manually delete cookies from your hard drive; (3) download and install a cookie-management program; or (4) use anonymous browsing programs that don't accept cookies. (For example, *www.anonymizer.com* offers anonymous surfing services; by switching on the Anonymizer privacy button or going through its Web site, you can hide your identity from nosy Web sites.) However, an increasing number of Web sites lock visitors out unless they allow cookies to be deposited on their hard drives.

Unfortunately, no major browsers let you track or block beacons without installing additional software (called plug-ins), such as Abine, Better Privacy, or Ghostery.

Personalization Software

In addition to using cookies to track consumer data, online marketers use **personalization software** to optimize the number, frequency, and mixture of their ad placements and to evaluate how visitors react to new ads. The goal is to turn first-time visitors to a site into paying customers and to facilitate greater cross-selling activities.

There are several types of personalization software. For example, rules-based personalization software uses business rules tied to customer-supplied preferences or online behavior to determine the most appropriate page views and product information to display when a user visits a Web site. For instance, if you use a Web site to book airline tickets to a popular vacation spot, rules-based software might ensure that you are shown ads for rental cars. Collaborative filtering offers consumer recommendations based on the types of products purchased by other people with similar buying habits. For example, if you bought a book by Dean Koontz, a company might recommend Stephen King books to you, based on the fact that a significant percentage of other customers bought books by both authors.

Demographic filtering is another form of personalization software. It augments click-stream data and user-supplied data with demographic information associated with user zip codes to make product suggestions. Microsoft has captured age, sex, and location information for years through its various Web sites, including MSN and Hotmail. It has accumulated a vast database on tens of millions of people, each assigned a global user ID. Microsoft has also developed a technology based on this database that enables marketers to target one ad to men and another to women. Additional information such as age and location can be used as ad-selection criteria.

Yet another form of personalization software, contextual commerce, associates product promotions and other e-commerce offerings with specific content a user may receive in a news story online. For example, as you read a story about white-water rafting, you may be offered a deal on rafting gear or a promotion for a white-water rafting vacation in West Virginia. Instead of simply bombarding customers at every turn with standard sales promotions that result in tiny response rates, marketers are getting smarter about where and how they use personalization. They are also taking great care to measure whether personalization is paying off. The intended result is that effective personalization increases online sales and improves consumer relationships.

Online marketers cannot capture personal information, such as names, addresses, and Social Security numbers, unless people provide them. Without this information, companies can't contact individual Web surfers who visit their sites. Data gathered about a user's Web browsing through the use of cookies is anonymous, as long as the network advertiser doesn't link the data with personal information. However, if a Web site visitor volunteers personal information, a Web site operator can use it to find additional personal information that the visitor may not want to disclose. For example, a name and address can be used to find a corresponding phone number, which can then lead to obtaining even more personal data. All this information becomes extremely valuable to the Web site operator, who is trying to build a relationship with Web site visitors and turn them into customers. The operator can use this data to initiate contact or sell it to other organizations with which they have marketing agreements.

Consumer Data Privacy

Consumer data privacy has grown into a major marketing issue. Companies that can't protect or don't respect customer information often lose business and some become defendants in class action lawsuits stemming from privacy violations. Privacy groups spoke out vigorously to protest the proposed merger of Web ad server DoubleClick and database marketing company Abacus Direct. The groups were concerned that the information stored in cookies would be combined with data from mailing lists, thus revealing the Web users' identities. This would enable a network advertiser to identify and track the habits of unsuspecting consumers. Public outrage and the threat of lawsuits forced DoubleClick to back off its plan to merge the data of the two companies.

Opponents of consumer profiling are also concerned that personal data is being gathered and sold to other companies without the permission of consumers who provide the data. After the data has been collected, consumers have no way of knowing how it is used or who is using it.

Treating Consumer Data Responsibly

When dealing with consumer data, strong measures are required to avoid customer relationship problems. A widely accepted approach to treating consumer data responsibly is for a company to adopt the Fair Information Practices and the 1980 OECD privacy guidelines. Under these guidelines, an organization collects only personal information that is necessary to deliver its product or service. The company ensures that the information is carefully protected and accessible only by those with a need to know, and that consumers can review their own data and make corrections. The company informs customers if it intends to use customer information for research or marketing, and it provides a means for them to opt out.

The Federal Trade Commission, which is in charge of protecting the privacy of U.S. consumers, admits that "its efforts to address privacy through self-regulation have been too slow, and up to now have failed to provide adequate and meaningful protection."[70] A preliminary staff report recommended a series of measures to improve the protection of consumer privacy, including: (1) implementating a Do Not Track option for Web browsing software that would keep online services from collecting surfing or ad-targeting data, (2) requiring that companies collect only the data they need and then delete it after it has met its purpose, and (3) requiring that online companies simplify their privacy polices so that consumers can read and understand them. These recommendations form the basis for a proposed framework of policies the FTC plans to put in place in 2011. The FTC asked interested parties to submit feedback on these proposals by January 31, 2011.[71]

The cross-industry Self-Regulatory Program for Online Behavioral Advertising is made up of several advertising industry associations, including the American Association of Advertising Agencies, the American Advertising Federation, the Association of National Advertisers, the Better Business Bureau, the Direct Marketing Association, the Interactive

Advertising Bureau, and the National Advertising Initiative. This group developed a set of seven principles that address consumer concerns about the use of personal information and interest-based advertising. Table 4-5 summarizes these seven principles.[72]

TABLE 4-5 Principles of the Self-Regulatory Program for Online Behavioral Advertising

Principle	Description
The Education Principle	Calls on organizations to educate individuals and organizations about online behavioral advertising
The Transparency Principle	Calls for clearer and easily accessible disclosures to consumers about data collection and use practices
The Consumer Control Principle	Provides consumers with an expanded ability to choose whether data is collected and used for online behavioral advertising purposes
The Data Security Principle	Calls for organizations to provide adequate security for, and limited retention of, data collected
The Material Changes Principle	Calls for obtaining consumer consent before a material change is made to a consumer's data collection and use policies
The Sensitive Data Principle	Recognizes that data collected from children and used for online behavioral advertising merits heightened protection and requires parental consent
The Accountability Principle	Calls for development of programs to further advance these principles

Source Line: Digital Advertising Alliance.

As a result of increased focus on the topic of data privacy, many companies recognize the need to establish corporate data privacy policies, and an increasing number of companies are appointing executives to oversee their data privacy policies and initiatives. Some companies are appointing a **chief privacy officer (CPO)**. A CPO is a senior manager within an organization whose role is to ensure that the organization does not violate government regulations while reassuring customers that their privacy will be protected. In order for the CPO to be effective, the organization must give him or her the power to stop or modify major company marketing initiatives if necessary. The CPO's general duties include training employees about privacy; checking the company's privacy policies for potential risks; figuring out if gaps exist and, if so, how to fill them; and developing and managing a process for customer privacy disputes.

The CPO should be briefed on marketing programs, information systems, and databases involving the collection or dissemination of consumer data while these projects are still in the planning phase. The rationale for early involvement in such initiatives is to ensure that potential problems can be identified at the earliest stages, when it is easier and cheaper to fix them. Some organizations fail to address privacy issues early on, and it takes a negative experience to make them appoint a CPO.

As Google tried to deal with the negative publicity over the wireless network scandal described in the opening vignette, Google CEO Eric Schmidt came under criticism for failure to have a strong CPO in place. Indeed, a strong CPO might have been able to avoid some public embarrassment for the firm. For example, when CNN *Parker Spitzer* cohost Kathleen Parker raised questions about her concerns over how much the Internet giant knows about her personally, the topic of Street View came up. Gaffe-prone Schmidt commented, "Street View. We drive exactly once. So, you can just move, right?" Other Schmidt gaffes on the topic of privacy include the following:

- "We know where you are. We know where you've been. We can more or less know what you're thinking about."
- "In order to escape the wide reach of Google's SEO (Search Engine Optimization), we'll soon have to change our names, as if covering up a digital scarlet letter."
- "There's such an overwhelming amount of information now, we can search where you are, see what you're looking at if you take a picture with your camera. One way to think about this is we're trying to make people better people, literally give them better ideas—augmenting their experience. Think of it as augmented humanity."
- "We can suggest what you should do next, what you care about. Imagine: We know where you are, we know what you like."[73]

RapLeaf is a personal information aggregation service capable of profiling users by name. It is able to create comprehensive and detailed databases about people by including data from voter registration files, shopping histories, social networking data, and real estate records. RapLeaf's goal is to help online companies better understand their consumers, engage them more meaningfully, and deliver the right message at the right time. The company recently developed software that can provide information about home ownership status, occupation, and income. In late 2010, the firm hired a new employee to serve as general counsel and chief privacy officer and to deal with the number of data privacy issues the software is likely to raise with concerned consumers.[74]

Table 4-6 provides useful guidance for ensuring that organizations treat consumer data responsibly. The preferred answer to each question is *yes*.

TABLE 4-6 Manager's checklist for treating consumer data responsibly

Question	Yes	No
Does your company have a written data privacy policy that is followed?		
Can consumers easily view your data privacy policy?		
Are consumers given an opportunity to opt in or opt out of your data policy?		
Do you collect only the personal information needed to deliver your product or service?		
Do you ensure that the information is carefully protected and accessible only by those with a need to know?		
Do you provide a process for consumers to review their own data and make corrections?		
Do you inform your customers if you intend to use their information for research or marketing and provide a means for them to opt out?		
Have you identified a person who has full responsibility for implementing your data policy and dealing with consumer data issues?		

Source Line: Course Technology/Cengage Learning.

Workplace Monitoring

There is plenty of data to support the conclusion that many workers waste large portions of their work time doing non-work-related activity at work. One source estimates that U.S. employers lose $4 billion dollars in productivity each year due to employees spending as much as two to three hours per day on personal Internet use.[75] Another source estimates that, on average, workers spend about four or five hours per week on personal matters. In a recent survey by an IT staffing firm, 54 percent of companies reported they were banning the use of social networking sites such as Facebook, Twitter, MySpace, and LinkedIn to help reduce waste at work.[76]

As discussed in Chapter 2, many organizations have developed policies on the use of IT in the workplace in order to protect against employee abuses that reduce worker productivity or that expose the employer to harassment lawsuits. For example, an employee may sue his or her employer for creating an environment conducive to sexual harassment if other employees are viewing pornography online while at work and the organization takes no measures to stop such viewing. (Email containing crude jokes and cartoons or messages that discriminate against others based on sex, race, or national origin can also spawn lawsuits.) The institution and communication of an IT usage policy establishes boundaries of acceptable behavior and enables management to take action against violators. Table 4-7 summarizes the extent of workplace monitoring.

TABLE 4-7 Extent of workplace monitoring

Subject of workplace monitoring	Percent of employers that monitor workers	Percent of companies that have fired employees for abuse or violation of company policy
Email	43%	28%
Web surfing	66%	30%
Time spent on phone as well as phone numbers called	45%	6%

Source Line: American Management Association Press Room, "2007 Electronic Monitoring and Surveillance Survey," © February 28, 2008, http://press.amanet.org/press-releases/177/2007-electronic-monitoring-surveillance-survey.

The potential for decreased productivity and increased legal liabilities has led many employers to monitor workers to ensure that corporate IT usage policies are being followed. Many U.S. firms find it necessary to record and review employee communications and activities on the job, including phone calls, email, and Web surfing. Some are even videotaping employees on the job. In addition, some companies employ random drug testing and psychological testing. With few exceptions, these increasingly common (and many would say intrusive) practices are perfectly legal.

The Fourth Amendment to the Constitution protects citizens from unreasonable government searches and is often invoked to protect the privacy of government employees. Public-sector workers can appeal directly to the "reasonable expectation of privacy" standard established by the 1967 Supreme Court ruling in *Katz v. United States*.

However, the Fourth Amendment cannot be used to limit how a private employer treats its employees because such actions are not taken by the government. As a result, public-sector employees have far greater privacy rights than those in private industry. Although private-sector employees can seek legal protection against an invasive employer under various state statutes, the degree of protection varies widely by state. Furthermore, state privacy statutes tend to favor employers over employees. For example, to successfully sue an organization for violation of their privacy rights, employees must prove that they were in a work environment where they had a reasonable expectation of privacy. As a result, courts typically rule against employees who file privacy claims for being monitored while using company equipment. A private organization can defeat a privacy claim simply by proving that an employee had been given explicit notice that email, Internet use, and files on company computers were not private and that their use might be monitored. When an employer engages in workplace monitoring, though, it must ensure that it treats all types of workers equally. For example, a company could get into legal trouble for punishing an hourly paid employee more seriously for visiting inappropriate Web sites than it punished a monthly paid employee.

Society is struggling to define the extent to which employers should be able to monitor the work-related activities of employees. On the one hand, employers want to be able to guarantee a work environment that is conducive to all workers, ensure a high level of worker productivity, and limit the costs of defending against privacy-violation lawsuits filed by disgruntled employees. On the other hand, privacy advocates want federal legislation that keeps employers from infringing on the privacy rights of employees. Such legislation would require prior notification to all employees of the existence and location of all electronic monitoring devices. Privacy advocates also want restrictions on the types of information collected and the extent to which an employer may use electronic monitoring. As a result, many privacy bills are being introduced and debated at the state and federal levels. As the laws governing employee privacy and monitoring continue to evolve, business managers must stay informed in order to avoid enforcing outdated usage policies. Organizations with global operations face an even greater challenge because the legislative bodies of other countries also debate these issues.

Advanced Surveillance Technology

A number of advances in information technology—such as surveillance cameras and satellite-based systems that can pinpoint a person's physical location—provide exciting new data-gathering capabilities. However, these advances can also diminish individual privacy and complicate the issue of how much information should be captured about people's private lives.

Advocates of advanced surveillance technology argue that people have no legitimate expectation of privacy in a public place. Critics raise concerns about the use of surveillance to secretly store images of people, creating a new potential for abuse, such as intimidation of political dissenters or blackmail of people caught with the "wrong" person or in the "wrong" place. Critics also raise the possibility that such technology may not identify people accurately.

Camera Surveillance

Surveillance cameras are used in major cities around the world in an effort to deter crime and terrorist activities. Critics believe that such scrutiny is a violation of civil liberties and are concerned about the cost of the equipment and people required to monitor the video feeds. Surveillance camera supporters offer anecdotal data that suggests the cameras are effective in preventing crime and terrorism. They can provide examples where cameras helped solve crimes by corroborating the testimony of witnesses and helping to trace suspects.

London has created a surveillance network consisting of more than 12,000 cameras continuously watching for problems. Two-thirds of the streets that used to lead into London have been closed to traffic and the remaining streets have been narrowed and are monitored by police in sentry boxes. In addition, cars are monitored by cameras—one aimed at the driver's face and one at the car's license plate.[77]

Washington, D.C.'s Homeland Security and Emergency Management Agency (HSEMA) receives video feeds from more than 4,500 surveillance cameras embedded in and around its schools and public transportation system hubs. HSEMA is evaluating the addition of thousands of more video feeds from private businesses such as banks, corner stores, and gas stations around the city. According to a HSEMA spokesperson, the cameras are designed to raise "situational awareness" during "developing significant events."[78]

The city of Chicago is estimated to have over 10,000 surveillance cameras located throughout the city. These cameras are technically sophisticated and have the capability to automatically identify and track specified individuals. They can also zoom in and magnify small details as well as view objects at great distances.[79]

More than 3,000 surveillance cameras in the New York subway system are being monitored by the city's police department as part of its counterterrorism efforts. About one-third of these cameras are networked to a computer system that runs sophisticated software capable of detecting suspicious activity such as a package sitting in one place for a long time. Approximately 500 live-feed cameras are installed at the transit system's three busiest hubs—Times Square, Pennsylvania Station, and Grand Central Station (see Figure 4-3). The cost of the system is nearly $200 million, but Mayor Michael Bloomberg

FIGURE 4-3 Surveillance cameras monitor commuters at Grand Central Station
Credit: Image copyright MalibuBooks, 2009. Used under license from Shutterstock.com.

and Police Commissioner Raymond Kelly fully support the system because subways remain at the top of the list of terrorist targets.[80]

GPS Chips

From automobiles to cell phones, Global Positioning System (GPS) chips are being placed in many devices to precisely locate users. The FCC has asked cell phone companies to implement methods for locating users so that police, fire, and medical personnel can be accurately dispatched to assist 911 callers. Similar location-tracking technology is also available for personal digital assistants, laptop computers, trucks, and boats. Parents can place one of these chips in their teenager's car and then use software to track the car's whereabouts.

Over a four-month period in 2007, U.S. Drug Enforcement Agency agents repeatedly monitored the movements of a drug suspect using GPS tracking devices attached to the undercarriage of his jeep. Information gathered from the tracking led to the suspect's eventual arrest and indictment. In court, the suspect claimed that the agents had violated his Fourth Amendment rights against unreasonable search because they had planted the GPS devices on a vehicle parked in his driveway without obtaining a warrant. However, two members of a three-judge panel of the U.S. Court of Appeals for the Ninth Circuit ruled that the defendant's constitutional rights were not violated. Their reasoning was that he had failed to take specific actions to exclude passersby from his driveway, such as installing a gate or posting no trespassing signs. Therefore, he could not claim reasonable privacy expectations. The third member of the panel dissented, arguing that most people don't expect that a car parked in their driveway "invites people to crawl under it and attach a [tracking] device. There is something creepy and un-American about such clandestine and underhanded behavior." The case is likely to be appealed to the Supreme Court.[81]

Banks, retailers, and airlines are eager to gain real-time access to consumer location data and have already devised a number of new services they want to provide—sending consumers digital coupons for stores that are nearby, providing the location of the nearest ATM, and updating travelers on flight and hotel information. Airlines are now using wireless devices both to enable passengers to check in for flights when they are close to the gate and to monitor when each person passes through the gate.

Businesses claim that they will respect the privacy of wireless users and allow them to opt in or opt out of marketing programs that are based on their location data. Wireless spamming is a distinct possibility—a user might continuously receive wireless ads, notices for local restaurants, and shopping advice while walking down the street. Another concern is that the data could be used to track people down at any time or to figure out where they were at some particular instant.

Summary

- The use of information technology in business requires balancing the needs of those who use the information that is collected against the rights and desires of the people whose information is being used. A combination of approaches—new laws, technical solutions, and privacy policies—is required to balance the scales.
- The Fourth Amendment reads, "The right of the people to be secure in their persons, houses, papers, and effects, against unreasonable searches and seizures, shall not be violated, and no Warrants shall issue, but upon probable cause, supported by Oath or affirmation, and particularly describing the place to be searched, and the persons or things to be seized." The courts have ruled that without a reasonable expectation of privacy, there is no privacy right to protect.
- Few laws provide privacy protection from private industry.
- There is no single, overarching national data privacy policy.
- There are a number of federal laws that provide protection for personal financial data, including the following:
    - Fair Credit Reporting Act—regulates operations of credit-reporting bureaus
    - Fair and Accurate Credit Transaction Act—allows consumers to request and obtain a free credit report from each of the three consumer credit reporting agencies
    - Right to Financial Privacy Act—protects the financial records of financial institution customers from unauthorized scrutiny by the federal government
    - Gramm–Leach–Bliley Act—establishes guidelines for the collection and disclosure of personal financial information; requires financial institutions to document their data security plan; and encourages institutions to implement safeguards against pretexting
- The Health Insurance Portability and Accountability Act (HIPAA) defined numerous standards to improve the portability and continuity of health insurance coverage; reduce fraud, waste, and abuse in health insurance care and healthcare delivery; and simplify the administration of health insurance.
- The American Recovery and Reinvestment Act included strong privacy provisions for electronic health records, including banning the sale of health information, promoting the use of audit trails and encryption, and providing rights of access for patients. It also mandated that each individual whose health information has been exposed be notified within 60 days after discovery of a data breach.
- The Children's Online Privacy Protection Act requires Web sites that cater to children to offer comprehensive privacy policies, notify parents or guardians about their data collection practices, and receive parental consent before collecting any personal information from children under the age of 13.
- The Family Educational Rights and Privacy Act provides students with specific rights regarding the release of their student records.
- The Communications Act of 1934 established the Federal Communications Commission and gave it responsibility for regulating all non-federal-government use of radio, television, and interstate telecommunications as well as all international communications that originate or terminate in the United States.

- Title III of the Omnibus Crime Control and Safe Streets Act regulates the interception of wire (telephone) and oral communications.
- The Foreign Intelligence Surveillance Act (FISA) describes procedures for the electronic surveillance and collection of foreign intelligence information between foreign powers and agents of foreign powers.
- The Electronic Communications Privacy Act (ECPA) deals with the protection of communications while in transit from sender to receiver; the protection of communications held in electronic storage; and the prohibition of devices to record dialing, routing, addressing, and signaling information without a search warrant.
- The Communications Assistance for Law Enforcement Act (CALEA) requires the telecommunications industry to build tools into its products that federal investigators can use—after gaining a court order—to eavesdrop on conversations and intercept electronic communications.
- The USA PATRIOT Act modified 15 existing statutes and gave sweeping new powers both to domestic law enforcement and to international intelligence agencies, including increasing the ability of law enforcement agencies to eavesdrop on telephone communication, intercept email messages, and search medical, financial, and other records; the act also eased restrictions on foreign intelligence gathering in the United States.
- Various organizations have defined guidelines to protect the transborder data flow of personal data. The Organisation for Economic Co-operation and Development created the Fair Information Practices—privacy guidelines that are often held up as the model for organizations to adopt for the ethical treatment of consumer data.
- The European Union Data Protection Directive requires member countries to ensure that data transferred to non–European Union countries is protected. It also bars the export of data to countries that do not have data privacy protection standards comparable to those of the European Union.
- The United States relies on self-regulation overseen by the Department of Commerce and organizations such as the Better Business Bureau Online (BBB*OnLine*) and TRUSTe.
- The Freedom of Information Act (FOIA) grants citizens the right to access certain information and records of the federal government upon request.
- The Privacy Act prohibits U.S. government agencies from concealing the existence of any personal data record-keeping system.
- Identity theft occurs when someone steals key pieces of personal information to impersonate a person; it is the fastest growing form of fraud in the United States.
- Identity thieves often create data breaches, purchase personal data, employ phishing, and install spyware to capture personal data.
- The number of data breaches is alarming, as is the lack of initiative by some companies in informing the people whose data is stolen.
- Discovery is part of the pretrial phase of a lawsuit in which each party can obtain evidence from the other party by various means, including requests for the production of documents. E-discovery is the collection, preparation, review, and production of electronically stored information for use in criminal and civil actions and proceedings.

- Companies use many different methods to collect personal data about visitors to their Web sites, including depositing cookies on visitors' hard drives and capturing click-stream, GET, and POST data. Beacons are small pieces of software that run on a Web page and are able to track what a viewer is doing on the page, such as what is being typed or where the mouse is moving.
- Marketers use personalization software to optimize the number, frequency, and mixture of their ad placements.
- Consumer data privacy has become a major marketing issue—companies that cannot protect or do not respect customer information have lost business and have become defendants in class actions stemming from privacy violations.
- One approach to treating consumer data responsibly is to adopt the Fair Information Practices; some companies also appoint a chief privacy officer.
- The Federal Trade Commission is in charge of protecting the privacy of U.S. consumers. It is changing its stance from a policy of self-regulation to a more proactive approach.
- Several advertising industry associations have banned together to create a set of principles that address consumer concerns about the use of personal information and interest-based advertising.
- Many organizations have developed IT usage policies to protect against employee abuses that can reduce worker productivity and expose employers to harassment lawsuits.
- Many U.S. firms record and review employee communications and activities on the job, including phone calls, email, Web surfing, and computer files.

Key Terms

affiliated Web site
American Recovery and Reinvestment Act of 2009
beacon
chief privacy officer (CPO)
Children's Online Privacy Protection Act (COPPA)
click-stream data
Communications Act of 1934
Communications Assistance for Law Enforcement Act (CALEA)
cookie
Electronic Communications Privacy Act of 1986 (ECPA)
electronic discovery (e-discovery)
electronically stored information (ESI)
European Union Data Protection Directive
Fair and Accurate Credit Transactions Act
Fair Credit Reporting Act
Fair Information Practices
Family Educational Rights and Privacy Act (FERPA)
foreign intelligence
Foreign Intelligence Surveillance Act (FISA)
Freedom of Information Act (FOIA)
Gramm–Leach–Bliley Act (GLBA)
Health Insurance Portability and Accountability Act of 1996 (HIPAA)
identity theft
information privacy
opt in

opt out

pen register

personalization software

Privacy Act of 1974

Right to Financial Privacy Act of 1978

spyware

sunset provision

Title III of the Omnibus Crime Control and Safe Street Act

trap and trace

USA PATRIOT Act

Self-Assessment Questions

The answers to the Self-Assessment Questions can be found in Appendix E.

1. The U.S. Supreme Court has ruled that the concept of privacy is protected by the ____________.
2. ____________ is part of the pretrial phase of a lawsuit in which each party can obtain evidence from the other part by various means.
3. During e-discovery, one party often sends the other party a litigation hold notice to save relevant data and to suspend data that might be due to be destroyed based on normal data retention rules. True or False?
4. An act designed to promote accuracy, fairness, and privacy of information in the files of credit-reporting companies is the:
   a. Gramm–Leach–Bliley Act
   b. Fair Credit Reporting Act
   c. HIPAA
   d. USA PATRIOT Act
5. If someone refuses to give an institution the right to collect and share personal data about himself or herself, he or she is said to ____________.
6. According to the Children's Online Privacy Protection Act, a Web site that caters to children must:
   a. offer comprehensive privacy policies
   b. notify parents or guardians about its data collection practices
   c. receive parental consent before collecting any personal information from preteens
   d. all of the above
7. The ____________ offers protection for victims of data breaches that go unreported for months after discovery.
8. ____________ *v. United States* is a famous court ruling that helped form the basis for the requirement that there be a reasonable expectation of privacy for the Fourth Amendment to apply.
9. The Right to Financial Privacy Act of 1978 protects the financial records of banking institutions from unauthorized scrutiny by the state governments. True or False?

10. Which of the following identifies the numbers dialed for outgoing calls?
    a. pen register
    b. wiretap
    c. trap and trace
    d. all of the above
11. ____________ gave sweeping new powers to law enforcement agencies to search telephone, email, medical, financial, and other records; it also eased restrictions on foreign intelligence gathering in the United States.
12. Under the USA PATRIOT Act, the FBI can issue a(n) ____________ to compel banks, Internet service providers, and credit-reporting companies to turn over information about their customers without a court order simply on the basis that the information is needed for an ongoing investigation.
13. The European philosophy of addressing privacy concerns employs strict government regulation, including enforcement by a set of commissioners; it differs greatly from the U.S. philosophy of having no federal privacy policy. True or False?
14. Which of the following is *not* a technique frequently employed by identity thieves?
    a. hacking databases
    b. spyware
    c. phishing
    d. trap and trace
15. ____________ is used by marketers to optimize the number, frequency, and mixture of their ad placements.
16. In a recent survey, over 25 percent of employers reported firing workers for violating or abusing their corporate email policy. True or False?

Discussion Questions

1. Go to the Web sites of the U.S. Department of Commerce and the Federal Trade Commission to gain an understanding of their roles in safeguarding consumer privacy. Are the roles of these two organizations redundant and overlapping, or are they complementary?
2. Prepare a set of arguments that would support the contention that the USA PATRIOT Act was overreaching in both its scope and its approach. Then prepare a set of arguments that support the USA PATRIOT Act as the proper and appropriate way to protect the United States from further terrorist acts.
3. Do research to find the nine Freedom of Information Act statutory exceptions under which an agency may refuse to disclose an agency record. Document two recent examples where an agency invoked one or more exemptions and refused to provide agency records.
4. Are surveillance cameras worth the cost in resources and loss of privacy, given the role that they play in deterring or solving crimes?

5. The American Civil Liberties Union (ACLU) has challenged the use of National Security Letters by the FBI in court several times. These lawsuits are in various stages of hearings and appeals. In one lawsuit, *Doe v. Holder*, the Court of the Southern District of New York and, upon appeal, the Second Circuit Court of Appeals ruled that the NSL gag provision—which prohibits NSL recipients from informing anyone, even the person who is the subject of the NSL request, that the government has secretly requested his or her records—violates the First Amendment. Do research to document the status of this lawsuit.
6. What benefits can consumer profiling provide to you as a consumer? Do these benefits outweigh the loss of your privacy?
7. How much effort should be required of a Web site operator to prevent preteen visitors who lie about their age from visiting the operator's adult-oriented Web site? Should Web site operators be prosecuted under COPPA if preteens provide false information in order to gain access to such sites?
8. An FOIA exemption prevents disclosure of records if it would invade someone's personal privacy. Develop a hypothetical example in which a person's privacy interests are clearly outweighed by competing public interests. Develop another hypothetical example in which a person's privacy interests are not outweighed by competing public interests.
9. Summarize the Fourth Amendment to the U.S. Constitution. Does it apply to the actions of private industry?
10. Do you feel that information systems to fight terrorism should be developed and used even if they infringe on privacy rights or violate the Privacy Act of 1974 and other such statutes?
11. Why do employers monitor workers? Do you think they should be able to do so? Why or why not?
12. Do you think that law enforcement agencies should be able to use advanced surveillance technology, such as surveillance cameras combined with facial recognition software? Why or why not?

What Would You Do?

1. Your friend is considering using an online service to identify people with compatible personalities and attractive physical features who would be interesting to date. Your friend must first submit some basic personal information, then complete a five-page personality survey, and finally provide several recent photos. Would you advise your friend to do this? Why or why not?
2. You are a recent college graduate with only a year of experience with your employer. You were recently promoted to manager of email services. You are quite surprised to receive a phone call at home on a Saturday from the Chief Financial Officer of the firm asking that you immediately delete all email from all email servers, including the archive and back-up servers, that is older than six months. He states that the reason for his request is that there have been an increasing number of complaints about the slowness of email services. In addition, he says he is concerned about the cost of storing so much email. This does not sound right to you because you have taken several measures that have actually speeded up email services. You recall a brief paragraph in the paper last week that stated

several suppliers were upset about what they called "price fixing" among the companies in your industry. What do you say to the Chief Financial Officer?

3. You have 15 years of experience in sales and marketing with three different organizations. You currently hold a middle management position and have been approached by a headhunter to make a move to a company seeking a chief privacy officer. The headhunter claims the position would represent a good move for you in your career path and lead to an eventual vice president of marketing position. You are not sure about the exact responsibilities of the CPO position or what authority the person in this position would have at this particular company. How would you go about evaluating this opportunity to determine if it actually is a good career move or a dead end?

4. As the information systems manager for a small manufacturing plant, you are responsible for anything relating to the use of information technology. A new inventory control system is being implemented to track the quantity and movement of all finished products stored in a local warehouse. Each time a forklift operator moves a case of product, he or she must first scan the product identifier code on the case. The product information is captured, as is the day, time, and forklift operator identification number. This data is transmitted over a local area network (LAN) to the inventory control computer, which then displays information about the case and where it should be placed in the warehouse.

   The warehouse manager is excited about using case movement data to monitor worker productivity. He will be able to tell how many cases per shift each operator moves, and he plans to use this data to provide performance feedback that could result in pay increases or termination. He has asked you if there are any potential problems with using the data in this manner and, if so, what should be done to avoid them. How would you respond?

5. You have been asked to help develop a company policy on what should be done in the event of a data breach, such as unauthorized access to your firm's customer database, which contains some 1.5 million records. What sort of process would you use to develop such a policy? What resources would you call on?

6. You are a new brand manager for a product line of Coach purses. You are considering purchasing customer data from a company that sells a large variety of women's products online. In addition to providing a list of names, mailing addresses, and email addresses, the data includes an approximate estimate of customers' annual income based on the zip code in which they live, census data, and highest level of education achieved. You could use the estimate of annual income to identify likely purchasers of your high-end purses, and use email addresses to send emails announcing the new product line and touting its many features. List the advantages and disadvantages of such a marketing strategy. Would you recommend this means of promotion in this instance? Why or why not?

7. Your company is rolling out a training program to ensure that everyone is familiar with the company's IT usage policy. As a member of the Human Resources department, you have been asked to develop a key piece of the training relating to why this policy is needed. What kind of concerns can you expect your audience to raise? How can you deal with this anticipated resistance to the policy?

8. You are the CPO of a midsized manufacturing company that has sales of more than $250 million per year and almost $50 million from online sales. You have been challenged

by the vice president of sales to change the company's Web site data privacy policy from an opt-in policy to an opt-out policy and to allow the sale of customer data to other companies. The vice president has estimated that this change would bring in at least $5 million per year in added revenue with little additional expense. How would you respond to this request?

Cases

1. The Use of Electronic Discovery in *Viacom v. YouTube*

In March 2007, entertainment giant Viacom International filed a lawsuit against YouTube, the popular video-sharing Web site on which users can upload and view video clips. Viacom accused YouTube of "massive intentional copyright infringement" and sought over $1 billion in damages. Viacom charged that approximately 160,000 clips copied from Viacom's products had been posted to YouTube and viewed over 1.5 billion times.[82]

YouTube, which was founded in February 2005 by Jawed Karim, Steve Chen, and Chad Hurley, claimed that it was legally protected by the Digital Millennium Copyright Act (DMCA) of 1998. According to the "safe harbor" provision in the DMCA, an online service provider is protected from copyright infringement liability as long as the provider does not have knowledge of the infringement and removes the copyrighted material once notified of the infringement.[83]

Viacom's lawsuit came just months after deep-pocketed Google purchased YouTube for $1.65 billion. Just prior to the purchase, YouTube had announced three separate deals—with CBS Corporation, Sony BMG Music Entertainment, and Universal Music Group—that would allow the use of copyrighted material from these media groups on YouTube.[84] The agreements allowed YouTube to avoid copyright infringement lawsuits from these companies. Such negotiations, however, did not prove fruitful with Viacom. A month before filing the suit, Viacom asked that YouTube remove 100,000 clips, thus making official the DMCA takedown requests.[85] YouTube complied with the request.[86]

The DMCA places the burden of proof entirely on the rights holders. Rights holders have to police any new Internet site that might post their content, document any unauthorized use, and then notify the site's owners. The DMCA requires rights holders to engage in detailed and costly electronic discovery. In July 2008, Viacom won a significant victory when the federal district court of New York issued an electronic discovery order that forced YouTube to turn over copies of the millions of infringing video clips that it had removed from its site. In addition, YouTube was ordered to supply Viacom with YouTube's "logging" database, which contained information about the login ID of users who watched each video, the number of times each video was watched, and the IP address of the device used to connect to the video.[87] This database would allow Viacom to determine exactly how many times each video for which it owned the copyright had been viewed on YouTube.

The discovery ruling raised significant privacy concerns, with bloggers and consumer advocates warning that Viacom could track people who watched the copyrighted clips and take legal action against them. If Viacom took such measures, all Internet companies whose success was predicated on user participation could be jeopardized.[88] Viacom immediately responded by issuing a press release stating that any personally identifying information would be stripped from the data before it was transferred to Viacom and that it would preserve the confidentiality of the YouTube data.[89]

A few months prior to the U.S. district court's final ruling in 2010, the court ordered the release of the case's documents, and both Google's and Viacom's briefs were made public.[90] Google pointed out in one brief that as early as December 2005, users were posting over 6,000 video clips a day. Google cited numerous measures it had taken to try to combat copyright infringement among this vast growing pool of uploaded videos. Those efforts including blocking the accounts of users who repeatedly posted copyrighted materials. Google also described the increasingly technical tools YouTube had developed to aid rights holders—from its search and report infringement tool to a customized fingerprinting technology that allows YouTube to scan all posted clips and compare them with a library provided by copyright holders.

Viacom briefs made extensive use of electronic discovery. One brief argued that email exchanges dating from early in 2005 clearly showed that the YouTube founders were aware that YouTube was a magnet for copyright infringing content and that they purposely exploited this type of content to expand their user base. When Steve Chen noted in an email to Jawed Karim and Chad Hurley that their Internet service provider was complaining that YouTube was violating its contract because "we're hosting copyrighted content," Chen commented, "I'm not about to take down content because our ISP is giving us [****]." One Viacom brief also claimed that when YouTube actually began to remove copyrighted content from its site in the summer of 2005, the company devised a strategy to remove only some but not all this material. The brief quotes from Chen's email: "That way, the perception is that we are concerned about this type of material and we're actively monitoring it. [But the] actual removal of this content will be in varying degrees. That way … YOU can find truckloads of … copyrighted content … [if] you [are] actively searching for it."[91]

The Viacom brief further argued that the founders themselves posted copyrighted material. Chen admonishes Karim in an email, writing, "We're going to have a tough time defending the fact that we're not liable for the copyrighted material on the site because we didn't put it up when one of the cofounders is blatantly stealing content from other sites and trying to get everyone to see it." Yet when Hurley expresses concerns about Chen advising that they steal certain other movies, Chen replies, "We need to attract traffic…."[92] Presuming the email quotes in Viacom's brief were accurate, Viacom's lawyers effectively used electronic discovery to show that the founders of YouTube had knowledge of and actively participated in copyright violation.

In July 2010, after nearly three and a half years of legal battles, the U.S. district court ruled that YouTube qualified for protection under the DMCA safe harbor provision. Judge Louis Stanton explained that proving that YouTube had a general knowledge of copyrighted videos that had been uploaded to its site was insufficient. Viacom would have had to have shown that YouTube had knowledge of specific Viacom videos that had been uploaded and had refused or neglected to remove them.[93] Because YouTube had complied with all DMCA takedown requests, he ruled in YouTube's favor.

Google saw the ruling as a victory for "the billions of people around the world who use the web to communicate and share experiences with each other."[94] Viacom was disappointed, and on December 3, 2010, the company filed an appeal with the U.S. Court of Appeals. Viacom argued that America's economic future is tied to the growth of trade in intellectual property and that this future is jeopardized if ideas are not protected under U.S. law.[95]

As the battle rages on, however, one thing is clear. Electronic discovery played a key role in this case, allowing the court to gain insight not only into the defendants' and the plaintiffs' actions, but also into their intent. Electronic discovery clearly has the potential to create a more fully informed judicial decision-making process.

Discussion Questions

1. What types of electronic discovery were used in this court case?
2. How was electronic discovery vital in supporting Viacom's argument?
3. How does this case highlight how privacy concerns and electronic discovery rulings can come into conflict?

2. Albert Gonzalez and Identity Theft

On a summer night in 2003, a plainclothes police officer in upper Manhattan noticed a young man wearing a woman's wig and a nose ring. Deciding to follow the young man into an ATM lobby, he watched as the man pulled out debit card after debit card, withdrawing hundreds of dollars. When questioned, the young man, Albert Gonzalez, gave the police officer his real name (although as investigators would discover, Gonzales frequently employed several aliases).[96]

Gonzalez was arrested for fraudulent ATM withdrawals, but when the Secret Service found out that he was none other than "Cumbajohhny," an administrator on Shadowcrew.com, they gave him the option of working undercover for the agency to avoid prison time. Shadowcrew was a major Web site for cybercriminals who bought and sold stolen bank card data, and Gonzalez could be a great help in catching this ring of cybercriminals. Gonzalez agreed, and the following year, the government launched the highly successful "Operation Firewall," which brought down the Shadowcrew site and resulted in the coordinated arrests of 28 of its leading members around the world.[97]

Gonzalez was highly intelligent and had a gift for explaining and sharing his expertise in credit card fraud with the government agents. Gonzales continued not only to help with further investigations once he was relocated to Miami (due to concerns about his safety) but also to speak at conferences.[98] He was a valuable resource at a time when identity and bank card theft was rising steeply despite government measures to stop it.

In late December 2006, lawyers from TJX, a leading U.S. retailer of apparel and home fashions (ranked 119 in the *Fortune* 500 listing),[99] contacted the U.S. Attorney's office in Massachusetts. TJX reported that data from a vast number of credit cards used at its T.J. Maxx and Marshalls stores had been stolen. A credit card company had alerted TJX to the situation, and when TJX examined its servers, it discovered that card data for well over half of the transactions conducted at its North American stores over the past 18 months had been stolen.[100] Eventually, it was discovered that thieves parked outside a retail store had managed to hack into TJX's network by penetrating the store's wireless system. They then installed a "sniffer" on the network that captured card data in real time as it traveled over TJX's network. TJX did not notice that thieves had downloaded 80 GBytes of data off its network.[101] TJX ultimately determined that data from 11.2 million cards had been stolen.[102] Between 2005 and 2007, other companies also reported extensive data breaches that suggested that an organized ring of bank card data thieves was at work.

During this same time, an undercover Secret Service agent in San Diego had been purchasing large amounts of stolen credit card data from a Ukranian man named Maksym

Yastremskiy. While meeting with Yastremskiy in Dubai, the agent secretly copied his hard drive, and the Secret Service discovered that Yastremskiy often chatted online with the person who appeared to be his main source of bank card data. However, the only information they could find on the hard drive was the person's Instant Messenger registration number.[103]

In the summer of 2007, Yastemskiy was arrested in Turkey and charged with selling stolen credit card information via online forums. Federal investigators learned that Yastemskiy had come into possession of a significant amount of credit card information stolen from TJX's servers.[104] Moreover, they discovered that Yastemskiy's source had asked Yastemskiy to furnish him with a fake passport for one of his cashers (a person who goes to the ATM to withdraw cash with fake bank cards), who had been arrested in United States. After several more weeks of digging, investigators found one Jonathan Williams, who had been arrested in North Carolina with $200,000 in cash and 80 blank debit cards. On a USB drive found in Williams's possession was a photo of Gonzalez and his sister's address. At about this time, Secret Service investigators received the Instant Messenger registration data on Yastemskiy's chat partner. There was only an email address: soupnazi@efnet.ru. As investigators knew by now, "soup nazi" was a favorite alias for Albert Gonzalez.[105]

Federal investigators ultimately discovered that while Gonzalez was earning $75,000 a year working for them as an undercover agent,[106] he had organized and run an international cybercrime operation that was responsible not only for the TJX heist, but for bank card data thefts from over 250 institutions, including Hannaford Brothers, Barnes & Noble, 7 Eleven, OfficeMax, Boston Market, and BJ's Wholesale Club.[107]

Cybercriminals and cyber identity thieves are notoriously hard to catch. They tend to hide their own identities fairly well. Additionally, state and federal governments have to rush to catch up with new and developing cybercriminal techniques and the resultant cybercrime waves that follow these developments. Although 44 states have now passed laws that require firms to report data breaches,[108] the Identity Theft Resource Center estimates that for nearly half of the breaches, the number of records potentially affected is not reported. In addition, for more than one-third of data breaches reported, the organizations do not specify what caused the breach.[109] Neither the government nor IT security companies can investigate crimes that are not reported. But companies have a significant incentive not to report data breaches, as they can be held financially liable. As part of a 2010 settlement related to a security breach, Countrywide Financial Corporation agreed to reimburse customers for up to $50,000 in losses for each time their identity was stolen; the data of approximately 17 million customers was exposed by the data breach. In addition, Countrywide had to pay $3.5 million in plaintiffs' legal fees.[110] There has been an increase in the number of data breaches reported in recent years. However, it is unclear whether this is due to the increasing number of breaches or compliance with the new notification laws.[111]

Although the federal government is expected to spend over $13 billion on cybersecurity annually by the year 2015, the government struggles to find professionals with the needed expertise. The FBI has recently approved two large contracts to outsource much of this work to two companies.[112] Facing a challenge in investigative expertise, the federal government has concentrated on prevention. The Identity Theft Red Flags Rule, passed in 2007 pursuant to the Fair and Accurate Credit Transactions Act of 2003, requires financial institutions to adopt identity theft prevention programs.[113] However, the Federal Trade Commission (FTC) has pushed back the implementation of this regulation five times as it tries to work with businesses and professional organizations to reach agreements on the specifics of the requirements. This most recent delay was announced after the American Medical Association filed a lawsuit demanding that physicians be excluded from the Red Flags Rule.[114] Thus, although the federal government has made great strides toward identity theft prevention and investigation, it is understandable why Secret Service agents might feel that they have a lot to gain by turning cybercriminals such as Albert Gonzalez over to their side of the law. However, federal agents will need to take the harsh lessons learned from the Gonzalez case with them as they investigate future cases.

Discussion Questions

1. How did the authorities catch Gonzalez the first time?
2. How did the authorities catch Gonzalez the second time?
3. What types of measures do you think should be taken so that federal, state, and local investigators can catch cybercriminals?

3. How Secure Is Our Healthcare Data?

The American Recovery and Reinvestment Act (ARRA) of 2009 funneled billions of dollars into the development of health information technology, promoting the use of electronic health record (EHR) systems. EHR systems are intended to reduce medical errors and make the healthcare system more efficient. At the time the Act was passed, only 17 percent of U.S. physicians and less than 10 percent of U.S. hospitals had even a basic EHR system.

ARRA was originally advanced as part of a wider effort to stimulate job creation. However, by encouraging the swift adoption of EHR systems, will the initiative improve or endanger patient privacy? Security breaches into medical records are already commonplace. Some breaches result in serious consequences:

- In December 2006, federal prosecutor Susan Harrison discovered that thousands of dollars had disappeared from her checking account. Two former employees of the Virginia Mason Medical Center in Seattle had snuck into the center at night using fake IDs. They accessed medical records and stole the personal information of 42 individuals, including Harrison's husband, and used it to withdraw money from their bank accounts.
- Susan Pugh White, a diabetes education nurse at Randolph Hospital in North Carolina, stole the personal information of 12 patients from their medical records and used that information to run up credit card bills in their names. She bought $60,000 worth of merchandise, including a large-screen television and a lawn mower. She was sentenced to nearly three years in prison.
- Between 2006 and 2007, Richard Yaw Adjei stole the identities of over 400 patients treated in Midwestern hospitals that used a Delaware-based billing company. Adjei had paid a claims processor at the company to pass him the names, birth dates, Social Security numbers, and medical information of these individuals. He then used this data to submit 163 tax returns and consequently obtained hundreds of thousands of dollars in tax rebates. Adjei was arrested and sentenced to six years and three months in jail.

As healthcare organizations rush to implement new EHR systems, will these types of breaches become more commonplace? Will new security threats emerge?

In 2008, the Healthcare Information and Management Systems Society (HIMSS) reported that between 2006 and 2007, over 1.5 million patient records were subject to security breaches in hospitals alone.[115] However, not all breaches are committed by individuals with malicious intent. Some commit security breaches out of curiosity—such as the 39 employees of New York City's hospital system who peeked at the records of a seven-year-old girl whose death had made her a tabloid sensation.[116] Others commit breaches out of carelessness. Television news investigators in Houston found computer printouts, prescription labels, and pill bottles with more than 200 patients' personal information in unlocked dumpsters outside 24 local pharmacies.[117]

But while an identity thief digging through a dumpster might recover personal information on 10, 20, or maybe 100 individuals, a thief that steals a laptop or an electronic storage device has access to thousands of medical records. In 2006 laptops containing medical records were stolen in Ohio, California, Pennsylvania, Michigan, New York, and Minnesota—most commonly from offices and vehicles of health provider employees. Sometimes these records were password protected and encrypted; sometimes they were not. These data breaches affect hundreds of thousands of people.[118] And although healthcare institutions often notify individuals when their EHRs have been compromised, it is the individuals who bear the responsibility for protecting their own financial assets and credit once their file has been breached. Title II of HIPAA, commonly called the Administrative Simplification Provisions, mandates the creation of standards to protect the confidentiality of electronic transactions within the healthcare system. The HIPAA Security Rule safeguards the confidentiality of electronic patient information. In implementing the Security Rule, the U.S. Department of Health and Human Services (HHS) regulated enforcement tasks to the Centers for Medicare and Medicaid Services (CMS). CMS has the authority to interpret the provisions, conduct compliance reviews, and impose monetary penalties on organizations that do not comply with HIPAA security regulations.

Although CMS was awarded these far-reaching authorities, for the first decade after HIPAA was enacted, the agency chose not to conduct compliance reviews. Instead, it relied entirely on individuals filing complaints against institutions as a means to ensure compliance. By January 31, 2009, CMS had received a total of 1,044 complaints, but not a single monetary penalty had been imposed.[119]

HHS's Office of Inspector General (OIG) launched an investigation into CMS enforcement of the HIPAA Security Rule. It began auditing hospitals in 2006 to determine whether CMS reliance on complaints was sufficient to promote compliance with the HIPAA privacy provisions. OIG first inspected Piedmont Hospital in Atlanta, Georgia, where it evaluated access to patient data, the monitoring of information system activities, and security violations. As a result of the OIG inspection, some healthcare institutions began to take a much closer look at how they could ensure compliance.

In 2007, a number of large healthcare corporations—such as CVS Caremark, Johnson & Johnson, and Philips Healthcare—and technology giant Cisco Systems established the Health Information Trust Alliance, an organization dedicated to developing a common security framework for electronic medical data. A HITRUST survey in 2008 found that 85 percent of health information technology executives supported such an effort.[120] Already, organizations such as HIMSS engaged in research and professional development to support EHR security. The common security framework went beyond this to provide shared standards and procedures for access control, human resource security, physical and environmental security, information security incident management, and many other areas directly or indirectly related to privacy breaches.

By 2008, OIG released its report, which found that CMS reliance on complaints was an ineffective means to ensure HIPAA compliance. Although CMS officially refuted OIG's conclusions, it agreed to begin compliance reviews. In January 2008, CMS conducted its first review at a hospital about which it had received a complaint regarding a lost laptop. As a result of the review, the hospital developed a plan that included the implementation of procedures requiring laptops to be secured to the workstation where they are used, the training of individuals working with mobile hardware and electronic storage devices, and 24-hour surveillance systems. The hospital was also required to make sure employees did not have greater access to EHRs than their jobs required. In addition, the hospital instituted measures that assured that the IT department would be notified when system users were terminated so that they could no longer access the EHRs.[121]

The question remains, however, as to whether these efforts can forestall a major proliferation in security breaches as the United States rushes to implement EHR technology in the wake of ARRA. In addition to providing monetary incentives for healthcare organizations to adopt electronic health records, ARRA expands the privacy provisions under HIPAA. Specifically, it extends the Security Rule to business associates of healthcare institutions. In the past, security breaches have been committed not only by these institutions but also by outside IT contractors. ARRA also requires healthcare institutions to notify individuals who are affected by security breaches, provides for increased fines for noncompliance, and authorizes state attorneys general to prosecute institutions violating HIPAA regulations.

The clear intent of these provisions is to increase enforcement of HIPAA, but what will be the actual outcome of these new policies? The development of procedures, standards, and technology is left to the private sector, which clearly recognizes that there are major challenges ahead and has begun to organize to meet these challenges. But will the private sector be able to produce secure solutions in time? Will suitable technology become accessible and affordable to smaller and rural healthcare institutions that are targeted by ARRA? Will CMS and newly authorized government bodies institute effective enforcement policies? These issues will determine in large part how secure health data will be in the coming years.

Discussion Questions

1. What type of security breaches of medical records are common today?
2. What measures are being taken by the government and private industry to safeguard EHRs?
3. How do you think the implementation of ARRA will affect the privacy of our healthcare and personal data? What breaches do you foresee? How can they be forestalled?

End Notes

[1] Justice Louis Brandeis, *Olmstead v. United States*, June 1928, http://scholar.google.com/scholar_case?case=5577544660194763070&hl=en&as_sdt=2&as_vis=1&oi=scholarr.

[2] Jeremy Kirk, "Germany Launches Criminal Investigation of Google," *PCWorld*, May 20, 2010, www.pcworld.com/article/196765/germany_launches_criminal_investigation_of_google.html.

[3] Andrew Orlowski, "Google Street View Logs WiFi Networks, Mac Addresses," *The Register*, April 22, 2010, www.theregister.co.uk/2010/04/22/google_streetview_logs_wlans.

[4] Google "Data Collected by Google Cars," *European Public Policy Blog*, April 27, 2010, http://googlepolicyeurope.blogspot.com/2010/04/data-collected-by-google-cars.html.

[5] Andrew Orlowski, "Google Street View Logs WiFi Networks, Mac Addresses," *The Register*, April 22, 2010, www.theregister.co.uk/2010/04/22/google_streetview_logs_wlans/.

[6] Google, "WiFi Data Collection: An Update," *The Official Google Blog*, May 14, 2010, http://googleblog.blogspot.com/2010/05/wifi-data-collection-update.html.

[7] Kevin O'Brien, "In Europe, Google Faces New Inquiries on Privacy," *New York Times*, May 20, 2010, www.nytimes.com/2010/05/21/technology/21streetview.html.

[8] Robert McMillan, "Google WiFi Uproar: Six Class Action Lawsuits Filed," *TechWorld*, June 4, 2010, http://news.techworld.com/networking/3225722/google-wifi-uproar-six-class-action-lawsuits-filed.

[9] David Kravets, "Packet-Sniffing Laws Murky as Open Wi-Fi Proliferates," *Wired*, June 22, 2010, www.wired.com/threatlevel/2010/06/packet-sniffing-laws-murky/.

[10] Tom Krazit, "Deciphering Google's Wi-Fi Headache (FAQ)," *CNET*, June 1, 2010, http://news.cnet.com/8301-30684_3-20006342-265.html.

[11] Google, "WiFi Data Collection: An Update," *The Official Google Blog*, May 14, 2010, http://googleblog.blogspot.com/2010/05/wifi-data-collection-update.html.

[12] Gregg Keizer, "Google Wants to Patent Technology used to 'Snoop' Wi-Fi Networks," *Computerworld*, June 3, 2010, www.computerworld.com/s/article/9177634/Google_wants_to_patent_technology_used_to_snoop_Wi_Fi_networks.

[13] Matt McGee, "FTC Ends Google WiFi Inquiry, No Penalties Announced," *Search Engine Land* (blog), October 27, 2010, http://searchengineland.com/ftc-ends-google-wifi-inquiry-no-penalties-54058.

[14] Tom Krazit, "Connecticut Heads Up 30-State Google Wi-Fi Probe," *CNET*, June 21, 2010, http://news.cnet.com/8301-30684_3-20008332-265.html.

[15] Alan Eustace, "Creating Stronger Privacy Controls Inside Google," *Google Public Policy Blog*, October 22, 2010, http://googlepublicpolicy.blogspot.com/2010/10/creating-stronger-privacy-controls.html.

[16] Chloe Albanesius, "FCC Investigating Google Street View Wi-Fi Data Collection," *PCMag.com*, November 10, 2010, www.pcmag.com/article2/0,2817,2372498,00.asp.

[17] Tom Krazit, "Deciphering Google's Wi-Fi Headache (FAQ)," *CNET*, June 1, 2010, http://news.cnet.com/8301-30684_3-20006342-265.html.

[18] Tomasz Zukowski and Irwin Brow, "Examining the Influence of Demographic Factors on Internet Users Information Privacy Concerns," ACM International Conference Proceedings Series, vol. 226, 2007 (accessed February 4, 2011).

[19] Privacy Protection Study Commission, "Personal Privacy in an Information Society: The Report of the Privacy Protection Study Commission," July 12, 1997, http://aspe.hhs.gov/datacncl/1977privacy/toc.htm.

[20] *Olmstead v. United States*, 277 U.S. 438 (1928), www.law.cornell.edu/supct/html/historics/USSC_CR_0277_0438_ZS.html (accessed February 5, 2011).

[21] Roger Clarke, "Introduction to Dataveillance and Information Privacy and Definition of Terms," August 15, 1997, www.rogerclarke.com/DV/Intro.html#Priv.

[22] Fair Credit Reporting Act, 15 U.S.C. § 1681, www.accessreports.com/statutes/FCRA.htm (accessed February 5, 2011).

[23] Federal Trade Commission, "Press Release: Provisions of New Fair and Accurate Credit Transaction Act Will Help Reduce Identity Theft and Help Victims Recover: FTC," June 15, 2004, www.ftc.gov/opa/2004/06/factaidt.shtm.

[24] The Right to Financial Privacy Act, 12 U.S.C. §§ 3401-342, http://epic.org/privacy/rfpa (accessed February 4, 2011).

[25] David Leonhardt, "Washington's Invisible Hand," *New York Times*, September 28, 2008.

[26] George V. Hulme, "Protecting Privacy," *InformationWeek*, April 16, 2001.

[27] U.S. Department of Health & Human Services, "Rite Aid Agrees to Pay $1 Million to Settle HIPAA Privacy Case," June 7, 2010, www.hhs.gov/ocr/privacy/hipaa/enforcement/examples/riteaidresagr.html.

[28] American Investment and Recovery Act of 2009, Title XIIII Health Information Technology, www.justlawlinks.com/ACTS/arara/araraA_title-XIII.php#subD (accessed February 6, 2011).

[29] Lucas Mearian, "Health Net Says 1.5M Medical Records Lost in Data Breach," *Computerworld*, November 19, 2009, www.computerworld.com/s/article/9141172/Health_Net_says_1.5M_medical_records_lost_in_data_breach.

[30] Jaikumar Vijayan, "Hospital Appeals $250,000 Fine for Late Breach Disclosure," *Computerworld*, September 10, 2010.

[31] Tamar Lewin, "If Your Kids Are Awake, They're Probably Online," *New York Times*, January 20, 2010, www.nytimes.com/2010/01/20/education/20wired.html.

[32] PogoWasRight.org, "Echometrix Agrees to Stop Collecting and Selling Kids' Chats," September 16, 2010, www.pogowasright.org/?tag=echometrix.

[33] Federal Trade Commission, "Frequently Asked Questions about the Children's Online Privacy Protection Rule," www.ftc.gov/privacy/coppafaqs.shtm (accessed January 26, 2011).

[34] William R. Van Dusen Jr., J.D., "FERPA: Basic Guidelines for Faculty and Staff—A Simple Step by Step Approach to Compliance," www.nacada.ksu.edu/Resources/FERPA-Overview.htm (accessed January 26, 2011).

[35] University of Washington, "FERPA for Faculty and Staff," www.washington.edu/students/reg/ferpafac.html (accessed February 26, 2011).

[36] Melissa B. Taboada, "Teacher Resigns After Posting Blog with Personal Student Info," *Homeroom, Statesman.com* (blog), September 27, 2010, www.statesman.com/blogs/content/shared-gen/blogs/austin/education/entries/2010/09/24/teacher_resigns_after_posting.html.

[37] U.S. Department of Justice, "Privacy & Civil Liberties:Title III of the Omnibus of the Crime Control and Safe Streets Act of 1968 (Wiretap Act)" www.it.ojp.gov/default.aspx?area=privacy&page=1284 (accessed February 6, 2011).

[38] *Katz v. United States*, 389 U.S. 347 (1967), http://supreme.justia.com/us/389/347/case.html (accessed February 6, 2011).

[39] Ryan Singel, "Police Wiretapping Jumps 26 Percent," *Wired*, April 30, 2010, www.wired.com/threatlevel/2010/04/wiretapping.

[40] U.S. Code, Chapter 36—Foreign Intelligence Surveillance, www.law.cornell.edu/uscode/html/uscode50/usc_sup_01_50_10_36.html (accessed February 6, 2011).

[41] Electronic Privacy Information Center, "Foreign Intelligence Surveillance Act Court Orders 1979–2009," http://epic.org/privacy/wiretap/stats/fisa_stats.html (accessed February 6, 2009).

[42] 18 U.S.C. § 2510, www.justice.gov/criminal/cybercrime/wiretap2510_2522.htm (accessed February 6, 2011).

[43] Communications Assistance for Law Enforcement Act, H.R. 4922, http://epic.org/privacy/wiretap/calea/calea_law.html (accessed February 6, 2011).

[44] "National Security Letters, American Civil Liberties Union, January 10, 2011, www.aclu.org/national-security-technology-and-liberty/national-security-letters.

[45] Electronic Privacy Information Center, "Foreign Intelligence Surveillance Act Court Orders 1979–2009," http://epic.org/privacy/wiretap/stats/fisa_stats.html (accessed February 6, 2009).

[46] Export.gov, "Welcome to the U.S.-EU & U.S.-Swiss Safe Harbor Frameworks," www.export.gov/safeharbor (accessed February 6, 2011).

[47] Andreas Udo de Haes, "Europeans Concerned Over Ongoing Privacy Fraud in U.S.," *Computerworld*, December 1, 2010, www.computerworld.com/s/article/9199019/Europeans_concerned_over_ongoing_privacy_fraud_in_U.S.

[48] BBB Online Web site, www.bbb.org/online (accessed February 6, 2011).

[49] Sharon Theimer, "Obama's Broken Promise: Federal Agencies Not More Transparent Under Obama Administration," *Huffington Post*, March 16, 2010.

[50] The Reporters Committee for Freedom of the Press, "Federal Open Government Guide,10th edition," www.rcfp.org/fogg/index.php?i=pt2#c (accessed February 7, 2011).

[51] The Reporters Committee for Freedom of the Press, "Federal Open Government Guide 10th edition," www.rcfp.org/fogg/index.php?i=pt2#c (accessed February 7, 2011).

[52] The Reporters Committee for Freedom of the Press, "Federal Open Government Guide 10th edition," www.rcfp.org/fogg/index.php?i=pt2#c (accessed February 7, 2011).

[53] Jaikumar Vijayan, "EPIC Files FOIA Request Over Reported Google, NSA Partnership," *Computerworld*, www.computerworld.com/s/article/9152438/EPIC_files_FOIA_request_over_reported_Google_NSA_partnership, (accessed February 4, 2011).

[54] Tom Burghardt, "Obama's National Cybersecurity Initiative: Privacy and Civil Liberties are Dammed," Global Research, March 8, 2010, www.globalresearch.ca/index.php?context=va&aid=17993.

[55] Debbie Turner, "Identity Theft: Statistics On Why We should Be Worried," *Online Social Media*, January 30, 2011, www.onlinesocialmedia.net/20110130/identity-theft-statistics-on-why-we-should-be-worried.

56 Thomas Claburn, "The Cost of Data Loss Rises, *InformationWeek*, November 28, 2007.

57 "Hannaford Bros. Faces Class Action Over Data Breach," *ConsumerAffairs.com*, March 21, 2008, www.consumeraffairs.com/news04/2008/03/hannaford_data2.html.

58 "HIPAA Compliance Strategies," *AISHealth*, February 25, 2008, www.aishealth.com.

59 Matthew Schwartz, "ID Theft Monitoring Services: What You Need to Know," *Information Week*, May 9, 2008, www.informationweek.com/news/security/privacy/showArticle.jhtml?articleID=207501091.

60 Jacob Leibenluft, "Credit Card Numbers for Sale," *Slate*, April 24, 2008, www.slate.com/id/2189902.

61 Gregg Keizer, "FBI Planted Spyware on Teen's PC to Trace Bomb Threats," *Computerworld*, June 19, 2007, www.computerworld.com/s/article/9027418/FBI_planted_spyware_on_teen_s_PC_to_trace_bomb_threats?taxonomyId=84&pageNumber=1.

62 Steven Levy and Brad Stone, "Grand Theft Identity," *Newsweek*, July 4, 2005.

63 Barbara J. Rothstein, Ronald J. Hedges, and Elizabeth C. Wiggins, *Managing Discovery of Electronic Information: A Pocket Guide for Judges*, Federal Judicial Center 2007, www.fjc.gov/public/pdf.nsf/lookup/eldscpkt.pdf/$file/eldscpkt.pdf (accessed February 5, 2011).

64 Frank Hayes, "One Less Patent Threat to Worry About," *Frankly Blogging, Computerworld* (blog), March 23, 2007, http://blogs.computerworld.com/node/5240.

65 Jeremy Kirk, "Broadcom to Pursue Injunction Against Qualcomm," *Computerworld*, November 26, 2007, www.computerworld.com/s/article/9048818/Broadcom_to_pursue_injunction_against_Qualcomm.

66 Dean Gonsowski, "What You Can Learn from Qualcomm v. Broadcom," *e-discovery 2.0* (blog), April 20, 2010, www.clearwellsystems.com/e-discovery-blog/2010/04/20/what-you-can-learn-from-qualcomm-v-broadcom.

67 "Roundtable Discussion: Changing Ethical Expectations," Navigating the Changing Ethical and Practical Expectations for E-Discovery presented at the Northern Kentucky University Chase College of Law Northern Kentucky Law Review Spring Symposium, February 28, 2009.

68 Julia Angwin, "The Web's New Gold Mine: Your Secrets," *Wall Street Journal*, July 30, 2010, http://online.wsj.com/article/SB10001424052748703940904575395073512989404.html.

69 Julia Angwin, "The Web's New Gold Mine: Your Secrets," *Wall Street Journal*, July 30, 2010, http://online.wsj.com/article/SB10001424052748703940904575395073512989404.html.

70 AFP, "US Regulator Wants 'Do Not Track' Button on Internet," *Google news*, December 1, 2010, www.google.com/hostednews/afp/article/ALeqM5gsDxgwFnQov7zDxSuOPk8rl0Ayig?docId=CNG.fa0914aaf88efbfc94d9b1b4fc7fecae.541.

71 AFP, "US Regulator Wants 'Do Not Track' Button on Internet," *Google news*, December 1, 2010, www.google.com/hostednews/afp/article/ALeqM5gsDxgwFnQov7zDxSuOPk8rl0Ayig?docId=CNG.fa0914aaf88efbfc94d9b1b4fc7fecae.541.

72 The Self-Regulatory Program for Online Behavioral Advertising "Self-Regulatory Principles Overview," www.aboutads.info/principles, (accessed February 11, 2011).

73 Austin Carr, "7 Creepy Faux Pas of Google CEO Eric Schmidt," *Fast Company*, October 6, 2010, www.fastcompany.com/1693384/google-ceo-eric-schmidt-gaffes-creepy-privacy-faux-pas#.

74 Mike Melanson, "Personal Info Aggregator Rapleaf Hires Chief Privacy Officer," *ReadWriteWeb*, November 29, 2010, www.readwriteweb.com/archives/personal_info_aggregator_rapleaf_hires_chief_priva.php.

75 Ashley Dively, "How Much Time Do Your Employees Waste on the Internet," Sozo Firm Inc., October 21, 2010, www.andrewjensen.net/do-your-employees-waste-time-on-the-internet.

76 24/7 Wall St., "The Top Ten Ways Workers Waste Time Online," September 30, 2010, http://247wallst.com/2010/09/30/the-top-ten-ways-workers-waste-tme-online/.

77 Kieran Long, "So Can the Secret Ring of Steel Save the City from Terrorism?," *London Evening Standard*, October 15, 2010, www.thisislondon.co.uk/lifestyle/article-23888163-so-can-the-secret-ring-of-steel-save-the-city-from-terrorism.do.

78 "Dramatic Expansion of DC Surveillance Camera Network," *Homeland Security Newswire*, February 1, 2011, http://homelandsecuritynewswire.com/dramatic-expansion-dc-surveillance-camera-network.

79 ACLU of Illinois, "Press Release: ACLU Report Details 'Unregulated' Nature of Chicago's Growing Surveillance Camera System, Calls for Moratorium on New Cameras," PRNews-Wire, February 8, 2011, www.prnewswire.com/news-releases/aclu-report-details-unregulated-nature-of-chicagos-growing-surveillance-camera-system-calls-for-moratorium-on-new-cameras-115537319.html.

80 Colleen Long, "Some New York Subway Cameras Now Transmitting in Real Time," *Huffington Post*, September 20, 2010, www.huffingtonpost.com/2010/09/20/new-york-subway-cameras-real-time-transmission_n_732242.html.

81 Jaikumar Vijayan, "Appeals Court OKs Warrantless GPS Tracking by Feds," *Computerworld*, August 27, 2010, www.computerworld.com/s/article/9182499/Appeals_court_OKs_warrantless_GPS_tracking_by_feds.

82 Anne Broache and Greg Sandoval, "Viacom Sues Google over YouTube Clips," *CNET*, March 13, 2007, http://news.cnet.com/Viacom-sues-Google-over-YouTube-clips/2100-1030_3-6166668.html.

83 Richard Keyt, "Web Site Owners: How to Protect Yourself from Claims of Copyright Infringement for Users' Conduct," Keytlaw, www.keytlaw.com/Copyrights/dmca.htm (accessed February 6, 2011).

84 Associated Press, "Google Buys YouTube for $1.65 Billion," *MSNBC*, October 10, 2006, www.msnbc.msn.com/id/15196982/ns/business-us_business.

85 Viacom, "Viacom Takedown Statement," February 2, 2007, http://news.viacom.com/news/Pages/youtubetakedownstatement.aspx.

[86] Candace Lombardi, "Viacom to YouTube: Take Down Pirated Clips," *CNET*, February 2, 2007, http://news.cnet.com/Viacom-to-YouTube-Take-down-pirated-clips/2100-1026_3-6155771.html.

[87] "*Viacom International, et al., v. YouTube, YouTube LLC, and Google Inc.*," United States District Court, Southern District of New York, Case 1:07-cv-02103-LLS, Document 117, July 2, 2008, http://beckermanlegal.com/Documents/viacom_youtube_080702DecisionDiscoveryRulings.pdf.

[88] Catherine Holahan, "Viacom vs. YouTube: Beyond Privacy," *Bloomberg Businessweek*, July 3, 2008, www.businessweek.com/technology/content/jul2008/tc2008073_435740.htm.

[89] Viacom, "Viacom Statement on Confidentiality of YouTube Data," July 7, 2008, http://news.viacom.com/news/Pages/youtubeconfidentiality.aspx.

[90] Sam Gustin, "Viacom and Google Trade Barbs As $1 Billion Lawsuit Heats Up," *Daily Finance*, March 18, 2010, www.dailyfinance.com/story/viacom-and-google-trade-barbs-as-1-billion-lawsuit-heats-up/19406030.

[91] Viacom, "Memorandum of Law in Support of Viacom's Motion for Partial Summary Judgment on Liability and Inapplicability of the Digital Millennium Copyright Act Safe Harbor Defense," www.viacom.com/news/Viacom%20Summary%20Judgment%20Motion/Viacom%20Summary%20Judgment%20Motion.pdf (accessed February 7, 2011).

[92] Viacom, "Memorandum of Law in Support of Viacom's Motion for Partial Summary Judgment on Liability and Inapplicability of the Digital Millennium Copyright Act Safe Harbor Defense," www.viacom.com/news/Viacom%20Summary%20Judgment%20Motion/Viacom%20Summary%20Judgment%20Motion.pdf (accessed February 7, 2011).

[93] Opinion and Order, *Viacom International, et al., v. YouTube, YouTube LLC, and Google Inc.*, United States District Court Southern District of New York, June 23, 2010, http://i.zdnet.com/blogs/viacom-youtube-ruling.pdf.

[94] Sam Diaz, "Google Prevails in Viacom-YouTube Copyright Lawsuit; Appeals on Deck," *ZDNet*, June 23, 2010, www.zdnet.com/blog/btl/google-prevails-in-viacom-youtube-copyright-lawsuit-appeals-on-deck/36229.

[95] Viacom, "Viacom Files Brief with U.S. Court of Appeals," Viacom Statements, December 12, 2010, .http://news.viacom.com/news/Pages/ytstatement.aspx.

[96] James Verini, "The Great Cyberheist," *New York Times*, November 10, 2010, www.nytimes.com/2010/11/14/magazine/14Hacker-t.html?pagewanted=all.

[97] Kim Zetter, "TJX Hacker Gets 20 Years in Prison," *Wired*, March 25, 2010, www.wired.com/threatlevel/2010/03/tjx-sentencing.

[98] James Verini, "The Great Cyberheist," *New York Times*, November 10, 2010, www.nytimes.com/2010/11/14/magazine/14Hacker-t.html?pagewanted=all.

[99] The TJX Companies, "About Us," Web site, www.tjx.com/about.asp (accessed February 24, 2011).

[100] James Verini, "The Great Cyberheist," *New York Times*, November 10, 2010, www.nytimes.com/2010/11/14/magazine/14Hacker-t.html?pagewanted=all.

101 Kim Zetter, "TJX Failed to Notice Thieves Moving 80-GBytes of Data on Its Network," *Wired*, October 26, 2007, www.wired.com/threatlevel/2007/10/tjx-failed-to-n.

102 Kim Zetter, "TJX Hacker Gets 20 Years in Prison," *Wired*, March 25, 2010, www.wired.com/threatlevel/2010/03/tjx-sentencing.

103 James Verini, "The Great Cyberheist," *New York Times*, November 10, 2010, www.nytimes.com/2010/11/14/magazine/14Hacker-t.html?pagewanted=all.

104 Don Gooding, "Ukrainian Jet Setter in World's Largest Cyber Heist?" *The Register*, August 22, 2007, www.theregister.co.uk/2007/08/22/possible_break_in_tjx_investigation.

105 James Verini, "The Great Cyberheist," *New York Times*, November 10, 2010, www.nytimes.com/2010/11/14/magazine/14Hacker-t.html?pagewanted=all.

106 Kim Zetter, "TJX Hacker Gets 20 Years in Prison," *Wired*, March 25, 2010, www.wired.com/threatlevel/2010/03/tjx-sentencing.

107 "Hacker Pleads Guilty in Credit Card Fraud Case," *RedOrbit*, December 30, 2009, www.redorbit.com/news/technology/1803786/hacker_pleads_guilty_in_credit_card_fraud_case/.

108 Kristin M. Finklea, "Identity Theft: Trends and Issues," Congressional Research Service, January 5, 2010, www.fas.org/sgp/crs/misc/R40599.pdf.

109 Mathew J. Schwartz, "ID Thefts Go Unreported Despite Notification Laws," *InformationWeek*, July 9, 2010, www.informationweek.com/news/security/vulnerabilities/showArticle.jhtml?articleID=225702822&queryText=ID%20theft.

110 Brett Barrouquere, "Judge approves Countrywide ID Theft Settlement," *Bloomberg Businessweek*, August 23, 2010, www.businessweek.com/ap/financialnews/archives/august2010/D9HPBMDG1.htm.

111 Kristin M. Finklea, "Identity Theft: Trends and Issues," Congressional Research Service, January 5, 2010, www.fas.org/sgp/crs/misc/R40599.pdf.

112 Elizabeth Montalbano, "Federal Cybersecurity Spending to Hit $13.3B By 2015, *InformationWeek*, December 1, 2010, www.informationweek.com/news/government/security/showArticle.jhtml?articleID=228500061&queryText=ID%20theft.

113 Kristin M. Finklea, "Identity Theft: Trends and Issues," Congressional Research Service, January 5, 2010, www.fas.org/sgp/crs/misc/R40599.pdf.

114 Jaikumar Vijayan, FTC Pushes Back Identity Theft Rules Deadline—For Fifth Time," *Computerworld*, June 1, 2010, www.computerworld.com/s/article/9177572/FTC_pushes_back_identity_theft_rules_deadline_for_fifth_time.

115 "2008 HIMSS Analytics Report: Security of Patient Data," April 2008, www.mmc.com/knowledgecenter/Kroll_HIMSS_Study_April2008.pdf.

116 Milt Freudenheim and Robert Pear, "Health Hazard: Companies Spilling Your History," *New York Times*, December 3, 2006.

117 "Pharmacies Dump Medical Information in the Trash," KPRC Local 2 (Houston), November 15, 2006.

118 Health Privacy Stories, Health Privacy Project, www.healthprivacy.org/use_doc/Privacystories.pdf.

[119] "OCR Receives 700 HIPAA Complaints in January 2009," Health Information Privacy/ Security Alerts: HIPAA Enforcement Statistics for March 2009, www.melamedia.com/ HIPAA.stats.home.html.

[120] George Holme, "The Security and Privacy of Healthcare Data, *InformationWeek*, August 20, 2008.

[121] HIPAA Compliance Review and Analysis and Summary of Results Reviews 2008, www.hhs.gov/ocr/privacy/hipaa/enforcement/cmscompliancerev08.pdf (accessed March 10, 2010).

CHAPTER 5

FREEDOM OF EXPRESSION

QUOTE

If the freedom of speech is taken away then dumb and silent we may be led, like sheep to the slaughter.

—George Washington

VIGNETTE

WikiLeaks

In October 2010, WikiLeaks (a nonprofit organization dedicated to "bringing important news and information to the public"[1]) released approximately 400,000 top-secret U.S. Army documents. Media sources claimed that the leak was the largest in U.S. history. These documents purportedly uncovered instances in which American troops ignored evidence that the Iraqi Shiite–dominated security forces were torturing Sunni prisoners. In addition, the documents disclosed 15,000 civilian deaths during the Iraq War that had allegedly gone unreported.[2]

This "document dump" was in fact the third major leak of U.S. military secrets by WikiLeaks in 2010. In April, the organization had posted a video of a U.S. army helicopter carrying out an operation in which civilians and two Reuters reporters were killed in Iraq. Then in July, WikiLeaks posted 92,000 military memos that suggested, among other things, that representatives of Pakistan's intelligence agency regularly met with Taliban fighters.[3]

The U.S. government, meanwhile, had been feverishly looking into whether it could prosecute WikiLeaks and prevent future leaks. The First Amendment guarantees citizens freedom of the press, and very few restrictions on this right have been permitted by the U.S. Supreme Court. One of the most notable examples of the Supreme Court affirming the right to freedom of the press was its refusal in 1971 to grant an injunction against the publication of the Pentagon Papers, which contained military secrets from the Vietnam War.[4] In 2010, the Congressional Research Service issued a report in which it concluded that because of the First Amendment's implications, no publisher of leaked information has ever been prosecuted for publishing the material. In July 2010, the U.S. government did charge Private First Class Bradley Manning with violating the Espionage Act for purportedly supplying WikiLeaks with the video of the helicopter and other classified materials. In order to prosecute WikiLeaks itself, however, the U.S. government would have to show that WikiLeaks convinced or pressured informants to release information.[5]

Shortly after the third leak in October 2010, several major Internet companies began to shut off services to WikiLeaks. These included PayPal and Moneybookers, two sites that supporters had used to contribute funds to the organization.[6] After massive denial-of-service attacks on the WikiLeaks site in November, the organization moved the site to Amazon servers.[7] Within a couple of days, however, Amazon refused to continue to host the site.[8] The following day, December 3, 2010, the American domain name system provider EveryDNS.net took the WikiLeaks domain offline. Supporters of WikiLeaks responded right away, and within two days, there were 208 WikiLeaks mirror sites up and operating. (Mirror sites provide an exact copy of another site's content.) [9] In addition, a loosely organized group of hackers calling themselves "Anonymous" counterattacked, launching denial-of-service attacks on many of the companies that had pulled the plug on WikiLeaks.[10] Then, on December 7, Julian Assange, the editor-in-chief of WikiLeaks, was arrested in London on rape charges issued from Sweden.[11]

As this online and offline battle raged, the public and the media expressed a range of views. Many argued that WikiLeaks had endangered national security; others staunchly defended WikiLeaks and its freedom to publish leaked information. Following the release of the July documents, however, WikiLeaks came under criticism from human rights organization and the international free press group Reporters Without Borders. Thousands of documents leaked by WikiLeaks contained the names of Afghan civilians who served as informants against the Taliban. With their identities now exposed, there was concern that these informants could be targeted by the Taliban in reprisal.

While many believe that WikiLeaks should have vetted the Afghanistan documents more fully to avoid releasing the names of informers, Assange insists that there is no evidence that any informer has been harmed or murdered.[12] His supporters argue that WikiLeaks has been a strong force for increased openness at a time when government and the traditional news media reporting are mired in ambiguity. They believe that to charge Assange with treason for publishing documents provided by whistle-blowers would be a dreadful precedent contrary to government transparency.[13]

Julian Assange, WikiLeaks' very public leader, has stated that his objective is to establish a new standard of scientific journalism. He publishes WikiLeaks' analysis of source material along with the source material itself, so that the readers themselves can come to their own conclusions.[14] At the top of its home page, WikiLeaks posts this quote from *Time* magazine: "Could become as important a journalistic tool as the Freedom of Information Act."

Clearly, WikiLeaks would like to portray itself as an advocate for the civil liberties of individuals. However, the political ramifications of its leaks are unclear, and it is not yet known whether the leaks will create serious concerns for national security.

Questions to Consider

1. How does the First Amendment protect WikiLeaks from prosecution?
2. What limits, if any, should be placed on WikiLeaks' right to post government or corporate secrets?
3. Could WikiLeaks set a journalistic precedent? What might happen if other groups began to post leaked information?

LEARNING OBJECTIVES

As you read this chapter, consider the following questions:

1. What is the basis for the protection of freedom of expression in the United States, and what types of expression are not protected under the law?
2. What are some key federal laws that affect online freedom of expression, and how do they impact organizations?
3. What important freedom of expression issues relate to the use of information technology?

FIRST AMENDMENT RIGHTS

The Internet enables a worldwide exchange of news, ideas, opinions, rumors, and information. Its broad accessibility, open discussions, and anonymity make the Internet a remarkable communications medium. It provides an easy and inexpensive way for a speaker to send a message to a large audience—potentially thousands or millions of people worldwide. In addition, given the right email addresses, a speaker can aim a message with laser accuracy at a select subset of powerful and influential people.

People must often make ethical decisions about how to use such incredible freedom and power. Organizations and governments have attempted to establish policies and laws to help guide people, as well as to protect their own interests. Businesses, in particular, have sought to conserve corporate network capacity, avoid legal liability, and improve worker productivity by limiting the nonbusiness use of IT resources.

The right to freedom of expression is one of the most important rights for free people everywhere. The First Amendment to the U.S. Constitution (shown in Figure 5-1) was adopted to guarantee this right and others. Over the years, a number of federal, state, and local laws have been found unconstitutional because they violated one of the tenets of this amendment.

FIGURE 5-1 The U.S. Constitution
Credit: Image copyright Kasia, 2009. Used under license from Shutterstock.com.

The First Amendment reads as follows:

> Congress shall make no law respecting an establishment of religion, or prohibiting the free exercise thereof; or abridging the freedom of speech, or of the press; or the right of the people peaceably to assemble, and to petition the government for a redress of grievances.

In other words, the First Amendment protects Americans' rights to freedom of religion and freedom of expression. This amendment has been interpreted by the Supreme Court as applying to the entire federal government, even though it only expressly refers to Congress.

Numerous court decisions have broadened the definition of speech to include nonverbal, visual, and symbolic forms of expression, such as flag burning, dance movements, and hand gestures. Sometimes the speech at issue is unpopular or highly offensive to a majority of people; however, the Bill of Rights provides protection for minority views. The Supreme Court has also ruled that the First Amendment protects the right to speak anonymously as part of the guarantee of free speech.

However, the Supreme Court has held that the following types of speech are not protected by the First Amendment and may be forbidden by the government: perjury, fraud, defamation, obscene speech, incitement of panic, incitement to crime, "fighting words," and sedition (incitement of discontent or rebellion against a government). Two of these types of speech—obscene speech and defamation—are particularly relevant to information technology.

Obscene Speech

Miller v. California is the 1973 Supreme Court case that established a test to determine if material is obscene and therefore not protected by the First Amendment. After conducting a mass mailing campaign to advertise the sale of adult material, Marvin Miller was convicted of violating a California statute prohibiting the distribution of obscene material.

Some unwilling recipients of Miller's brochures complained to the police, initiating the legal proceedings. Although the brochures contained some descriptive printed material, they primarily consisted of pictures and drawings explicitly depicting men and women engaged in sexual activity. In ruling against Miller, the Supreme Court determined that speech can be considered obscene and not protected under the First Amendment based on the following three questions:

- Would the average person, applying contemporary community standards, find that the work, taken as a whole, appeals to the prurient interest?
- Does the work depict or describe, in a patently offensive way, sexual conduct specifically defined by the applicable state law?
- Does the work, taken as a whole, lack serious literary, artistic, political, or scientific value?

These three tests have become the U.S. standard for determining whether something is obscene. The requirement that a work be assessed by its impact on an average adult in a community has raised many questions:

- Who is an average adult?
- What are contemporary community standards?
- What is a community? (This question is particularly relevant in cases in which potentially obscene material is displayed worldwide via the Internet.)

Defamation

The right to freedom of expression is restricted when the expressions, whether spoken or written, are untrue and cause harm to another person. Making either an oral or a written statement of alleged fact that is false and that harms another person is **defamation**. The harm is often of a financial nature, in that it reduces a person's ability to earn a living, work in a profession, or run for an elected office. An oral defamatory statement is **slander**, and a written defamatory statement is **libel**. Because defamation is defined as an untrue statement of fact, truth is an absolute defense against a charge of defamation. Although people have the right to express opinions, they must exercise care in their online communications to avoid possible charges of defamation. Organizations must also be on their guard and prepared to take action in the event of libelous attacks against them.

FREEDOM OF EXPRESSION: KEY ISSUES

Information technology has provided amazing new ways for people to communicate with others around the world, but with these new methods come new responsibilities and new ethical dilemmas. This section discusses a number of key issues related to freedom of expression, including controlling access to information on the Internet, anonymity on the Internet, defamation and hate speech, corporate blogging, and pornography.

Controlling Access to Information on the Internet

Although there are clear and convincing arguments to support freedom of speech online, the issue is complicated by the ease with which children can access the Internet. Even some advocates of free speech acknowledge the need to restrict children's Internet access, but it is difficult to restrict their access without also restricting adults' access. In attempts to address this issue, the U.S. government has passed laws and software manufacturers have invented special software to block access to objectionable material. The following sections summarize these approaches.

Communications Decency Act (CDA) (1996)

The Telecommunications Act became law in 1996. Its purpose was to allow freer competition among phone, cable, and TV companies. Embedded in the Telecommunications Act was the **Communications Decency Act (CDA)**, aimed at protecting children from pornography. The CDA imposed $250,000 fines and prison terms of up to two years for the transmission of "indecent" material over the Internet.

In February 1996, the American Civil Liberties Union (ACLU) and 18 other organizations filed a lawsuit challenging the criminalization of so-called indecency on the Web under the CDA. After a three-judge federal panel ruled unanimously that the law unconstitutionally restricted free speech, the government appealed to the Supreme Court (see Figure 5-2) in a case that became known as *Reno v. ACLU*. Examples of indecency identified as potentially criminal by government witnesses included Web postings of a photo of actress Demi Moore, naked and pregnant on the cover of *Vanity Fair*, and any online use of the infamous "seven dirty words." In addition to the ACLU, the plaintiffs of the original suit included Planned Parenthood, Stop Prisoner Rape, Human Rights Watch, and the Critical Path AIDS Project. Many of these organizations feared that much of their online material could be classified as indecent, because examples cited by government witnesses included speech about abortion, prisoner rape, safe-sex practices, and other sexually related topics. The plaintiffs argued that much of this information was important to both minors and adults.

FIGURE 5-2 U.S. Supreme Court building
Credit: Image copyright Johnny Kuo, 2009. Used under license from Shutterstock.com.

The problem with the CDA was its broad language and vague definition of *indecency*, a standard that was left to individual communities to determine. In June 1997, the Supreme Court ruled the law unconstitutional and declared that the Internet must be afforded the highest protection available under the First Amendment.[15] The Supreme Court said in its ruling that "the interest in encouraging freedom of expression in a democratic society outweighs any theoretical but unproven benefit of censorship." The ruling also said that "the growth of the Internet has been and continues to be phenomenal. As a matter of constitutional tradition, and in the absence of evidence to the contrary, we presume government regulation of the content of speech is more likely to interfere with the free exchange of ideas than to encourage it."[16] The ruling applied essentially the same free-speech protections to communication over the Internet as exist for print communication.

If the CDA had been judged constitutional, it would have opened all aspects of online content to legal scrutiny. Many current Web sites would probably either not exist or would look much different today had the law not been overturned. Web sites that might have been deemed indecent under the CDA would be operating under an extreme risk of liability.

Child Online Protection Act (COPA) (1998)

In October 1998, the **Child Online Protection Act (COPA)** was signed into law. (This act is not to be confused with the Children's Online Privacy Protection Act [COPPA], discussed in Chapter 4. COPPA is directed at Web sites that wish to gather personal information from children under the age of 13.) COPA states that "whoever knowingly and with knowledge of the character of the material, in interstate or foreign commerce by means of the World Wide Web, makes any communication for commercial purposes that is available to any minor and that includes any material that is harmful to minors shall be fined not more than $50,000, imprisoned not more than 6 months, or both." (Subsequent sections of the act allow for penalties of up to $150,000 for each day of violation.)[17]

The law became a rallying point for proponents of free speech. Not only could it affect sellers of explicit material online and their potential customers, but it could ultimately set standards for Internet free speech. Supporters of COPA (primarily the Department of Justice) argued that the act protected children from online pornography while preserving the rights of adults. However, privacy advocacy groups—such as the Electronic Privacy Information Center, the ACLU, and the Electronic Frontier Foundation—claimed that the language was overly vague and limited the ability of adults to access material protected under the First Amendment.

Following a temporary injunction as well as numerous hearings and appeals, in June 2004 the Supreme Court ruled in *Ashcroft v. American Civil Liberties Union* that there would be "a potential for extraordinary harm and a serious chill upon protected speech" if the law went into effect.[18] The ruling made it clear that COPA was unconstitutional and could not be used to shelter children from online pornography.

Internet Filtering

An **Internet filter** is software that can be used to block access to certain Web sites that contain material deemed inappropriate or offensive. The best Internet filters use a combination of URL, keyword, and dynamic content filtering. With URL filtering, a particular URL or domain name is identified as belonging to an objectionable site, and the user is not allowed access to it. Keyword filtering uses keywords or phrases—such as *sex*, *Satan*, and *gambling*—to block Web sites. With dynamic content filtering, each Web site's content is evaluated immediately before it is displayed, using techniques such as object analysis and image recognition.

The negative side of Internet filters is that they can "over block" Internet content and keep users from accessing useful information. Studies by various organizations (Kaiser Family Foundation, Consumer Reports, the Free Expression Policy Project, and the San Jose Public Library) found that filters block Web sites containing useful information about civil rights, health, sex, and politics as well as online databases and online book catalogs.[19]

Organizations may direct their network administrators to install filters on employees' computers to prevent them from viewing sites that contain pornography or other objectionable material. Employees unwillingly exposed to such material would have a strong case for sexual harassment. The use of filters can also ensure that employees do not waste their time viewing non-business-related Web sites.

According to TopTenREVIEWS, the top rated Internet filters for home users for 2011 include Net Nanny Parental Controls, Pure Sight PC, CYBERsitter, Safe Eyes®, and CyberPatrol.[20]

Safe Eyes® from InternetSafety.com™ is Internet filtering software that filters videos on YouTube, manages the viewing of online TV by choosing the age-appropriate ratings (e.g., TV-G and TV-PG), and blocks the use of media sharing applications used to download pirated music and videos. See Figure 5-3.

FIGURE 5-3 Screenshot of Safe Eyes® from Internet Safety
Source Line: Used with permission from InternetSafety.com, part of McAfee Inc.

Internet software filters have also been developed to run on mobile devices such as the iPhone and other smartphones, the iPod Touch, and the iPad.

A filtering tool designed for use by Web site owners is available through ICRA (formerly the Internet Content Rating Association), which is part of the nonprofit Family Online Safety Institute (FOSI), whose members as of February 2010 included such Internet industry leaders as AOL, AT&T, Facebook, Google, Microsoft, MySpace, Verizon, and Yahoo!.[21] FOSI's mission is to enable the public to make informed decisions about electronic media through the open and objective labeling of content. Its goals are to protect children from potentially harmful material while safeguarding free speech online.

The ICRA rating system requires that Web authors fill out an online questionnaire to describe the content of their site. The questionnaire covers such broad topics as the presence of chat rooms or other user-generated content; the language used; the presence of nudity and sexual content; and the depiction of violence, alcohol, drugs, gambling, and suicide. Within each broad category, the Web author is asked whether specific items or features are present on the site. Based on the author's responses, ICRA generates a content label (a short piece of computer code) that the author adds to the site. This label conforms to an Internet industry standard known as the Platform for Internet Content Selection (PICS). Internet users can then set their browsers to allow or disallow access to Web sites based on the information declared in the content label and their own preferences.

Note that ICRA does not rate Web content—the content providers do. However, relying on Web site authors to do their own ratings has its weaknesses. For one, many hate sites and sexually explicit sites don't have ICRA ratings. Thus, these sites won't be blocked unless a browser is set to block *all* unrated sites, which would make Web surfing a virtually useless activity in that it would block many acceptable sites, as well. Site labeling also depends on the honesty with which Web site authors rate themselves. If authors lie when completing the ICRA questionnaire, their site could receive a content label that doesn't accurately reflect the site's content. For these reasons, site labeling is at best a complement to other filtering techniques.

Another approach to restricting access to Web sites is to subscribe to an Internet service provider (ISP) that performs the blocking. The blocking occurs through the ISP's server rather than via software loaded onto each user's computer. One ISP, ClearSail/Family.NET, prevents access to known Web sites that address such topics as bomb making, gambling, hacking, hate, illegal drugs, pornography, profanity, public chat, satanic activities, and suicide. ClearSail employees search the Web daily to uncover new sites to add to ClearSail's block list. The ISP blocks specific URLs and known pornographic hosting services, as well as other sites based on certain keywords. ClearSail's filtering blocks millions of Web pages. Newsgroups are also blocked because of the potential for pornography within them.[22]

Children's Internet Protection Act (CIPA) (2000)

In another attempt to protect children from accessing pornography and other explicit material online, Congress passed the **Children's Internet Protection Act (CIPA)** in 2000. The act required federally financed schools and libraries to use some form of technological protection (such as an Internet filter) to block computer access to obscene material, pornography, and anything else considered harmful to minors. Congress did not specifically define what content or Web sites should be forbidden or what measures should be used—these decisions were left to individual school districts and library systems. Any school or library that failed to comply with the law would no longer be eligible to receive federal money through the E-Rate program, which provides funding to help pay for the cost of Internet connections. The following points summarize CIPA:

- Under CIPA, schools and libraries subject to CIPA will not receive the discounts offered by the E-Rate program unless they certify that they have certain Internet safety measures in place. These include measures to block or filter pictures that (1) are obscene, (2) contain child pornography, or (3) are harmful to minors (for computers used by minors).
- Schools subject to CIPA are required to adopt a policy to monitor the online activities of minors.
- Schools and libraries subject to CIPA are required to adopt a policy addressing (1) access by minors to inappropriate matter online; (2) the safety and security of minors when using email, chat rooms, and other forms of direct electronic communications; (3) unauthorized access, including hacking and other unlawful activities by minors online; (4) unauthorized disclosure, use, and dissemination of personal information regarding minors; and (5) restricting minors' access to materials harmful to them. CIPA does not require the tracking of Internet use by minors or adults.[23]

Opponents of the law were concerned that it transferred power over education to private software companies who develop the Internet filters and define what sites to block. Furthermore, opponents felt that the motives of these companies were unclear—for example, some filtering companies track students' Web-surfing activities and sell the data to market research firms. Opponents also pointed out that some versions of these filters were ineffective, blocking access to legitimate sites and allowing access to objectionable ones. Yet another objection was that penalties associated with the act could cause schools and libraries to lose federal funds from the E-Rate program, which is intended to help bridge the digital divide between rich and poor, urban and rural. Loss of federal funds would lead to a less capable version of the Internet for students at poorer schools, which have the fewest alternatives to federal aid.

CIPA's proponents contended that shielding children from drugs, hate, pornography, and other topics was a sufficient reason to justify filters. They argued that Internet filters are highly flexible and customizable and that critics exaggerated the limitations. Proponents pointed out that schools and libraries could elect not to implement a children's Internet protection program; they just wouldn't receive federal money for Internet access.

Many school districts implemented programs consistent with CIPA. Acceptance of an Internet filtering system is more meaningful if the system and its rationale are first discussed with parents, students, teachers, and administrators. Then the program can be refined, taking into account everyone's feedback. An essential element of a successful program is to require that students, parents, and employees sign an agreement outlining the school district's acceptable-use policies for accessing the Internet. Controlling Internet access via a central district-wide network rather than having each school set up its own filtering system reduces administrative effort and ensures consistency. Procedures must be defined to block new objectionable sites as well as remove blocks from Web sites that should be accessible.

Implementing CIPA in libraries is much more difficult because a library's services are open to people of all ages, including adults who have First Amendment rights to access a broader range of online materials than are allowed under CIPA. One county library was sued for filtering, while another was sued for not filtering enough. At least one federal court has ruled that a local library board may not require the use of filtering software on all library computers connected to the Internet. A possible compromise for public libraries with multiple computers would be to allow unrestricted Internet use for adults but to provide computers with only limited access for children.

The ACLU filed a suit to challenge CIPA. In May 2002, a three-judge panel in eastern Pennsylvania held that "we are constrained to conclude that the library plaintiffs must prevail in their contention that CIPA requires them to violate the First Amendment rights of their patrons, and accordingly is facially invalid" under the First Amendment. The ruling instructed the government not to enforce the act. This ruling, however, was reversed in June 2003 by the U.S. Supreme Court in *United States v. American Library Association*. The Supreme Court, in a 6-3 decision, held that public libraries must purchase filtering software and comply with all portions of CIPA.[24]

Rather than deal with all the technical and legal complications, some librarians say they wish they could simply focus on training students and adults to use the Internet safely and wisely.

Internet Censorship

"**Internet censorship** is the control or suppression of the publishing or accessing of information on the Internet."[25] Censorship can take many forms—such as limiting access to certain Web sites, allowing access to only some content or modified content at certain Web sites, rejecting the use of certain keywords in search engine searches, tracking and monitoring the Internet activities of individuals, and harassing or even jailing individuals for their Internet use.

China has the largest online population in the world, with over 420 million Internet users (see Table 5-1, which depicts the top five countries in terms of number of Internet users). However, Internet censorship in China is perhaps the most rigorous in the world. Table 5-2 provides examples of Internet censorship in several countries.

TABLE 5-1 The top five countries with the highest number of Internet users

Country	Population (million)	Internet users (million)	% of country's population
China	1,330	420	31.6%
United States	310	240	77.3%
Japan	127	99	78.2%
India	1,173	81	6.9%
Brazil	201	76	37.8%

Source Line: "Top 20 Countries with the Highest Number of Internet Users," Internet World Stats, © June 30, 2010, www.internetworldstats.com/top20.htm.

TABLE 5-2 Internet censorship examples by selected country

Country	Form of censorship
Brazil[26]	Brazilian government demands have closed more Google Gmail accounts and more blogger sites than in any other country. In Brazil, filing a lawsuit to demand that Internet content be taken down is relatively easy and inexpensive. The ability of litigants to challenge content and demand that anonymous sources be revealed stifles Brazilian journalists and Internet bloggers.
Burma[27]	Dissemination of information via by the Internet is tightly monitored and controlled. Two high-ranking government officials were sentenced to death for emailing documents abroad.
China[28]	The government blocks access to Web sites that discuss any of a long list of topics that are considered objectionable—including the Buddhist leader the Dalai Lama, anything to do with the government crackdown on the 1989 Tiananmen Square protests, and the banned spiritual movement Falun Gong. Chinese Web sites also employ censors who monitor and delete objectionable content. The government hires workers to post comments favorable to the government.
Cuba[29]	The ability to go on the Internet requires special permission of the ruling Communist Party, and when a user does get connected, it is only to a highly censored version of the Internet.
Egypt[30]	Although Egypt had not set up an Internet filtering system under the regime of Hosni Mubarak, it did track Internet posters and arrest bloggers who made posts that were unacceptable to the Mubarak government. The government also disabled access to the Internet completely for periods of time.
United States[31]	Many in the United States feel that U.S. laws about interception of online communications do not provide sufficient privacy guarantees for users. There are also concerns that some U.S. companies are selling equipment and technology to the Chinese government that enable it to censor Internet content for users in China.

Anonymity on the Internet

Anonymous expression is the expression of opinions by people who do not reveal their identity. The freedom to express an opinion without fear of reprisal is an important right of a democratic society. Anonymity is even more important in countries that don't allow free speech. However, in the wrong hands, anonymous communication can be used as a tool to commit illegal or unethical activities.

Anonymous political expression played an important role in the early formation of the United States. Before and during the American Revolution, patriots who dissented against British rule often used anonymous pamphlets and leaflets to express their opinions. England had a variety of laws designed to restrict anonymous political commentary, and people found guilty of breaking these laws were subject to harsh punishment—from whippings to hangings. A famous case in 1735 involved a printer named John Zenger, who was prosecuted for seditious libel because he wouldn't reveal the names of anonymous authors whose writings he published. The authors were critical of the governor of New York. The British were outraged when the jurors refused to convict Zenger, in what is considered a defining moment in the history of freedom of the press in the United States.

Other democracy supporters often authored their writings anonymously or under pseudonyms. For example, Thomas Paine was an influential writer, philosopher, and statesman of the Revolutionary War era. He published a pamphlet called *Common Sense*, in which he criticized the British monarchy and urged the colonies to become independent by establishing a republican government of their own. Published anonymously in 1776, the pamphlet sold more than 500,000 copies when the population of the colonies was estimated to have been less than 4 million; it provided a stimulus to produce the Declaration of Independence six months later.

Despite the importance of anonymity in early America, it took nearly 200 years for the Supreme Court to render rulings that addressed anonymity as an aspect of the Bill of Rights. One of the first rulings was in the 1958 case of *National Association for the Advancement of Colored People (NAACP) v. Alabama*, in which the court ruled that the NAACP did not have to turn over its membership list to the state of Alabama. The court believed that members could be subjected to threats and retaliation if the list were disclosed and that disclosure would restrict a member's right to freely associate, in violation of the First Amendment.

Another landmark anonymity case involved a sailor threatened with discharge from the U.S. Navy because of information obtained from AOL. In 1998, following a tip, a Navy investigator asked AOL to identify the sailor, who used a pseudonym to post information in an online personal profile that suggested he might be gay. Thus, he could be discharged under the military's "don't ask, don't tell" policy on homosexuality. AOL admitted that its representative violated company policy by providing the information. A federal judge ruled that the Navy had overstepped its authority in investigating the sailor's sexual orientation and had also violated the Electronic Communications Privacy Act, which limits how government agencies can seek information from email or other online data. The sailor received undisclosed monetary damages from AOL and, in a separate agreement, was allowed to retire from the Navy with full pension and benefits.[32]

Maintaining anonymity on the Internet is important to some computer users. They might be seeking help in an online support group, reporting defects about a manufacturer's goods or services, taking part in frank discussions of sensitive topics, expressing a minority or antigovernment opinion in a hostile political environment, or participating in chat rooms. Other Internet users would like to ban Web anonymity because they think its use increases the risks of defamation, fraud, libel, and the exploitation of children.

When data is sent over the Internet, a computer's IP address (a numeric identification for a computer attached to the Internet) is logged by the ISP. The IP address can be used to identify the sender of an email or an online posting. Internet users who want to remain anonymous can send email to an **anonymous remailer** service, which uses a computer program to strip the originating IP number from the message. It then forwards the message to its intended recipient—an individual, a chat room, or a newsgroup—with either no IP address or a fictitious one. This ensures that no header information can identify the author. Some remailers route messages through multiple remailers to provide a virtually untraceable level of anonymity.

The use of a remailer keeps communications anonymous; what is communicated, and whether it is ethical or legal, is up to the sender. The use of remailers by people committing unethical or even illegal acts in some states or countries has spurred controversy. Remailers are frequently used to send pornography, to illegally post copyrighted material to Usenet newsgroups, and to send unsolicited advertising to broad audiences (spamming). An organization's IT department can set up a firewall to prohibit employees from accessing remailers or to send a warning message each time an employee communicates with a remailer.

John Doe Lawsuits

Businesses must monitor and respond to both the public expression of opinions that might hurt their reputations and the public sharing of confidential company information. When anonymous employees reveal harmful information online, the potential for broad dissemination is enormous, and it can require great effort to identify the people involved and stop them.

An aggrieved party can file a **John Doe lawsuit** against a defendant whose identity is temporarily unknown because he or she is communicating anonymously or using a pseudonym. Once the John Doe lawsuit is filed, the plaintiff can request court permission to issue subpoenas to command a person to appear under penalty. If the court grants permission, the plaintiff can serve subpoenas on any third party—such as an ISP or a Web site hosting firm—that may have information about the true identity of the defendant. When, and if, the identity becomes known, the complaint is modified to show the correct name(s) of the defendant(s).

A company may file a John Doe lawsuit because it is upset by anonymous email messages that criticize the company or reveal company secrets. For example, TheFunded is a Web site that allows startup founders to leave anonymous comments about venture capital firms. EDF Ventures, a small venture capital firm, sued a John Doe for defamation for anonymously posting a comment on TheFunded's Web site implying that the people running the firm were dishonest. EDF Ventures subpoenaed the operator of the Web site seeking the identity of the poster. The management of TheFunded cooperated with the subpoena. However, the company was unable to produce any information about the identity of the poster because TheFunded takes careful measures to safeguard the anonymity of its posters, including not storing IP addresses, email addresses, or any other personal information associated with a member account.[33]

U.S. courts have ruled that under the Federal Communications Decency Act, Web sites are not liable for comments posted if they take no active role in screening those comments. Instead, the people posting the comments are legally responsible for their content. Defending oneself against a libelous post can be extremely expensive and time consuming.

AOL, EarthLink, NetZero, Verizon Online, and other ISPs receive more than a thousand subpoenas per year directing them to reveal the identity of John Does. Free-speech advocates argue that if someone charges libel, the anonymity of the Web poster should be preserved until the libel is proved. Otherwise, the subpoena power can be used to silence anonymous, critical speech.

Proponents of such lawsuits point out that most John Doe cases are based on serious allegations of wrongdoing, such as libel or disclosure of confidential information. For example, stock price manipulators can use chat rooms to affect the share price of stocks—especially those of very small companies that have just a few outstanding shares. In addition, competitors of an organization might try to create the feeling that the organization is a miserable place to work, which could discourage job candidates from applying, investors from buying stock, or consumers from buying company products. Proponents of John Doe lawsuits argue that perpetrators should not be able to hide behind anonymity to avoid responsibility for their actions.

Anonymity on the Internet is not guaranteed. By filing a lawsuit, companies gain immediate subpoena power, and many message board hosts release information as soon as it is requested, often without notifying the poster. Everyone who posts comments in a public place on the Web should consider the consequences if their identities were to be exposed. Furthermore, everyone who reads anonymous postings online should think twice about believing what they read.

The California State Court in *Pre-Paid Legal v. Sturtz et al.*[34] set another legal precedent that refined the criteria the courts apply when deciding whether or not to approve subpoenas requesting the identity of anonymous Web posters. The case involved a subpoena issued by Pre-Paid Legal Services (PPLS), which requested the identity of eight anonymous posters on Yahoo!'s Pre-Paid message board. Attorneys for PPLS argued that it needed the posters' identities to determine whether they were subject to a voluntary injunction that prevented former sales associates from revealing PPLS's trade secrets.

The Electronic Frontier Foundation (EFF) represented two of the John Does whose identities were subpoenaed. EFF attorneys argued that the message board postings cited by PPLS revealed no company secrets but were merely disparaging the company and its treatment of sales associates. They argued further that requiring the John Does to reveal their identities would let the company punish them for speaking out and set a dangerous precedent that would discourage other Internet users from voicing criticism. Without proper safeguards on John Doe subpoenas, a company could use the courts to uncover its critics.

EFF attorneys urged the court to apply the four-part test adopted by the federal courts in *Doe v. 2TheMart.com, Inc.* [35] to determine whether a subpoena for the identity of the Web posters should be upheld. In that case, the federal court ruled that a subpoena should be enforced only when the following occurs:

- The subpoena was issued in good faith and not for any improper purpose.
- The information sought related to a core claim or defense.
- The identifying information was directly and materially relevant to that claim or defense.
- Adequate information was unavailable from any other source.

In August 2001, a judge in Santa Clara County Superior Court invalidated the subpoena to Yahoo! requesting the posters' identities. He ruled that the messages were not obvious violations of the injunctions invoked by PPLS and that the First Amendment protection of anonymous speech outweighed PPLS's interest in learning the identity of the speakers.

Defamation and Hate Speech

In the United States, speech that is merely annoying, critical, demeaning, or offensive enjoys protection under the First Amendment. Legal recourse is possible only when hate speech turns into clear threats and intimidation against specific citizens. Persistent or malicious harassment aimed at a specific person can be prosecuted under the law, but general, broad statements expressing hatred of an ethnic, racial, or religious group cannot. A threatening private message sent over the Internet to a person, a public message displayed on a Web site describing intent to commit acts of hate-motivated violence against specific individuals, and libel directed at a particular person are all actions that can be prosecuted.

Although ISPs do not have the resources to prescreen content (and they do not assume any responsibility for content provided by others), many ISPs do reserve the right to remove content that, in their judgment, does not meet their standards. The speed at which content may be removed depends on how quickly such content is called to the attention of the ISP, how egregious the content is, and the general availability of ISP resources to handle such issues.

For example, AOL has documented a set of standards it calls the AOL Community Guidelines. AOL states clearly that it is not responsible for any failure or delay in removing material that violates these standards. To become an AOL subscriber, you must agree to follow these guidelines and acknowledge that AOL has the right to enforce them in its sole discretion. Thus, if you or anyone using your account violates the AOL Community Guidelines, AOL may take action against your account, ranging from issuing a warning to terminating your account.

The AOL Community Guidelines state the following in regard to the use of hate speech:

> Do not use hate speech. Hate speech is unacceptable, and AOL reserves the right to take appropriate action against any account using the service to post, transmit, promote, distribute or facilitate distribution of content intended to victimize, harass, degrade or intimidate an individual or group on the basis of age, disability, ethnicity, gender, race, religion or sexual orientation.[36]

Because such prohibitions are included in the service contracts between a private ISP and its subscribers and do not involve the federal government, they do not violate the subscribers' First Amendment rights. Of course, ISP subscribers who lose an account for violating the ISP's regulations may resume their hate speech by simply opening a new account with some other, more permissive ISP.

Although they may implement a speech code, public schools and universities are legally considered agents of the government and therefore must follow the First Amendment's prohibition against speech restrictions based on content or viewpoint. Corporations, private schools, and private universities, on the other hand, are not part of state or federal government. As a result, they may prohibit students, instructors, and other employees from engaging in offensive speech using corporate-, school-, or university-owned computers, networks, or email services.

Most other countries do not provide constitutional protection for hate speech. For example, promoting Nazi ideology is a crime in Germany, and denying the occurrence of the Holocaust is illegal in many European countries. Authorities in Britain, Canada, Denmark, France, and Germany have charged people for crimes involving hate speech on the Web.

Thousands of people faced the potential of criminal charges after posting hate messages and threats to the Facebook account of Brendan Sokaluk, who is accused of setting bushfires that killed 21 people in Victoria, Australia, in February 2009. "It's the cyber-world equivalent of angry mobs forming outside court, hurling abuse," said Michael Pearce, president of Liberty Victoria (one of Australia's leading civil liberties organizations). "There is a clear risk that these people are going to imperil a fair trial for the accused and also that they are in contempt of court."[37]

A U.S. citizen who posts material on the Web that is illegal in a foreign country can be prosecuted if he subjects himself to the jurisdiction of that country—for example, by visiting there. As long as the person remains in the United States, he is safe from prosecution because U.S. laws do not allow a person to be extradited for engaging in an activity protected by the U.S. Constitution, even if the activity violates the criminal laws of another country.

Corporate Blogging

A growing number of organizations allow employees to create their own personal blogs relating to their employment. They see blogging as a new way to reach out to partners, customers, and other employees and to improve their corporate image. Under the best conditions, individual employees use their blogs to ask other employees for help with work-related problems, to share work-related information in a manner that invites conversation, or to invite others to refine or build on a new idea. However, most organizations are well aware that such blogs can also provide an outlet for uncensored commentary and interaction. Employees can use their blogs to criticize corporate policies and decisions. Employee blogging also involves the risk that employees might reveal company secrets or breach federal security disclosure laws.

Mark Jen, an associate product manager for Google, began a blog chronicling his experiences at Google after his first day on the job. He thought his entries might be of interest to his family and friends. Little did Mark realize that his blog would attract the attention of thousands of people curious about life inside Google. His entries candidly discussed his first day on the job, a global sales meeting, and Google's compensation package. His comments also included information about Google's future products and economic performance. Within a couple of days, Mark's audience had grown into the tens of thousands. The following week, Mark's blog was offline for a couple days. In his next posting, Mark revealed that he had been asked to take down sensitive information about the company. Then the entries stopped altogether, and rumors were rampant as the number of visitors to his blog approached 100,000 per day. A few weeks later, Mark finally checked back in to let his readers know that he had been fired. Within the blogosphere, Mark had become a cause célèbre, and Google's reputation suffered. The incident sent a shock wave through the IT industry, forcing companies to evaluate and establish their own blogging policies.[38]

Within a few months of Mark's dismissal, companies such as Sun, IBM, and Yahoo! began to formulate and publish employee blogging policies. Many guidelines suggested that employees use their common sense: Don't reveal company secrets. Don't insult people that you interact with for eight hours a day. Check with the Public Relations Department if you are not sure whether you should include specific information. Other guidelines provided suggestions that might improve the quality of a blog and make it more appealing to potential readers: Be interesting. Don't use pen names or ghost writers. Be authentic.

Pornography

Many people, including some free-speech advocates, believe that there is nothing illegal or wrong about purchasing adult pornographic material made by and for consenting adults. They argue that the First Amendment protects such material. On the other hand, most parents, educators, and other child advocates are concerned that children might be exposed to pornography. They are deeply troubled by its potential impact on children and fear that increasingly easy access to pornography encourages pedophiles and sexual predators.

Clearly, the Internet has been a boon to the pornography industry by providing fast, cheap, and convenient access to well over 4.2 million porn Web sites worldwide. Access via the Internet enables pornography consumers to avoid offending others or being embarrassed by others observing their purchases. There is no question that adult pornography on the Web is big business (however, revenue estimates vary widely between $1 billion and $97 billion)[39] and generates a lot of traffic; it is estimated that there are 72 million visitors to pornographic Web sites monthly.[40]

Pornography purveyors are free to produce and publish whatever they want; however, if what they distribute or exhibit is judged obscene, they are subject to prosecution under the obscenity laws. The precedent-setting *Miller v. California* ruling on obscenity discussed earlier in the chapter predates the Internet. The judges in that case ruled that contemporary community standards should be used to judge what is obscene. The judges allowed that different communities could have different norms.

The key question in deciding what Internet material is obscene is: "Whose community standards are used?" Because Internet content publishers cannot easily direct their content into or away from a particular geographic area, one answer to this question is that the Internet content publisher must conform to the norms of the most restrictive community. However, this line of reasoning was challenged by the Third Circuit Court of Appeals in the *Ashcroft v. American Civil Liberties Union* case, which involved a challenge to the 1998 Child Online Protection Act (COPA). The Supreme Court reversed the circuit court's ruling in this case—but with five different opinions and no clear consensus on the use of local or national community standards.[41] In *United States v. Kilbride*, the Ninth Circuit Court of Appeals ruled that "a national community standard must be applied in regulating obscene speech on the Internet, including obscenity disseminated via email."[42] In *United States v. Little*, the Eleventh Circuit Court of Appeals rejected the national community standard and adopted the older, local community standard.[43] Currently there is no clear agreement within the courts on whether local or national community standards are to be used to judge obscenity.

U.S. organizations must be very careful when dealing with issues relating to pornography in the workplace. By providing computers, Internet access, and training in how to use those computers and the Internet, companies could be seen by the law as purveyors of pornography because they have enabled employees to store pornographic material and retrieve it on demand. A Nielsen survey on the viewing of pornography in the workplace revealed that 21 million Americans accessed porn from their work computers in March 2010—29 percent of the workforce.[44] In addition, if an employee sees a coworker viewing porn on a workplace computer, that employee may be able to claim that the company has created a hostile work environment. Such a claim opens the organization to a sexual harassment lawsuit that can cost hundreds of thousands of dollars and tie up managers and executives in endless depositions and court appearances.

Many companies believe that they have a duty to stop the viewing of pornography in the workplace. As long as they can show that they took reasonable steps and determined actions to prevent it, they have a valid defense if they become the subject of a sexual harassment lawsuit. If it can be shown that a company made only a half-hearted attempt to stop the viewing of pornography in the workplace, then its defense in court would be weak. Reasonable steps include establishing a computer usage policy that prohibits access to pornography sites, identifying those who violate the policy, and taking action against those users—regardless of how embarrassing it is for the users or how harmful it might be for the company.

The Texas Workforce Commission has posted a sample Internet, email, and computer usage policy at *www.texasworkforce.org/news/efte/internetpolicy.html*. Table 5-3 shows portions of the sample policy that pertain to the topics being discussed in this chapter.

TABLE 5-3 Portions of a sample Internet, email, and computer usage policy

"Use of Company computers, networks, and Internet access is a privilege granted by management and may be revoked at any time for inappropriate conduct carried out on such systems, including, but not limited to: • Violating the laws and regulations of the United States or any other nation or any state, city, province, or other local jurisdiction in any way; • Engaging in unlawful or malicious activities; • Using abusive, profane, threatening, racist, sexist, or otherwise objectionable language in either public or private messages; • Sending, receiving, or accessing pornographic materials; • Becoming involved in partisan politics; • Maintaining, organizing, or participating in nonwork-related Web logs ('blogs'), Web journals, 'chat rooms,' or private/personal/instant messaging."

Source Line: "Internet, E-Mail, And Computer Usage Policy" Texas Workforce Commission, www.twc.state.tx.us/news/efte/internetpolicy.html.

A few companies take the opposite viewpoint—that they cannot be held liable if they don't know employees are viewing, downloading, and distributing pornography. Therefore, they believe the best approach is to ignore the problem by never investigating it, thereby ensuring that they can claim that they never knew it was happening. Many people would consider such an approach unethical and would view management as shirking an important responsibility to provide a work environment free of sexual harassment. Employees unwillingly exposed to pornography would have a strong case for sexual harassment because they could claim that pornographic material was available in the workplace and that the company took inadequate measures to control the situation.

There are numerous federal laws addressing child pornography. Possession of child pornography is a federal offense punishable by up to five years in prison. The production and distribution of such materials carry harsher penalties; decades or even life in prison is not an unusual sentence. In addition to these federal statutes, all states have enacted laws against the production and distribution of child pornography, and all but a few states have outlawed the possession of child pornography. At least seven states have passed laws that require computer technicians who discover child pornography on clients' computers to report it to law enforcement officials.[45]

Sexting—sending sexual messages, nude or seminude photos, or sexually explicit videos over a cell phone—is a fast-growing trend among teens and young adults. According to a survey by the National Campaign to Prevent Teen and Unplanned Pregnancy, one in five teenagers has sent or posted nude or seminude photos of himself/herself, including 22 percent of teen girls, 18 percent of teen boys, and 11 percent of young teen girls aged 13 to 16. Increasingly, teens are suffering the consequences of this new fad.[46]

Controlling the Assault of Non-Solicited Pornography and Marketing (CAN-SPAM) Act (2003)

The **Controlling the Assault of Non-Solicited Pornography and Marketing (CAN-SPAM) Act** specifies requirements that commercial emailers must follow when sending messages that advertise or promote a commercial product or service. The key requirements of the law include:

- The "From" and "To" fields in the email, as well as the originating domain name and email address, must be accurate and identify the person who initiated the email.
- The subject line of the email cannot mislead the recipient as to the contents or subject matter of the message.
- The email must be identified as an advertisement and include a valid physical postal address for the sender.
- The emailer must provide a return email address or some other Internet-based response procedure to enable the recipient to request no future emails, and the emailer must honor such requests to opt out.
- The emailer has 10 days to honor the opt-out request.
- Additional rules prohibit the harvesting of email addresses from Web sites, using automated methods to register for multiple email accounts, or relaying email through another computer without the owner's permission.

Each violation of the provisions of the CAN-SPAM Act can result in a fine of up to $250 for each unsolicited email, and fines can be tripled in certain cases.[47] The Federal Trade Commission (FTC) is charged with enforcing the act. In addition, the FTC maintains a consumer complaint database relating to the law. Consumers can submit complaints online at *www.ftc.gov* or forward email to the FTC at *spam@use.gov*.

Asis Internet Services is a small ISP with just four employees. Asis was awarded $2.6 million in a CAN-SPAM lawsuit. The firm claimed it received 25,000 spam messages over an 18-month period from a business called Find a Quote. The judge in the case determined that Asis was entitled to damages of $865,340, but decided to triple the fine because the spammers used automatic scripts to send their messages.[48] It is unlikely that Asis will ever collect the award because the owner of Find a Quote fled to Panama. Interestingly, a few months later, in another CAN-SPAM lawsuit against AzoogleAds.com, a federal magistrate ordered Asis to pay over $800,000 to Azoogle for "filing groundless claims" that had mired Azoogle in years of costly litigation. Asis also had sued at least two other firms for violation of the CAN-SPAM act.[49]

The CAN-SPAM Act can also be used in the fight against the dissemination of pornography. For example, two men were indicted by an Arizona grand jury in August 2005 for violating the CAN-SPAM Act by sending massive amounts of unsolicited email advertising pornographic Web sites. They had amassed an email database of 43 million people and used it to send emails containing pornographic images. AOL stated it received over

660,000 complaints from people who received spam from the two. Their operation was highly profitable and enabled them to earn $1.4 million in 2003. The defendants ran afoul of the CAN-SPAM Act by sending messages with false return addresses and for using domain names registered using false information. They were convicted of multiple counts of spamming and criminal conspiracy, which carry a maximum sentence of five years each plus a fine of $500,000 and up to 20 years for money laundering. This is believed to be the first conviction involving CAN-SPAM Act violations.[50]

There is considerable debate over whether the CAN-SPAM Act has helped control the growth of spam. After all, the act clearly defines the conditions under which the sending of spam is legal, and as long as mass emailers meet these requirements, they cannot be prosecuted. In addition, the FTC has done little to enforce the act. It is estimated that more than 100 billion spam messages are sent each year, yet the FTC brought only 30 law-enforcement actions in the first five years of the act.[51] Some suggest that the act could be improved by penalizing the companies that use spam to advertise, as well as ISPs who support the spammers.

Table 5-4 is a manager's checklist for dealing with issues of freedom of expression in the workplace. In each case, the preferred answer is *yes*.

TABLE 5-4 Manager's checklist for handling freedom of expression issues in the workplace

Question	Yes	No
Do you have a written data privacy policy that is followed?		
Does your corporate IT usage policy discuss the need to conserve corporate network capacity, avoid legal liability, and improve worker productivity by limiting the nonbusiness use of information resources?		
Did the developers of your policy consider the need to limit employee access to non-business-related Web sites (e.g., Internet filters, firewall configurations, or the use of an ISP that blocks access to such sites)?		
Does your corporate IT usage policy discuss the inappropriate use of anonymous remailers?		
Has your corporate firewall been set to detect the use of anonymous remailers?		
Has your company (in cooperation with legal counsel) formed a policy on the use of John Doe lawsuits to identify the authors of libelous, anonymous email?		
Does your corporate IT usage policy make it clear that defamation and hate speech have no place in the business setting?		
Does your corporate IT usage policy prohibit the viewing and sending of pornography?		
Does your policy communicate that employee email is regularly monitored for defamatory, hateful, and pornographic material?		
Does your corporate IT usage policy tell employees what to do if they receive hate mail or pornography?		

Source Line: Course Technology/Cengage Learning.

Summary

- The First Amendment protects Americans' rights to freedom of religion and freedom of expression. The Supreme Court has ruled that the First Amendment also protects the right to speak anonymously.
- Obscene speech, defamation, incitement of panic, incitement to crime, "fighting words," and sedition are not protected by the First Amendment and may be forbidden by the government.
- The Internet enables a worldwide exchange of news, ideas, opinions, rumors, and information. Its broad accessibility, open discussions, and anonymity make it a powerful communications medium. People must often make ethical decisions about how to use such remarkable freedom and power.
- Organizations and governments have attempted to establish policies and laws to help guide Internet use as well as protect their own interests. Businesses, in particular, have sought to conserve corporate network capacity, avoid legal liability, and improve worker productivity by limiting the nonbusiness use of IT resources.
- Although there are clear and convincing arguments to support freedom of speech on the Internet, the issue is complicated by the ease with which children can use the Internet to gain access to material that many parents and others feel is inappropriate. The conundrum is that it is difficult to restrict children's Internet access without also restricting adults' access.
- The U.S. government has passed several laws to attempt to address this issue, including the Communications Decency Act (CDA) (aimed at protecting children from online pornography) and the Child Online Protection Act (COPA), which prohibited making harmful material available to minors via the Internet. Both laws were ultimately ruled unconstitutional.
- Software manufacturers have developed Internet filters, which are designed to block access to objectionable material through a combination of URL, keyword, and dynamic content filtering.
- The Children's Internet Protection Act (CIPA) requires federally financed schools and libraries to use filters to block computer access to any material considered harmful to minors.
- Internet censorship is the control or suppression of the publishing or accessing of information on the Internet. There are many forms of Internet censorship. Many countries practice some form on Internet censorship.
- Maintaining anonymity on the Internet is important to some computer users. Such users sometimes use an anonymous remailer service, which strips the originating IP number from the message and then forwards the message to its intended recipient.
- Many businesses monitor the Web for the public expression of opinions that might hurt their reputations. They also try to guard against the public sharing of company confidential information.
- Organizations may file a John Doe lawsuit to enable them to gain subpoena power in an effort to learn the identity of anonymous Internet users who have caused some form of harm through their postings.

- In the United States, speech that is merely annoying, critical, demeaning, or offensive enjoys protection under the First Amendment. Legal recourse is possible only when hate speech turns into clear threats and intimidation against specific citizens.
- Some ISPs have voluntarily agreed to prohibit their subscribers from sending hate messages using their services. Because such prohibitions can be included in the service contracts between a private ISP and its subscribers and do not involve the federal government, they do not violate subscribers' First Amendment rights.
- Numerous organizations allow employees to create their own personal blogs relating to their employment as a means to reach out to partners, customers, and other employees and to improve their corporate image.
- Organizations are advised to formulate and publish employee blogging policies to avoid potential negative consequences from employee criticism of corporate policies and decisions, or the revelation of company secrets or confidential financial data.
- Many adults and free-speech advocates believe there is nothing illegal or wrong about purchasing adult pornographic material made by and for consenting adults. However, organizations must be very careful when dealing with pornography in the workplace. As long as companies can show that they were taking reasonable steps to prevent pornography, they have a valid defense if they are subject to a sexual harassment lawsuit.
- Reasonable steps include establishing a computer usage policy that prohibits access to pornography sites, identifying those who violate the policy, and taking action against those users—regardless of how embarrassing it is for the users or how harmful it might be for the company.
- The key question in deciding what Internet material is obscene is: "Whose community standards are used?"
- The CAN-SPAM Act specifies requirements that commercial emailers must follow in sending out messages that advertise a commercial product or service. The CAN-SPAM Act can also be used in the fight against the dissemination of pornography.

Key Terms

anonymous expression
anonymous remailer
Child Online Protection Act (COPA)
Children's Internet Protection Act (CIPA)
Communications Decency Act (CDA)
Controlling the Assault of Non-Solicited Pornography and Marketing (CAN-SPAM) Act
defamation
Internet censorship
Internet filter
John Doe lawsuit
libel
Miller v. California
sexting
slander

Self-Assessment Questions

The answers to the Self-Assessment Questions can be found in Appendix E.

1. The most basic legal guarantee to the right of freedom of expression in the United States is contained in the:
   a. Bill of Rights
   b. Fourth Amendment
   c. First Amendment
   d. U.S. Constitution
2. Both the Communications Decency Act and the Child Online Protection Act were aimed at protecting children from harmful material, and both were found to be unconstitutional. True or False?
3. An important Supreme Court case that established a three-part test to determine if material is obscene and therefore not protected speech was ___________.
4. An oral statement that is false and that harms another person is called:
   a. a lie
   b. slander
   c. libel
   d. freedom of expression
5. The ___________ Act, which was passed in 2000, required federally financed schools and libraries to use some form of technological protection to block computer access to obscene material, pornography, and anything else considered harmful to minors.
6. The IRCA rating system requires Web authors to fill out an online questionnaire to describe the content of their site. True or False?
7. The U.S. Supreme Court in *United States v. American Library Association* ruled that public libraries must install Internet filtering software to comply with all portions of:
   a. the Children's Internet Protection Act
   b. the Child Online Protection Act
   c. the Communications Decency Act
   d. the Internet Filtering Act
8. The best Internet filters rely on the use of:
   a. URL filtering
   b. keyword filtering
   c. dynamic content filtering
   d. all of the above
9. Anonymous expression, or the expression of opinions by people who do not reveal their identities, has been found to be unconstitutional. True or False?

10. A lawsuit in which the true identity of the defendant is temporarily unknown is called a ____________.
11. The California State Court in *Pre-Paid Legal v. Sturtz et al.* set a precedent that courts apply when deciding whether to approve subpoenas requesting the identity of ____________.
12. In the United States, speech that is merely annoying, critical, demeaning, or offensive enjoys protection under the First Amendment. Legal recourse is possible only when hate speech turns into clear threats and intimidation against specific citizens. True or False?
13. Which of the following statements about Internet censorship is NOT true?
    a. China has the largest online population in the world and perhaps the most rigorous censorship in the world.
    b. The Brazilian government places little, if any, censorship on the use of the Internet by its citizens.
    c. Access to the Internet in Cuba requires the permission of the Communist Party.
    d. There are concerns over Internet censorship in the United States.
14. The ____________ Act specifies requirements that commercial emailers must follow in sending out messages that advertise or promote a commercial product or service.

Discussion Questions

1. Outline a scenario in which you might be acting ethically but might still want to remain anonymous while using the Internet. How might someone learn your identity even if you attempt to remain anonymous?
2. Research the use of the Internet by the people in Tunisia, Egypt, Bahrain, and Libya during the uprisings in these countries in early 2011. To what degree did communications over the Internet stimulate the start of the uprisings? To what degree was the Internet used to communicate the news of what was happening to people outside these countries?
3. Discuss to what degree you believe the United States is free from Internet censorship. Do you feel the amount of censorship is appropriate or overly restrictive?
4. What can an ISP do to limit the distribution of hate email? Why would such actions not be considered a violation of the subscriber's First Amendment rights?
5. Do research to identify the current countries that are identified as "Enemies of the Internet" by the group Reporters Without Borders. What criteria did Reporters Without Borders use when deciding which countries to place on this list?
6. Go to the FOSI Web site at *www.fosi.org.* After exploring this site, briefly describe how its ICRA rating process works. What are the advantages and disadvantages of this system?
7. Find out if your school or employer has a set of policies that cover the use of corporate blogging. Do you consider these policies to be overly restrictive?
8. Do research on the Web to locate an anonymous remailer. Find out what is required to sign up for this service and what fees are involved.

9. Look carefully at the email you receive over the next few days. Are any of the emails advertisements for a commercial product or service that violate the CAN-SPAM Act? If so, what might you do to stop receiving such email in the future?
10. What is a John Doe lawsuit? Do you think that a corporation should be allowed to use a subpoena to identify a John Doe before proving that the person has done damage to the company? Why or why not? Under what conditions will the courts execute a John Doe lawsuit?
11. Do you think further efforts to limit the dissemination of pornography on the Internet are appropriate? Why or why not?
12. How did the Children's Internet Protection Act escape from being ruled unconstitutional? Talk to your local librarian and find out if the library has implemented filtering. If so, is it experiencing any problems enforcing the use of filters? Write a short paragraph summarizing your findings.

What Would You Do?

1. A former high school classmate of yours who has returned to China emails you that he has been offered a part-time job monitoring a Chinese Web site and posting comments favorable to the government. How would you respond?
2. Your 15-year-old nephew exclaims "Oh wow!" and proceeds to tell you about a very revealing photo attachment he just received in a text message from his 14-year-old girlfriend of three months. He can't wait to forward the message to others in his school. What would you say to your nephew? Are further steps needed besides a discussion on sexting?
3. A college friend of yours approaches you about an idea to start a PR firm that would specialize in monitoring the Internet for "bad PR" about a company and "fixing" it. One tactic the firm would use is to threaten negative posters with a libel lawsuit unless they remove their posting. Should that fail, the PR firm will generate dozens of positive postings to outweigh the negative posting. What would you say to your friend about her idea?
4. You are the computer technical resource for county's public library system. The library is making plans to install Internet filtering software so that it will conform to the Children's Internet Protection Act and be eligible for federal funding. What sort of objections can you expect regarding implementation of Internet filters? How might you deal with such objections?
5. Imagine that you receive a hate email at your school or job. What would you do? Does your school or workplace have a policy that covers such issues?
6. A coworker confides to you that he is going to begin sending emails to your employer's internal corporate blog site, which serves as a suggestion box. He plans to use an anonymous remailer and sign the messages "Anonymous." Your friend is afraid of retribution from superiors but wishes to call attention to instances of racial and sexual discrimination observed during his five years as an employee with the firm. What would you say to your friend?

7. You are a member of your company's computer support group and have just helped someone from the company's board of directors upgrade his computer. As you run tests after making the upgrade, you are shocked to find dozens of downloads from adult porn sites on the hard drive. What would you do?
8. A friend contacts you about joining his company, Anonymous Remailers Anonymous. He would like you to lead the technical staff and offers you a 20 percent increase in salary and benefits over your current position. Your initial project would be to increase protection for users of the company's anonymous remailer service. After discussing the opportunity with your friend, you suspect that some of the firm's customers are criminal types and purveyors of pornography and hate mail. Although your friend cannot be sure, he admits it is possible that hackers and terrorists may use his firm's services. Would you accept the generous job offer? Why or why not?
9. You are representing the IT organization in a meeting with representatives from Legal, Human Resources, and Finance to discuss the advisability of allowing employees to start their own corporate blogs. It is your turn to speak. What would you say?
10. You are a member of the human resources department and are working with a committee to complete your company's computer usage policy. What advice would you offer the committee regarding how to address online pornography? Would you suggest that the policy be laissez-faire, or would you recommend that the committee require strict enforcement of tough corporate guidelines? Why?

Cases

1. Facebook, Freedom of Speech, Defamation, and Cyberbullying

In February 2009, Oceanside High School graduate Denise Finkel sued Facebook, four former classmates, and their parents for false and defamatory statements made against her.[52] The classmates, part of a private Facebook group called "90 Cents Short of a Dollar," had posted comments to a password-protected page alleging that Finkel used intravenous drugs, had AIDS, and had engaged in inappropriate sexual behavior.[53]

Many legal commentators found the case surprising. Why would an attorney sue Facebook, when the company is obviously protected by Section 230 of the Communications Decency Act (CDA)? The CDA protects Web site owners from liability for content posted by a third party. Thus, Facebook should be immune from defamation lawsuits that arise from user-posted content.[54] However, Finkel claimed that Facebook's Terms of Use agreement established Facebook's proprietary rights to the site's content. In support of this argument, she pointed to the following clause in the agreement: "All content on the site … [is] the proprietary property of the Company, its users or licensors with all rights reserved."[55] As a result of this unique argument, the case attracted national attention because the consequences of the ruling could have serious implications for other social media sites. In September 2009, however, the Supreme Court of New York ruled that the Terms of Use did not disqualify Facebook from immunity under the CDA.[56]

This ruling provoked commentary by legal professionals, but no one was very surprised. What was less predictable was whether or not the court would hold Finkel's classmates liable.

To what extent do students have the right to exercise their freedom of speech through social media outlets and on the Internet? What restrictions have been placed on this freedom?

In July 2010, the New York Supreme Court dismissed the case against Finkel's former classmates. The court determined that teenage members of the "90 Cents Short of a Dollar" Facebook group had simply acted childishly and were not guilty of defamation.[57] Finkel did not have any recourse against her former classmates via an antibullying statute because although the state of New York has a law prohibiting bullying, it only prohibits bullying on school property. The law does not apply to cyberbullying. Furthermore, the law merely requires that the schools take disciplinary action, and it does not provide for criminal sanctions.[58]

In 1969, in a landmark case, *Tinker v. Des Moines School District*, the U.S. Supreme Court established that public schools cannot curtail students' freedom of speech unless this speech would cause a substantial disruption to school activities or violate the rights of other students.[59] Hence, a federal judge overturned the suspension of a high school senior who created a fake MySpace profile of his principal; the profile said the principal took drugs and kept alcohol at his desk. However, in a similar case, a U.S. district court ruled against a Pennsylvania junior high student who had created a fake MySpace profile of her principal claiming that he was a sex addict and a pedophile. She had been suspended for 10 days, and her parents sued the school. Teachers at the school, however, testified that the profile had caused a disruption in class because students were too busy talking about the profile to pay attention in class. The court determined that the talking had constituted a "substantial disruption." Both of these decisions were upheld on appeal in February 2010.[60]

Many states have recently enacted bullying laws that restrict written and symbolic speech on social media sites. In 2005, a Florida honors student committed suicide after three years of teasing at school and online bullying. As a result, in 2008, the state enacted a tough law called the "Jeffrey Johnston Stand Up for All Students Act." [61] The law prohibits the teasing; social exclusion; threat; intimidation; stalking; sexual, religious, or racial harassment; or public humiliation of any public school student or employee on or offline. The law is limited in that it only applies to behavior during school, on a school bus, during any school-related or school-sponsored program or activity, or from computers that are part of a K–12 system.[62]

In January 2011, two Florida teenagers, Taylor Wynn and McKenzie Barker, were arrested for allegedly setting up a false Facebook account for a classmate that included nude photos. The teenage girls are accused of doctoring photographs, placing the classmate's face on the bodies of naked men and women, and posting them to the site. Although the victim told the school resource officer that the teasing would eventually "go away," a parent of another student notified authorities. Investigators traced the IP addresses to Wynn and Barker and collected incriminating text messages and emails linking them to the false Facebook page. Wynn and Barker were charged with stalking a minor under the new Jeffrey Johnston Stand Up for All Students Act.[63]

On January 19, 2011, 14-year old Kameron Jacobsen, a freshman at Monroe-Woodbury High School in Orange County, New York, committed suicide after being taunted on Facebook for his supposed sexual orientation.[64]

Although freedom of speech is a right guaranteed by the U.S. Constitution, it can be restricted where it violates the other rights of individuals—as established by state or federal laws. To date, 31 states have antibullying laws that include electronic forms of bullying.[65]

Discussion Questions

1. Why is Facebook protected from liability for content posted by third parties? Should Facebook be held liable?
2. How is a student's freedom of speech restricted on social media sites such as Facebook?
3. Should Taylor Wynn and McKenzie Barker have been prosecuted despite the victim's attitude that the teasing would pass? How should cyberbullying laws be implemented?

2. Google Problems with China

China has the largest online population in the world, with over 420 million Internet users—nearly twice as many users as the United States, which is in second place. However, China also imposes stringent Internet censorship by blocking access to Web sites that discuss a wide range of topics considered objectionable, including the Dalai Lama, the Tiananmen Square protests, and pornography. The Chinese government also blocks access to Web sites run by human rights organization groups, Western news groups, Facebook, YouTube, Twitter, and WikiLeaks.[66]

The Beijing government erected a firewall consisting of proxy servers that form connections and relay information between millions of individual PCs and the various Web site hosts on the global Internet.[67] These servers are configured to intercept requests for objectionable Web sites and, instead of returning information, simply return an "error" message to the end user. The firewall is sometimes referred to as The Great Firewall of China, and it came online in 2003.[68]

For Google, offering access and service to Chinese users from a Web site outside of China meant that all search requests and Gmail email had to pass through the Great Firewall of China. Not only were searches and results censored, but performance was slow. Executives at Google understood that while operating outside of China, the company would never compete effectively with the Chinese search engine Baidu. The size of the Chinese search market was appealing to Google; it was estimated to be worth $1.7 billion in 2010 with an expected 50 percent growth rate each year for the next four years.[69] Thus Google obtained the Google.cn domain and a license to conduct business in China in January 2006. As a condition of doing business in China, Google agreed to obey Chinese laws, including self-censorship of its search results consistent with the Chinese government censorship. Google justified this decision as a means to provide the Chinese people with better, faster search results.[70]

Many people did not agree with Google's decision to work with a government that has one of the worst records for human rights and censorship just so Google could grow its business. As it turns out, Google inside of China was still unable to compete with Baidu, and after four years of operation in China, Google's share of the search engine market was just 14 percent compared to the Baidu's 62 percent.[71]

In January 2010, Google executives released information about a "highly sophisticated and targeted attack" on their corporate computing and network infrastructure. The attacks occurred in December 2009, and Google concluded that they originated from China. While one of the goals of the attacks apparently was to gather information about Chinese human rights activists, it appears that the attacks had very limited success. Only two Google Gmail accounts were partially compromised, with limited account information accessed; the attacks were not successful in accessing the content of emails.[72] In the course of its investigation, Google also discovered that third parties frequently accessed the Gmail accounts of dozens of Chinese human rights advocates, not only in China, but in the United States and Europe. These attacks were separate from the December 2010 attack and involved phishing scams and/or the downloading of malware onto unsuspecting users' computers.[73]

David Drummond, chief legal officer at Google, said that the attacks had led the company to reconsider the feasibility of its Chinese operations. "We recognize that this may well mean having to shut down Google.cn, and potentially our offices in China." Drummond also revealed that Google would stop censoring its search results on Chinese Google sites.[74]

Privacy advocates expressed their support for Google's reversal on its censorship activities. Leslie Harris, president of the Center of Democracy and Technology, stated that Google had "taken a bold and difficult step for Internet freedom in support of fundamental human rights. No company should be forced to operate under government threat to its core values or to the rights and safety of its users."[75]

In March 2010, Google decided it would no longer censor its search results, but it wanted to give the appearance of abiding by Chinese law. Its solution was to automatically redirect all Chinese users who went to the *Google.cn* Web site to *Google.com.hk*, an uncensored Chinese language Web site administered in Hong Kong. This tactic upset Chinese officials, so in June 2010, Google modified its approach by creating a revised Web page that required users in mainland China to make a choice rather than automatically directing them to the Hong Kong site. The move to require a double-click to get to the Hong Kong site was a key concession, enabling Google to continue to operate within China.[76] Edward Yu of Analysys International explains: "In China, it is very common that you need to give the government face if you want to do business here. The double-click rule shows that Google can compromise and give them face."[77]

In July 2010, the Beijing government renewed Google's license to operate in mainland China. David Drummond stated: "We are very pleased that the government renewed our Internet Content Provider license, and we look forward to continuing to provide web search and local products to our users in China."[78]

As early as December 2009, Google began a partnership with the National Security Agency (NSA) to investigate the cyberattacks. An NSA spokeswoman stated, "NSA is not able to comment on specific relationships we may or may not have with U.S. companies. NSA works with a broad range of commercial partners and research associates" to safeguard the government's national security computer networks. Google would not comment on the partnership other than to say that it was "working with the relevant U.S. authorities."[79]

Many companies have been reluctant to work with the NSA since it was revealed that the agency conducted domestic surveillance without warrants during the Bush administration. There were also allegations in 2006 that AT&T provided NSA with full access to its customers' phone calls and routed its Internet traffic to secret data mining equipment in a San Francisco switching center.[80] As a result, there is concern that the partnership could lead to NSA gaining access to Google users' personal information.

Discussion Questions

1. Enter *Google.cn* as a URL. When you click anywhere on the page, you can access an "English" option, which will allow search results to be displayed in English. List the results as you enter various search terms such as *Dalai Lama*, *Tiananmen Square protests*,

Facebook, and *Twitter*. Compare these results to the search results found when you access Google's U.S. site.

2. With the benefit of hindsight, how do you think Google should have gone about starting its business in China that could have avoided some of the problems and criticism the company has received?
3. How concerned should we be that the NSA and Google have formed an alliance? What is your worst-case scenario as to where this alliance may lead? What is your best-case scenario?

3. Sexting

Over the years, the courts have made many decisions regarding what kinds of communication are obscene and whether or not they are protected by our First Amendment rights to freedom of expression. The line between what is legal and what is not has been in constant flux. Now comes a new test for our society: sexting.

Jessie Logan was a good kid—lively, artistic, and fun. But in her last year at Sycamore High School, she made a terrible mistake. She used her cell phone to take nude photos of herself and then sent them to her boyfriend. After the couple broke up, Jessie's ex-boyfriend forwarded the photos to several other teenage girls; eventually, the pictures were sent to hundreds of teens in the Cincinnati area. Classmates, and even kids she did not know, started teasing her. They called her a tramp and worse; some even threw things at her. Instead of attending classes, she began sleeping in her car in the school parking lot or hiding in the bathroom—skipping classes to avoid further embarrassment. She stopped interacting with her friends, and her grades dropped. Finally, in the summer of 2008, Jessie was so full of despair that she hung herself.[81]

Jessie's mother, Cynthia Logan, wondered why school officials and authorities didn't do more to help Jessie. The school superintendent, Adrienne James, said that the sexting problem had been addressed at a parents' meeting. No other action could be taken, James said, because some of the students involved attended other school districts and because Jessie had taken the photos at home—not at school. A school resource officer said that he reprimanded students who harassed Jessie. The officer also spoke to a prosecutor who told him that nothing could be done because Jessie was 18 years old.[82]

But were Jessie's ex-boyfriend's actions legal? Certainly, if they involved exposing underage children to the nude photos, the ex-boyfriend would have been violating laws protecting children from exploitation and pornography. However, the ex-boyfriend and other students who distributed the photos were never arrested or charged with any crime.

Authorities, however, are starting to come down hard on teens under the age of 18 who engage in sexting:

- Seventeen-year-old Alex Phillips of La Crosse, Wisconsin, received nude photos from his 16-year-old girlfriend. In 2008, he posted the pictures on MySpace with obscene captions as a means of "venting" after the two broke up. When police asked Phillips to remove the photos from the Web site, he refused. Police then charged Phillips with possession of child pornography, sexual exploitation of a child, and defamation. These charges were eventually dropped, and he was charged with causing mental harm to a child.[83] As part of a plea bargain, Phillips was eventually sentenced to three years of probation and 100 hours of community service.[84]

- In Middletown, Ohio, a 13-year-old boy was arrested after a photo of an eighth-grade girl involved in sexual activity was found on his cell phone by school officials. He had shared the photo with other students at a skating party.[85]
- Phillip Albert sent nude pictures of his 16-year-old ex-girlfriend to 70 people, including her parents and grandparents, after she taunted him. The 18-year-old was charged with transmitting child pornography and was sentenced to five years of probation and will be registered as a sex offender until he reaches the age of 43.[86]

Teens are also increasingly being charged for merely exchanging nude photos of themselves over their cell phones. A group of Pennsylvania teenagers were charged with disseminating and possessing child pornography after three 13-year-old girls sent nude or seminude images of themselves to three 16- and 17-year-old boys.[87] Two Ohio teenagers, four Alabama middle-school students, and many teens in other states have been arrested on similar charges for taking and sharing nude or seminude photos of themselves.

Across the nation, lawmakers and citizens have begun to debate whether teenagers involved in sexting should be charged with such serious child-porn-related offenses. Some people argue that applying child pornography laws to teens is too harsh—that the purpose of these laws is to protect children, not prosecute them. In Utah, the legislature recently changed sexting from a felony to a misdemeanor.[88] Authorities in Montgomery County, Ohio, have created a program for teens arrested for sexting. This program will prevent some first-time offenders from being registered as sex offenders—a designation that can stay with them for up to 20 years.[89] In other instances, authorities have not prosecuted teens under child pornography laws, but rather brought them up on lesser charges.

The crackdown is part of a greater effort to send a message: Sexting can be dangerous. The practice can have serious consequences, which those who engage in it may not initially be aware of—creating a situation that can get out of hand very easily. In 2009, the National Center for Missing and Exploited Children reported that of the 2,100 children the center had identified as being victims of online child pornography, 24 percent took the photos themselves.[90] Sometimes these photos are used for purposes other than just distribution. In Wisconsin, a teenage boy enticed other boys at his school to sext him by pretending to be a girl. He then blackmailed seven of them into performing sexual favors for him in exchange for not distributing their photos.[91]

When photos depict a person who is over 18, the recipient often has the right to share these photos with others who are over 18—even when sharing is not the "right" thing to do. In these cases, there's no legal recourse for anyone to take to stop that person. In the meantime, underage participants—those who have the least life experience to help them make good decisions—are being held accountable. Until we learn how to deal with this new phenomenon appropriately in homes, in schools, and in the courts, many teens and young adults will likely continue to get hurt.

Discussion Questions

1. Does sexting represent a form of expression protected by the First Amendment?
2. What can be done to protect people from the dangers of sexting while still safeguarding our First Amendment rights?
3. What key points would you make if you had to speak about sexting to a group of parents and teenagers?

End Notes

1 WikiLeaks, http://213.251.145.96 (accessed March 19, 2011).

2 Associated Press, "WikiLeaks Releases Largest Leak in U.S. History," *ABC WTVG*, October 22, 2010, http://abclocal.go.com/wtvg/story?section=news/national_world&id=7741331.

3 "A WikiLeaks Timeline," *National Post*, November 28, 2010, http://news.nationalpost.com/2010/11/28/a-wikileaks-timeline.

4 Susan Welch, John Gruhl, John Comer, Susan Rigdon, and Margery Abrosius, *Understanding American Government* (Boston: Wadsworth, 2001).

5 Declan McCullagh, "GOP Senator Proposes Law Targeting WikiLeaks 'Cowards,'" *CNET*, October 26, 2010, http://news.cnet.com/8301-13578_3-20020738-38.html.

6 David Leigh and Rob Evans, "WikiLeaks Says Funding Has Been Blocked After Government Blacklisting," *Guardian*, October 14, 2010, www.guardian.co.uk/media/2010/oct/14/wikileaks-says-funding-is-blocked.

7 Gregg Keizer, "WikiLeaks Moves to Amazon Servers after DoS Attacks," *Computerworld*, November 29, 2010, www.computerworld.com/s/article/9198418/WikiLeaks_moves_to_Amazon_servers_after_DoS_attacks.

8 Patrick Thibodeau, "With WikiLeaks, Amazon Shows Its Power Over Customers," *Computerworld*, December 2, 2010, www.computerworld.com/s/article/9199258/With_WikiLeaks_Amazon_shows_its_power_over_customers.

9 Ravi Somaiya, "Hundreds of WikiLeaks Mirror Sites Appear," *New York Times*, December 5, 2010, www.nytimes.com/2010/12/06/world/europe/06wiki.html?_r=2.

10 Leslie Horn, "'Anonymous' Launches DDoS Attacks Against WikiLeaks Foes," *PCMagazine*, December 8, 2010, www.pcmag.com/article2/0,2817,2374023,00.asp.

11 Arthur Bright, "WikiLeaks' Julian Assange Arrested in London on Rape Charges," *Christian Science Monitor*, December 7, 2010, www.csmonitor.com/World/terrorism-security/2010/1207/WikiLeaks-Julian-Assange-arrested-in-London-on-rape-charges.

12 Carole Cadwalladr and Paul Harris, "WikiLeaks Founder Accuses US Army of Failing to Protect Afghan Informers," *Guardian*, August 1, 2010, www.guardian.co.uk/media/2010/aug/01/julian-assange-wikileaks-afghanistan-us.

13 Aloysius Agendia, "In Defense of Freedom of Expression: Journalists Worldwide Rally Behind WikiLeaks," posted November 8, 2010, http://agendia.jigsy.com/entries/international/in-defence-of-freedom-of-expression-journalists-worldwide-rally-behind-wikileaks.

14 Raffi Khatchadourian, "No Secrets: Julian Assange's Mission For Total Transparency," *New Yorker*, June 7, 2010, www.newyorker.com/reporting/2010/06/07/100607fa_fact_khatchadourian?currentPage=1.

15 Courtney Macavinta, "The Supreme Court Today Rejected the Communications Decency Act," *CNET*, June 26, 1997, http://news.cnet.com/High-court-rejects-CDA/2009-1023_3-200957.html.

[16] *Reno, Attorney General of the United States v. American Civil Liberties Union, et al.*, 521 U.S. 844 (1997), Legal Information Institute, Cornell University Law School, www.law.cornell.edu/supct/html/96-511.ZS.html (accessed February 17, 2011).

[17] Title XIV—Child Online Protection Act, Electronic Privacy Information Center, http://epic.org/free_speech/censorship/copa.html (accessed February 18, 2011).

[18] *Ashcroft v. American Civil Liberties Union* (03-218), 542 U.S. 656 (2004), Legal Information Institute, Cornell University Law School, www.law.cornell.edu/supct/html/03-218.ZS.html (accessed February 19, 2011).

[19] Joel Engardio, "Internet Filters, Voluntary OK, Not Government Mandate," *Blog of Rights*, American Civil Liberties Union, January 26, 2009, www.aclu.org/2009/01/26/internet-filters-voluntary-ok-not-government-mandate.

[20] "2011 Best Internet Filter Software and Comparisons," *Top Ten Reviews*, http://internet-filter-review.toptenreviews.com, February 17, 2011.

[21] Family Online Safety Institute, "FOSI and Member Companies Promote Careful Clicking on Safer Internet Day," February 09, 2010 www.fosi.org/fosi-press-releases-2010/57-pr-2010/571-february-09-2010.html.

[22] "About Clearsail/Family.net, www.clearsail.net/about.htm (accessed February 17, 2011).

[23] Federal Communications Commission, "Children's Internet Protection Act: FCC Consumer Facts," www.fcc.gov/cgb/consumerfacts/cipa.html (accessed February 18, 2011).

[24] *United States v. American Library Association*, Supreme Court Online, Duke Law, www.law.duke.edu/publiclaw/supremecourtonline/editedcases/univame.html (accessed March 17, 2011).

[25] "Internet Censorship: Enemies of the Internet," *OnlyInfoGraphic*, February 8, 2011, www.onlyinfographic.com/2011/internet-censorship-enemies-of-the-internet/.

[26] Danny O'Brien, "Is Brazil the Censorship Capital of the Internet? Not Yet," *CPJ Blog*, Committee to Protect Journalists, April 28, 2010, www.cpj.org/blog/2010/04/is-brazil-the-censorship-capital-of-the-internet.php.

[27] Abraham Hyatt, "Enemies of the Internet: Not Just for Dictators Anymore," *ReadWriteWeb*, March 11, 2010, www.readwriteweb.com/archives/enemies_of_the_internet_not_just_for_dictators_anymore.php.

[28] Jonathan Zittrain and Benjamin Edelman, "Empirical Analysis of Internet Filtering in China," Berkman Center for Internet & Society, Harvard Law School, http://cyber.law.harvard.edu/filtering/china/ (accessed March 17, 2011).

[29] Lance Whitney, "Report Names Enemies of the Internet," *CNET*, March 15, 2010, http://news.cnet.com/8301-13578_3-10468332-38.html.

[30] Abraham Hyatt, "Enemies of the Internet: Not Just for Dictators Anymore," *ReadWriteWeb*, March 11, 2010, www.readwriteweb.com/archives/enemies_of_the_internet_not_just_for_dictators_anymore.php.

[31] Lance Whitney, "Report Names Enemies of the Internet," *CNET*, March 15, 2010, http://news.cnet.com/8301-13578_3-10468332-38.html.

[32] Frank Rich, "Journal; The 2 Tim McVeighs," *New York Times*, January 17, 1998, www.nytimes.com/1998/01/17/opinion/journal-the-2-tim-mcveighs.html?n=Top/Reference/Times%20Topics/Subjects/H/Homosexuality.

[33] Jason Kincaid, "VC Firm Subpoenas TheFunded for Negative Review," *TechCrunch* (blog), August 12, 2008, http://techcrunch.com/2008/08/12/vc-firm-confirms-that-its-clueless-subpoenas-thefunded-for-negative-review/#.

[34] "PrePaid Legal v. Sturtz Case Archive." Electronic Frontier Foundation. http://w2.eff.org/Censorship/SLAPP/Discovery_abuse/PrePaid_Legal_v_Sturtz/?f=20010712_proposed_order.html (accessed March 17, 2011).

[35] *John Doe v. 2TheMart.com Inc.,* Berkman Center for Internet & Society, Harvard Law School, http://cyber.law.harvard.edu/stjohns/2themart.html (accessed March 17, 2011).

[36] AOL, "AOL Member Community Guidelines," http://help.aol.com/help/viewContent.do?externalId=221226 (accessed February 19, 2011).

[37] Selma Milovanovic, "Online Hate Mail Threat to Arson Case," *The Age*, February 17, 2009, www.theage.com.au/national/online-hate-mail-threat-to-arson-case-20090216-899r.html.

[38] Evan Hansen, "Google Blogger: 'I Was Terminated,'" *CNET*, February 11, 2005, http://news.cnet.com/Google-blogger-I-was-terminated/2100-1038_3-5572936.html.

[39] Gilbert Wondracek, Thorsten Holz, Christian Platzer, Engin Kirda, and Christopher Kruegel, "Is the Internet for Porn? An Insight Into the Online Adult Industry," International Secure Systems Lab, www.iseclab.org/papers/weis2010.pdf (accessed March 19, 2011).

[40] Jerry Ropelato, "Internet Pornography Statistics," *TopTenREVIEWS*, http://internet-filter-review.toptenreviews.com/internet-pornography-statistics.html#anchor2 (accessed February 13, 2011).

[41] *Ashcroft v. American Civil Liberties Union* (03-218), 542 U.S. 656 (2004), Legal Information Institute, Cornell University Law School, www.law.cornell.edu/supct/html/03-218.ZS.html (accessed February 19, 2011).

[42] Eric Goldman, "Internet Obscenity Conviction Requires Assessment of National Community Standards—US v. Kilbride" *Technology & Marketing Law Blog*, October 30, 2009, blog.ericgoldman.org/archives/2009/10/internet_obscen.htm.

[43] David D. Johnson, "U.S. v. Little: Emerging Circuit Split on Whether National Community Standards Should Be Applied in Internet Obscenity Cases," *Internet and E-commerce Law Blog*, February 11, 2010, www.internetecommercelaw.com/2010/02/articles/internet-decency-and-defamatio/us-v-little-emerging-circuit-split-on-whether-national-community-standards-should-be-applied-in-internet-obscenity-cases.

[44] Stanley Holdritch, "Porn in the Workplace on the Rise," *InternetSafety.com* (blog), April 28, 2010, http://blog.internetsafety.com/2010/04/28/porn-in-the-workplace-on-the-rise.

[45] National Conference of State Legislatures, "Child Pornography Reporting Requirements (ISPs and IT Workers)," December 31, 2008.

[46] "Sex and Tech: Results from a Survey of Teens and Young Adults," National Campaign to Prevent Teen and Unplanned Pregnancy, www.thenationalcampaign.org/sextech/PDF/SexTech_Summary.pdf (accessed February 25, 2011).

47 U.S. House of Representatives, 15 USC Chapter 103—Controlling The Assault of Non-Solicited Pornography and Marketing Office of the Law Revision Counsel, http://uscode.house.gov/download/pls/15C103.txt (accessed March 24, 2011).

48 Dan Goodin, "Spammers Ordered to Pay Tiny ISP Whopping $2.4M," *The Register*, May 6, 2010, www.theregister.co.uk/2010/05/06/spam_judgment/print.html.

49 Dan Goodin, "Veteran Spam Suit Troll Plaintiff Calls It Quite," *The Register*, September 14, 2010, at www.theregister.co.uk/2010/09/14/asis_calls_it_quits/print.html.

50 Ken Magill, "Porn CAN-SPAM Conviction Upheld," *Direct*, September 11, 2007, http://directmag.com/email/news/porn_can-spam_conviction/#.

51 Scott Bradner, "The CAN-SPAM Act as a Warning," *Network World*, January 6, 2009, www.networkworld.com/columnists/2009/010609bradner.html.

52 David Ardia, "Finkel v. Facebook: Court Rejects Defamation Claim Against Facebook Premised on 'Ownership' of User Content," Citizen Media Law Project (blog post), October 21, 2009, www.citmedialaw.org/blog/2009/finkel-v-facebook-court-rejects-defamation-claim-against-facebook-premised-ownership-user-.

53 Chris Matyszczyk, "Teen Sues Facebook, Classmates over Cyberbullying," *CNET*, March 3, 2009, http://news.cnet.com/8301-17852_3-10187531-71.html.

54 Title 47, Chapter 5, Subchapter II, Part I, § 230, "Protection for Private Blocking and Screening of Offensive Material," Legal Information Institute, Cornell University Law School, www.law.cornell.edu/uscode/html/uscode47/usc_sec_47_00000230----000-.html (accessed February 22, 2011).

55 Supreme Court of the State of the New York Count of New York, *Denise E. Finkel, Plaintiff against Facebook, Inc., Michael Dauber, Jeffrey Schwarz, Melinda Danowitz, Leah Herz, Richard Dauber, Amy Schwartz, Elliot Schwartz, Martin Danowitz, Bari Danowitz, Alan Herz, and Ellen Herz, Defendants*, Affirmation in Opposition, Index No. 101578-09, www.citmedialaw.org/sites/citmedialaw.org/files/2009-03-26-Finkel%20Opposition%20to%20Facebook%20Motion%20to%20Dismiss.pdf (accessed February 22, 2011).

56 *Finkel v. Facebook,* Citizen Media Law Project, www.citmedialaw.org/threats/finkel-v-facebook (accessed February 22, 2011).

57 *Finkel v. Facebook,* Citizen Media Law Project, www.citmedialaw.org/threats/finkel-v-facebook (accessed February 22, 2011).

58 Sameer Hinduja and Justin W. Patchin, "State Cyberbullying Laws: A Brief Review of State Cyberbullying Laws and Policies," Cyberbullying Research Center, www.cyberbullying.us/Bullying_and_Cyberbullying_Laws.pdf (accessed February 23, 2011).

59 *Tinker v. Des Moines School Dist.*, 393 U.S. 503 (1969), First Amendment Center, www.firstamendmentschools.org/freedoms/case.aspx?id=404 (accessed February 23, 2011).

60 David Kravets, "Rulings Leave Online Student Speech Rights Unresolved." *Wired*, February 4, 2010, www.wired.com/threatlevel/2010/02/rulings-leave-us-student-speech-rights-unresolved.

61 "I Couldn't Get Them to Stop," Polk County Public Schools, www.polk-fl.net/parents/general information/documents/bullying_jeffreyjohnstonstory.pdf (accessed February 24, 2011).

[62] The 2010 Florida Statutes, 1006.147 "Bullying and Harassment Prohibited," Online Sunshine, The Florida Legislature, www.leg.state.fl.us/Statutes/index.cfm?App_mode=Display_Statute&Search_String=&URL=1000-1099/1006/Sections/1006.147.html (accessed February 24, 2011).

[63] Amar Toor, "Teens Arrested for Cyberbullying Classmate with Fake Facebook Profile, Nude Photos," *Switched*, January 14, 2011, www.switched.com/2011/01/14/fake-facebook-profile-cyberbullying-gets-teens-arrested/.

[64] Lisa Evers, "Sources: Teenager Kills Himself After Facebook Taunts," *Fox 5 News*, January 19, 2011, www.myfoxny.com/dpp/news/local_news/new_york_state/sources-teenager-kills-himself-after-facebook-taunts-20110119.

[65] Sameer Hinduja and Justin W. Patchin, "State Cyberbullying Laws: A Brief Review of State Cyberbullying Laws and Policies," Cyberbullying Research Center, www.cyberbullying.us/Bullying_and_Cyberbullying_Laws.pdf (accessed February 25, 2011).

[66] Alexis Madrigal, "Chinese Paper Takes Down Story on Great Firewall's Creator," *Atlantic*, February 21, 2011, www.theatlantic.com/technology/archive/2011/02/chinese-paper-takes-down-story-on-great-firewalls-creator/71517.

[67] Sarah Lai Stirland, "Cisco Leak 'Great Firewall of China' Was A Chance to Sell More Routers," *Wired*, May 20, 2008, www.wired.com/threatlevel/2008/05/leaked-cisco-do.

[68] Alexis Madrigal, "Chinese Paper Takes Down Story on Great Firewall's Creator," *The Atlantic*, February 21, 2011, www.theatlantic.com/technology/archive/2011/02/chinese-paper-takes-down-story-on-great-firewalls-creator/71517.

[69] Melanie Lee, "Analysis: A Year After China Retreat, Google Plots New Growth," *Reuters*, January 13, 2011, www.reuters.com/article/2011/01/13/us-google-china-idUSTRE70C1X820110113.

[70] Tom Krazit, "Dynasty Denied, Google Rethinks China," *CNET*, January 12, 2010, http://news.cnet.com/8301-30684_3-10433745-265.html.

[71] Tom Krazit, "Dynasty Denied, Google Rethinks China," *CNET*, January 12, 2010, http://news.cnet.com/8301-30684_3-10433745-265.html.

[72] Michael Arrington, "Google Defends Against Large Scale Chinese Cyber Attack: May Cease Chinese Operations," *TechCrunch*, January 12, 2010, http://techcrunch.com/2010/01/12/google-china-attacks/.

[73] Ellen Nakashima, Steven Mufson, and John Pomfret, "Google Threatens to Leave China After Attacks on Activists' E-Mail," *Washington Post*, January 13, 2010, www.washington post.com/wp-dyn/content/article/2010/01/12/AR2010011203024.html.

[74] Ellen Nakashima, Steven Mufson, and John Pomfret, "Google Threatens to Leave China After Attacks on Activists' e-Mail," *Washington Post*, January 13, 2010, www.washington post.com/wp-dyn/content/article/2010/01/12/AR2010011203024.html.

[75] Ellen Nakashima, Steven Mufson, and John Pomfret, "Google Threatens to Leave China After Attacks on Activists' e-Mail," *Washington Post*, January 13, 2010, www.washington post.com/wp-dyn/content/article/2010/01/12/AR2010011203024.html.

[76] "Internet Censorship in China," *New York Times*, July 9, 2010, http://topics.nytimes.com/topics/news/international/countriesandterritories/china/internet_censorship/index.html.

[77] John C. Abell, "China Renews Google's License," *Wired*, July 9, 2010, www.wired.com/epicenter/2010/07/china-renews-googles-license.

[78] John C. Abell, "China Renews Google's License," *Wired*, July 9, 2010, www.wired.com/epicenter/2010/07/china-renews-googles-license.

[79] Siobhan Gorman and Jessica E. Vascellaro, "Google Working with NSA to Investigate Cyber Attack," *Wall Street Journal*, February 4, 2010, http://online.wsj.com/article/SB10001424052748704041504575044920905689954.html.

[80] Ryan Singel, "Whistle Blower Outs NSA Spy Room," *Wired*, April 7, 2006, www.wired.com/science/discoveries/news/2006/04/70619.

[81] Mike Celizic, "Her Teen Committed Suicide Over 'Sexting,'" *TodayShow.com*, March 6, 2009, www.msnbc.msn.com/id/29546030.

[82] Sheree Paolello, "Mom Loses Daughter over 'Sexting,' Demands Accountability," *WLWT.com, March 5*, 2009, www.wlwt.com/news/18866515/detail.html.

[83] "Teen Nabbed for Naked MySpace Photos," *The Smoking Gun*, May 21, 2008, www.thesmokinggun.com/archive/years/2008/0521081myspace1.html.

[84] "UPDATE: Teen Put Girl's Nude Pics on MySpace," *nbc15.com*, March 6, 2009, www.nbc15.com/state/headlines/38449989.html.

[85] Jennifer Baker, "Sex Images Found on Boy's Phone," *Cincinnati.com*, March 19, 2009, http://news.cincinnati.com/article/20090319/NEWS0107/303190036.

[86] Bianca Prieto, "Teens Learning There Are Consequences to 'Sexting,'" *Seattle Times*, March 11, 2009, http://seattletimes.nwsource.com/html/nationworld/2008845324_sexting12.html.

[87] Dahlia Lithwick, "Teens, Nude Photos and the Law," *Newsweek*, February 14, 2009, www.newsweek.com/2009/02/13/teens-nude-photos-and-the-law.html.

[88] Wendy Koch, "Teens Caught 'Sexting' Face Porn Charges," *USA Today*, March 11, 2009, www.usatoday.com/tech/wireless/2009-03-11-sexting_N.htm.

[89] Lou Grieco, "County Eases 'Sexting' Penalty," *Dayton Daily News*, March 5, 2009, www.daytondailynews.com/n/content/oh/story/news/local/2009/03/05/ddn030509sexting.html.

[90] Wendy Koch, "Teens Caught 'Sexting' Face Porn Charges," *USA Today*, March 11, 2009, www.usatoday.com/tech/wireless/2009-03-11-sexting_N.htm.

[91] "Boy Posing as Girl on Facebook Extorts Sex," *CBS News*, February 5, 2009, www.cbsnews.com/stories/2009/02/05/national/main4777194.shtml?source=RSSattr=HOME_4777194.

CHAPTER 6

INTELLECTUAL PROPERTY

QUOTE

The advancement of the arts, from year to year, taxes our credulity, and seems to presage the arrival of that period when human improvement must end. (This quote is often misstated as—everything which can be invented has already been patented).[1]

—Henry Ellsworth, Commissioner of Patents, 1843

VIGNETTE

Smartphone Patent Wars

In January 2011, the research firm Canalys announced that Google's Android had finally surpassed Nokia's proprietary Symbian operating system as the most popular smartphone operating system worldwide.[2] It was not only a victory for the open source software movement, but a turning point in the highly competitive race to capture market share in the smartphone industry. The race began in 2007 when Apple released its iPhone, the first highly successful smartphone, with a touchscreen and top-notch Web-browsing capabilities. Less than a year later, Google released Android (an open source operating system for mobile devices) to the entire mobile industry. The release presented a challenge to Apple, Nokia, and Blackberry producer RIM, who were big movers in the market and who each had their own proprietary operating systems. In 2009, Motorola delivered the Droid, the first major Android success. The Droid sold over 1 million units in its first few months on the market.[3] Then in 2010, Taiwan-based smartphone manufacturer HTC came out with the Android-based EVO 4G, which ran on Sprint, the fastest wireless network in the United States. Google released its own Nexus One that same year.[4] Suddenly, Apple, Nokia, and RIM faced a great deal of competition.

Perhaps not surprisingly, in 2010, the competition for market share disintegrated into an all-out patent war in the U.S. court system. The battle focused not on operating systems, but rather on other technologies involved in the development of smartphones. On March 2, Apple filed a lawsuit against HTC, claiming that HTC had infringed on 20 of Apple's patents.[5] The patents involve the touchscreens, graphical user interfaces, multitouch touch pads, signal processing, and other features developed for Apple's iPhone. Simultaneously, Apple filed a complaint with the International Trade Commission (ITC) against HTC in the hopes that the commission would block imports of HTC handsets that violated Apple patents. Analysts speculated that this was an indirect attack on Google and its Android operating system.[6] Because both Google and Microsoft provided the underlying operating system technology on HTC's smartphones, these companies might be forced to defend their own products if Apple were to be successful in its legal battle.[7] Originally caught off guard by Apple's lawsuit, HTC responded two months later by filing a complaint against Apple, accusing it of infringing on five HTC patents pertaining to power management and to hardware and software needed to implement directories used in mobile phones. HTC asked the ITC to block the import of Apple's iPhone, iPad, and iPod.[8]

Then, in October, Motorola filed four lawsuits alleging that Apple had violated 18 of its patents. In a separate move shortly after, Motorola asked a federal judge to invalidate Apple's multitouch patents, including those named in Apple's lawsuit against HTC. Apple countered later that month with lawsuits against Motorola, claiming the Droid violated two patents: one pertaining to the Apple's multitouch feature and the other to its user interface.[9]

That same month, Microsoft filed a patent infringement lawsuit against Motorola. Microsoft alleged that Motorola's Android violated patents related to how the smartphone synchronizes emails, contacts, and calendars; how it schedules meetings; and how it alerts applications to changes in battery power or signal strength.[10] By the end of 2010, every major player in the smartphone market was engaged in a patent war with at least one competitor.[11]

Today, Nokia holds a leading position in the mobile industry, with almost 30 percent of the market share, primarily because Apple, RIM, and others have focused on smartphone products that target advanced markets. However, in the fourth quarter of 2010, over 50 percent of all global smartphone sales were limited to Western Europe and North America. As smartphone prices continue to decline and mobile networks improve around the world, smartphones will grab an increasing share of the market. Already in 2010, Nokia's market share fell by 7.5 points compared to the previous year. Apple, meanwhile, saw an 87.2 percent growth in 2009. This resulted largely from its expansion into new countries, but the increase also came as a result of the expiration of exclusive sales agreements with certain communication service providers, including AT&T. The iPhone is now available through 185 communication service providers, including Verizon Wireless. This means that the communication service providers will compete to offer the Apple iPhone more cheaply.[12]

Some analysts argue that the patent infringement battles won't have an enormous impact on market share. The major competitors have carefully built up portfolios of intellectual property, with tens of thousands of patents. They are spending time and research to uncover where the other companies might be infringing on their patents. "But, at the end of the day," one *Wired* analyst writes, "all the major players in the market will end up signing licensing agreements with one another. Motorola could pocket a couple of bucks from each iPhone sale, while Apple might rake in a small amount each time an HTC Touch is sold."[13]

Yet, in March 2006, RIM was forced to hand over $612.5 million to patent holding company NTP to avoid being shut out of the U.S. market. NTP is sometimes derogatorily referred to as a "patent troll" because it holds patents but does not develop technology using these patents. It frequently sues other companies that do use its intellectual property.[14] However, NPT also arranges license agreements with technology companies, such as Nokia, to use its intellectual property.[15] In 2001, NPT sued RIM for patent infringement in the development of its popular Blackberry mobile phone. In 2006, U.S. district court Judge James Spencer threatened to initiate an injunction on the U.S. sale and support of RIM's Blackberry if it did not reach a settlement agreement with NTP.[16] Within days, RIM agreed to pay NTP $612.5 million. The litigation had gone on for nearly five years, and the settlement agreement was a setback for RIM, but the company saw it as the only way to ensure Blackberry's future.[17]

If the current spate of patent infringement cases follows a similar course, companies could have to pay out significant royalties or face being shut out of the American market. These patent wars may, therefore, have a significant impact on the race to capture smartphone market share.

Questions to Consider

1. Why are there so many patent disputes among the manufacturers of smartphones?
2. To what degree do patents and related lawsuits slow down innovation in the information technology industry?

LEARNING OBJECTIVES

As you read this chapter, consider the following questions:

1. What does the term *intellectual property* encompass, and why are organizations so concerned about protecting intellectual property?
2. What are the strengths and limitations of using copyrights, patents, and trade secret laws to protect intellectual property?
3. What is plagiarism, and what can be done to combat it?
4. What is reverse engineering, and what issues are associated with applying it to create a lookalike of a competitor's software program?
5. What is open source code, and what is the fundamental premise behind its use?
6. What is the essential difference between competitive intelligence and industrial espionage, and how is competitive intelligence gathered?
7. What is cybersquatting, and what strategy should be used to protect an organization from it?

WHAT IS INTELLECTUAL PROPERTY?

Intellectual property is a term used to describe works of the mind—such as art, books, films, formulas, inventions, music, and processes—that are distinct and owned or created by a single person or group. Intellectual property is protected through copyright, patent, and trade secret laws.

Copyright law protects authored works, such as art, books, film, and music; patent law protects inventions; and trade secret law helps safeguard information that is critical to an organization's success. Together, copyright, patent, and trade secret legislation forms a complex body of law that addresses the ownership of intellectual property. Such laws can also present potential ethical problems for IT companies and users—for example, some innovators believe that copyrights, patents, and trade secrets stifle creativity by making it harder to build on the ideas of others. Meanwhile, the owners of intellectual property want to control and receive compensation for the use of their intellectual property. Should the need for ongoing innovation or the rights of property owners govern how intellectual property is used?

Defining and controlling the appropriate level of access to intellectual property are complex tasks. For example, protecting computer software has proven to be difficult because it has not been well categorized under the law. Software has sometimes been treated as the expression of an idea, which can be protected under copyright law. In other cases, has been treated as a process for changing a computer's internal structure, making it eligible for protection under patent law. At one time, software was even judged to be a series of mental steps, making it inappropriate for ownership and ineligible for any form of protection.

COPYRIGHTS

Copyright and patent protection was established through the U.S. Constitution, Article I, section 8, clause 8, which specifies that Congress shall have the power "to promote the Progress of Science and useful Arts, by securing for limited Times to Authors and Inventors the exclusive Rights to their respective Writings and Discoveries."

A **copyright** is the exclusive right to distribute, display, perform, or reproduce an original work in copies or to prepare derivative works based on the work. Copyright protection is granted to the creators of "original works of authorship in any tangible medium of expression, now known or later developed, from which they can be perceived, reproduced, or otherwise communicated, either directly or with the aid of a machine or device."[18] The author may grant this exclusive right to others. As new forms of expression develop, they can be awarded copyright protection. For example, in the Copyright Act of 1976, audiovisual works were added and computer programs were assigned to the literary works category.

Copyright infringement is a violation of the rights secured by the owner of a copyright. Infringement occurs when someone copies a substantial and material part of another's copyrighted work without permission. The courts have a wide range of discretion in awarding damages—from $200 for innocent infringement to $100,000 for willful infringement.

Copyright Term

Copyright law guarantees developers the rights to their works for a certain amount of time. Since 1960, the term of copyright has been extended 11 times from its original limit of 28 years. The Copyright Term Extension Act, also known as the Sonny Bono Copyright Term Extension Act (after the legislator, and former singer/entertainer, who was one of the co-sponsors of the bill in the House of Representatives), signed into law in 1998, established the following time limits:

- For works created after January 1, 1978, copyright protection endures for the life of the author plus 70 years.
- For works created but not published or registered before January 1, 1978, the term endures for the life of the author plus 70 years, but in no case expires earlier than December 31, 2004.
- For works created before 1978 that are still in their original or renewable term of copyright, the total term was extended to 95 years from the date the copyright was originally secured.[19]

These extensions were primarily championed by movie studios concerned about retaining rights to their early films. Opponents argued that lengthening the copyright period made it more difficult for artists to build on the work of others, thus stifling creativity and innovation. The Sonny Bono Copyright Term Extension Act was legally challenged by Eric Eldred, a bibliophile who wanted to put digitized editions of old books online. The case went all the way to the Supreme Court, which ruled the act constitutional in 2003.[20]

Eligible Works

The types of work that can be copyrighted include architecture, art, audiovisual works, choreography, drama, graphics, literature, motion pictures, music, pantomimes, pictures, sculptures, sound recordings, and other intellectual works, as described in Title 17 of the U.S. Code. To be eligible for a copyright, a work must fall within one of the preceding categories, and it must be original. Copyright law has proven to be extremely flexible in covering new technologies; thus, software, video games, multimedia works, and Web pages can all be protected. However, evaluating the originality of a work is not always a straightforward process, and disagreements over whether or not a work is original sometimes lead to litigation. For example, former Beatles member George Harrison was entangled for decades in litigation over similarities between his hit "My Sweet Lord," released in 1970, and "He's So Fine," composed by Ronald Mack and recorded by the Chiffons in 1962.[21]

Some works are not eligible for copyright protection, including those that have not been fixed in a tangible form of expression (such as an improvisational speech) and those that consist entirely of common information that contains no original authorship, such as a chart showing conversions between European and American units of measure.

Fair Use Doctrine

Copyright law tries to strike a balance between protecting an author's rights and enabling public access to copyrighted works. The fair use doctrine was developed over the years as courts worked to maintain that balance. The **fair use doctrine** allows portions of copyrighted materials to be used without permission under certain circumstances. Title 17, section 107, of the U.S. Code established that courts should consider the following four factors when deciding whether a particular use of copyrighted property is fair and can be allowed without penalty:

- The purpose and character of the use (such as commercial use or nonprofit, educational purposes)
- The nature of the copyrighted work
- The portion of the copyrighted work used in relation to the work as a whole
- The effect of the use on the value of the copyrighted work[22]

The concept that an idea cannot be copyrighted but the expression of an idea can be is key to understanding copyright protection. For example, an author cannot copy the exact words that someone else used to describe his feelings during a World War II battle, but he can convey the sense of horror that the other person expressed. Also, there is no copyright infringement if two parties independently develop a similar or even identical work. For example, if two writers happened to use the same phrase to describe a key historical figure, neither would be guilty of infringement. Of course, independent creation can be extremely difficult to prove or disprove.

Software Copyright Protection

The use of copyrights to protect computer software raises many complicated issues of interpretation. For example, a software manufacturer can observe the operation of a competitor's copyrighted program and then create a program that accomplishes the same result and performs in the same manner. To prove infringement, the copyright holder must show a striking resemblance between its software and the new software that could be explained only by copying. However, if the new software's manufacturer can establish that it developed the program on its own, without any knowledge of the existing program, there is no infringement. For example, two software manufacturers could conceivably develop separate programs for a simple game such as tic-tac-toe without infringing the other's copyright.

One area that holds the potential for software copyright infringement involves the sale of refurbished consumer computer supplies, such as toner and inkjet cartridges. Lexmark International, a manufacturer and supplier of printers and associated supplies, objected to the use of refurbished cartridges. In 2002, Lexmark filed suit against Static Control Components (SCC), a producer of components used to make refurbished printer cartridges. The suit alleged that SCC's Smartek chips included Lexmark software in violation of copyright law. (The software is necessary to allow the refurbished toner cartridges to work with Lexmark's printers.) In February 2003, Lexmark was granted an injunction that prevented SCC from selling the chips until the case could be resolved at trial. However, the ruling was overturned in October 2003 by the Sixth Circuit Court of Appeal, which said that copyright law should not be used to inhibit interoperability between the products of rival vendors. The appeals court upheld its own decision in February 2005.[23]

The Prioritizing Resources and Organization for Intellectual Property (PRO-IP) Act of 2008

The Prioritizing Resources and Organization for Intellectual Property (PRO-IP) Act of 2008 increased trademark and copyright enforcement and substantially increased penalties for infringement. For example, the penalty for infringement of a 10-song album was raised from $7,500 to $1.5 million. The law also created the Office of the United States Intellectual Property Enforcement Representative within the U.S. Justice Department. The agency was charged with creating and overseeing a Joint Strategic Plan against counterfeiting and privacy and coordinating the efforts of the many government agencies that deal with these issues.[24] Tom Donohue, the president of the U.S. Chamber of Commerce, stated, "The PRO-IP Act sends the message to [intellectual property] criminals everywhere that the [United States] will go the extra mile to protect American innovation." [25] Meanwhile, opponents of the act proclaimed that "its penalties were far too harsh and that it didn't balance users' rights and concerns over those of major software, media and pharmaceutical companies."[26]

General Agreement on Tariffs and Trade (GATT)

The original General Agreement on Tariffs and Trade (GATT) was signed in 1947 by 150 countries. Since then, there have been eight rounds of negotiations addressing various trade issues. The Uruguay Round, completed in December 1993, resulted in a trade agreement among 117 countries. This agreement also created the World Trade Organization (WTO) in Geneva, Switzerland, to enforce compliance with the agreement. GATT includes a section covering copyrights called the Agreement on Trade-Related Aspects of Intellectual Property Rights (TRIPS), discussed in the following section. U.S. law was amended to be essentially consistent with GATT through both the Uruguay Round Agreements Act of 1994 and the Sonny Bono Copyright Term Extension Act of 1998. Despite GATT, copyright protection varies greatly from country to country, and extreme caution must be exercised on all international usage of any intellectual property.

The WTO and the WTO TRIPS Agreement (1994)

The World Trade Organization (WTO) deals with rules of international trade based on WTO agreements that are negotiated and signed by representatives of the world's trading nations. The WTO is headquartered in Geneva, Switzerland, and had 153 member nations as of March 2011. Its goal is to help producers of goods and services, exporters, and importers conduct their business.[27]

Many nations recognize that intellectual property has become increasingly important in world trade, yet the extent of protection and enforcement of intellectual property rights varies around the world. As a result, the WTO developed the **Agreement on Trade-Related Aspects of Intellectual Property Rights**, also known as the TRIPS Agreement, to establish minimum levels of protection that each government must provide to the intellectual property of all WTO members. This binding agreement requires member governments to ensure that intellectual property rights can be enforced under their laws and that penalties for infringement are tough enough to deter further violations. Table 6-1 provides a brief summary of copyright, patent, and trade secret protection under the TRIPS Agreement.

TABLE 6-1 Summary of the WTO TRIPS Agreement

Form of intellectual property	Key terms of agreement
Copyright	Computer programs are protected as literary works. Authors of computer programs and producers of sound recordings have the right to prohibit the commercial rental of their works to the public.
Patent	Patent protection is available for any invention—whether a product or process—in all fields of technology without discrimination, subject to the normal tests of novelty, inventiveness, and industrial applicability. It is also required that patents be available and patent rights enjoyable without discrimination as to the place of invention and whether products are imported or locally produced.
Trade secret	Trade secrets and other types of undisclosed information that have commercial value must be protected against breach of confidence and other acts that are contrary to honest commercial practices. However, reasonable steps must have been taken to keep the information secret.

Source Line: World Trade Organization, "Overview: The TRIPS Agreement," *www.wto.org/english/tratop_e/trips_ e/intel2_e.htm*.

The World Intellectual Property Organization (WIPO) Copyright Treaty (1996)

The World Intellectual Property Organization (WIPO), headquartered in Geneva, Switzerland, is an agency of the United Nations established in 1967. WIPO is dedicated to developing "a balanced and accessible international intellectual property (IP) system, which rewards creativity, stimulates innovation and contributes to economic development while safeguarding the public interest."[28] It has 184 member nations and administers 24 international treaties. Since the 1990s, WIPO has strongly advocated for the interests of intellectual property owners. Its goal is to ensure that intellectual property laws are uniformly administered.

The WIPO Copyright Treaty, adopted in 1996, provides additional copyright protections to address electronic media. The treaty ensures that computer programs are protected as literary works and that the arrangement and selection of material in databases is also protected. It provides authors with control over the rental and distribution of their work, and prohibits circumvention of any technical measures put in place to protect the works. The WIPO Copyright Treaty is implemented in U.S. law through the Digital Millennium Copyright Act (DMCA), which is discussed in the next section.

The Digital Millennium Copyright Act (1998)

The **Digital Millennium Copyright Act (DMCA)** was signed into law in 1998 and implements two 1996 WIPO treaties: the WIPO Copyright Treaty and the WIPO Performances and Phonograms Treaty. The act is divided into the following five sections:[29]

- Title I (WIPO Copyright and Performances and Phonograms Treaties Implementation Act of 1998)—This section implements the WIPO treaties by making certain technical amendments to U.S. law in order to provide appropriate references and links to the treaties. It also creates two new prohibitions in the Copyright Act (Title 17 of the U.S. Code)—one on circumvention of technological measures used by copyright owners to protect their works and one on tampering with copyright management information. Title I also adds civil remedies and criminal penalties for violating the prohibitions.
- Title II (Online Copyright Infringement Liability Limitation Act)—This section amends the Copyright Act by adding a new section to create new limitations on liability for copyright infringement by online service providers.
- Title III (Computer Maintenance Competition Assurance Act)—This section permits the owner or lessee of a computer to make or authorize the making of a copy of a computer program in the course of maintaining or repairing that computer. The new copy cannot be used in any other manner and must be destroyed immediately after the maintenance or repair is completed.
- Title IV (Miscellaneous provisions)—This section adds language to the Copyright Act confirming the Copyright Office's authority to continue to perform the policy and international functions that it has carried out for decades under its existing general authority.
- Title V (Vessel Hull Design Protection Act)—This section creates a new form of protection for the original design of vessel hulls.

The portion of Title I dealing with anticircumvention provisions makes it an offense to do any of the following:

- Circumvent a technical protection
- Develop and provide tools that allow others to access a technologically protected work
- Manufacture, import, provide, or traffic in tools that enable others to circumvent protection and copy a protected work

Violations of these provisions carry both civil and criminal penalties, including up to five years in prison, a fine of up to $500,000 for each offense, or both. Unlike traditional copyright law, the statute does not govern copying; instead, it focuses on the distribution of tools and software that can be used for copyright infringement as well as for legitimate noninfringing use. Although the DMCA explicitly outlaws technologies that can defeat copyright protection devices, it does permit reverse engineering for encryption, interoperability, and computer security research.

Several cases brought under the DMCA have dealt with the use of software to enable the copying of DVD movies. For example, motion picture companies supported the development and worldwide licensing of the Content Scramble System (CSS), which enables a DVD player (shown in Figure 6-1) or a computer drive to decrypt, unscramble, and play back motion pictures on DVDs, but not copy them.

FIGURE 6-1 Several cases brought under the DMCA have dealt with the use of software to enable the copying of DVD movies

Credit: Image copyright Polat, 2009. Used under license from Shutterstock.com

However, a software program called DeCSS can break the encryption code and enable users to copy DVDs. The posting of this software on the Web in January 2000 led to a lawsuit by major movie studios against its author. After a series of cases, courts finally ruled that the use of DeCSS violated the DMCA's anticircumvention provisions.

Title II provides "safe harbors" for ISPs whose customers/subscribers might be breaking copyright laws by downloading, posting, storing, or sending copyrighted material via its services. These safe harbors are granted only if the ISP complies with clearly defined "notice and takedown" procedures that grant copyright holders a quick and simple way to halt access to allegedly infringing content. Copyright holders are also granted the right to issue subpoenas to alleged copyright holders identified through their ISP. Title II also provides defined procedures for ISP users to challenge improper takedowns.

The takedown procedure works as follows. The owners of copyrighted material who allege that their material has been infringed send a notice to the ISP hosting the content. The ISP forwards the notice to whoever was responsible for uploading material. That individual is given a chance to respond. If there is no response, the ISP must ensure that the material is no longer accessible.

Because many copyright infringers take measures to conceal their true identity, copyright owners must take additional steps if they wish to sue for copyright infringement. Provided a copyright owner has sent a DMCA notice, a John Doe subpoena can be obtained from a court clerk without even commencing a lawsuit. The subpoena compels the ISP to reveal the identity of the anonymous poster. The ISP is unlikely to resist the subpoena due to the associated legal costs.

During 2010, there were several copyright infringement lawsuits on behalf of adult film makers. Three separate copyright infringement lawsuits were filed against a total of more than 4,000 John Doe defendants by attorney Evan Stone on behalf of Larry Flynt Publishing for allegedly illegally sharing the movie "This Ain't Avatar XXX."[30] Separately, Ken Ford filed a copyright infringement lawsuit on behalf of Axel Braun Productions, a well-known adult film studio, alleging that some 7,098 defendants illegally shared the adult film "Batman XXX: A Porn Parody."[31] Ford filed eight additional mass lawsuits against over 15,000 alleged file swappers of adult-oriented films.[32] Overall, more than 100,000 peer-to-peer file sharers were sued for sharing movies during 2010.[33] Although each complaint specified thousands of John Doe defendants, the suits have not gained much traction due to jurisdiction issues and ISPs' reluctance to deal with these sorts of mass lawsuits.

The typical process for such lawsuits is that the IP addresses are collected for the alleged copyright violators. Attorneys then file a John Doe complaint in federal court and request the court to issue subpoenas to all ISPs used by the defendants. The subpoenas compel the ISPs to provide the defendants' name and other contact information. The attorneys then contact the defendants to offer them the opportunity to settle out of court and thus avoid embarrassment and legal fees.

In an unusual move, Time Warner Cable (TWC), a major ISP, refused to fully cooperate with a request to provide the identities of customers tied to the IP addresses that were identified as being used to illegally share film in the Larry Flynt Publishing case. According to Stone, TWC has agreed to provide only ten names per month.[34]

In another setback for the plaintiffs in these cases, a federal judge hearing the Ken Ford cases in West Virginia dismissed every defendant except one and demanded that each case be tried separately. This would require Ford to pay a $350 per case filing fee, which could amount to millions of dollars. The judge also instructed Ford to only re-submit those IP addresses likely to be connected to Internet users living in his jurisdiction of West Virginia. The judge stated that it is not proper to join that many defendants simply because they allegedly used the same peer-to-peer software and infringed the same movie. Each defendant might each have a completely different defense. In addition, the judge stated that the defendants should be tried separately in a court that has jurisdiction where the defendant resides.[35]

In response to these setbacks, The US Copyright Group (also known as Dunlap, Grubb & Weave or DGW), a copyright law firm, is joining forces with other law firms throughout the country in an effort that will enable them to "file the cases in the local court without any one law firm having to travel the entire country, seeking the right jurisdiction."[36]

Some see the DMCA as a boon to the growth of the Internet and its use as a conduit for innovation and freedom of expression. Without the safe harbors that the DMCA provides, the risk of copyright liability would be so great as to seriously discourage ISPs from hosting and transmitting user-generated content. Others see the DMCA as extending too much power to copyright holders. They share the viewpoint of Verizon General Counsel William P. Barr, who stated in testimony before Congress that the "broad and promiscuous subpoena procedure" of the DMCA grants "truly breathtaking powers to anyone who can claim to be or represent a copyright owner; powers that Congress has not even bestowed on law enforcement and national security personnel."[37]

PATENTS

A **patent** is a grant of a property right issued by the United States Patent and Trademark Office (USPTO) to an inventor. A patent permits its owner to exclude the public from making, using, or selling a protected invention, and it allows for legal action against violators. Unlike a copyright, a patent prevents independent creation as well as copying. Even if someone else invents the same item independently and with no prior knowledge of the patent holder's invention, the second inventor is excluded from using the patented device without permission of the original patent holder. The rights of the patent are valid only in the United States and its territories and possessions. Figure 6-2 shows the number of patents applied for and granted in recent years.

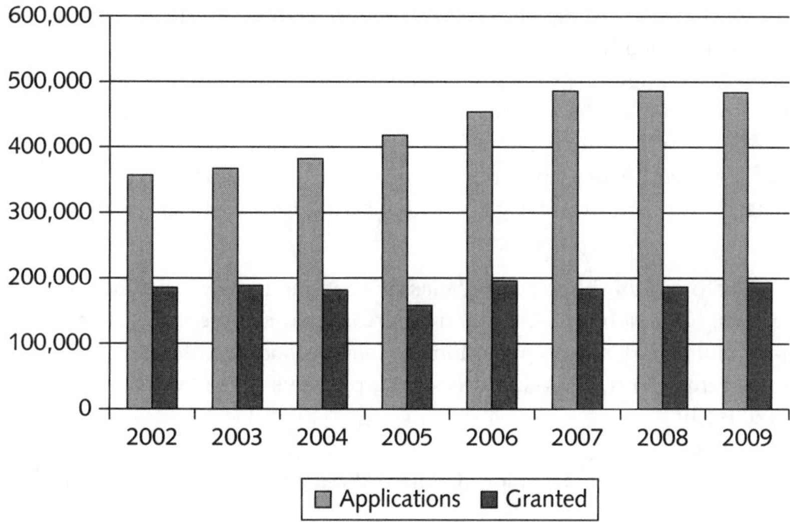

FIGURE 6-2 Patents applied for and granted
Source Line: U.S. Patent Statistics Calendar Years 1963–2010, www.uspto.gov/web/offices/ac/ido/oeip/taf/us_stat.pdf.

The value of patents to a company cannot be underestimated. IBM obtained 4,914 patents in 2009, the seventeenth consecutive year it received more patents than any other company in the United States.[38] It has been estimated that IBM's licensing of patents and technologies generates several hundred million dollars in annual revenue.[39] Table 6-2 lists the organizations that have been granted the most patents over the 40-year period from 1969 to 2009.

To obtain a U.S. patent, an application must be filed with the USPTO according to strict requirements. As part of the application, the USPTO searches the **prior art**—the existing body of knowledge available to a person of ordinary skill in the art—starting with patents and published material that have already been issued in the same area. The

TABLE 6-2 Organizations that received the most patents since 1986

Organization	Number of patents granted
IBM	58,121
Canon	38,202
Samsung	31,155
Toshiba	29,195
Hitachi	28,596
Sony	24,976

Source Line: U.S. Patent and Trademark Office, "All Technologies Report," © March 2011, www.uspto.gov/web/offices/ac/ido/oeip/taf/all_tech.pdf.

USPTO will not issue a patent for an invention whose professed improvements are already present in the prior art. Although the USPTO employs 6,225 patent examiners to research the originality of each patent application, the average time from filing until the application is issued as a patent or abandoned by the applicant has increased from 31.1 months in 2006 to 35.3 months in 2010. Such delays can be costly for companies that want to bring patented products to market quickly.[40] As a result, in many cases, people trained in the patent process, rather than the inventors themselves, prepare patent applications.

The main body of law that governs patents is contained in Title 35 of the U.S. Code, which states that an invention must pass the following four tests to be eligible for a patent:

- It must fall into one of five statutory classes of items that can be patented: (1) processes, (2) machines, (3) manufactures (such as objects made by humans or machines), (4) compositions of matter (such as chemical compounds), and (5) new uses in any of the previous four classes.
- It must be useful.
- It must be novel.
- It must not be obvious to a person having ordinary skill in the same field.[41]

The U.S. Supreme Court has ruled that three classes of items cannot be patented: abstract ideas, laws of nature, and natural phenomena. Standing on its own, mathematical subject matter is also not entitled to patent protection. Thus, Pythagoras could not have patented his formula for the length of the hypotenuse of a right triangle ($c^2 = a^2 + b^2$). The statute does not identify computer software, gene sequences, or genetically modified bacteria as patentable subject matter. However, these items have subsequently been determined to be patentable.

Patent infringement, or the violation of the rights secured by the owner of a patent, occurs when someone makes unauthorized use of another's patent. Unlike copyright infringement, there is no specified limit to the monetary penalty if patent infringement is found. In fact, if a court determines that the infringement is intentional, it can award up to three times the amount of the damages claimed by the patent holder. The most common defense against patent infringement is a counterattack on the claim of infringement and the validity of the patent itself. Even if the patent is valid, the plaintiff must still prove that every element of a claim was infringed and that the infringement caused some sort of damage.

Software Patents

A software patent claims as its invention some feature or process embodied in instructions executed by a computer. Prior to 1981, the courts regularly turned down requests for such patents, giving the impression that software could not be patented. In the 1981 *Diamond v. Diehr* case, the Supreme Court granted a patent to Diehr, who had developed a process control computer and sensors to monitor the temperature inside a rubber mold. The USPTO interpreted the court's reasoning to mean that just because an invention used software did not mean that the invention could not be patented. Based on this ruling, courts have slowly broadened the scope of protection for software-related inventions.[42]

The creation of the U.S. Court of Appeals for the Federal Circuit in 1982 further improved the environment for the use of patents in software-related inventions. This court is charged with hearing all patent appeals and is generally viewed as providing stronger enforcement of patents and more effective punishment for willful infringement.

Since the early 1980s, the USPTO has granted as many as 20,000 software-related patents per year.[43] Applications software, business software, expert systems, and system software have been patented, along with such software processes as compilation routines, editing and control functions, and operating system techniques. Many patents have been granted for business methods implemented in software. The question of whether business methods are patentable is a separate issue from the question of whether software is.

Some software experts think that too many software patents are being granted, and they believe that this inhibits new software development. For example, in September 1999, Amazon.com obtained a patent for "one-click shopping," based on the use of a shopping-cart purchase system for electronic commerce. In October 1999, Amazon.com sued Barnes & Noble for allegedly infringing this patent with its Express Lane feature. The filing of the suit prompted many complaints about the issuing of patents to business methods, which critics deride as being overly broad and unoriginal concepts that do not merit patents. Some critics considered one-click shopping little more than a simple combination of existing Web technologies. Following preliminary court hearings and the discovery that others had used the one-click technology before Amazon.com even began business, Amazon.com and Barnes & Noble settled out of court in March 2002.

Bilski v. Kappos was a business method patent case that provided the Supreme Court an opportunity in 2010 to set a clear precedent on how such patents should be judged in the future. Bernard Bilski and Rand Warsaw filed an application for a patent on certain methods of hedging risks associated with commodities trading. The Patent Office rejected the application due to Section 101 of the Copyright Right Act (Title 35 of the U.S. Code). Although this section does not definitely exclude business methods from being patented, it does exclude laws of nature, physical phenomena, and abstract ideas. The Patent Office felt that the risk management method represented nothing more than an abstract idea and so was not patentable. Bilski appealed the decision to the U.S. Court of Appeals for the Federal Circuit, which affirmed the Patent Office's rejection of the application. Bilski then appealed to the Supreme Court, which also affirmed the rejection. Unfortunately, the Supreme Court failed to set any clear criteria for when a business process can be patented and when it cannot.[44] According to Justice Anthony Kennedy, "The patent application here can be rejected under our precedents on the unpatentability of abstract ideas. The court, therefore, need not define further what constitutes a patentable process." In his ruling, Justice Kennedy rejected reasoning advanced by the U.S. Court of Appeals for the Federal Circuit that defined what types of nonphysical processes can be patented. Kennedy stated that such a change "would create uncertainty as to the patentability of software" and "inventions based on linear programming, data compression, and the manipulation of digital signals."[45] As a result, there is still no clear, definitive test for when software can be patented and when it cannot.

Cross-Licensing Agreements

Many large software companies have cross-licensing agreements in which each party agrees not to sue the other over patent infringements. For example, since 2003 Microsoft has put in place more than 600 licensing agreements with companies including Amazon, Apple, Brother, Cisco, IBM, Hewlett-Packard, Nikon, Panasonic, Pioneer, and SAP.[46,47] This strategy to obtain the rights to technologies that it might use in its products provides a tremendous amount of development freedom to Microsoft without risk of expensive litigation.

Major IT firms usually have little interest in cross-licensing with smaller firms, so small businesses often have no choice but to license patents if they use them. As a result, small businesses must pay an additional cost from which many larger companies are exempt. Furthermore, small businesses are generally unsuccessful in enforcing their patents against larger companies. Should a small business bring a patent infringement suit against a large firm, the larger firm can overwhelm the small business with multiple patent suits, whether they have merit or not. Considering that the average patent lawsuit costs $3 to $10 million and takes two to three years to litigate, a small firm often simply cannot afford to fight; instead, it usually settles and licenses its patents to the large company.[48]

Defensive Publishing and Patent Trolls

Inventors sometimes employ a tactic called defensive publishing as an alternative to filing for patents. Under this approach, a company publishes a description of its innovation in a bulletin, conference paper, trade journal, or on a Web site. Although this obviously provides competitors with access to the innovation, it also establishes the idea's legal existence as prior art. Therefore, competitors cannot patent the idea or charge licensing fees to other users of the technology or technique. This approach costs mere hundreds of dollars, requires no lawyers, and is fast.

A **patent troll** is a firm that acquires patents with no intention of manufacturing anything, instead licensing the patents to others. As discussed in the opening vignette, NTP is a patent holding company that filed patent infringement lawsuits against Research in Motion, Apple, Google, HTC, LG, Microsoft, Motorola, and others. Many would say that NTP is a clear example of a patent troll.[49] The outcome of the lawsuits is very much undecided. The U.S. Patent Office recently completed a re-examination of eight NTP wireless email patents and found fault with all eight, rejecting some entirely and partially rejecting others. NTP is fighting the recent findings.[50]

Submarine Patents and Patent Farming

A **standard** is a definition that has been approved by a recognized standards organization or accepted as a de facto standard within a particular industry. Standards exist for communication protocols, programming languages, operating systems, data formats, and electrical interfaces. Standards are extremely useful because they enable hardware and software from different manufacturers to work together.

A technology, process, or principle that has been patented may be embedded—knowingly or unknowingly—within a standard. If so, the patent owner can either demand a royalty payment from any party that implements the standard or refuse to permit certain parties to use the patent, thus effectively blocking them from using the standard. A patented process or invention that is surreptitiously included within a standard without being made public until after the standard is broadly adopted is called a **submarine patent**. A devious patent holder might influence a standards organization to make use of its patented item without revealing the existence of the patent. Later, the patent holder might demand royalties from all parties that use the standard. This strategy is known as **patent farming**.

In 1995, the term for a U.S. patent was modified from 17 years from issuance to 20 years from filing. This change makes it unlikely that patents will remain pending for decades and should reduce the incidence of submarine patents. However, in 2010, TiVo was granted a patent titled "Data Storage Management and Scheduling System" that was first applied for in 1999, meaning that the patent was pending at the U.S. Patent and Trademark Office for 10 years. The patent applies to any and all DVRs currently on the market and will likely generate a fortune in royalty payments to TiVo.[51] Other manufacturers of similar recording devices will be forced to pay royalties to TiVo if their design infringes on the TiVo patent or they will need to perform further innovation to create a new design.

TRADE SECRETS

In Chapter 2, a trade secret was defined as business information that represents something of economic value, has required effort or cost to develop, has some degree of uniqueness or novelty, is generally unknown to the public, and is kept confidential.

Trade secret protection begins by identifying all the information that must be protected—from undisclosed patent applications to market research and business plans—and developing a comprehensive strategy for keeping the information secure. Trade secret law protects only against the *misappropriation* of trade secrets. If competitors come up with the same idea on their own, it is not misappropriation; in other words, the law doesn't prevent someone from using the same idea if it was developed independently.

Trade secret law has several key advantages over the use of patents and copyrights in protecting companies from losing control of their intellectual property, as summarized in the following list:

- There are no time limitations on the protection of trade secrets, as there are with patents and copyrights.
- There is no need to file an application, make disclosures to any person or agency, or disclose a trade secret to outsiders to gain protection.

(After the USPTO issues a patent, competitors can obtain a detailed description of it.)

- Although patents can be ruled invalid by the courts, meaning that the affected inventions no longer have patent protection, this risk does not exist for trade secrets.
- No filing or application fees are required to protect a trade secret.

Because of these advantages, trade secret laws protect more technology worldwide than patent laws do.

Trade Secret Laws

Trade secret protection laws vary greatly from country to country. For example, the Philippines provides no legal protection for trade secrets. In some European countries, pharmaceuticals, methods of medical diagnosis and treatment, and information technology cannot be patented. Many Asian countries require foreign corporations operating there to transfer rights to their technology to locally controlled enterprises. (Coca-Cola reopened its operations in India in 1993 after halting sales for 16 years to protect the "secret formula" for its soft drink, even though India's vast population represented a huge potential market.) American businesses that seek to operate in foreign jurisdictions or enter international markets must take these differences into account.

Uniform Trade Secrets Act (UTSA)

The Uniform Trade Secrets Act (UTSA) was drafted in the 1970s to bring uniformity to all states in the area of trade secret law. The first state to enact the UTSA was Minnesota in 1981, followed by 39 more states and the District of Columbia. The UTSA defines a trade secret as "information, including a formula, pattern, compilation, program, device, method, technique, or process, that:

- Derives independent economic value, actual or potential, from not being generally known to, and not being readily ascertainable by, persons who can obtain economic value from its disclosure or use, and
- Is the subject of efforts that are reasonable under the circumstances to maintain its secrecy."[52]

Under these terms, computer hardware and software can qualify for trade secret protection by the UTSA.

The Economic Espionage Act (EEA) (1996)

The Economic Espionage Act (EEA) of 1996 imposes penalties of up to $10 million and 15 years in prison for the theft of trade secrets. Before the EEA, there was no specific criminal statute to help pursue economic espionage; the FBI was investigating nearly 800 such cases in 23 countries when the EEA was enacted.[53] The Office of the National Counterintelligence Executive has estimated that the "combined costs of foreign and domestic economic espionage, including the theft of intellectual property, [are] as high as $300 billion per year and rising."[54] As with the UTSA, information is considered a trade secret under the EEA only if companies take steps to protect it.

One of the earliest convictions under this act was in 2008 and involved a software engineer born in China and working in California. The trade secrets involved software products of Quantum 3D that provide advanced visual simulations for military training. The engineer was convicted of economic espionage and sentenced to a prison term of 24 months and a three-year term of supervised release and was fined $10,000. The trade secrets were used in a demonstration in China with the intent to benefit the Peoples Republic of China Naval Research Center.[55] In a subsequent case (*United States v. Lee and Ge*), a California federal judge ruled that prosecution under the act requires specific intent to benefit a foreign government. This ruling meant that theft of trade secrets for benefit of a foreign corporation is not punishable under the Economic Espionage Act.[56]

Employees and Trade Secrets

Employees are the greatest threat to the loss of company trade secrets—they might accidentally disclose trade secrets or steal them for monetary gain. Organizations must educate employees about the importance of maintaining the secrecy of corporate information. Trade secret information should be labeled clearly as confidential and should only be accessible by a limited number of people. Most organizations have strict policies regarding nondisclosure of corporate information.

Losing customer information to competitors is a growing concern in industries in which companies compete for many of the same clients. There are numerous cases of employees making unauthorized use of their employer's customer list. For example, the Ohio State Supreme Court upheld a verdict against a man who left a financial services firm and recruited former clients to start his own firm. His former employer sued him, even though the former employee had not stolen a client list. "This ruling says, it doesn't matter if the confidential list is on paper or in your memory if it qualifies as a trade secret," said Susan Guerette, a specialist in restrictive covenant and trade secret law at Pennsylvania-based Fisher & Phillips LLP.[57]

Legally, a customer list is not automatically considered a trade secret. If a company doesn't treat the list as valuable, confidential information internally, neither will the court. The courts must consider two main factors in making this determination. First, did the firm take prudent steps to keep the list secret? Second, did the firm expend money or effort to develop the customer list? The more the firm invested to build its customer list and the more that the list provides the firm with a competitive advantage, the more likely the courts are to accept the list as a trade secret.

Because organizations can risk losing trade secrets when key employees leave, they often try to prohibit employees from revealing secrets by adding **nondisclosure clauses** to employment contracts. Thus, departing employees cannot take copies of computer programs or reveal the details of software owned by the firm. Defining reasonable nondisclosure agreements can be difficult, as seen in the following example involving Apple. In addition to filing hundreds of patents on iPhone technology, the firm put into place a restrictive nondisclosure agreement to provide an extra layer of protection. Many iPhone developers complained bitterly about the tough restrictions, which prohibited them from talking about their coding work with anyone not on the project team and even prohibited them from talking about the restrictions themselves. Eventually, Apple admitted that its nondisclosure terms were overly restrictive and loosened them for iPhone software that was already released.[58]

Another option for preserving trade secrets is to have an experienced member of the Human Resources Department conduct an exit interview with each departing employee. A key step in the interview is to review a checklist that deals with confidentiality issues. At the end of the interview, the departing employee is asked to sign an acknowledgment of responsibility not to divulge any trade secrets.

Employers can also use noncompete agreements to protect intellectual property from being used by competitors when key employees leave. A **noncompete agreement** prohibits an employee from working for any competitors for a period of time, often one to two years. When courts are asked to settle disputes over noncompete agreements, they must weigh several factors. First, they must consider the reasonableness of the restriction and how it protects confidential and trade secret information of the former employer. Second, they must weigh the employee's right to work and seek employment in the area where the employee has gained skill, experience, and business contacts. The courts also consider geographic area and the length of time of the restriction in relation to the pace of change in the industry.

Most states only enforce such noncompete agreements to the extent required to shelter the employer's legitimate confidential business interests. However, there is a wide range of treatment on noncompete agreements among the various states. Ohio is highly supportive of former employers enforcing noncompete agreements while noncompete agreements are seldom enforced in California.[59]

The following is an example of a noncompete agreement:

> The employee agrees as a condition of employment that in the event of termination for any reason, he or she will not engage in a similar or competitive business for a period of two years, nor will he or she contact or solicit any customer with whom Employer conducted business during his or her employment. This restrictive covenant shall be for a term of two years from termination, and shall encompass the geographic area within a 100-mile radius of Employer's place of business.

Giovanni Visentin was General Manager of IBM's Integrated Technology Services business services. In early 2011, Visentin informed IBM that he was resigning to go to work at Hewlett-Packard (HP). The very next day, IBM filed suit against Visentin seeking to enforce a noncompete clause in his employment contract. The court ruled against IBM and denied its motion to prevent Visentin from working for HP for one year. The judge's ruling was based on the following facts and reasoning:

- IBM's witnesses "failed to provide specific examples of confidential or trade secret information that could actually be used to IBM's detriment if Mr. Visentin were allowed to assume his new position at HP."
- IBM's claim that it took sensible measures to protect its trade secrets was undermined because other IBM employees who worked with trade secrets were not bound by noncompete or confidentiality agreements.
- One IBM employee testified that the noncompete agreements were used as "retention devices" to pressure IBM's employees from leaving their jobs.[60]

KEY INTELLECTUAL PROPERTY ISSUES

This section discusses several issues that apply to intellectual property and information technology, including plagiarism, reverse engineering, open source code, competitive intelligence, trademark infringement, and cybersquatting.

Plagiarism

Plagiarism is the act of stealing someone's ideas or words and passing them off as one's own. The explosion of electronic content and the growth of the Web have made it easy to cut and paste paragraphs into term papers and other documents without proper citation or quotation marks. To compound the problem, hundreds of online "paper mills" enable users to download entire term papers. Although some sites post warnings that their services should be used for research purposes only, many users pay scant heed. As a result, plagiarism has become an issue from elementary schools to the highest levels of academia.

Plagiarism is also common outside academia. Popular literary authors, playwrights, musicians, journalists, and even software developers have been accused of it:

- Katie Couric gave a *CBS News* commentary about the joys of getting her first library card that plagiarized portions of a column in the *Wall Street Journal*; the "commentary" that Couric read was actually written by a CBS producer who was subsequently fired.[61]
- Reporter Jayson Blair resigned from the *New York Times* after he was accused of plagiarism and fabricating quotes and other information in news stories. Executive Editor Howell Raines and Managing Editor Gerald Boyd also resigned in the fallout from the scandal.[62]
- Science fiction writer Harlan Ellison successfully sued movie director James Cameron for taking key elements from two different episodes of the TV series *The Outer Limits*—written by Ellison—and using them in the 1984 classic movie *The Terminator*.[63]
- A plagiarism lawsuit was brought by the estate of Adrian Jacobs against Scholastic, the U.S. publisher of J. K. Rowling's *Harry Potter and the Goblet of Fire*. The lawsuit alleged that the *Goblet of Fire* plagiarized from Jacobs' book entitled *The Adventures of Willy the Wizard*. The judge in the case ruled that "the contrast between the total concept and feel of the works [was] so stark that any serious comparison on the two strains credulity." The lawsuit was dismissed.[64]

Despite school codes of ethics that clearly define plagiarism and prescribe penalties ranging from no credit on a paper to expulsion, many students still do not understand what constitutes plagiarism. Some students believe that all electronic content is in the public domain, while other students knowingly commit plagiarism because they either feel pressure to achieve a high GPA or are too lazy or pressed for time to do original work.

Some instructors say that being familiar with a student's style of writing, grammar, and vocabulary enables them to determine if the student actually wrote a paper. In addition, plagiarism detection systems (see Table 6-3) allow teachers, corporations, law firms, and publishers to check for matching text in different documents as a means of identifying potential plagiarism.

TABLE 6-3 Partial list of plagiarism detection services and software

Name of service	Web site	Provider
iThenticate	www.ithenticate.com	iParadigms
Turnitin	www.turnitin.com	iParadigms
SafeAssign	www.safeassign.com	Blackboard
Glatt Plagiarism Services	www.plagiarism.com	Glatt Plagiarism Services
EVE Plagiarism Detection	www.canexus.com/eve	CaNexus

Source Line: Course Technology/Cengage Learning.

These systems work by checking submitted material against one or more of the following databases of electronic content:

- More than 5 billion pages of publicly accessible electronic content on the Internet
- Millions of works published in electronic form, including newspapers, magazines, journals, and electronic books
- A database of papers submitted to the plagiarism detection service from participating institutions

Turnitin has been available since 1999 and is used in over 9,500 academic institutions around the world. It can check documents in 30 languages. iThenticate is available from the same company that created Turnitin but is designed to meet the needs of members of the information industry, such as publishers, research facilities, legal firms, government agencies, and financial institutions.[65]

Interestingly, four high school students brought a lawsuit against iParadigms, accusing the firm of copyright infringement. The basis of their lawsuit was that the firm's primary product, Turnitin, used archived student papers without their permission to assess the originality of newly submitted papers. However, both a district court and a court of appeals ruled that the use of student papers for purposes of plagiarism detection constitutes a fair use and is therefore not a copyright infringement. A U.S. court of appeals ruled that such use of student papers "has a protective effect" on the future marketability of the students' works and "provides a substantial public benefit through the network of institutions using Turnitin."[66]

The following list shows some of the actions that schools can take to combat student plagiarism:

- Help students understand what constitutes plagiarism and why they need to cite sources properly.
- Show students how to document Web pages and materials from online databases.
- Schedule major writing assignments so that portions are due over the course of the term, thus reducing the likelihood that students will get into a time crunch.
- Make clear to students that instructors are aware of Internet paper mills.

- Ensure that instructors both educate students about plagiarism detection services and make them aware that they know how to use these services.
- Incorporate detection software and services into a comprehensive antiplagiarism program.

Reverse Engineering

Reverse engineering is the process of taking something apart in order to understand it, build a copy of it, or improve it. Reverse engineering was originally applied to computer hardware but is now commonly applied to software as well. Reverse engineering of software involves analyzing it to create a new representation of the system in a different form or at a higher level of abstraction. Often, reverse engineering begins by extracting design-stage details from program code. Design-stage details about an information system are more conceptual and less defined than the program code of the same system.

One frequent use of reverse engineering for software is to modify an application that ran on one vendor's database so that it can run on another's (for example, from Access to Oracle). Database management systems use their own programming language for application development. As a result, organizations that want to change database vendors are faced with rewriting existing applications using the new vendor's database programming language. The cost and length of time required for this redevelopment can deter an organization from changing vendors and deprive it of the possible benefits of converting to an improved database technology.

Using reverse engineering, a developer can use the code of the current database programming language to recover the design of the information system application. Next, code-generation tools can be used to take the design and produce code (forward engineer) in the new database programming language. This reverse-engineering and code-generating process greatly reduces the time and cost needed to migrate the organization's applications to the new database management system. No one challenges the right to use this process to convert applications developed in-house. After all, those applications were developed and are owned by the companies using them. It is quite another matter, however, to use this process on a purchased software application developed and licensed by outside parties. Most IT managers would consider this action unethical because the software user does not actually own the right to the software. In addition, a number of intellectual property issues would be raised, depending on whether the software was licensed, copyrighted, or patented.

Other reverse-engineering issues involve tools called compilers and decompilers. A compiler is a language translator that converts computer program statements expressed in a source language (such as COBOL, Pascal, or C) into machine language (a series of binary codes of 0s and 1s) that the computer can execute. When a software manufacturer provides a customer with its software, it usually provides the software in machine-language form. Tools called reverse-engineering compilers, or decompilers, can read the machine language and produce the source code. For example, REC (Reverse Engineering Compiler) is a decompiler that reads an executable, machine-language file and produces a C-like representation of the code used to build the program.

Decompilers and other reverse-engineering techniques can be used to reveal a competitor's program code, which can then be used to develop a new program that either duplicates the original or interfaces with the program. Thus, reverse engineering provides a way to gain access to information that another organization may have copyrighted or classified as a trade secret.

The courts have ruled in favor of using reverse engineering to enable interoperability. In the early 1990s, video game maker Sega developed a computerized lock so that only Sega video cartridges would work on its entertainment systems. This essentially shut out competitors from making software for the Sega systems. *Sega Enterprises Ltd. v. Accolade, Inc.* dealt with rival game maker Accolade's use of a decompiler to read the Sega software source code. With the code, Accolade could create new software that circumvented the lock and ran on Sega machines. An appeals court ultimately ruled that if someone lacks access to the unprotected elements of an original work and has a "legitimate reason" for gaining access to those elements, disassembly of a copyrighted work is considered to be a fair use under section 107 of the Copyright Act. The unprotected element in this case was the code necessary to enable software to interoperate with the Sega equipment. The court reasoned that to refuse someone the opportunity to create an interoperable product would allow existing manufacturers to monopolize the market, making it impossible for others to compete. This ruling had a major impact on the video game industry, allowing video game makers to create software that would run on multiple machines.[67]

Software license agreements increasingly forbid reverse engineering. As a result of the increased legislation affecting reverse engineering, some software developers are moving their reverse-engineering projects offshore to avoid U.S. rules.

The reverse engineering of copyrighted or patented hardware or software items done for the purposes of interoperability (e.g., to support undocumented file formats or undocumented hardware peripherals) is often considered to be legal. However, copyright and patent owners often contest this in an attempt to stifle reverse engineering of their products for any reason.

The ethics of using reverse engineering are debated. Some argue that its use is fair if it enables a company to create software that interoperates with another company's software or hardware and provides a useful function. This is especially true if the software's creator refuses to cooperate by providing documentation to help create interoperable software. From the consumer's standpoint, such stifling of competition increases costs and reduces business options. Reverse engineering can also be a useful tool in detecting software bugs and security holes.

Others argue strongly against the use of reverse engineering, saying it can uncover software designs that someone else has developed at great cost and taken care to protect. Opponents of reverse engineering contend it unfairly robs the creator of future earnings and significantly reduces the business incentive for software development.

Open Source Code

Historically, the makers of proprietary software have not made their source code available, but not all developers share that philosophy. **Open source code** is any program whose source code is made available for use or modification, as users or other developers see fit. The basic premise behind open source code is that when many programmers can read, redistribute, and modify a program's code, the software improves. Programs with open source code can be adapted to meet new needs, and bugs can be rapidly identified and fixed. Open source code advocates believe that this process produces better software than the traditional closed model.

A considerable amount of open source code is available, and an increasing number of organizations use open source code. For example, Walmart, Ticketmaster, McGraw-Hill Education, Cardinal Health, Symantec, and Wikipedia all use the open source database management system software MySQL. A common use of open source software is to move data from one application to another and to extract, transform, and load business data into large databases. Two frequently cited reasons for using open source software are that it provides a better solution to a specific business problem and that it costs less.[68] Open source software is used in applications developed for smartphones and other mobile devices, such as Apple's iPhone, Palm's Treo, and Research In Motion's BlackBerry. See Table 6-4 for a listing of commonly used open source software.

TABLE 6-4 Commonly used open source software

Open source software	Purpose
7-Zip	File compression
Ares Galaxy	Peer-to-peer file sharing
Audacity	Sound editing and special effects
Azureus	Peer-to-peer file sharing
Blender 3D	3D modeling and animation
eMule	Peer-to-peer file sharing
Eraser	Erase data completely
Firefox	Internet browser
OpenOffice	Word processing, spreadsheets, presentations, graphics, and databases
Video Dub	Video editing

Source Line: Course Technology/Cengage Learning.

Why would firms or individual developers create open source code if they do not receive money for it? Here are several reasons:

- Some people share code to earn respect for solving a common problem in an elegant way.
- Some people have used open source code that was developed by others and feel the need to pay back.
- A firm may be required to develop software as part of an agreement to address a client's problem. If the firm is paid for the employees' time spent to develop the software rather than for the software itself, it may decide to license the code as open source and use it either to promote the firm's expertise or as an incentive to attract other potential clients with a similar problem.

- A firm may develop open source code in the hope of earning software maintenance fees if the end user's needs change in the future.
- A firm may develop useful code but may be reluctant to license and market it, and so might donate the code to the general public.

There are various definitions of what constitutes open source code, each with its own idiosyncrasies. The GNU General Public License (GPL) was a precursor to the open source code defined by the Open Source Initiative (OSI). GNU is a computer operating system composed entirely of free software; its name is a recursive acronym for GNU's Not Unix. The GPL is intended to protect GNU software from being made proprietary, and it lists terms and conditions for copying, modifying, and distributing free software. The OSI is a nonprofit organization that advocates for open source and certifies open source licenses. Its certification mark, "OSI Certified," may be applied only to software distributed under an open source license that meets OSI criteria, as described at its Web site, *www.opensource.org*.

A software developer could attempt to make a program open source simply by putting it into the public domain with no copyright. This would allow people to share the program and their improvements, but it would also allow others to revise the original code and then distribute the resulting software as their own proprietary product. Users who received the program in the modified form would no longer have the freedoms associated with the original software. Use of an open source license avoids this scenario.

Competitive Intelligence

Competitive intelligence (as defined in Chapter 3) is legally obtained information that is gathered to help a company gain an advantage over its rivals. For example, some companies have employees who monitor the public announcements of property transfers to detect any plant or store expansions of competitors. An effective competitive intelligence operation requires the continual gathering, analysis, and evaluation of data with controlled dissemination of useful information to decision makers. Competitive intelligence is often integrated into a company's strategic plan and decision making. Many companies, such as Eastman Kodak, Monsanto, and United Technologies, have established formal competitive intelligence departments. Some companies have even employed former CIA analysts to assist them.

Competitive intelligence is not the same as **industrial espionage**, which is the use of illegal means to obtain business information not available to the general public. In the United States, industrial espionage is a serious crime that carries heavy penalties.

Almost all the data needed for competitive intelligence can be collected from examining published information or interviews, as outlined in the following list:

- 10-K or annual reports
- An SC 13D acquisition—a filing by shareholders who report owning more than 5 percent of common stock in a public company
- 10-Q or quarterly reports
- Press releases
- Promotional materials
- Web sites

- Analyses by the investment community, such as a Standard & Poor's stock report
- Dun & Bradstreet credit reports
- Interviews with suppliers, customers, and former employees
- Calls to competitors' customer service groups
- Articles in the trade press
- Environmental impact statements and other filings associated with a plant expansion or construction
- Patents

By coupling this data with analytical tools and industry expertise, an experienced analyst can make deductions that lead to significant information. According to Avinash Kaushik, self-described Analytics Evangelist for Google, "The Web is the best competitive intelligence tool in the world." Kaushik likens the failure to use such data to driving a car 90 miles an hour with the windshield painted black, then scraping off the paint and realizing "you're going 90 but everyone else is going 220 and you're going to die."[69]

Competitive intelligence gathering has become enough of a science that over two dozen colleges and universities offer courses or even entire programs in this subject. Also, the Strategic and Competitive Intelligence Professionals organization (*www.scip.org*) offers ongoing training programs and conferences.

Without proper management safeguards, the process of gathering competitive intelligence can cross over to industrial espionage and dirty tricks. One frequent trick is to enter a bar near a competitor's plant or headquarters, strike up a conversation, and ply people for information after their inhibitions have been weakened by alcohol.

Competitive intelligence analysts must avoid unethical or illegal actions, such as lying, misrepresentation, theft, bribery, or eavesdropping with illegal devices. Table 6-5 provides a manager's checklist for running an ethical competitive intelligence operation. The preferred answer to each question in the checklist is *yes*.

TABLE 6-5 A manager's checklist for running an ethical competitive intelligence operation

Question	Yes	No
Has the competitive intelligence organization developed a mission statement, objectives, goals, and a code of ethics?		
Has the company's legal department approved the mission statement, objectives, goals, and code of ethics?		
Do analysts understand the need to abide by their organization's code of ethics and corporate policies?		
Is there a rigorous training and certification process for analysts?		
Do analysts understand all applicable laws—domestic and international—including the Uniform Trade Secrets Act and the Economic Espionage Act, and do they understand the critical importance of abiding by them?		
Do analysts disclose their true identity as well as the name of their organization prior to any interviews?		

(*Continued*)

TABLE 6-5 A manager's checklist for running an ethical competitive intelligence operation (*Continued*)

Question	Yes	No
Do analysts understand that everything their firm learns about the competition must be obtained legally?		
Do analysts respect all requests for anonymity and confidentiality of information?		
Has the company's legal department approved the processes for gathering data?		
Do analysts provide honest recommendations and conclusions?		
Is the use of third parties to gather competitive intelligence carefully reviewed and managed?		

Source Line: Course Technology/Cengage Learning.

Failure to act prudently when gathering competitive intelligence can get analysts and companies into serious trouble. For example, Procter & Gamble (P&G) admitted in 2001 that it unethically gained information about Unilever, its competitor in the multibillion-dollar hair-care business. Unilever markets brands such as Salon Selectives, Finesse, and Thermasilk, while P&G manufactures Pantene, Head & Shoulders, and Pert. Competitive intelligence managers at P&G hired a contractor, who in turn hired several subcontractors to spy on P&G's competitors. Unilever was the primary target.

In at least one instance, the espionage included going through dumpsters on public property outside Unilever corporate offices in Chicago. In addition, competitive intelligence operatives were alleged to have misrepresented themselves to Unilever employees, suggesting that they were market analysts. (P&G confirms the dumpster diving, but it denies that misrepresentation took place.) The operatives captured critical information about Unilever's brands, including new-product rollouts, selling prices, and operating margins.

When senior P&G officials discovered that the firm hired by the company was operating unethically, P&G immediately stopped the campaign and fired the three managers responsible for hiring the firm. P&G then did something unusual—it blew the whistle on itself, confessed to Unilever, returned stolen documents to Unilever, and started negotiations with them to set things straight. P&G's chairman of the board was personally involved in ensuring that none of the information obtained would ever be used in P&G business plans. Several weeks of high-level negotiations between P&G and Unilever executives led to a secret agreement between the two companies. P&G is believed to have paid tens of millions of dollars to Unilever. In addition, several hair-care product executives were transferred to other units within P&G.

Experts in competitive intelligence agree that the firm hired by P&G crossed the line of ethical business practices by sorting through Unilever's garbage. However, they also give P&G credit for going to Unilever quickly after it discovered the damage. Such prompt action was seen as the best approach. If no settlement had been reached, of course, Unilever could have taken P&G to court, where embarrassing details might have been revealed, causing more bad publicity for a company that is generally perceived as highly ethical. Unilever also stood to lose from a public trial. The trade secrets at the heart of the case may have been disclosed in depositions and other documents during a trial, which could have devalued proprietary data.[70]

Trademark Infringement

A **trademark** is a logo, package design, phrase, sound, or word that enables a consumer to differentiate one company's products from another's. Consumers often cannot examine goods or services to determine their quality or source, so instead they rely on the labels attached to the products. The Lanham Act of 1946 (also known as the Trademark Act, Title 15, Chapter 22 of the U.S. Code) defines the use of a trademark, the process for obtaining a trademark from the Patent and Trademark Office, and the penalties associated with trademark infringement. The law gives the trademark's owner the right to prevent others from using the same mark or a confusingly similar mark on a product's label.

The United States has a federal system that stores trademark information; merchants can consult this information to avoid adopting marks that have already been taken. Merchants seeking trademark protection apply to the USPTO if they are using the mark in interstate commerce or if they can demonstrate a true intent to do so. Trademarks can be renewed forever—as long as a mark is in use.

It is not uncommon for an organization that owns a trademark to sue another organization over the use of that trademark in a Web site or a domain name. The court rulings in such cases are not always consistent and are impossible to judge in advance.

Nominative fair use is a defense often employed by the defendant in trademark infringement cases where a defendant uses a plaintiff's mark to identify the plaintiff's products or services in conjunction with its own product or services. To successfully employ this defense, the defendant must show three things:[71]

- that the plaintiff's product or service cannot be readily identifiable without using the plaintiff's mark,
- that it uses only as much of plaintiff's mark as necessary to identify defendant's product or service, and
- that the defendant does nothing with plaintiff's mark that suggests endorsement or sponsorship by the plaintiff.

This defense was first applied to Web sites in *Playboy Enterprises, Inc. v. Terri Welles*. Welles was the Playboy™ Playmate of the Year™ in 1981. In 1997, she created a Web site to offer free photos of herself, advertise the sale of additional photos, solicit memberships in her photo club, and promote her spokeswoman services. Welles used the trademarked terms *Playboy* model and *Playmate of the Year* to describe herself on her Web site. The Ninth Circuit Court of Appeals determined that the former Playboy model's use of trademarked terms was permissible, nominative use. By using the nominative fair use defense, Welles avoided a motion for preliminary injunction, which would have restrained her from continuing to use the trademarked terms on her Web site.[72]

Farzad and Lisa Tabaris operated an auto broker business that assisted consumers in buying Toyota Lexus™ automobiles at the lowest possible price. They created two Web sites, *buy-a-lexus.com* and *buyorleaselexus.com*, which were the subject of a trademark lawsuit with Toyota. A district court agreed with Toyota and forbid the use of the trademark Lexus in any domain name. However, in July 2010, the Ninth Circuit Court of Appeals disagreed and reversed the decision based on the nominative fair use doctrine and First Amendment concerns.[73]

Cybersquatting

Companies that want to establish an online presence know that the best way to capitalize on the strengths of their brand names and trademarks is to make the names part of the domain names for their Web sites. When Web sites were first established, there was no procedure for validating the legitimacy of requests for Web site names, which were given out on a first-come, first-served basis. **Cybersquatters** registered domain names for famous trademarks or company names to which they had no connection, with the hope that the trademark's owner would eventually buy the domain name for a large sum of money.

The main tactic organizations use to circumvent cybersquatting is to protect a trademark by registering numerous domain names and variations as soon as the organization knows it wants to develop a Web presence (for example, UVXYZ.com, UVXYZ.org, and UVXYZ.info). In addition, trademark owners who rely on non-English-speaking customers should register their names in multilingual form. Registering additional domain names is far less expensive than attempting to force cybersquatters to change or abandon their domain names.

Other tactics can also help curb cybersquatting. For example, the Internet Corporation for Assigned Names and Numbers (ICANN) is a nonprofit corporation responsible for managing the Internet's domain name system. Prior to 2000, eight generic Top-Level Domain names were in existence: .com, .edu, .gov, .int, .mil, .net, .org, and .arpa. In 2000, ICANN introduced seven more: .aero, .biz, .coop, .info, .museum, .name, and .pro. In 2004, ICANN introduced .asia, .cat, .mobi, .tel, and .travel. The generic Top-Level Domain .xxx was approved in 2011. With each new round of generic Top-Level Domains, current trademark holders are given time to assert rights to their trademarks in the new top-level domains before registrations are opened up to the general public.

ICANN also has a Uniform Domain Name Dispute Resolution Policy, under which most types of trademark-based domain name disputes must be resolved by agreement, court action, or arbitration before a registrar will cancel, suspend, or transfer a domain name. The ICANN policy is designed to provide for the fast, relatively inexpensive arbitration of a trademark owner's complaint that a domain name was registered or used in bad faith.

The Anticybersquatting Consumer Protection Act (ACPA), enacted in 1999, allows trademark owners to challenge foreign cybersquatters who might otherwise be beyond the jurisdiction of U.S. courts. Also under this act, trademark holders can seek civil damages of up to $100,000 from cybersquatters that register their trade names or similar-sounding names as domain names. The act also helps trademark owners challenge the registration of their trademark as a domain name even if the trademark owner has not created an actual Web site.

OnlineNIC was one of the very first domain registrars licensed by ICANN. During 2008, Verizon Communications, Microsoft, and Yahoo! each filed separate lawsuits against OnlineNIC because that firm registered hundreds of domain names identical or similar to their trademark names (e.g., *verizon-cellular.com, encarta.com,* and *yahoozone.com*). In December 2008, Verizon was awarded damages of $31 million. OnlineNIC was prohibited from registering any additional names containing Verizon trademarks, and it was ordered to transfer the disputed domain names to Verizon.[74]

Summary

- *Intellectual property* is a term used to describe works of the mind—such as art, books, films, formulas, inventions, music, and processes—that are distinct and owned or created by a single person or group.
- Copyrights, patents, trademarks, and trade secrets provide a complex body of law relating to the ownership of intellectual property, which represents a large and valuable asset to most companies. If these assets are not protected, other companies can copy or steal them, resulting in significant loss of revenue and competitive advantage.
- A copyright is the exclusive right to distribute, display, perform, or reproduce an original work in copies; prepare derivative works based on the work; and grant these exclusive rights to others.
- Copyright law has proven to be extremely flexible in covering new technologies, including software, video games, multimedia works, and Web pages. However, evaluating the originality of a work can be difficult and can lead to litigation.
- Copyrights provide less protection for software than patents; software that produces the same result in a slightly different way may not infringe a copyright if no copying occurred.
- The fair use doctrine establishes four factors for courts to consider when deciding whether a particular use of copyrighted property is fair and can be allowed without penalty: the purpose and character of the use, the nature of the copyrighted work, the portion of the copyrighted work used, and the effect of the use on the value of the copyrighted work.
- The use of copyright to protect computer software raises many complicated issues of interpretation.
- The Prioritizing Resources and Organization for Intellectual Property (PRO-IP) Act of 2008 increased trademark and copyright enforcement; it also substantially increased penalties for infringement.
- The original General Agreement on Tariffs and Trade (GATT) was signed in 1947 by 150 countries. It created the World Trade Organization (WTO) in Geneva, Switzerland, to enforce compliance with the agreement. GATT includes a section covering copyrights called the Agreement on Trade-Related Aspects of Intellectual Property Rights (TRIPS).
- The WTO deals with rules of international trade based on WTO agreements that are negotiated and signed by representatives of the world's trading nations. Its goal is to help producers of goods and services, exporters, and importers conduct their business.
- The World Intellectual Property Organization (WIPO) is an agency of the United Nations dedicated to developing "a balanced and accessible international intellectual property (IP) system, which rewards creativity, stimulates innovation and contributes to economic development while safeguarding the public interest."
- The Digital Millennium Copyright Act (DMCA) implements two WIPO treaties in the United States. It also makes it illegal to circumvent a technical protection or develop and provide tools that allow others to access a technologically protected work. It also limits the liability of online service providers for copyright infringement by their subscribers or customers.

- Some view the DMCA as a boon to the growth of the Internet and its use as a conduit for innovation and freedom of expression. Others believe that the DMCA has given excessive powers to copyright holders.
- A patent enables an inventor to sue people who manufacture, use, or sell the invention without permission while the patent is in force. A patent prevents copying as well as independent creation (which is allowable under copyright law).
- For an invention to be eligible for a patent, it must fall into one of five statutory classes of items that can be patented; it must be useful; it must be novel; and it must not be obvious to a person having ordinary skill in the same field.
- Unlike copyright infringement, for which monetary penalties are limited, if the court determines that a patent has been intentionally infringed, it can award up to triple the amount of the damages claimed by the patent holder.
- To qualify as a trade secret, information must have economic value and must not be readily ascertainable. In addition, the trade secret's owner must have taken steps to maintain its secrecy. Trade secret laws do not prevent someone from using the same idea if it was developed independently or from analyzing an end product to figure out the trade secret behind it.
- Trade secret law has three key advantages over the use of patents and copyrights in protecting companies from losing control of their intellectual property: (1) There are no time limitations on the protection of trade secrets, unlike patents and copyrights; (2) there is no need to file any application or otherwise disclose a trade secret to outsiders to gain protection; and (3) there is no risk that a trade secret might be found invalid in court.
- To plagiarize is to steal someone's ideas or words and pass them off as one's own. Plagiarism detection systems enable people to check the originality of documents and manuscripts.
- Reverse engineering is the process of breaking something down in order to understand it, build a copy of it, or improve it. Reverse engineering was originally applied to computer hardware but is now commonly applied to software.
- In some situations, reverse engineering might be considered unethical because it enables access to information that another organization may have copyrighted or classified as a trade secret.
- Recent court rulings and software license agreements that forbid reverse engineering, as well as restrictions in the DMCA, have made reverse engineering a riskier proposition in the United States.
- Open source code refers to any program whose source code is made available for use or modification, as users or other developers see fit. The basic premise behind open source code is that when many programmers can read, redistribute, and modify it, the software improves. Open source code can be adapted to meet new needs, and bugs can be rapidly identified and fixed.
- Competitive intelligence is legally obtained information that is gathered to help a company gain an advantage over its rivals. Competitive intelligence is not the same as industrial

espionage, which is the use of illegal means to obtain business information that is not readily available to the general public. In the United States, industrial espionage is a serious crime that carries heavy penalties.

- Competitive intelligence analysts must take care to avoid unethical or illegal behavior, including lying, misrepresentation, theft, bribery, or eavesdropping with illegal devices.
- A trademark is a logo, package design, phrase, sound, or word that enables a consumer to differentiate one company's products from another's. Web site owners who sell trademarked goods or services must take care to ensure they are not sued for trademark infringement.
- Cybersquatters register domain names for famous trademarks or company names to which they have no connection, with the hope that the trademark's owner will eventually buy the domain name for a large sum of money.
- The main tactic organizations use to circumvent cybersquatting is to protect a trademark by registering numerous domain names and variations as soon as they know they want to develop a Web presence.

Key Terms

Agreement on Trade-Related Aspects of Intellectual Property Rights
copyright
copyright infringement
cybersquatters
Digital Millennium Copyright Act (DMCA)
fair use doctrine
industrial espionage
intellectual property
noncompete agreement
nondisclosure clauses
open source code
patent
patent farming
patent infringement
patent troll
plagiarism
prior art
reverse engineering
standard
submarine patent
trademark

Self-Assessment Questions

The answers to the Self-Assessment Questions can be found in Appendix E.

1. Which of the following is an example of intellectual property?
   a. a work of art
   b. a computer program
   c. a trade secret of an organization
   d. all of the above
2. The courts may award up to triple damages for copyright infringement. True or False?

3. The ____________ doctrine established four factors for courts to consider when deciding whether a particular use of copyrighted property is fair and can be allowed without penalty.
4. Two software manufacturers develop separate but nearly identical programs for playing a game. Even though the second manufacturer can establish that it developed the program on its own, without knowledge of the existing program, that manufacturer can be found guilty of copyright infringement. True or False?

5. Title II of the ____________ amends the Copyright Act by adding a new section to create new limitations on liability for copyright infringement by online service providers.
6. A ____________ is a logo, package design, phrase, sound, or word that enables a consumer to differentiate one company's products from another's.
7. Many large software companies have cross-licensing agreements in which each agrees not to sue the other over ____________.
8. A ____________ is a form of protection for intellectual property that does not require any disclosures or filing of an application.
   a. copyright
   b. patent
   c. trade secret
   d. trademark
9. A customer list can be considered a trade secret if an organization treats the information as valuable and takes measures to safeguard it. True or False?
10. ____________ established minimum levels of protection that each government must provide to the intellectual property of all WTO members.
11. Plagiarism is an issue only in academia. True or False?
12. The process of taking something apart in order to understand it, build a copy of it, or improve it is called ____________.
13. As part of the patent application, the USPTO searches the existing body of knowledge that is available to a person of ordinary skill in the art. This existing body of knowledge is also called ____________.
14. Almost all the data needed for competitive intelligence can be collected either through carefully examining published information or through interviews. True or False?
15. The main tactic used to circumvent cybersquatting is to register numerous domain name variations as soon as an organization thinks it might want to develop a Web presence. True or False?

Discussion Questions

1. Explain the concept that an idea cannot be copyrighted, but the expression of an idea can be, and why this distinction is a key to understanding copyright protection.

2. Some view the DMCA as a boon to the growth of the Internet and its use as a conduit for innovation and freedom of expression. Others believe that the DMCA has given excessive powers to copyright holders. What is your opinion, and why?
3. What is a cross-licensing agreement? How do large software companies use such agreements? Do you think their use is fair to small software development firms? Why or why not?
4. What is the role of the WTO and what is the scope and intent of its TRIPS agreement?
5. Discuss the use of a patent by a software manufacturer to protect against the unauthorized use of its software. Is this an effective approach? Which would you recommend—the copyright or patent approach—to safeguard software? Why?
6. Discuss the conditions under which a company's customer list can be considered a trade secret.
7. What is a submarine patent? Do you think that the use of a submarine patent is an ethical practice by information technology manufacturers? Why or why not?
8. Identify and briefly discuss three key advantages that trade secret law has over the use of patents and copyrights in protecting intellectual property. Are there any drawbacks with the use of trade secrets to protect intellectual property?
9. What problems can arise in using nondisclosure and noncompete agreements to protect intellectual property?
10. Outline an approach that a university might take to successfully combat plagiarism among its students.
11. Under what conditions is the use of reverse engineering an acceptable business practice?
12. How might a corporation use reverse engineering to convert to a new database management system? How might it use reverse engineering to uncover the trade secrets behind a competitor's software?
13. Why might an organization elect to use open source code instead of proprietary software?
14. Compare the key issues in the *Sega v. Accolade* and *Lexmark v. Static Control Components* reverse-engineering lawsuits.

What Would You Do?

1. Your friend is a two-time winner of the Ironman™ Arizona Triathlon (2.4-mile swim, 112-mile bike, and 26.2-mile run). He is also a popular and well-known marathon runner throughout the Southwest. He has asked you to design a Web page to promote the sale of a wide variety of health products, vitamins, food supplements, and clothing targeted at the athletes training to participate in the triathlon. The products will carry his personal trademark. However, much of the information on the Web page will include discussion of his personal success in various triathlons and marathons in which he has competed. Many of these events have corporate sponsors and carry their own trademark. He has asked you if there are any potential trademark issues with his marketing plans. What would you do?
2. Because of the amount of the expense, your company's CFO had to approve a $500,000 purchase order for hardware and software needed to upgrade the servers used to store

data for the Product Development Department. Everyone in the department had expected an automatic approval, and they were disappointed when the purchase order request was turned down. Management said that the business benefits of the expenditure were not clear. Realizing that she needs to develop a more solid business case for the order, the vice president of product development has come to you for help. Can you help her identify arguments related to protecting intellectual property that might strengthen the business case for this expenditure?

3. You are interviewing for the role of human resources manager for a large software developer. Over the last year, the firm has lost a number of high-level executives who left the firm to go to work for competitors. During the course of your interview, you are asked what measures you would put in place to reduce the potential loss of trade secrets from executives leaving the firm. How would you respond?
4. You have been asked by the manager of software development to lead a small group of software developers in an attempt to reengineer the latest release of the software by your leading competitor. The goal of the group is to identify features that could be implemented into the next few releases of your firm's software. You are told that the group would relocate from the United States to the island of Antigua, in the Caribbean Sea, to "reduce the risk of the group being distracted by the daily pressures associated with developing fixes and enhancements with the current software release." What sort of legal and/or ethical questions might be raised by this reengineering effort? Would you consider taking this position?
5. You have procrastinated too long and now your final paper for your junior English course is due in just five days—right in the middle of final exam week! The paper counts for half your grade for the term and would probably take you at least 20 hours to research and write. Your roommate, an English major with a 3.8 GPA, has suggested two options: (1) He will write an original paper for you for $100, or (2) he will show you two or three "paper mill" Web sites, from which you can download a paper for less than $35. You want to do the right thing, but writing the paper will take away from the time you have available to study for your final exam in three other courses. What would you do?
6. You are beginning to feel very uncomfortable in your new position as a computer hardware salesperson for a firm that is the major competitor of your previous employer. Today, for the second time, someone has mentioned to you how valuable it would be to know what the marketing and new product development plans were of your ex-employer. You stated that you are unable to discuss such information under the nondisclosure contract signed with your former employer, but you know your response did not satisfy your new coworkers. You fear that the pressure to reveal information about the plans of your former company is only going to increase over the next few weeks. What do you do?
7. You have been asked to lead your company's new competitive intelligence organization. What would you do to ensure that members of the new organization obey applicable laws and the company's own ethical policies?
8. You are the vice president for software development at a small, private firm. Sales of your firm's products have been strong, but you recently detected a patent infringement by one of your larger competitors. Your in-house legal staff has identified three options: (1) ignore the infringement out of fear that your larger competitor will file numerous countersuits; (2) threaten to file suit, but try to negotiate an out-of-court settlement for an amount of money

that you feel your larger competitor would readily pay; or (3) point out the infringement and negotiate aggressively for a cross-licensing agreement with the competitor, which has numerous patents you had considered licensing. Which option would you pursue and why?

Cases

1. Google Book Search Library Project

In 2005, Google announced the Google Book Search Library Project, a highly ambitious plan to scan and digitize books from various libraries, including the New York Public Library and the libraries at Harvard University, Oxford University, Stanford University, and the University of Michigan.[75] Google's goal is to "work with publishers and libraries to create a comprehensive, searchable, virtual card catalog of all books in all languages that helps users discover new books and publishers discover new readers."[76]

Because many of the books are protected under copyright law, Google needed a way to avoid problems with copyright infringement. Therefore, Google established a process requiring publishers and copyright holders to opt out of the program if they did not want their books to be searchable. Publishers and copyright holders were incensed and argued that they should control who can view and search their books. In October 2005, the Authors Guild and the Association of American Publishers (on behalf of McGraw-Hill, Simon & Schuster, John Wiley & Sons, Pearson Education, and the Penguin Group) filed suit against Google to stop the program. They argued that making a full copy of a copyright-protected book does not fit into the narrow exception to the law defined by fair use.

After more than two years of discussions, the parties negotiated a settlement in October 2008. The settlement did not resolve the legal dispute over whether Google's project is permissible as a fair use; however, it concluded the litigation, enabling the parties to avoid the cost and risk of a trial.[77] The proposed settlement would give Google the right to display up to 20 percent of a book online and to profit from it by selling access to all or part of it. Google would also sell subscriptions to its entire collection to universities and other institutions, but offer free portals to public libraries where users could pay a per-page fee to print parts of the book.[78] In addition, Google would set aside $125 million to compensate authors and publishers for originally infringing on their copyrights, to pay the legal fees of the authors and publishers, and to establish a Book Rights Registry where rights holders can register their works to receive a share of ad revenue and digital book sales.[79]

Google, as well as many authors and publishers, defended the settlement, saying the project would benefit authors, publishers, and the public and renew access to millions of out-of-print books.[80]

However, in a further complication, the U.S. Department of Justice (DOJ) began an inquiry in April 2009 into the proposed settlement. In September, the DOJ urged the court to reject the settlement. The DOJ concluded that the settlement violated copyright, antitrust, and class action laws on three grounds. First, one goal of the settlement was to offer copyrighted materials to the public electronically while compensating copyright holders. However, the DOJ concluded that the current system does not require copyright owners to register. Moreover, the project includes many "orphan books"—those whose copyright holders are unknown or cannot be located. In addition, the DOJ argued that the settlement should result in a marketplace in which consumers have a choice of outlets from which they can obtain the access and in which prices are kept competitive. Finally, the DOJ harshly criticized the settlement because, as a class action, it failed to protect the rights of absent class members. The DOJ generally questioned whether a class action lawsuit was an appropriate method of dealing with the issues that arise from such a large-scale project to provide public electronic access to copyrighted material. A more appropriate venue, the DOJ suggested, would be the legislature.[81]

The parties in the case quickly responded by working out a new agreement. Through the revised agreement, Google's book registry would actively seek out authors and rights holders and Google would only scan books in English-speaking countries. In addition, the settlement limited ways that Google could make money from the project.[82]

In February 2010, however, the DOJ rejected the amended settlement for violating class action, antitrust, and copyright laws. The DOJ made specific suggestions to help avoid copyright infringement, such as arranging for authors to opt-in rather than opt-out and listing a book in the registry for two years prior to making it available online. But, from an antitrust perspective, the arrangement was still extremely problematic, the DOJ noted, as there are no serious competitors in the market. Amazon has approximately three million to Google's tens of millions of books.[83]

However, this time the parties did not rush to develop a new agreement. Instead, New York Federal District Judge Denny Chin postponed a ruling on the agreement a few weeks later.[84] The judge wanted to give all parties involved time to submit comments on the amended agreement.[85] The court issued no ruling during 2010. Then in December 2010, Google launched its own online bookstore of eBooks. Of its over three million titles, only 200,000 had been licensed through publishers. The remaining 2.8 million were texts that Google had scanned from libraries that were no longer covered by copyright law in the United States.[86]

Discussion Questions

1. Do you think that Google should have taken a different approach that would have allowed it to avoid litigation and a lengthy delay in implementing its Book Search Library Project? Please explain your answer.
2. As a potential user, are you in favor of or do you oppose the Book Search Library Project? Please explain your answer.
3. Do you think that the proposed settlement gives Google an unfair advantage to profit from creating an online service that allows people to access and search millions of books?

2. Applied Materials Indicted for Industrial Espionage

Samsung Electronics manufactures a wide range of consumer and industrial electronic equipment and products such as semiconductors, personal computers, peripherals, monitors, televisions, and home appliances. The company has been in business for over 79 years[87] and is considered South Korea's top electronics company. Hynix Semiconductor Inc. is a major South Korean semiconductor manufacturer and a direct competitor to Samsung. The firms are the two top producers of dynamic random access memory (DRAMM), used primarily in computers. They are also intense rivals in the manufacture of NAND flash memory chips, which are used in smartphones, cameras, and music players.

Applied Materials, Inc. is a Santa Clara, California–based company that develops, manufactures, markets, and services semiconductor wafer fabrication equipment and related spare parts for the worldwide semiconductor industry. The firm is under contract with both Samsung and Hynix to provide semiconductor manufacturing equipment and services.

In early 2010, it became public knowledge that employees of Applied Materials were under investigation for having leaked certain proprietary Samsung semiconductor technology to its domestic rival Hyrix.[88] Certain employees in Applied Materials' South Korean operation had routine access to Samsung's "core technology" through their ongoing role installing and maintaining the company's chip manufacturing equipment. According to prosecutors, Applied Materials employees obtained the technology from Samsung employees and then delivered it to Hynix.[89] A former Samsung employee who is now working for Applied Materials in the United States is suspected of playing a key role in the leaking of the technology.[90]

Applied Materials confirmed that the former head of its South Korea unit, who is currently a vice president of the firm, along with other employees of Applied Materials, had been indicted in a Seoul Eastern District court.[91] Eventually it was learned that ten employees from Applied Materials, four from Samsung, and five from Hynix were indicted for stealing the technology.[92]

In a filing to the U.S. Securities and Exchange Commission, Applied Materials stated that it had strict policies in place to protect the intellectual property of its customers, suppliers, and other third parties and took violation of these policies seriously.[93] A spokeswoman for Hynix stated that none of the stolen technology was ever used by Hynix in development or mass production.[94]

To avoid several lawsuits related to the alleged industrial espionage and to get the incident behind it, Applied Materials agreed to sell equipment to Samsung at a discounted rate for three years. This agreement does not affect the ongoing criminal proceedings against individuals, which are being heard in a South Korean court. These charges are still in the process of resolution.[95]

Discussion Questions

1. Do you believe that Applied Materials' people resources and money should be used to defend its employees who were charged with industrial espionage? Why or why not?
2. Do you think that it is possible for nineteen people from three different companies to work in collusion to steal secrets from Samsung and pass them to Hynix without Applied Materials senior management knowledge and support? Why or why not?
3. Although Applied Materials had strict policies in place to protect the intellectual property of its customers, suppliers, and other third parties and took violation of these policies seriously, these policies failed. Are there additional measures, controls, or audits that the company should put into place to prevent a repeat incident?

3. Intellectual Property and the War over Software Maintenance

In late 2006, Oracle Corporation noticed unusually heavy activity on its Customer Connection Web site—the password-protected site through which Oracle supplies customer support for its PeopleSoft and JD Edwards products. In November 2006, more than 10,000 copyrighted technical support items were downloaded by unauthorized users in what Oracle called "corporate theft on a grand scale."[96]

Oracle is one of the largest providers of business software in the world, with over 370,000 customers in over 145 countries using its applications.[97] In 2005, the company purchased PeopleSoft (which had acquired JD Edwards & Company in 2003) after a federal judge ruled that the purchase would not violate antitrust laws. PeopleSoft offered high-quality human resource management, customer relations, finance, and enterprise performance systems.

JD Edwards provided similar products, but for smaller companies that could not afford PeopleSoft. After the acquisition, Oracle announced that it planned to continue to support PeopleSoft and JD Edwards products and that it would eventually integrate them into Oracle's enterprise resource planning (ERP) product.

When Oracle made its acquisition, it gained not only the sales revenue generated by the PeopleSoft and JD Edwards products, but also the sizable maintenance contract revenue that came with those product lines. In return for a substantial annual fee, Oracle provides telephone and online support as well as access to instruction manuals, patches, code, and other documents through its Customer Connection Web site.

Oracle and other software vendors typically charge about 20 percent of the price of the original software for maintenance and support. So, if a package cost $1,000,000, a company might spend $200,000 each year to fix bugs that arose during and after installation or add enhancements released by the software developer. If the software provider continues to add value to the product each year, the client may not see these maintenance expenses as excessive. However, if the bugs are quickly ironed out and the software provider is doing little to maintain or improve the software package, a company might begin to look for other ways to obtain support for the product. Third-party vendors are one option, but the level of support they can provide depends on the level of access that the software provider is willing to offer these third parties. Not surprisingly, software providers are generally unwilling to cooperate with third parties that siphon off maintenance revenue. As a result, some business analysts argue that clients are held hostage by these software providers and are forced into expensive contracts.[98]

By the fourth or fifth year of operation, a business may not need a great deal of support for its software systems. In this case, a third-party vendor that only charges 10 to 15 percent of the original cost of the software package may be a very good option. At least, that is what Honeywell, Lockheed Martin, Coors, and others thought when they switched their business to TomorrowNow Inc., a company formed by former PeopleSoft executives. TomorrowNow originally supported PeopleSoft products and then branched out to JD Edwards and other applications. At the time Oracle acquired PeopleSoft and the JD Edwards product line, TomorrowNow already had about 100 clients. As PeopleSoft clients became nervous about Oracle's plans for the software package, traffic on TomorrowNow's site jumped 300 percent, and TomorrowNow pulled some maintenance support clients away from Oracle.[99]

In 2005, just days after the acquisition, Oracle's chief competitor, SAP, bought TomorrowNow. Many analysts believe that SAP procured TomorrowNow to counter Oracle's acquisition of PeopleSoft. With this purchase, SAP could erode Oracle's maintenance contract revenue while encouraging PeopleSoft and JD Edwards customers to migrate to SAP's ERP system through its Safe Harbor migration program. The Safe Harbor initiative was aimed at clients who were already using a combination of software, such as PeopleSoft human resource and SAP financial systems. Clients would receive a 75 percent discount on SAP's ERP software, and TomorrowNow would continue to support PeopleSoft and JD Edwards products through the migration. Any clients who migrated to the SAP software would pay an annual maintenance fee based on the full value of the ERP system.[100] Although some analysts believed that TomorrowNow was too small to accommodate the potential client base that would migrate over, Oracle clearly saw this move as a threat.

Oracle filed a lawsuit on March 22, 2007, in U.S. district court alleging that SAP had violated the federal Computer Fraud and Abuse Act as well as California law through the actions of TomorrowNow. The complaint accused TomorrowNow of downloading thousands of documents from Oracle's Customer Connection Web site in late 2006 and early 2007. TomorrowNow purportedly downloaded copyrighted material containing software updates, patches, bug fixes, and instructions for PeopleSoft and JD Edwards products.

During late 2006, Oracle customers using expired and soon-to-be-expiring passwords were retrieving information and code from every library on the Web site with much greater frequency than usual. One of these "customers," who had previously averaged 20 downloads per month, averaged over 1,800 downloads per day over the course of four days. Although paying customers can access support materials for any PeopleSoft or JD Edwards product by entering a username and password, they have no legal rights to support documents for software products for which they have not purchased a maintenance contract. In this sweep of the Customer Connection Web site, many of these people downloaded resources unrelated to the systems they had paid Oracle to support. Oracle claims these "customers" had one thing in common: They were or would soon be clients of TomorrowNow. Moreover, Oracle claimed that the downloads all originated from an Internet protocol (IP address) in Bryan, Texas—the headquarters of TomorrowNow. Oracle believed that through its subsidiary, TomorrowNow, SAP downloaded an extensive collection of Oracle's copyrighted software and support materials onto SAP computers.[101]

Oracle Corporation claimed that TomorrowNow was engaged in unethical business practices as far back as 2005. Oracle purported that TomorrowNow used "non-production" copies of PeopleSoft software to develop support solutions. Oracle alleged that SAP knew of these improprieties when it purchased TomorrowNow.[102] Oracle sought $1 billion in damages from SAP.[103]

SAP responded that TomorrowNow was acting on behalf of its clients who were authorized to make such downloads. SAP admitted, however, that in doing so, TomorrowNow also made some inappropriate downloads. In November 2007, SAP announced a change in management at TomorrowNow, and in July 2008, SAP decided to close the subsidiary. It helped TomorrowNow's customers transition back to Oracle or to third-party vendor Rimini Street for maintenance and support.[104]

Rimini Street is a third-party software maintenance company established by TomorrowNow cofounder Seth Ravin, a former PeopleSoft manager who left TomorrowNow three months after SAP acquired the company. In August 2007, when asked whether the Oracle lawsuit was scaring clients off third-party software support companies such as Rimini Street, Ravin replied, "The lawsuit actually opened up business—because some people didn't even know there was a choice. A lot of people read the lawsuit and said, 'My God, Merck and Honeywell, and all these large companies are using third-party support.' And instead of saying, 'Wow, look what happened here!' The issue really is, 'Holy gee, am I the only guy paying full price?'"[105]

So, as two software giants wrestle, third-party vendors remain an option. The questions remain as to how and whether these businesses can open up the maintenance contract market to competition without violating federal law.

In August 2010, SAP accepted liability for copyright infringement on the part of TomorrowNow but argued that the $1 billion in damages claimed by Oracle was greatly exaggerated. SAP states that it was misconduct on the part of TomorrowNow and wants to move to put the lawsuits behind it.[106]

Discussion Questions

1. According to Oracle, how was TomorrowNow violating intellectual property law?
2. Why do you think Oracle sued SAP?
3. What do you think should be done, if anything, to open the maintenance contract market to third-party contractors?

End Notes

[1] The Patent Station, "Famous Quotes," The Law Office of Mark D. Machtinger, LTD, www.patentstation.com/mdm/p142.htm#quotes (accessed March 11, 2011).

[2] Tarmo Virki, "Google Topples Symbian From Smart Phones Top Spot," *The Globe and Mail*, January 31, 2011, www.theglobeandmail.com/news/technology/mobile-technology/google-topples-symbian-from-smartphones-top-spot/article1888557/?cmpid=rss1.

[3] Brad Reed, "A Brief History of Smartphones," *PCWorld*, June 18, 2010, www.pcworld.com/article/199243/a_brief_history_of_smartphones.html.

[4] Brad Reed, "A Brief History of Smartphones," *PCWorld*, June 18, 2010, www.pcworld.com/article/199243/a_brief_history_of_smartphones.html.

[5] Marguerite Reardon, "Apple Sues HTC Over iPhone Patents," *CNET*, March 2, 2010, http://news.cnet.com/8301-1035_3-10462116-94.html.

[6] Brian X. Chen, "Apple Fires at HTC, But the Target Is Google," *Wired*, March 2, 2010, www.wired.com/epicenter/2010/03/apple-fires-at-htc.

[7] Marguerite Reardon, "Apple Sues HTC over iPhone patents," *CNET*, March 2, 2010, http://news.cnet.com/8301-1035_3-10462116-94.html.

[8] Erica Ogg, "HTC Fires Back at Apple with Patent Complaint," *CNET*, May 12, 2010, http://news.cnet.com/8301-31021_3-20004809-260.html.

[9] Eric Bangeman, "Apple Tries Short-Circuiting Droid With Patent Lawsuit," *Wired*, October 31, 2010, www.wired.com/epicenter/2010/10/apple-tries-short-circuiting-droid-with-patent-lawsuit.

[10] Michelle Kessler, "Microsoft Sues Motorola over Android Phone Patents," *USA Today*, October 1, 2010, http://content.usatoday.com/communities/technologylive/post/2010/10/microsoft-sues-motorola-over-android-phone-patents/1.

[11] Chris Foresman, "Motorola Asks ITC, Two Federal Courts to Throw Book at Apple," *ARS Technica*, http://arstechnica.com/apple/news/2010/10/motorola-asks-itc-two-federal-courts-to-throw-book-at-apple.ars.

[12] Gartner, "Press Release: Gartner Says Worldwide Mobile Device Sales to End Users Reached 1.6 Billion Units in 2010; Smartphone Sales Grew 72 Percent in 2010," February 9, 2011, www.gartner.com/it/page.jsp?id=1543014.

[13] Eric Bangeman, "Apple Tries Short-Circuiting Droid With Patent Lawsuit," *Wired*, October 31, 2010, www.wired.com/epicenter/2010/10/apple-tries-short-circuiting-droid-with-patent-lawsuit.

[14] Dan Frommer, "Patent Troll NTP Sues Apple, Google, HTC, Microsoft, and Motorola Over Email Patents," *Business Insider SAI*, July 9, 2010, www.businessinsider.com/ntp-sues-apple-google-htc-microsoft-and-motorola-over-email-patents-2010-7#ixzz1GacLtkjZ.

15 Heather Green, "NTP Sticks It to RIM," *Bloomberg Businessweek*, December 15, 2005, www.businessweek.com/technology/content/dec2005/tc20051215_806425.htm.

16 Tom Krazit, "Judge Faces 'Reality' in BlackBerry Case," *CNET*, February 24, 2006, http://news.cnet.com/Judge-faces-reality-in-BlackBerry-case/2100-1041_3-6043212.html.

17 Eric Bangeman, "It's Over: RIM and NTP Settle BlackBerry Dispute," *Ars Technica*, March 3, 2006, http://arstechnica.com/old/content/2006/03/6314.ars.

18 U.S. Code, Title 17, § 102(a).

19 United States Copyright Office, "Copyright Law of the United States of America and Related Laws Contained in Title 17 of the United States Code," www.copyright.gov/title17/92chap3.html (accessed April 19, 2011).

20 "Sonny Bono Copyright Term Extension Act," *EconomicExpert.com,* www.economicexpert.com/a/Sonny:Bono:Copyright:Term:Extension:Act.htm (accessed April 19, 2011).

21 UCLA Law and Columbia Law School Copyright Infringement Project, *Bright Tunes Music v. Harrisongs Music*, http://cip.law.ucla.edu/cases/case_brightharrisongs.html (accessed April 19, 2011).

22 U.S. Code, Title 17, § 107.

23 James Niccolai, "Court Won't Block Low-Cost Cartridges," *PC World*, February 22, 2005, www.pcworld.com/article/119747/court_wont_block_lowcost_cartridges.html.

24 K. C. Jones, "Groups Weigh in on Intellectual Property Protection Act," *InformationWeek*, October 14, 2008.

25 "IP Enforcement Bill Becomes Law," *National Journal*, October 13, 2008, http://techdailydose.nationaljournal.com/2008/10/ip-enforcement-bill-becomes-la.php.

26 Steven Schwankert, "Bush Enacts PRO-IP Anti-piracy Law," *PC World*, October 14, 2008.

27 World Trade Organization, "What Is the WTO?," www.wto.org (accessed March 11, 2011).

28 World Intellectual Property Organization, "What Is WIPO?," www.wipo.int/about-wipo/en/what_is_wipo.html, (accessed March 11, 2011).

29 United States Copyright Office, "The Digital Millennium Copyright Act of 1998, U.S. Copyright Office Summary," December 1998, www.copyright.gov/legislation/dmca.pdf.

30 Greg Sandoval, "ISP Won't Reveal Names of Alleged Porn Pirates," *CNET*, December 27, 2010, http://news.cnet.com/8301-31001_3-20026654-261.html.

31 Greg Sandoval, "Porn Maker Sues 7,098 Alleged Film Pirates," *CNET*, November 2, 2010, http://news.cnet.com/8301-31001_3-20021438-261.html.

32 Nate Anderson, "Judge Kills Massive P2P Porn Lawsuit, Kneecaps Copyright Troll," *Ars Technica*, December 17, 2010, http://arstechnica.com/tech-policy/news/2010/12/judge-kills-massive-p2p-porn-lawsuit-kneecaps-copyright-troll.ars.

33 Eriq Gardner, "More Than 100,000 People Have Been Sued for Sharing Movies in Past Year," *Hollywood Reporter*, February 1, 2011, www.hollywoodreporter.com/blogs/thr-esq/100000-people-sued-sharing-movies-95095.

[34] Greg Sandoval, "ISP Won't Reveal Names of Alleged Porn Pirates," *CNET*, December 27, 2010, http://news.cnet.com/8301-31001_3-20026654-261.html.

[35] Nate Anderson, "Judge Kills Massive P2P Porn Lawsuit, Kneecaps Copyright Troll," *Ars Technica*, December 17, 2010, http://arstechnica.com/tech-policy/news/2010/12/judge-kills-massive-p2p-porn-lawsuit-kneecaps-copyright-troll.ars.

[36] Sean F., "Law Firms Join Forces to Fight Online Piracy," *Digital Digest* (blog), January 12, 2011, www.digital-digest.com/news-62835-Law-Firms-Join-Forces-To-Fight-Online-Piracy.html.

[37] *Pornography, Technology, and Process: Problems and Solutions on Peer-to-Peer Networks, Hearing Before the Senate Judiciary Committee*, Testimony of William Barr, Executive Vice President and General Counsel, Verizon Communications, September 17, 2003, http://judiciary.senate.gov/hearings/testimony.cfm?id=902&wit_id=2563 (accessed March 19, 2011).

[38] IBM "Press Release: 2009 IBM Patent Leadership," January 12, 2010, www-03.ibm.com/press/us/en/presskit/29164.wss.

[39] Joff Wild, "More on the IBM $1 Billion Patent Licensing Urban Legend," *Intellectual Asset Management*, March 27, 2008, www.iam-magazine.com/blog/Detail.aspx?g=9be3f156-79b1-49f4-abf1-9bee7e788501.

[40] "The Patent Filing Process," Essortment, www.essortment.com/all/patentfilingpr_rrgx.htm (accessed March 25, 2011).

[41] U.S. Code, Title 35, § 103–4.

[42] *Diamond v. Diehr*, 450 U.S. 175(1981), BitLaw, www.bitlaw.com/source/cases/patent/Diamond_v_Diehr.html (accessed March 25, 2011).

[43] "Patents and Innovation: Trends and Policy Challenges," Organisation for Economic Cooperation and Development, 2004 www.oecd.org/dataoecd/48/12/24508541.pdf (accessed March 26, 2011).

[44] Professor Dennis Crouch, *Bilski v. Kappos*, Patently-O, June 28, 2010, www.patentlyo.com/patent/2010/06/bilski-v-kappos-business-methods-out-software-still-patentable.html.

[45] Stephen Shankland and Declan McCullagh, "Supreme Court Sidesteps Software Patent Issue," *CNET*, June 28, 2010, http://news.cnet.com/8301-30685_3-20009019-264.html.

[46] Ina Fried, "Microsoft—License to Deal," *CNET*, November 8, 2004, http://news.cnet.com/Microsoft–license-to-deal/2100-1012_3-5440881.html.

[47] Nicholas Kolakowski, "Microsoft, Panasonic Announce Patent Cross-Licensing Agreement", *eWeek*, February 25, 2010, www.eweek.com/c/a/Windows/Microsoft-Panasonic-Announce-Patent-CrossLicensing-Agreement-671192.

[48] "Patent Litigation Costs, How Much Does It Cost to Protect a Patent?," www.inventionstatistics.com/Patent_Litigation_Costs.html (accessed March 27, 2011).

[49] Steve Lohr, "Smartphone Patent Suits Challenge Big Makers," *New York Times*, July 9, 2010.

[50] "US Appeals Court to Consider NTP's Wireless-Email Patents," Fox Business, February 9, 2011, www.foxbusiness.com/markets/2011/02/09/appeals-court-consider-ntps-wireless-email-patents/#.

51 Gene Quinn, "Submarine Patents Alive and Well: Tivo Patents DVR Scheduling," *IPWatchdog*, February 19, 2020, http://ipwatchdog.com/tag/submarine-patent.

52 National Conference of Commissioners on Uniform State Laws, "Uniform Trade Secrets Act," http://euro.ecom.cmu.edu/program/law/08-732/TradeSecrets/utsa.pdf (accessed March 28, 2011).

53 Gerald J. Mossinghoff, J. Derek Mason, Ph.D., and David A. Oblon, "The Economic Espionage Act: A New Federal Regime of Trade Secret Protection," Oblon Spivak, www.oblon.com/publications/economic-espionage-act-new-federal-regime-trade-secret-protection (accessed March 28, 2011).

54 Ryan Averbeck and Gregory A. Gaddy, "Protecting Your Organization's Innovations," *CSO Online*, February 22, 2009, www.csoonline.com/article/481815/protecting-your-organization-s-innovations.

55 "First Conviction Under the Economic Espionage Act," *Homeland Security News Wire*, June 20, 2008, http://homelandsecuritynewswire.com/first-conviction-under-economic-espionage-act-1996.

56 Robert Abrams and Gregory L. Baker, "International Trade Secret Theft Not Violation of the Economic Espionage Act Unless Defendants Intended to Benefit a Foreign Government, Federal Court Holds," martindale.com, June 7, 2010, www.martindale.com/criminal-law/article_Howrey-LLP_1044122.htm.

57 Barbara Rose, "Non-Compete Clause Tying Hands of Employees," *Chicago Tribune*, February 25, 2008, http://articles.chicagotribune.com/2008-02-25/business/0802230096_1_non-competes-employees-supreme-court/2.

58 Thomas Claburn, "Apple's Controversial iPhone Developer Agreement Published," *InformationWeek*, October 28, 2008, www.informationweek.com/news/personal_tech/iphone/showArticle.jhtml?articleID=211601121.

59 Bill Nolan, "Noncompete Agreements: Critical IP and Employment Protection," *Columbus CEO*, February 14, 2011, www.columbusceo.com/parting_shots/article_56dd9cf4-3884-11e0-a106-0017a4a78c22.html.

60 Kenneth J. Vanko, "IBM Loses Preliminary Injunction Motion Over Executive's Departure to Hewlett Packard (IBM v Visetin)," February 24, 2011, *Legal Developments In Non-Competition Agreements* (blog) www.non-competes.com/2011/02/ibm-loses-preliminary-injunction-motion.html.

61 Howard Kurtz, "'Katie's Notebook' Item Cribbed from *W.S. Journal*," *Washington Post*, April 11, 2007, www.washingtonpost.com/wp-dyn/content/article/2007/04/10/AR2007041001537.html.

62 Jacques Steinberg, "Times's 2 Top Editors Resign After Furor on Writer's Fraud," *New York Times*, June 6, 2003, www.nytimes.com/2003/06/06/business/media/06PAPE.html.

63 Pat Sierchio, "Feisty, Prolific SF Author Harlan Ellison Bares 'Sharp Teeth' in Bio-Pic," *Jewish Journal*, August 2, 2007, www.jewishjournal.com/arts/article/feisty_prolific_sf_author_harlan_ ellison_bares_sharp_teeth_in_biopic

64 Krista Westervelt, "Harry Potter Plagiarism Lawsuit Dismissed by US Judge," *Entertainment*, January 7, 2011, http://entertainment.gather.com/viewArticle.action?articleId=281474978887412.

65 iThenticate, "About Us—A History of Plagiarism Detection," www.ithenticate.com/about.html (accessed March 28, 2011).

66 Turnitin, "Press Release: US Court of .Appeals Unanimously Affirms Finding of 'Fair Use' for Turnitin," April 27, 2009, http://turnitin.com/static/resources/documentation/turnitin/sales/Turnitin_RELEASE_042709_Appeals_Court_Affirms_Fair_Use.pdf.

67 *Sega Enterprises Ltd. v. Accolade,* Inc., http://digital-law-online.info/cases/24PQ2D1561.htm (accessed April 4, 2011).

68 Ed Scannell, "1 in 3 IT Shops Uses Combo Proprietary, Open Source Software," *InformationWeek*, March 13, 2009.

69 Tom Smith, "5 Web Lessons from Google's Analytics Guru," *InformationWeek*, September 17, 2008.

70 "P&G, Unilever Reach Spying Settlement," *Business Courier of Cincinnati*, September 6, 2001.

71 "What Is Trademark Fair Use?," Trademark Education & Information, www.trademark-education.com/fairuse.html (accessed March 15, 2011).

72 *Playboy Enterprises, Inc. v. Terri Welles*, www.loundy.com/CASES/Playboy_v_Wells.html (accessed March 15, 2011).

73 Eric Goldman, "Funky Ninth Circuit Opinion on Domain Names and Nominative Use—Toyota v. Tabari," *Technology & Marketing Law Blog*, July 14, 2010, http://blog.ericgoldman.org/archives/2010/07/funky_ninth_cir.htm.

74 Peter Sayer, "Verizon Wins $31M Judgment in Cybersquatting Case," *Computerworld*, December 26, 2008.

75 Jonathan Band, "The Google Library Project: Both Sides of the Story," www.plagiary.org/Google-Library-Project.pdf (accessed March 21, 2011).

76 Google Book Search, "Google Books Library Project," http://books.google.com/googlebooks/library.html.

77 Association of Research Libraries, "A Guide for the Perplexed: Libraries and the Google Library Project Settlement," November 13, 2008, www.arl.org/bm~doc/google-settlement-13nov08.pdf.

78 Stephanie Condon, "Google Reaches $125 Million Settlement with Authors," *CNET*, October 28, 2008, http://news.cnet.com/8301-13578_3-10076948-38.html.

79 Chris Snyder, "Google, Authors and Publishers Settle Book-Scan Suit," *Wired*, October 28, 2008, www.wired.com/epicenter/2008/10/google-authors.

80 Miguel Helft, "Justice Dept. Opens Antitrust Inquiry into Google Books Deal," *New York Times*, April 28, 2009.

81 "Statement of Interest of the United States Regarding Proposed Amended Settlement," *The Authors Guild, Inc. et al., Plaintiffs, v. Google Inc., Defendant*; 05 Civ. 8136 (DC) ECF Case, Filed September 18, 2009, http://thepublicindex.org/docs/letters/usa.pdf (accessed April 20, 2011).

82 Elinor Mills, "Google Books Settlement Sets Geographic, Business Limits," *CNET*, November 13, 2009, http://news.cnet.com/8301-1023_3-10397787-93.html.

83 "Statement of Interest of the United States Regarding Proposed Class Settlement," *The Authors Guild, Inc. et al., Plaintiffs, v. Google Inc., Defendant*; 05 Civ. 8136 (DC) ECF Case, Filed February 4, 2010, http://thepublicindex.org/docs/amended_settlement/usa.pdf (accessed April 19, 2011).

84 Motoko Rich, "Judge Hears Arguments on Google Book Settlement," *New York Times*, February 18, 2010, www.nytimes.com/2010/02/19/technology/19google.html.

85 Greg Sandoval, "Google Book Settlement Draws Fire in Court," *CNET*, February 18, 2010, http://news.cnet.com/8301-31001_3-10456382-261.html.

86 Ryan Singel, "Google Launches Online Bookstore, Challenging Amazon," *Wired*, December 6, 2010, www.wired.com/epicenter/2010/12/google-bookstore.

87 Samsung, "About Samsung," www.samsung.com/us/aboutsamsung (accessed April 22, 2011).

88 "World's Top Computer Memory Chip Manufacture in Espionage Row," *domain-b*, February 2, 2010, www.domain-b.com/industry/Semiconductors/20100204_chip_manufacture.html.

89 "Chip Espionage Case Heats Up," *CBS News*, February 4, 2010, www.cbsnews.com/stories/2010/02/04/tech/main6172338.shtml.

90 "Chip Espionage Case Heats Up," *CBS News*, February 4, 2010, www.cbsnews.com/stories/2010/02/04/tech/main6172338.shtml.

91 "World's Top Computer Memory Chip Manufacture in Espionage Row," *domain-b*, February 2, 2010, www.domain-b.com/industry/Semiconductors/20100204_chip_manufacture.html.

92 "Defendents in Samsung Case Deny Industrial Espionage Charges," *PI Newswire*, March 12, 2010, www.pinewswire.net/2010/03/defendents-in-samsung-case-deny-industrial-espionage-charges.

93 "World's Top Computer Memory Chip Manufacture in Espionage Row," *domain-b*, February 2, 2010, www.domain-b.com/industry/Semiconductors/20100204_chip_manufacture.html.

94 "Chip Espionage Case Heats Up," *CBS News*, February 4, 2010, www.cbsnews.com/stories/2010/02/04/tech/main6172338.shtml.

95 Jun Yang and Victoria Batchelor, "Applied Materials, Samsung Settle Espionage Dispute," *Bloomberg Businessweek*, November 30, 2010.

96 "Complaint for Damages and Injunctive Relief," United States District Court, Northern District of California, March 22, 2007, www.oracle.com/sapsuit/complaint.pdf.

97 Oracle, "About Oracle," www.oracle.com/us/corporate/index.html (accessed April 22, 2011).

98 Linda Tucci, "Oracle/SAP Lawsuit Fuels Third-Party Maintenance Argument," *SearchCIO*, August 9, 2007, http://searchcio/techtarget.com.au/articles/21119-Oracle-SAP-lawsuit-fuels-third-party-maintenance-argument.

[99] Charles Babcock, "Support Comes from the Outside," *InformationWeek*, January 10, 2005, www.informationweek.com/news/global-cio/showArticle.jhtml?articleID=57300371.

[100] Renee Boucher Ferguson, "SAP Buy Targets PeopleSoft Migration," *eWeek.com*, January 19, 2005, www.eweek.com/c/a/Enterprise-Applications/SAP-Buy-Targets-People Soft-Migration.

[101] "Complaint for Damages and Injunctive Relief," United States District Court, Northern District of California, March 22, 2007, www.oracle.com/sapsuit/complaint.pdf.

[102] "Third Amended Complaint for Damages and Injunctive Relief," United States District Court, Northern District of California, October 8, 2008, www.oracle.com/sapsuit/third-ammended-complaint.pdf.

[103] Brandon Bailey, "Oracle Expands Theft Allegations Against SAP," *Trade Secrets Vault*, July 30, 2008, www.tradesecretsblog.info/2008/07/oracle_exapnds_theft_allegation.html.

[104] Rick Whiting, "SAP Shuttering Its TomorrowNow Services Subsidiary," *ChannelWeb*, July 21, 2008, www.cn.com/software/209102245.

[105] Linda Tucci, "Interview: Seth Ravin on Third-Party Software Maintenance," *SearchCIO*, August 28, 2007, http://searchcio.techtarget.com/news/interview/0,289202,sid182_gci1269359,00.html.

[106] Larry Dignan, "SAP Accepts TomorrowNow Liability, Says Oracle Damages Inflated," *ZDNet*, August 5, 2010.

CHAPTER 7

SOFTWARE DEVELOPMENT

QUOTE

It's not a bug—it's an undocumented feature.
—Author unknown

VIGNETTE

Software Errors Lead to Death

Medical linear accelerators have long been a critical piece of medical equipment in the fight against cancer. Linear accelerators deliver radiation therapy to cancer patients by accelerating electrons to create high-energy beams, which can kill cancer tumors without impacting surrounding healthy tissue. Tumors close to the skin can be treated with the accelerated electrons; however, for tumors that are more deeply embedded, the electron beam is converted into an X-ray photon beam, which is diffused using a beam spreader plate.

The Canadian firm Atomic Energy of Canada Limited (AECL) and a French company named CGR collaborated to build two models of medical linear accelerators. The Therac-6 was capable of producing only X-rays that could be used to kill tumors close to the skin. The Therac-20 was capable of producing both X-ray photons and electrons and thus could kill both shallow and deeply embedded tumors. Computer software was used to simplify the operation of the equipment but not to control and monitor its operation. Instead industry standard hardware safety features were built into both models.[1]

After the business relationship between the two firms failed, AECL went on to build the Therac-25 based on a new design concept. Unlike the Therac-6 and Therac-20, which operated without significant computer controls, computer software was used to both control and monitor the Therac-25 accelerator.[2]

The software for the Therac-25 was based on modified code from the Therac-6. The software monitored the machine, accepted technician input for specific patient treatment, initialized the machine to administer the defined treatment, and controlled the machine to execute the defined treatment. The machine was enclosed in the patient treatment room to prevent radiation exposure to the technicians. Audio and visual equipment allowed the patient to communicate with the technicians.[3]

Eleven Therac-25 machines were installed in the United States and Canada. Over a 19-month period from June 1985 to January 1987, six serious incidents involving the use of the device occurred. In each of the incidents, the patient received an overdose of radiation. Four of the patients died from the overdose, and another eventually had to have both breasts removed and lost use of her right arm as a result of the overdose. A final patient received burns but was able to fully recover after several years.[4]

Following each incident, AECL was contacted and asked to investigate the situation. However, AECL at first refused to believe that its machine could have been responsible for an overdose. Indeed, following the third incident, AECL responded, "After careful consideration, we are of the opinion that this damage could not have been produced by any malfunction of the Therac-25 or by any operator error." AECL made some minor changes to the equipment, but because the company did not address the root cause of the problem, additional incidents occurred.[5]

Finally, a physicist at a hospital where two incidents occurred was able to re-create the malfunction and show that the problem was due to a defect in the machine and its software. A failure occurred when a specific sequence of keystrokes was entered by the operator. Because this sequence of keystrokes was nonstandard, the problem rarely occurred and went undetected for a long time. Entry of this combination of keystrokes within a period of eight seconds did not allow time for the beam spreader plate to be rotated into place. The software did not recognize the error, and the patient was then hit with a high-powered electron beam roughly 100 times the intended dose of radiation.[6]

In early 1987, the Food and Drug Administration (FDA) and Health Canada (the Canadian counterpart to the FDA) insisted that all Therac-25 units be shut down. Within the next six months, the AECL implemented numerous code changes, installed independent hardware safety locks, and implemented other changes to correct the problem.[7] After these changes, the Therac-25 device continued to be safely used for many years. However, at least three lawsuits were filed against AECL and the hospitals involved in the earlier incidents. The lawsuits were settled out of court, and the results were never revealed.[8]

Questions to Consider

1. What additional measures must be taken in the development of software that, if it fails, can cause loss of human life?
2. What can organizations do to reduce the negative consequences of software development problems in the production of their products and the operation of their business processes and facilities?

LEARNING OBJECTIVES

As you read this chapter, consider the following questions:

1 Why do companies require high-quality software in business systems, industrial process control systems, and consumer products?

2 What potential ethical issues do software manufacturers face in making trade-offs between project schedules, project costs, and software quality?

3 What are the four most common types of software product liability claims?

4 What are the essential components of a software development methodology, and what are the benefits of using such a methodology?

5 How can the Capability Maturity Model Integration® improve an organization's software development process?

6 What is a safety-critical system, and what special actions are required during its development?

STRATEGIES FOR ENGINEERING QUALITY SOFTWARE

High-quality software systems are easy to learn and use because they perform quickly and efficiently; they meet their users' needs; and they operate safely and reliably so that system downtime is kept to a minimum. Such software has long been required to support the fields of air traffic control, nuclear power, automobile safety, health care, military and defense, and space exploration. Now that computers and software have become integral parts of almost every business, the demand for high-quality software is increasing. End users cannot afford system crashes, lost work, or lower productivity. Nor can they tolerate security holes through which intruders can spread viruses, steal data, or shut down Web sites. Software manufacturers face economic, ethical, and organizational challenges associated with improving the quality of their software. This chapter covers many of these issues.

A **software defect** is any error that, if not removed, could cause a software system to fail to meet its users' needs. The impact of these defects can be trivial; for example, a computerized sensor in a refrigerator's ice cube maker might fail to recognize that the tray is full and continue to make ice. Other defects could lead to tragedy—the control system for an automobile's antilock brakes could malfunction and send the car into an uncontrollable spin. The defect might be subtle and undetectable, such as a tax preparation package that makes a minor miscalculation; or the defect might be glaringly obvious, such as a payroll program that generates checks with no deductions for Social Security or other taxes. Here are some notable software bugs that have occurred over the past several years:

- In December 2005, a critical software bug was detected that affected 39 different Symantec software products—including both home and enterprise versions of its antivirus software. The bug could have been exploited to load unauthorized code onto a computer and to potentially take control of the computer.[9]
- Windows Vista—the operating system from Microsoft—became the subject of much criticism for issues relating to performance, security, and privacy. Indeed, the initial version of the operating system was so unpopular that in mid-2006, Microsoft began allowing PC manufacturers to offer a "downgrade" option to buyers who purchased machines with Vista but who wanted to switch to Windows XP.[10]
- Skype—which offers a service to enable users to make phone calls over the Internet—suffered a massive outage in August 2007, which affected the 220 million people who were registered Skype users at that time.[11] According to a Skype spokesperson, a "previously unseen software bug within the network resource algorithm" caused the problem.[12]
- Software problems in the automated baggage sorting system at Heathrow airport caused the system to go offline for almost two days in February 2008. As a result, carriers in the terminal were forced to sort baggage manually, and 6,000 passengers experienced delays, flight cancellations, and frustration. The breakdown reportedly occurred during a software upgrade, despite

pretesting of the software. The system continued to experience problems in subsequent months.[13]

- In January 2010, a software defect triggered by the change in the decade affected 30 million German debit and credit cards, leaving bank customers unable to charge purchases or make cash withdrawals.[14]

Software quality is the degree to which a software product meets the needs of its users. **Quality management** focuses on defining, measuring, and refining the quality of the development process and the products developed during its various stages. These products—including statements of requirements, flowcharts, and user documentation—are known as **deliverables**. The objective of quality management is to help developers deliver high-quality systems that meet the needs of their users. Unfortunately, the first release of any software rarely meets all its users' expectations. A software product does not usually work as well as its users would like it to until it has been used for a while, found lacking in some ways, and then corrected or upgraded.

A primary cause of poor software quality is that many developers do not know how to design quality into software from the very start; some simply do not take the time to do so. In order to develop high-quality software, developers must define and follow a set of rigorous engineering principles and be committed to learning from past mistakes. In addition, they must understand the environment in which their systems will operate and design systems that are as immune to human error as possible.

All software designers and programmers make mistakes in defining user requirements and turning them into lines of code. According to one study, even experienced software developers unknowingly inject an average of one design or implementation defect for every 7 to 10 lines of code.[15] The developers aren't incompetent or lazy—they're just human. Everyone makes mistakes, but in software, these mistakes can result in defects.

The Microsoft Windows 7 operating system contains more than 50 million lines of code.[16] Assume that the Microsoft software developers produced code at the accuracy rate mentioned above. Even if 99.9 percent of the defects were identified and fixed before the product was released to the public, there would still be about one bug per 10,000 lines of code, or roughly 5,000 bugs in Windows 7. Thus, software used daily by workers worldwide likely contains thousands of bugs.

Another factor that can contribute to poor-quality software is the extreme pressure that software companies feel to reduce the time to market for their products. They are driven by the need to beat the competition in delivering new functionality to users, to begin generating revenue to recover the cost of development, and to show a profit for shareholders. They are also driven by the need to meet quarterly earnings forecasts used by financial analysts to place a value on the stock. The resources and time needed to ensure quality are often cut under the intense pressure to ship a new product. When forced to choose between adding more user features and doing more testing, most software companies decide in favor of more features. They often reason that defects can be patched in the next release, which will give customers an automatic incentive to upgrade. Additional features make a release more useful and therefore easier to sell to customers. A major ethical dilemma for software development organizations is: "How much additional cost and effort should they expend to ensure that their products and services meet customers' expectations?"

Customers are stakeholders who are key to the success of a software application, and they may benefit from new features. However, they also bear the burden of errors that aren't caught or fixed during testing. Thus, many customers challenge whether the decision to cut quality in favor of feature enhancement is ethical.

As a result of the lack of consistent quality in software, many organizations avoid buying the first release of a major software product or prohibit its use in critical systems; their rationale is that the first release often has many defects that cause problems for users. Because of the defects in the first two popular Microsoft operating systems (DOS and Windows), including their tendency to crash unexpectedly, many believe that Microsoft did not have a reasonably reliable operating system until its third major variation—Windows NT.

Even software products that have been reliable over a long period can falter unexpectedly when operating conditions change. For instance, the software in the Cincinnati Bell telephone switch had been thoroughly tested and had operated successfully for months after it was deployed. Later that year, however, when the time changed from daylight saving time to standard time, the switch failed because it was overwhelmed by the number of calls to the local "official time" phone number from people who wanted to set their clocks. The large increase in the number of simultaneous calls to the same number was a change in operating conditions that no one had anticipated.

The Importance of Software Quality

A **business information system** is a set of interrelated components—including hardware, software, databases, networks, people, and procedures—that collects and processes data and disseminates the output. A common type of business system is one that captures and records business transactions. For example, a manufacturer's order-processing system captures order information, processes it to update inventory and accounts receivable, and ensures that the order is filled and shipped on time to the customer. Other examples include an airline's online ticket-reservation system and an electronic-funds transfer system that moves money among banks. The accurate, thorough, and timely processing of business transactions is a key requirement for such systems. A software defect can be devastating, resulting in lost customers and reduced revenue. How many times would bank customers tolerate having their funds transferred to the wrong account before they stopped doing business with that bank?

Another type of business information system is the **decision support system (DSS)**, which is used to improve decision making in a variety of industries. A DSS can be used to develop accurate forecasts of customer demand, recommend stocks and bonds for an investment portfolio, or schedule shift workers in such a way as to minimize cost while meeting customer service goals. A software defect in a DSS can result in significant negative consequences for an organization and its customers.

Software is also used to control many industrial processes in an effort to reduce costs, eliminate human error, improve quality, and shorten the time it takes to manufacture products. For example, steel manufacturers use process control software to capture data from sensors about the equipment that rolls steel into bars and about the furnace that heats the steel before it is rolled. Without process control computers, workers could react to defects only after the fact and would have to guess at the adjustments needed to correct the process. Process control computers enable the process to be monitored for variations from operating standards (e.g., a low furnace temperature or incorrect levels of iron ore) and to eliminate product defects *before* they affect product quality. Any defect in this software can lead to decreased product quality, increased waste and costs, or even unsafe operating conditions for employees.

Software is also used to control the operation of many industrial and consumer products, such as automobiles, medical diagnostic and treatment equipment, televisions, radios, stereos, refrigerators, and washers. A software defect could have relatively minor consequences, such as clothes not drying long enough, or it could cause serious damage, such as a patient being overexposed to powerful X-rays.

As a result of the increasing use of computers and software in business, many companies are now in the software business whether they like it or not. The quality of software, its usability, and its timely development are critical to almost everything businesses do. The speed with which an organization develops software can put it ahead of or behind its competitors. Software problems may have caused frustrations in the past, but mismanaged software can now be fatal to a business, causing it to miss product delivery dates, incur increased product development costs, and deliver products that have poor quality.

Business executives frequently face ethical questions of how much money and effort they should invest to ensure high-quality software. A manager who takes a short-term, profit-oriented view may feel that any additional time and money spent on quality assurance will only delay a new product's release, resulting in a delay in sales revenue and profits. However, a different manager may consider it unethical not to fix all known problems before putting a product on the market and charging customers for it.

Other key questions for executives are whether their products could cause damage and what their legal exposure would be if they did. Fortunately, software defects are rarely lethal, and few personal injuries are related to software failures. However, the use of software introduces product liability issues that concern many executives.

SOFTWARE PRODUCT LIABILITY

Software product litigation is certainly not new. One lawsuit in the early 1990s involved a financial institution that became insolvent because defects in a purchased software application caused errors in its integrated general ledger system, customers' passbooks, and loan statements. Dissatisfied depositors responded by withdrawing more than $5 million. In another case from 1992, a Ford truck stalled because of a software defect in the truck's fuel injector. In the ensuing accident, a young child was killed.[17] A state supreme court later affirmed an award of $7.5 million in punitive damages against the manufacturer. In October 2008, a faulty onboard computer caused a Qantas passenger flight traveling between Perth and Singapore to plunge some 8,000 feet in 10 seconds, injuring 46 passengers. Qantas moved quickly to compensate all passengers with a refund of their ticket prices, a $2,000 travel voucher, and a promise to pay all medical related expenses. Even so the Australian law firm of Slater & Gordon was engaged to represent a dozen of the passengers.[18]

The liability of manufacturers, sellers, lessors, and others for injuries caused by defective products is commonly referred to as **product liability**. There is no federal product liability law; instead, product liability is mainly covered by common law (made by state judges) and Article 2 of the Uniform Commercial Code, which deals with the sale of goods.

If a software defect causes injury or loss to purchasers, lessees, or users of the product, the injured parties may be able to sue as a result. Injury or loss can come in the form of physical mishaps and death, loss of revenue, or an increase in expenses due to a business disruption caused by a software failure. Software product liability claims are typically based on strict liability, negligence, breach of warranty, or misrepresentation—sometimes in combination with one another. Each of these legal concepts is discussed in the following sections.

Strict liability means that the defendant is held responsible for injuring another person, regardless of negligence or intent. The plaintiff must prove only that the software product is defective or unreasonably dangerous and that the defect caused the injury. There is no requirement to prove that the manufacturer was careless or negligent, or to prove who caused the defect. All parties in the chain of distribution—the manufacturer, subcontractors, and distributors—are strictly liable for injuries caused by the product and may be sued.

Defendants in a strict liability action may use several legal defenses, including the doctrine of supervening event, the government contractor defense, and an expired statute of limitations. Under the doctrine of supervening event, the original seller is not liable if the software was materially altered after it left the seller's possession and the alteration caused the injury. To establish the government contractor defense, a contractor must prove that the precise software specifications were provided by the government, that the software conformed to the specifications, and that the contractor warned the government of any known defects in the software. Finally, there are statutes of limitations for claims of liability, which means that an injured party must file suit within a certain amount of time after the injury occurs.

As discussed in Chapter 2, negligence is the failure to do what a reasonable person would do, or doing something that a reasonable person would not do. When sued for negligence, a software supplier is not held responsible for every product defect that causes customer or third-party loss. Instead, responsibility is limited to harmful defects that could have been detected and corrected through "reasonable" software development practices. Even when a contract is written expressly to protect against supplier negligence, courts may disregard such terms as unreasonable. Software manufacturers or organizations with software-intensive products are frequently sued for negligence and must be prepared to defend themselves.

The defendant in a negligence case may either answer the charge with a legal justification for the alleged misconduct or demonstrate that the plaintiffs' own actions contributed to their injuries (**contributory negligence**). If proved, the defense of contributory negligence can reduce or totally eliminate the amount of damages the plaintiffs receive. For example, if a person uses a pair of pruning shears to trim his fingernails and ends up cutting off a fingertip, the defendant could claim contributory negligence.

A **warranty** assures buyers or lessees that a product meets certain standards of quality. A warranty of quality may be either expressly stated or implied by law. Express warranties can be oral, written, or inferred from the seller's conduct. For example, sales contracts contain an implied warranty of merchantability, which requires that the following standards be met:

- The goods must be fit for the ordinary purpose for which they are used.
- The goods must be adequately contained, packaged, and labeled.
- The goods must be of an even kind, quality, and quantity within each unit.
- The goods must conform to any promise or affirmation of fact made on the container or label.
- The quality of the goods must pass without objection in the trade.
- The goods must meet a fair average or middle range of quality.

If the product fails to meet the terms of its warranty, the buyer or lessee can sue for **breach of warranty**. Of course, most dissatisfied customers will first seek a replacement, a substitute product, or a refund before filing a lawsuit.

Software suppliers frequently write warranties to attempt to limit their liability in the event of nonperformance. Although a certain software may be warranted to run on a given machine configuration, often no assurance is given as to what that software will do. Even if a contract specifically excludes the commitment of merchantability and fitness for a specific use, the court may find such a disclaimer clause unreasonable and refuse to enforce it or refuse to enforce the entire contract. In determining whether warranty disclaimers are unreasonable, the court attempts to evaluate if the contract was made between two "equals" or between an expert and a novice. The relative education, experience, and bargaining power of the parties and whether the sales contract was offered on a take-it-or-leave-it basis are considered in making this determination.

The plaintiff must have a valid contract that the supplier did not fulfill in order to win a breach-of-warranty claim. Because the software supplier writes the warranty, this claim can be extremely difficult to prove. For example, in 1993, M. A. Mortenson Company—one of the largest construction companies in the United States—had a new version of bid-preparation software installed for use by its estimators. During the course of preparing one new bid, the software allegedly malfunctioned several times, each time displaying the same cryptic error message. Nevertheless, the estimator submitted the bid and Mortenson won the contract. Afterward, Mortenson discovered that the bid was $1.95 million lower than intended and filed a breach-of-warranty suit against Timberline Software, makers of the bid software. Timberline acknowledged the existence of the bug. However, the courts ruled in Timberline's favor because the license agreement that came with the software explicitly barred recovery of the losses claimed by Mortenson.[19] Even if breach of warranty can be proven, the damages are generally limited to the amount of money paid for the product.

As mentioned in Chapter 2, intentional misrepresentation occurs when a seller or lessor either misrepresents the quality of a product or conceals a defect in it. For example, if a cleaning product is advertised as safe to use in confined areas and some users subsequently pass out from the product's fumes, they could sue the seller for intentional misrepresentation or fraud. Advertising, salespersons' comments, invoices, and shipping labels are all forms of representation. Most software manufacturers use limited warranties and disclaimers to avoid any claim of misrepresentation.

Software Development Process

Developing information system software is not a simple process; it requires completing many complex activities, with many dependencies among the various activities. System analysts, programmers, architects, database specialists, project managers, documentation specialists, trainers, and testers are all involved in large software projects. Each of these groups of workers has a role to play and has specific responsibilities and tasks. In addition, each group makes decisions that can affect the software's quality and the ability of an organization or an individual to use it effectively.

Many software companies have adopted a **software development methodology**—a standard, proven work process that enables systems analysts, programmers, project managers, and others to make controlled and orderly progress in developing high-quality software. A methodology defines activities in the software development process, and the individual and group responsibilities for accomplishing these activities. It also recommends specific techniques for accomplishing the various activities, such as using a flowchart to document the logic of a computer program. A methodology also offers guidelines for managing the quality of software during the various stages of development. See Figure 7-1. If an organization has developed such a methodology, it is applied to any software development that the company undertakes.

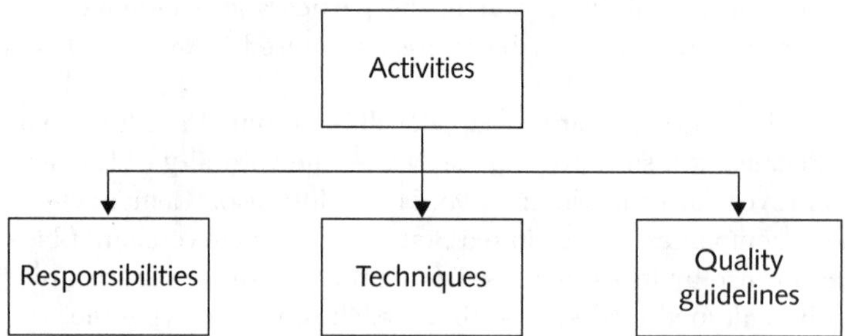

FIGURE 7-1 Software development methodology
Source Line: Course Technology/Cengage Learning.

As with most things, it is usually easier and cheaper to avoid software problems at the beginning than to attempt to fix the damages after the fact. Studies have shown that the cost to identify and remove a defect in an early stage of software development (requirements definition) can be up to 100 times less than removing a defect in a piece of software that has been distributed to customers (see Figure 7-2).[20,21] (Although these studies were conducted several years ago, their results still hold true today.)

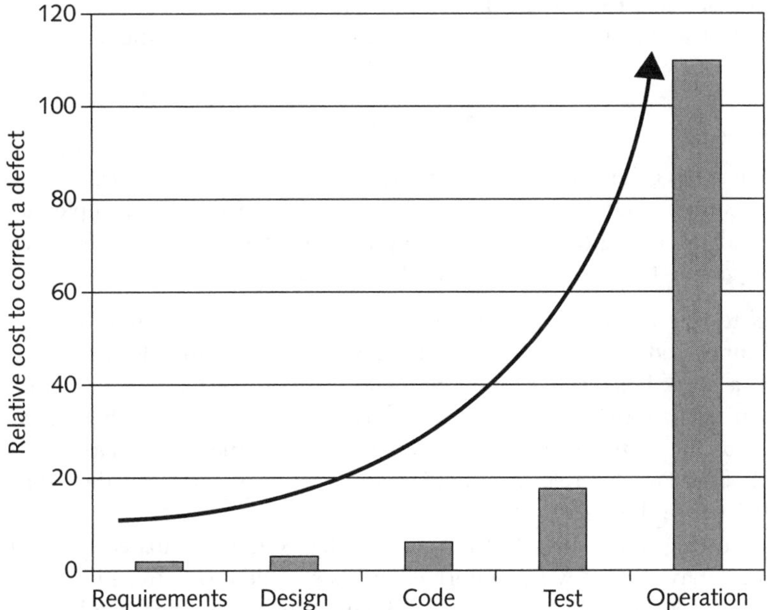

FIGURE 7-2 The cost of removing software defects
Source Line: Used with permission from LKP Consulting Group.

If a defect is uncovered during a later stage of development, some rework of the deliverables produced in preceding stages will be necessary. The later the error is detected, the greater the number of people who will be affected by the error. Thus, the greater the costs will be to communicate and fix the error. Consider the cost to communicate the details of a defect, distribute and apply software fixes, and possibly retrain end users for a software product that has been sold to hundreds or thousands of customers. Thus, most software developers try to identify and remove errors early in the development process not only as a cost-saving measure but as the most efficient way to improve software quality.

A product containing inherent defects that harm the user may be the subject of a product liability suit. The use of an effective methodology can protect software manufacturers from legal liability in two ways. First, an effective methodology reduces the number of software errors that might occur. Second, if an organization follows widely accepted development methods, negligence on its part is harder to prove. However, even a *successful* defense against a product liability case can cost hundreds of thousands of dollars in legal fees. Thus, failure to develop software carefully and consistently can be serious in terms of liability exposure.

Quality assurance (QA) refers to methods within the development cycle designed to guarantee reliable operation of a product. Ideally, these methods are applied at each stage of the development cycle. However, some software manufacturing organizations without a formal, standard approach to QA consider testing to be their only QA method. Instead of checking for errors throughout the development process, they rely primarily on testing just before the product ships to ensure some degree of quality.

Several types of tests are used in software development, as discussed in the following sections.

Dynamic Software Testing

Software is developed in units called subroutines or programs. These units, in turn, are combined to form large systems. One approach to QA is to test the code for a completed unit of software by actually entering test data and comparing the results to the expected results. This is called **dynamic testing**. There are two forms of dynamic testing:

- **Black-box testing** involves viewing the software unit as a device that has expected input and output behaviors but whose internal workings are unknown (a black box). If the unit demonstrates the expected behaviors for all the input data in the test suite, it passes the test. Black-box testing takes place without the tester having any knowledge of the structure or nature of the actual code. For this reason, it is often done by someone other than the person who wrote the code.
- **White-box testing** treats the software unit as a device that has expected input and output behaviors but whose internal workings, unlike the unit in black-box testing, are known. White-box testing involves testing all possible logic paths through the software unit with thorough knowledge of its logic. The test data must be carefully constructed so that each program statement executes at least once. For example, if a developer creates a program to calculate an employee's gross pay, the tester would develop data to test cases in which the employee worked less than 40 hours, exactly 40 hours, and more than 40 hours (to check the calculation of overtime pay).

Other Types of Software Testing

Other forms of testing include the following:

- **Static testing**—Special software programs called static analyzers are run against the new code. Rather than reviewing input and output, the static analyzer looks for suspicious patterns in programs that might indicate a defect.
- **Integration testing**—After successful unit testing, the software units are combined into an integrated subsystem that undergoes rigorous testing to ensure that the linkages among the various subsystems work successfully.
- **System testing**—After successful integration testing, the various subsystems are combined to test the entire system as a complete entity.
- **User acceptance testing**—Independent testing performed by trained end users to ensure that the system operates as they expect.

Capability Maturity Model Integration

Capability Maturity Model Integration (CMMI)—developed by the Software Engineering Institute at Carnegie Mellon—is a process improvement approach that defines the essential elements of effective processes. The model is general enough to be used to evaluate and improve almost any process, and a specific application of CMMI—**CMMI-Development (CMMI-DEV)**—is frequently used to assess and improve software development practices. CMMI defines five levels of software development maturity (see Table 7-1) and identifies the issues that are most critical to software quality and process improvement.

TABLE 7-1 Definition of CMMI maturity levels

Maturity level	Description
Initial	Process is ad hoc and chaotic; organization tends to over commit and processes are often abandoned during times of crisis
Managed	Projects employ processes and skilled people; status of work products is visible to management at defined points
Defined	Processes are well defined and understood and are described in standards, procedures, tools, and methods; processes are consistent across the organization
Quantitatively managed	Quantitative objectives for quality and process performance are established and are used as criteria in managing projects; specific measures of process performance are collected and statistically analyzed
Optimizing	Organization continually improves its processes based on a quantitative understanding of its business objectives and performance needs

Source Line: Used with permission from Carnegie Mellon University.

A maturity level consists of practices for a set of process areas that improve an organization's overall performance. Identifying an organization's current maturity level enables it to specify necessary actions to improve the organization's future performance. The model also enables an organization to track, evaluate, and demonstrate its progress over the years.

Table 7-2 shows the percentages of organizations at each CMMI maturity level, as reported in a recent survey of 5,346 organizations.

TABLE 7-2 Maturity level distribution across a large sample of organizations

Maturity level	Percent of 5,346 organizations surveyed
Not provided	6.2%
Initial	0.7%
Managed	26.7%
Defined	56.9%
Quantitatively managed	2.5%
Optimizing	7.0%

Source Line: Used with permission from Carnegie Mellon University.

CMMI-DEV is a set of guidelines for 22 process areas related to systems development. The premise of the model is that those organizations that do these 22 things well will have an outstanding software development process. The Software Engineering Institute has documented the following results from CMMI-DEV implementations:

- A 33 percent decrease in the cost to fix defects
- A reduction in the number of defects found, from 6.6 to 2.1 per thousand lines of code
- A 30 percent increase in productivity
- An increase in project-schedule milestones met, from 50 percent to 95 percent[22]

After an organization decides to adopt CMMI-DEV, it must conduct an assessment of its software development practices (using trained, outside assessors to ensure objectivity) and determine where they fit in the capability model. The assessment identifies areas for improvement and establishes action plans needed to upgrade the development process. Over the course of a few years, the organization can improve its maturity level by executing the action plan.

CMMI-DEV can also be used as a benchmark for comparing organizations. In the awarding of software contracts—particularly by the federal government—organizations that bid on a contract may be required to have adopted CMMI and to be performing at a certain level.

Achieving Maturity Level 5 is a significant accomplishment for any organization, and it can lead to substantial business benefits. It means that the organization is able to statistically evaluate the performance of its software development processes. This in turn leads to better control and continual improvement in the processes making it possible to deliver software products of high quality and within defined timeframes.

As the maturity level increases, the organization improves its ability to deliver good software on time and on budget. For example, Lockheed Martin Management and Data Systems converted to CMMI between 1996 and 2002, with outstanding improvements. Lockheed increased software productivity by 30 percent, reduced unit software costs by 20 percent, and cut the costs of finding and fixing software defects by 15 percent.[23]

KEY ISSUES IN SOFTWARE DEVELOPMENT

Although defects in any system can cause serious problems, the consequences of software defects in certain systems can be deadly. In these kinds of systems, the stakes involved in creating quality software are raised to the highest possible level. The ethical decisions involving a trade-off—if one must be considered—between quality and such factors as cost, ease of use, and time to market require extremely serious examination. The next sections discuss safety-critical systems and the special precautions companies must take in developing them.

Development of Safety-Critical Systems

A **safety-critical system** is one whose failure may cause injury or death. The safe operation of many safety-critical systems relies on the flawless performance of software; such systems control automobiles' antilock brakes, nuclear power plant reactors, airplane navigation, elevators, and numerous medical devices, to name just a few. The process of building software for such systems requires highly trained professionals, formal and rigorous methods, and state-of-the-art tools. Failure to take strong measures to identify and remove software errors from safety-critical systems "is at best unprofessional and at worst lead to disastrous consequences."[24] However, even with these precautions, the software associated with safety-critical systems is still vulnerable to errors that can lead to injury or death. Here are several examples of safety-critical system failures:

- The *Mariner I* space probe, which was intended to make a close flyby of the planet Venus, was ordered destroyed less than five minutes after launch in July 1962. Faulty software code caused the flight control computer to perform a series of unnecessary course corrections, which threw the spacecraft dangerously off course.[25]
- A Royal Air Force helicopter took off from Northern Ireland in June 1994 with 25 British intelligence officials who were heading to a security conference in Inverness. Just 18 minutes into its flight, the helicopter crashed on the peninsula of Kintyre in Argyll, Scotland, killing everyone aboard. The engine management software, which controlled the acceleration and deceleration of the engines, was suspected of causing the crash.[26]
- Between November 2000 and March 2002, therapy planning software at the National Oncology Institute in Panama City, Panama, miscalculated the proper dosage of radiation for patients undergoing therapy; at least eight patients died while another 20 received overdoses that caused significant health problems. Sadly, the developers of this software had not learned from the lessons of the Therac-25 tragedy discussed in the opening vignette.[27]
- Fire broke out on a Washington, D.C., six-car Metro train as it pulled out of the L'Enfant Plaza station in April 2007. Fire and smoke were seen underneath the last car, but thankfully, the flames did not penetrate the floor of the car. The train operator stopped and evacuated the passengers. It was determined that the train's brake resistor grid, which checks various subsystems and voltages, overheated and caught fire. Monitoring software failed to perform as expected in detecting and preventing excess power usage in equipment on the passenger rail cars, resulting in overheating and fire.[28]

When developing safety-critical systems, a key assumption must be that safety will *not* automatically result from following an organization's standard development methodology. Safety-critical software must go through a much more rigorous and time-consuming development process than other kinds of software. All tasks—including requirement definition, systems analysis, design, coding, fault analysis, testing, implementation, and change control—require additional steps, more thorough documentation, and vigilant checking and rechecking. As a result, safety-critical software takes much longer to complete and is much more expensive to develop.

The key to ensuring that these additional tasks are completed is to appoint a **project safety engineer**, who has explicit responsibility for the system's safety. The safety engineer uses a logging and monitoring system to track hazards from a project's start to its finish. The hazard log is used at each stage of the software development process to assess how it has accounted for detected hazards. Safety reviews are held throughout the development process, and a robust configuration management system tracks all safety-related documentation to keep it consistent with the associated technical documentation. Informal documentation is not acceptable for safety-critical system development; formal documentation is required, including verification reviews and signatures.

The increased time and expense of completing safety-critical software can draw developers into ethical dilemmas. For example, the use of hardware mechanisms to back up or verify critical software functions can help ensure safe operation and make the consequences of software defects less critical. However, such hardware may make the final product more expensive to manufacture or harder for the user to operate—potentially making the product less attractive than a competitor's. Companies must carefully weigh these issues to develop the safest possible product that also appeals to customers.

Another key issue is deciding when the QA staff has performed sufficient testing. How much testing is enough when you are building a product whose failure could cause loss of human life? At some point, software developers must determine that they have completed sufficient QA activities and then sign off to indicate their approval. Determining how much testing is sufficient demands careful decision making.

When designing, building, and operating a safety-critical system, a great deal of effort must be put into considering what can go wrong, the likelihood and consequences of such occurrences, and how risks can be averted or mitigated. One approach to answering these questions is to conduct a formal risk analysis. **Risk** is the probability of an undesirable event occurring times the magnitude of the event's consequences if it does happen. These consequences include damage to property, loss of money, injury to people, and death.

For example, if an undesirable event has a 1 percent probability of occurring and its consequences would cost $1,000,000, then the risk can be calculated as 0.01 × $1,000,000, or $10,000. The risk for this event would be considered greater than that of an event with a 10 percent probability of occurring, at a cost of $100 (0.10 × $100 = $10). Risk analysis is important for safety-critical systems but is useful for other kinds of software development as well.

Another key element of safety-critical systems is **redundancy**, the provision of multiple interchangeable components to perform a single function in order to cope with failures and errors. An example of a simple redundant system would be an automobile with a spare tire or a parachute with a backup chute attached. A more complex system used in IT is a redundant array of independent disks (RAID), which is commonly used in high-volume data storage for file servers. RAID systems use many small-capacity disk drives to store large amounts of data to provide increased reliability and redundancy. Should one of the drives fail, it can be removed and a new one inserted in its place. Because the data has also been stored elsewhere, data on the failed disk can be rebuilt automatically without the server ever having to be shut down.

N-version programming is a form of redundancy involving the simultaneous execution of a series of program instructions by two different systems. The two systems use different algorithms to execute instructions that accomplish the same result. The results from the two systems are then compared; if a difference is found, another algorithm is executed to determine which system yielded the correct result. In some cases, instructions for the two systems are written by programmers from two different companies and run on different hardware devices. The rationale behind N-version programming is that both systems are highly unlikely to fail at the same time under the same conditions. Thus, one of the two systems should yield a correct result. IBM employs N-version programming to reduce disk sector failures in data storage devices. Two pieces of code in the same application save a piece of data and then compare the data to ensure that no errors occurred.

During times of widespread disaster, lack of sufficient redundant systems can lead to major problems. For example, the designers of the reactors at Japan's Fukushima Daiichi Nuclear Power Plant anticipated that a strong earthquake and even a tsunami might hit the facility. So in addition to a main power supply, backup generators were put in place to ensure that coolant could be circulated to the nuclear reactors even if the main power supply was knocked out. When a 9.0 earthquake hit the area in early 2011, it knocked out the main power supply, but the backup power supply was still working until it was hit with a tsunami 10 meters high, twice the height of what had been anticipated in the design of the redundant power supplies of the plant.[29]

After an organization determines all pertinent risks to a system, it must decide what level of risk is acceptable. This decision is extremely difficult and controversial because it involves forming personal judgments about the value of human life, assessing potential liability in case of an accident, evaluating the surrounding natural environment, and estimating the system's costs and benefits. System modifications must be made if the level of risk in the design is judged to be too great. Modifications can include adding redundant components or using safety shutdown systems, containment vessels, protective walls, or escape systems. Another approach is to mitigate the consequences of failure by devising emergency procedures and evacuation plans. In all cases, organizations must ask how safe is safe enough if human life is at stake.

Manufacturers of safety-critical systems must sometimes decide whether to recall a product when data indicates a problem. For example, automobile manufacturers have been known to weigh the cost of potential lawsuits against that of a recall. Drivers and passengers in affected automobiles (and, in many cases, the courts) have not found this approach to be ethically sound. Manufacturers of medical equipment and airplanes have had to make similar decisions, which can be complicated if data cannot pinpoint the cause of the problem. For example, there was great controversy in 2000 over the use of Firestone tires on Ford Explorers after numerous tire blowouts and Explorer rollovers caused multiple injuries and deaths. However, it was difficult to determine if the rollovers were caused by poor automobile design, faulty tires, or improperly inflated tires. Consumers' confidence in both manufacturers and their products was nevertheless shaken.

Reliability is the probability of a component or system performing without failure over its product life. For example, if a component has a reliability of 99.9 percent, it has one chance in one thousand of failing over its lifetime. Although this chance of failure may seem low, remember that most systems are made up of many components. As you add more components, the system becomes more complex, and the chance of failure increases. For example, assume that you are building a complex system made up of seven components, each with 99 percent reliability. If none of the components has redundancy built in, the system has a 93.8 percent ($.99^7$) probability of operating successfully with no component malfunctions over its lifetime. If you build the same type of system using 10 components, each with 99 percent reliability, the overall probability of operating without an individual component failure falls to 90 percent. Thus, building redundancy into systems that are both complex and safety critical is imperative.

One of the most important and difficult areas of safety-critical system design is the human interface. Human behavior is not nearly as predictable as the reliability of hardware and software components in a complex system. The system designer must consider what human operators might do to make a system work less safely or effectively. The challenge is to design a system that works as it should and leaves little room for erroneous judgment on the part of the operator. For instance, a self-medicating pain-relief system must allow a patient to press a button to receive more pain reliever, but must also regulate itself to prevent an overdose. Additional risk can be introduced if a designer does not anticipate the information an operator needs and how the operator will react under the daily pressures of actual operation, especially in a crisis. Some people keep their wits about them and perform admirably in an emergency, but others may panic and make a bad situation worse.

Poor design of a system interface can greatly increase risk, sometimes with tragic consequences. For example, in July 1988, the guided missile cruiser USS *Vincennes* mistook an Iranian Air commercial flight for an enemy F-14 jet fighter and shot the airliner down over international waters in the Persian Gulf. All 290 people on board perished. Some investigators blamed the tragedy on the confusing interface of the $500 million Aegis radar and weapons control system. The Aegis radar on the *Vincennes* locked onto an Airbus 300, but it was misidentified as a much smaller F-14 by its human operators. The Aegis operators also misinterpreted the system signals and thought that the target was descending, even though the airbus was actually climbing. A third human error was made in determining the target altitude—it was off by 4,000 feet. As a result of this combination of human errors, the *Vincennes* crew thought the ship was under attack and shot down the plane.[30]

Quality Management Standards

The International Organization for Standardization (ISO), founded in 1947, is a worldwide federation of national standards bodies from 161 countries. The ISO issued its 9000 series of business management standards in 1988. These standards require organizations to develop formal quality-management systems that focus on identifying and meeting the needs, desires, and expectations of their customers.

The **ISO 9001 family of standards** serves as a guide to quality products, services, and management. ISO 9001:2008 provides a set of standardized requirements for a quality management system. It is the only standard in the family for which organizations can be certified. Approximately 1 million organizations in more than 175 countries have ISO 9001 certification.[31] Although companies can use the standard as a management guide for their own purposes in achieving effective control, the priority for many companies is having a qualified external agency certify that they have achieved ISO 9001 certification. Many businesses and government agencies both in the United States and abroad insist that a potential vendor or business partner have a certified quality management system in place as a condition of doing business. Becoming ISO 9001 certified provides proof of an organization's commitment to quality management and continuous improvement.

To obtain this coveted certificate, an organization must submit to an examination by an external assessor and must fulfill the following requirements:

- Have written procedures for all processes
- Follow those procedures
- Prove to an auditor that it has fulfilled the first two requirements; this proof can require observation of actual work practices and interviews with customers, suppliers, and employees

Many software development organizations are applying ISO 9001 to meet the special needs and requirements associated with the purchase, development, operation, maintenance, and supply of computer software.

Failure mode and effects analysis (FMEA) is an important technique used to develop ISO 9000–compliant quality systems by both evaluating reliability and determining the effects of system and equipment failures. Failures are classified according to their impact on a project's success, personnel safety, equipment safety, customer satisfaction, and customer safety. The goal of FMEA is to identify potential design and process failures early in a project, when they are relatively easy and inexpensive to correct. FMEA is used heavily in the manufacture of healthcare devices and the analysis of healthcare procedures. In addition, thousands of Ford Motor Company engineers use FMEA software to evaluate various system designs.[32]

A failure mode describes how a product or process could fail to perform the desired functions described by the customer. An effect is an adverse consequence that the customer might experience. Unfortunately, most systems are so complex that there is seldom a one-to-one relationship between cause and effect. Instead, a single cause may have multiple effects, and a combination of causes may lead to one effect or multiple effects. It is not uncommon for a FMEA of a system to identify 50 to 200 potential failure modes.

The use of FMEA helps to prioritize those actions necessary to reduce potential failures with the highest relative risks. The following steps are used to identify the highest priority actions to be taken[33]:

- Determine the severity rating: The potential effects of a failure are scored on a scale of 1 to 10 (or 1 to 5) with 10 assigned to the most serious consequence (9 or 10 are assigned to safety or regulatory related effects).
- Determine the occurrence rating: The potential causes of that failure occurring are also scored on a scale of 1 to 10, with 10 assigned to the cause with the greatest probability of occurring.
- Determine the criticality, which is the product of severity times occurrence.
- Determine the detection rating: The ability to detect the failure in advance of it occurring due to the specific cause under consideration is also scored on a scale of 1 to 10, with 10 assigned to the failure with the least likely chance of advance detection. For software, the detection rating would represent the ability of planned tests and inspections to remove the cause of a failure.
- Calculate the risk priority rating: The severity rating is multiplied by the occurrence rating and by the detection rating to arrive at the risk priority rating.

Table 7-3 shows a sample FMEA risk priority table.

TABLE 7-3 Sample FMEA risk priority table

Issue	Severity	Occurrence	Criticality	Detection	Risk priority
#1	3	4	12	9	108
#2	9	4	36	2	72
#3	4	5	20	4	80

Source Line: Course Technology/Cengage Learning.

Many organizations consider those issues with the highest criticality rating (severity × occurrence) as the highest priority issues to address. They may then go on to address those issues with the highest risk priority (severity × occurrence × detection). So although Issue #2 shown in Table 7-3 has the lowest risk priority, it may be assigned the highest priority because of its high criticality rating.

Table 7-4 provides a manager's checklist for upgrading the quality of the software an organization produces. The preferred answer to each question is *yes*.

TABLE 7-4 Manager's checklist for improving software quality

Question	Yes	No
Has senior management made a commitment to develop quality software?		
Have you used CMMI to evaluate your organization's software development process?		
Has your company adopted a standard software development methodology?		
Does the methodology place a heavy emphasis on quality management and address how to define, measure, and refine the quality of the software development process and its products?		
Are software project managers and team members trained in the use of this methodology?		
Are software project managers and team members held accountable for following this methodology?		
Is a strong effort made to identify and remove errors as early as possible in the software development process?		
Are both static and dynamic software testing methods used?		
Are white-box testing and black-box testing methods used?		
Has an honest assessment been made to determine if the software being developed is safety critical?		
If the software is safety critical, are additional tools and methods employed, and do they include the following: a project safety engineer, hazard logs, safety reviews, formal configuration management systems, rigorous documentation, risk analysis processes, and the FMEA technique?		

Source Line: Course Technology/Cengage Learning.

Summary

- High-quality software systems are easy to learn and use. They perform quickly and efficiently to meet their users' needs, operate safely and reliably, and have a high degree of availability that keeps unexpected downtime to a minimum.
- High-quality software has long been required to support the fields of air traffic control, nuclear power, automobile safety, health care, military and defense, and space exploration, among others.
- Now that computers and software have become integral parts of almost every business, the demand for high-quality software is increasing. End users cannot afford system crashes, lost work, or lower productivity. Nor can they tolerate security holes through which intruders can spread viruses, steal data, or shut down Web sites.
- Software developers are under extreme pressure to reduce the time to market of their products. They are driven by the need to beat the competition in delivering new functionality to users, to begin generating revenue to recover the cost of development, and to show a profit for shareholders.
- The resources and time needed to ensure quality are often cut under the intense pressure to ship a new software product. When forced to choose between adding more user features and doing more testing, many software companies decide in favor of more features. They often reason that defects can be patched in the next release, which will give customers an automatic incentive to upgrade. Additional features make the release more useful and therefore easier to sell to customers.
- Software product liability claims are typically based on strict liability, negligence, breach of warranty, or misrepresentation—sometimes in combination.
- A software development methodology defines the activities in the software development process, defines individual and group responsibilities for accomplishing objectives, recommends specific techniques for accomplishing the objectives, and offers guidelines for managing the quality of the products during the various stages of the development cycle.
- Using an effective development methodology enables a manufacturer to produce high-quality software, forecast project-completion milestones, and reduce the overall cost to develop and support software. It also helps protect software manufacturers from legal liability for defective software in two ways: it reduces the number of software errors that could cause damage, and it makes negligence more difficult to prove.
- CMMI defines five levels of software development maturity: initial, managed, defined, quantitatively managed, and optimizing. CMMI identifies the issues that are most critical to software quality and process improvement. Its use can improve an organization's ability to predict and control quality, schedule, costs, and productivity when acquiring, building, or enhancing software systems.
- CMMI also helps software engineers analyze, predict, and control selected properties of software systems.

- A safety-critical system is one whose failure may cause injury or death. In the development of safety-critical systems, a key assumption is that safety will *not* automatically result from following an organization's standard software development methodology.
- Safety-critical software must go through a much more rigorous and time-consuming development and testing process than other kinds of software; the appointment of a project safety engineer and the use of a hazard log and risk analysis are common.
- The International Organization for Standardization (ISO) issued its 9000 series of business management standards in 1988. These standards require organizations to develop formal quality management systems that focus on identifying and meeting the needs, desires, and expectations of their customers.
- The ISO 9001:2008 standard serves as a guide to quality products, services, and management. Approximately 1 million organizations in more than 175 countries have ISO 9001 certification. Many businesses and government agencies specify that a vendor must be ISO 9001 certified to win a contract from them.
- Failure mode and effects analysis (FMEA) is an important technique used to develop ISO 9001–compliant quality systems. FMEA is used to evaluate reliability and determine the effects of system and equipment failures.

Key Terms

black-box testing
breach of warranty
business information system
Capability Maturity Model Integration (CMMI)
CMMI-Development
contributory negligence
decision support system
deliverables
dynamic testing
failure mode and effects analysis (FMEA)
integration testing
ISO 9001 family of standards
N-version programming
product liability
project safety engineer
quality assurance (QA)
quality management
redundancy
reliability
risk
safety-critical system
software defect
software development methodology
software quality
static testing
strict liability
system testing
user acceptance testing
warranty
white-box testing

Self-Assessment Questions

The answers to the Self-Assessment Questions can be found in Appendix E.

1. The need for high quality software has only recently arisen. True or False?
2. ____________ is the degree to which a software product meets the needs of its users.
3. Which of the following is *not* a major cause of poor software quality?
   a. Many developers do not know how to design quality into software or do not take the time to do it.
   b. Programmers make mistakes in turning design specifications into lines of code.
   c. Software developers are under extreme pressure to reduce the time to market of their products.
   d. Many organizations avoid buying the first release of a major software product.
4. A type of system used to control many industrial processes in an effort to reduce costs, eliminate human error, improve quality, and shorten the time it takes to make products is called ____________.
5. The liability of manufacturers, sellers, lessors, and others for injuries caused by defective products is commonly referred to as ____________.
6. A standard, proven work process for the development of high-quality software is called a(n) ____________.
7. The cost to identify and remove a defect in an early stage of software development can be up to 100 times more expensive than the cost of removing a defect in an operating piece of software after it has been distributed to many customers. True or False?
8. Methods within the development cycle designed to guarantee reliable operation of the product are known as ____________.
9. A form of software testing that involves viewing a software unit as a device that has expected input and output behaviors but whose internal workings are unknown is known as:
   a. dynamic testing
   b. white-box testing
   c. integration testing
   d. black-box testing
10. An approach that defines the essential elements of an effective process and outlines a system for continuously improving software development is:
    a. ISO 9000
    b. FMEA
    c. CMMI
    d. DOD-178B
11. One of the most important and difficult areas of safety-critical system design is the human interface. True or False?

12. The provision of multiple interchangeable components to perform a single function to cope with failures and errors is called:
    a. risk
    b. redundancy
    c. reliability
    d. availability
13. A reliability evaluation technique that can determine the effect of system and equipment failures is ____________.

14. A set of standards requiring organizations to develop formal quality management systems that focus on identifying and meeting the needs, desires, and expectations of their customers are the ____________ standards.
15. In a lawsuit alleging ____________, responsibility is limited to harmful defects that could have been detected and corrected through "reasonable" software development practices.

Discussion Questions

1. Identify the three criteria you consider to be most important in determining whether or not a system is a quality system. Briefly discuss your rationale for selecting these criteria.
2. Explain why the cost to identify and remove a defect in the early stages of software development might be 100 times less than the cost of removing a defect in software that has been distributed to hundreds of customers. What are the implications for a software development organization?
3. Identify and briefly discuss two ways that the use of an effective software development methodology can protect software manufacturers from legal liability for defective software.
4. Your company is considering using N-version programming with two software development firms and two hardware devices for the navigation system of a guided missile. Briefly describe what this means, and outline several advantages and disadvantages of this approach.
5. Why is the human interface one of the most important but difficult areas of safety-critical systems? Do a search on the Internet and find three good sources of information relating to how to design an effective computer to human interface.
6. Identify and briefly discuss the implications to a project team of classifying a piece of software as safety critical.
7. Your organization develops accounting software for use by individuals to budget and forecast their expenses and pay their bills while keeping track of the amount in their savings and checking accounts. Develop a strong argument for the management of your firm as to why they must conduct as assessment of their current software development practices.
8. Discuss why an organization might elect to use a separate, independent team for quality testing rather than the group of people who originally developed the software.

9. You are considering contracting for the development of new software that is essential to the success of your midsized manufacturing firm. One candidate firm boasts that its software development practices are at level 4 of CMMI. Another firm claims that all its software development practices are ISO 9001 compliant. How much weight should you give to these certifications when deciding which firm to use? Do you think that a firm could lie or exaggerate its level of compliance with these standards?

What Would You Do?

1. Read the fictional Killer Robot case at the Web site for the Online Ethics Center for Engineering at *www.onlineethics.com/CMS/computers/compcases/killerrobot.aspx*. The case begins with the manslaughter indictment of a programmer for writing faulty code that resulted in the death of a robot operator. Slowly, over the course of many articles, you are introduced to several factors within the corporation that contributed to the accident. After reading the case, answer the following questions:
   a. Responsibility for an accident is rarely defined clearly and is often difficult to trace to one or two people or causes. In this fictitious case, it is clear that a large number of people share responsibility for the accident. Identify all the people you think were at least partially responsible for the death of Bart Matthews, and explain why you think so.
   b. Imagine that you are the leader of a task force assigned to correct the problems uncovered by this accident. Develop a list of the 10 most significant actions to take to avoid future problems. What process would you use to identify the most critical actions?
2. You used a tax preparation software package to prepare your federal tax return this year. Today you received a form letter from the software manufacturer informing all its customers that there was an error in the software that resulted in a substantial underestimation of the amount owed for both those who indicated that they were single and those who were married but filing separate tax returns. In the letter, the software manufacturer suggests that such users of its software promptly file an amended tax return before the IRS sends them a letter informing them that they discovered an error and requesting the payment of fines and interest. What do you do?
3. You are the project manager in charge of developing the latest release of your software firm's flagship product. The product release date is just two weeks away, and enthusiasm for the product is extremely high among your customers. Stock market analysts are forecasting sales of more than $25 million per month. If so, earnings per share will increase by nearly 50 percent. There is just one problem: two key features promised to customers in this release have several bugs that would severely limit the software's usefulness. You estimate that at least six weeks are needed to find and fix the problems. In addition, even more time is required to find and fix 15 additional, less severe bugs just uncovered by the QA team. What do you recommend to management?
4. You developed a spreadsheet program that helps you perform your role of inventory control manager at a small retail bakery. The software uses historical sales data to calculate

expected weekly sales for each of 250 baked goods carried by the store. Based on that forecast, you order the appropriate raw ingredients.

Your store is one of four bakeries in the city all owned by the same person. You sent a copy of the spreadsheet to each of the people responsible for inventory control at the other three retail stores, and they are all now using your software to help them do their jobs. You have started getting complaints that the software is not entirely accurate, and you notice that your own estimates are no longer as accurate as they used to be. What do you do?

5. You have been assigned to manage software that controls the shutdown of the new chemical reactors to be installed at a manufacturing plant. Your manager insists the software is not safety critical. The software senses temperatures and pressures within a 50,000-gallon stainless steel vat and dumps in chemical retardants to slow down the reaction if it gets out of control. In the worst possible scenario, failure to stop a runaway reaction would result in a large explosion that would send fragments of the vat flying and spray caustic liquid in all directions.

   Your manager points out that the stainless steel vat is surrounded by two sets of protective concrete walls and that the reactor's human operators can intervene in case of a software failure. He feels that these measures would protect the plant employees and the surrounding neighborhood if the shutdown software failed. Besides, he argues, the plant is already more than a year behind its scheduled start-up date. He cannot afford the additional time required to develop the software if it is classified as safety critical. How would you work with your manager and other appropriate resources to decide whether the software is safety critical?

6. You are a senior software development consultant with a major consulting firm. You have been asked to conduct a follow-up assessment of the software development process for ABCXYZ Corporation, a company for which you had performed an initial assessment using CMMI two years prior. At the initial assessment, you determined the company's level of maturity to be level 2. Since that assessment, the organization has spent a lot of time and effort following your recommendations to raise its level of process maturity. The organization appointed a senior member of its IT staff to be a process management guru and paid him $150,000 per year to lead the improvement effort. This senior member adopted a methodology for standard software development and required all project managers to go through a one-week training course at a total cost of more than $2 million.

   Unfortunately, these efforts did not significantly improve process maturity because senior management failed to hold project managers accountable for actually using the standard development methodology in their projects. Too many project managers convinced senior management that the new methodology was not necessary for their particular project and would just slow things down. You are concerned that when senior management learns that no real progress has been made, they will refuse to accept partial blame for the failure and instead drop all attempts at further improvement. You want senior management to ensure that the new methodology is used on all projects—with no exceptions. What would you do?

7. You are the CEO for a small, struggling software firm that produces educational software for high school students. Your latest software is designed to help students improve their SAT and ACT scores. To prove the value of your software, a group of 50 students who had taken the ACT test were retested after using your software for just two weeks. Unfortunately,

there was no dramatic increase in their scores. A statistician you hired to ensure objectivity in measuring the results claimed that the variation in test scores was statistically insignificant. You had been counting on touting the results in the promotion of your new software.

A small core group of educators and systems analysts will need at least six months to start again from scratch to design a viable product. Programming and testing could take another six months. Another option would be to go ahead and release the current version of the product and then, when the new product is ready, announce it as a new release. This would generate the cash flow necessary to keep your company afloat and save the jobs of 10 or more of your 15 employees. Given this information about your company's product, what would you do?

Cases

1. Apple Guidelines for App Approval

Apple's App Store has been a huge success ever since it was launched in 2008. As of the end of 2010, the App Store offered more than 250,000 applications available for sale to owners of Apple iPhone, iPad, and iPod devices, with more than 6.5 billion downloads since it opened.[34]

Before software applications can be sold through the App Store, they must go through a review process. Apple has been accused by some of using clandestine and capricious rules to reject some programs and thus blocking them from reaching the very large and growing market of iPhone, iPad, and iPod Touch users. One application developer complained: "If you submit an app, you have no idea what's going to happen. You have no idea when it's going to be approved or if it's going to be approved."[35] The developers of an app called "South Park" complained that their app was rejected because the content was deemed "potentially offensive," even though episodes of the award-winning animated sitcom are available at the Apple iTunes Store.[36] In September 2010, after more than two years of complaints, Apple finally provided application developers the guidelines it uses to review software.

Most guidelines seem to be aimed at ensuring that Apple users can only access high-quality and noncontroversial apps from its App Store. Some of the Apple guidelines are clear and their rationale is easy to understand, such as "apps that rapidly drain the device's battery or generate excessive heat will be rejected." However, other guidelines are unclear and highly subjective, such as "We will reject apps for any content or behavior that we believe is over the line." What line, you ask? Well, as a Supreme Court Justice once said, "'I'll know it when I see it.' And we think that you will also know it when you cross it."[37] ("I know it when I see it" was the phrase used by U.S. Supreme Court Justice Potter Stewart to describe his ability to recognize rather than to provide a precise definition of hard core pornography in his opinion in the case *Jacobellis v. Ohio* in 1964.)

The Electronic Frontier Foundation believes that while the guidelines are helpful, in some cases Apple is defining the content of third-party software and placing limits on what is available to customers of Apple's App Store.[38]

By way of comparison, Google places few restrictions on developers of software for its competing Android Marketplace. However, there have been many low-quality applications offered to Android Marketplace customers, including some that include unwanted malware. Indeed by early 2011, Google had pulled 21 Android applications from its Android Marketplace because, once downloaded, the applications not only stole users' information and device data, but also created a backdoor for even more harmful attacks.[39] Apple's decision to finally share its applications guidelines may have been an attempt to combat the rapidly increasing popularity of the Android.[40] It may also have been a response to a U.S. Federal Trade Commission investigation of a complaint from Adobe concerning Apple's banning of the Flash software from devices that run using Apple's iOS operating system. (Adobe® Flash® Player is a browser-based application that runs on many computer hardware/operating system combinations and supports the viewing of so called "rich, expressive applications," content, and videos across screens and browsers.)[41]

Discussion Questions

1. Should Apple conduct extensive screening of Apps before they are allowed to be sold on the App Store? Why or why not?
2. Do research to determine the current status of the FCC investigation of Apple for banning use of the Adobe Flash software on devices that use the iOS operating system.
3. What do you think of Apple's guideline that says it will reject an app for any content or behavior that they believe is over the line? Could such a statement be construed as a violation of the developer's freedom of speech? Why or why not?

2. Billing Errors at Verizon Upset Customers

Starting as early as November 2007, millions of Verizon Wireless customers noticed sporadic $1.99 charges for data usage on their monthly bills even though they had not signed up for a data plan.[42] When the billing problems were not promptly corrected, many consumers complained to the U.S. Federal Communications Commission (FCC), which initiated a 10-month investigation into the situation.[43]

It appears that software errors led to many of the billing mistakes. Some customers who did not have data usage plans were charged because of data transmissions initiated by software that was built into their phones. For example, if a customer tried a demo of a game that came preloaded on a phone, a data transmission from the phone could be triggered, without the customer realizing it. The customer was then charged by Verizon for the data exchange.[44] Another cause of the billing problems related to how Verizon pre-programs its smartphones. Many of them have a shortcut combination of physical keys that launch the device's Web browser. Users who accidently pressed the shortcut set of keys would launch the Web browser. Even if the customers cancelled out of the data session, Verizon charged them a data usage fee.[45]

In an October 2010 press release, a Verizon spokesperson said: "We made inadvertent billing mistakes. We accept responsibility for those errors and apologize to our customers who received accidental data charges on their bills."[46] Following review of their customer billing records, Verizon agreed to reimburse some 15 million of its 92 million customers a minimum of $52.5 million. The firm also agreed to pay a fine of $25 million to the U.S. Treasury, a punishment enforced by the FCC. Verizon stated that it had made changes to eliminate overbilling problems in the future.[47]

In response to these actions by Verizon, an FCC spokesperson issued the following statement: "While I appreciate that Verizon Wireless has acknowledged its billing errors, the refunds to millions of Americans have been a long time coming. It appears the company was first notified, more than two years ago, about certain billing errors. As I pointed out in December last year, the company's initial response to public reports of the phantom fees was that it does not charge customers for accidental launching of the Web browser. Yesterday's announcement clearly requires further explanation."[48]

People become quite upset when there are recurring errors in their bills, even if the amount is relatively small. Their anger can lead them to file complaints with regulatory agencies and can also cause them to leave their service provider, especially if the service provider is slow to resolve the issue. Customer ill will is created whether the billing errors were intentional or not.

Discussion Questions

1. In addition to the monetary costs identified above, what other negative impacts did Verizon suffer from this billing error incident? How could you attempt to quantify these losses?
2. Are you satisfied with Verizon's explanation of the billing errors? Why or why not? Should the FCC review how Verizon determined that "just" 15 million of its customers were overbilled by $52.5 million?
3. With the benefit of hindsight, how might Verizon have minimized the negative impact of these billing errors?

3. Prius Plagued by Programming Error

In May 2005, the National Highway Traffic Safety Administration (NHTSA) revealed that it had opened an investigation into problems with the Toyota Prius hybrid, following several complaints of engines stalling. Motorists reported that while they were driving or stuck in traffic, warning lights flashed on and the gas engine suddenly switched off. In all cases, the motorists were able to maneuver to the side of the road safely because the electric motor, steering, and brake system continued to function.[49]

As one of the most fuel-efficient cars in the United States, the Prius delivers improved mileage per gallon by shifting between its gas engine and its electric motor in different driving situations. For instance, when the car is stopped in traffic, the gas engine shuts off, and the car is powered by the electric motor. At high speeds, the gas engine takes over to deliver more power to the vehicle. An electronic control unit is designed to ensure a smooth transition between the gas engine and the electric motor. Toyota eventually determined that an error in the software used by the electronic control unit caused it to malfunction and the gas engine to stall.

Today, the average car is equipped with 30 to 40 microprocessors, so software quality is a critical issue for all automakers. With 35 million lines of code per car, the potential for error is enormous. IBM automotive software specialist Stavros Stefanis reports that as many as one-third of all warranty claims result from software glitches and electronic defects. "It's a big headache for the automakers," said Stefanis.[50]

The potential for headaches is high because software impacts engine performance and controls many safety-related systems, such as steering, antilock brakes, and air bags.

A programming error could have an enormous financial impact on an automaker in terms of major lawsuits and recalls.

When NHTSA opened its investigation, Toyota—a company with an established reputation for reliability—proved extremely cooperative. In fact, the company had already issued a service bulletin to owners of 2004 and 2005 Prius hybrids in October 2004. Toyota conducted an internal investigation, and in October 2005 it recalled 75,000 Prius hybrids sold in the United States; the company recalled an additional 85,000 units sold in Japan, Europe, and other markets.[51]

According to Toyota spokeswoman Allison Takahashi, NHTSA determined that passenger safety was not at issue: "We are voluntarily initiating a customer-service campaign to assure that this unusual occurrence does not cause inconvenience." Following the announcement of the recall campaign, NHTSA closed its investigation.[52]

The investigation, however, did not slow sales of the most popular hybrid on the market. Sales in 2005 were up 200 percent over the previous year. In fact, Toyota had to concentrate a good deal of its software development efforts on projects that would increase production to meet the demand for new hybrids in the United States.

Although the recall's effect on sales was probably negligible, the cost of the recall may be significant. Toyota had to pay for advertising and other costs involved in contacting Prius owners about the recall; the company also had to pay for the repairs. The recall serves as a warning to automakers about the costs of human errors that go undetected during software production.

Discussion Questions

1. Do you agree with NHTSA's assessment that the problem with the Prius was not a safety-critical issue? In such cases, who should decide whether a software bug creates a safety-critical issue—the manufacturer, consumers, government agencies, or some other group?
2. How would the issue have been handled differently if it had been a safety-critical matter? Would it have been handled differently if the costs involved hadn't been so great?
3. As the amount of hardware and software embedded in the average car continues to grow, what steps can automakers take to minimize warranty claims and ensure customer safety?

End Notes

[1] ComputingCases.org, "Therac History, Genesis of the Therac-25," http://computingcases.org/case_materials/therac/supporting_docs/levenson/Therac%20History.html (accessed April 17, 2011).

[2] ComputingCases.org, "Therac History, Genesis of the Therac-25," http://computingcases.org/case_materials/therac/supporting_docs/levenson/Therac%20History.html (accessed April 17, 2011).

[3] Troy Gallagher, "Therac-25 Computerized Radiation Therapy," www.kellyhs.org/itgs/ethics/reliability/THERAC-25.htm (accessed April 17, 2011).

[4] Nancy Leveson and Clark S. Turner, "An Investigation of the Therac-25 Accidents," *IEEE Computer*, http://courses.cs.vt.edu/cs3604/lib/Therac_25/Therac_1.html (accessed April 17, 2011).

[5] Nancy Leveson and Clark S. Turner, "An Investigation of the Therac-25 Accidents," *IEEE Computer*, http://courses.cs.vt.edu/cs3604/lib/Therac_25/Therac_1.html (accessed April 17, 2011).

[6] Troy Gallagher, "Therac-25 Computerized Radiation Therapy," www.kellyhs.org/itgs/ethics/reliability/THERAC-25.htm (accessed April 17, 2011).

[7] "A History of Introduction and Shut Down of Therac-25," www.computingcases.org/case_materials/therac/case_history/Case%20History.html (accessed May 20, 2011)

[8] Cunningham & Cunningham, Inc., "Therac Twenty-Five," http://c2.com/cgi/wiki?TheracTwentyFive (accessed May 20, 2011).

[9] Will Knight, "Critical Bug Found in Anti-virus Software," *New Scientist*, December 22, 2005, www.newscientist.com/article/dn8505-critical-bug-found-in-antivirus-software.html.

[10] Ina Fried, "The XP Alternative for Vista PCs," *CNET*, September 21, 2007, http://news.cnet.com/The-XP-alternative-for-Vista-PCs/2100-1016_3-6209481.html.

[11] Telecompaper, "Skype Grows FY Revenues 20%, Reaches 663 Mln Users," March 8, 2011, www.telecompaper.com/news/skype-grows-fy-revenues-20-reaches-663-mln-users.

[12] Wolfgang Gruener, "Software Bug Took Skype Out," *TG Daily*, August 20, 2007, www.tgdaily.com/content/view/33452/103.

[13] Michael Krigsman, "System Update Kicks Heathrow Baggage System Offline," *IT Project Failures ZDNet Business Network* (blog), February 21, 2008, www.zdnet.com/blog/projectfailures/system-update-kicks-heathrow-baggage-system-offline/610.

[14] James Wilson, "'Year 2010' Software Glitch Hits German Bank Cards," *The Financial Times*, January 6, 2010, www.ft.com/cms/s/0/00da0e24-fa63-11de-beed-00144feab49a.html#axzz1JsntiPi2.

[15] Noopur Davis and Julia Mullaney, "The Team Software Process in Practice: A Summary of Recent Results," Technical Report CMU/SEI-2003-TR-014, September 2003.

[16] Jason Hiner, "Five Super-Secret Features in Windows 7," *TechRepublic*, October 12, 2009, www.techrepublic.com/blog/hiner/five-super-secret-features-in-windows-7/3120.

[17] Cem Kaner, "Quality Cost Analysis: Benefits and Risks," *Software Quality Assurance* 3, no. 1 (1996), www.kaner.com/pdfs/Quality_Cost_Analysis.pdf (accessed May 20, 2011).

[18] Paul Bibby, "Qantas Exposed to Compo Claims," *WA Today*, October 9, 2008, www.watoday.com.au/national/qantas-exposed-to-compo-claims-20081009-4x6t.html.

[19] Martin Samson, "*M. A. Mortenson Co. v. Timberline Software Co. et al.*" *Internet Library of Law and Court Decisions*, www.internetlibrary.com/cases/lib_case206.cfm (accessed April 18, 2011).

[20] Barry W. Boehm, "Improving Software Productivity," *IEEE Computer* 20, no. 8 (1987): 43–58.

[21] Capers Jones, "*Software Quality in 2002: A Survey of the State of the Art*," Technical Report, Software Productivity Research, Inc., November 2002.

[22] Dennis R. Goldenson and Diane L. Gibson, "Demonstrating the Impact and Benefits of CMMI," CMU/SEI-2003-SR-009, October 2003, www.sei.cmu.edu/reports/03sr009.pdf.

[23] Dennis R. Goldenson, Diane L. Gibson, Robert W. Ferguson, "Why Make the Switch? Evidence About the Benefits of CMMI," Carnegie Mellon Software Engineering Institute, SEPG 2004, www.sei.cmu.edu/library/assets/evidence.pdf.

[24] Jonathan P. Bowen, "The Ethics of Safety-Critical Systems," *Communications of the ACM* 43 (2000): 91–97.

[25] "NASA, USAF, JPL Announce Mariner I Lost Because Flight Control Computer Generated Incorrect Steering Commands," *New York Times*, July 28, 1962.

[26] Peter B. Ladkin and Mike Beims, "The Chinook Crash," *The Risks Digest (blog)*, January 10, 2002, http://catless.ncl.ac.uk/Risks/21.20.html#subj7.

[27] Matt Lake, "Epic Failures: 11 Infamous Software Bugs," *New Zealand PCWorld*, September 27, 2010, http://pcworld.co.nz/pcworld/pcw.nsf/feature/epic-failures-11-infamous-software-bugs-p8.

[28] "Surge Caused Fire in Rail Car," *Washington Times*, April 27, 2007, www.washingtontimes.com/news/2007/apr/12/20070412-104206-9871r.

[29] Ryan Witt, "Map of U.S. Nuke Reactors Reveals What Happened in Japan Could Happen in America," *Examiner*, March 14, 2011, www.examiner.com/political-buzz-in-national/could-japan-s-nuclear-crisis-happen-the-united-states.

[30] George C. Wilson, "Navy Missile Downs Iranian Jetliner," *Washington Post*, July 4, 1988, www.washingtonpost.com/wp-srv/inatl/longterm/flight801/stories/july88crash.htm.

[31] International Organization for Standardization, "ISO 9000–Quality Management," www.iso.org/iso/iso_catalogue/management_and_leadership_standards/quality_management.htm.

[32] Byteworx, "Byteworx FMEA Software," www.byteworx.com (accessed April 20, 2011).

[33] Oliver Mackel, "Software FMEA: Opportunities and Benefits in the Development of Process of Software Intensive Technical Systems," www.fmeainfocentre.com/papers/mackel1.pdf (accessed May 20, 2011).

[34] "Apple Opens iOS to Third-Party Dev Tools, Reveals Approval Guidelines," *AppleInsider*, September 9, 2010, www.appleinsider.com/articles/10/09/09/apple_no_longer_banning_third_party_ios_development_tools.html.

[35] "Apple Publishes Guidelines for App Approval," *CBS News*, September 9, 2010, www.cbsnews.com/stories/2010/09/09/tech/main6850597.shtml.

[36] Fred von Lohmann, "Another iPhone App Banned: Apple Deems South Park 'Potentially Offensive,'" Electronic Frontier Foundation, February 17, 2009, www.eff.org/deeplinks/2009/02/south-park-iphone-app-denied.

[37] "Apple Publishes Guidelines for App Approval," CBS News, September 9, 2010, www.cbsnews.com/stories/2010/09/09/tech/main6850597.shtml.

[38] "Apple Publishes Guidelines for App Approval," CBS News, September 9, 2010, www.cbsnews.com/stories/2010/09/09/tech/main6850597.shtml.

[39] Lauren Acurantes, "Google Removes 21 Bad Apps from Android Market," *Manila Bulletin Online*, March 2, 2011, www.mb.com.ph/articles/307060/google-removes-21-bad-apps-android-market.

40 “Apple Publishes Guidelines for App Approval,” CBS News, September 9, 2010, www.cbsnews.com/stories/2010/09/09/tech/main6850597.shtml.

41 “Apple Opens iOS to Third-Party Dev Tools, Reveals Approval Guidelines,” AppleInsider, September 9, 2010, www.appleinsider.com/articles/10/09/09/apple_no_longer_banning_third_party_ios_development_tools.html.

42 W. David Gardner, “Verizon Paying FCC $25 M to Settle ‘Mystery Fees,’” *InformationWeek*, October 29, 2010, http://mobile.informationweek.com/10993/show/e2727722f4e8c35df41ecfaa842b51f3&t=53f39a3e32ca9d3a743e4e218e8fca9c.

43 Paul Muschick, “Verizon Wireless Paying Big Bucks for ‘Inadvertent’ Billing Errors,” *WatchDog* (blog), October 28, 2010, http://blogs.mcall.com/watchdog/2010/10/verizon-wireless-paying-big-bucks-for-billing-errors.html.

44 Joseph A. Giannone and Sinead Carew, “Verizon Wireless to Pay Refunds for Billing Errors,” *Reuters*, October 4, 2010, www.reuters.com/article/2010/10/05/us-verizonwireless-idUSTRE6922C420101005.

45 Eric Zeman, “Verizon Wireless to Refund $90M in False Charges,” *InformationWeek*, October 4, 2010, www.informationweek.com/news/hardware/handheld/227600149.

46 W. David Gardner, “Verizon Paying FCC $25 M to Settle ‘Mystery Fees,’” *InformationWeek*, October 29, 2010, http://mobile.informationweek.com/10993/show/e2727722f4e8c35df41ecfaa842b51f3&t=53f39a3e32ca9d3a743e4e218e8fca9c.

47 W. David Gardner, “Verizon Paying FCC $25 M to Settle ‘Mystery Fees,’” *InformationWeek*, October 29, 2010, http://mobile.informationweek.com/10993/show/e2727722f4e8c35df41ecfaa842b51f3&t=53f39a3e32ca9d3a743e4e218e8fca9c.

48 Eric Zeman, “Verizon Wireless to Refund $90M in False Charges,” *InformationWeek*, October 4, 2010, www.informationweek.com/news/hardware/handheld/227600149.

49 Matt Nauman, “Toyota to Fix Software Problem on 75,000 Prius Hybrids,” *San Jose Mercury News*, October 13, 2005.

50 Mel Duvall, “Software Bugs Threaten Toyota Hybrids,” *Baseline*, August 4, 2005.

51 Associated Press, “Toyota Recalls 160,000 Prius Hybrids,” *Washington Post*, October 14, 2005.

52 Associated Press, “Toyota Recalls 160,000 Prius Hybrids,” *Washington Post*, October 14, 2005.

CHAPTER 8

THE IMPACT OF INFORMATION TECHNOLOGY ON PRODUCTIVITY AND QUALITY OF LIFE

QUOTE

If you don't set a baseline standard for what you'll accept in life, you'll find it easy to slip into behaviors and attitudes or a quality of life that's far below what you deserve.
—Anthony Robbins, self-help guru[1]

VIGNETTE

Kaiser Permanente Implements Electronic Health Record (EHR) System

Kaiser Permanente is a not-for-profit integrated healthcare organization founded in 1945. The company operates one of the nation's largest not-for-profit health plans with 8.6 million health plan subscribers. Kaiser Permanente also includes Kaiser Foundation Hospitals—encompassing 35 hospitals in California and Hawaii—and The Permanente Medical Groups, with 431 medical offices in 9 states. The company employs nearly 170,000 people, including some 15,000 physicians and 40,000 nurses. Its annual operating revenue exceeds $40 billion.[2]

HealthConnect is Kaiser's comprehensive health information system, which includes an electronic health record (EHR) application that was fully implemented at all of its hospitals and clinics in March 2010. In 2003, Kaiser announced its intention to work with Epic Systems Corporation over a three-year period to build an integrated set of systems to support EHRs, computerized physician order entry, scheduling and billing, and clinical decision support at an estimated cost of $1.8 billion. This decision came after Kaiser had already made several unsuccessful attempts at clinical automation projects. The project eventually ballooned into a seven-year, $4.2 billion effort as the scope of the project was expanded time and again.[3] Training and productivity losses made up more than 50 percent of the cost of the project as Kaiser had to cut physician's hours at clinics during training and was forced to hire physicians temporarily to handle the workload.[4]

The HealthConnect system connects Kaiser healthcare plan subscribers to their healthcare providers and to their personal healthcare information. The system uses EHR to coordinate patient care among the physician's office, the hospital, testing labs, and pharmacies. The EHR is designed to ensure that all healthcare providers have access to current, accurate, and complete patient data.[5]

Physicians and nurses in hospitals, clinics, and private offices document treatment in the EHR system. Once physicians enter a diagnosis, they may receive a message informing them that there is a "best practice order set" available for treating the condition. When they enter a medication order, physicians receive alerts about potential allergic reactions or adverse drug reactions based on other medications a patient is already taking.[6] HealthConnect also provides capabilities to support barcoding for the safe administration of medicine. Under this system of administering medication, the nurse first scans the patient's bar-coded identification wristband. The nurse next scans a bar code on the medication container that identifies the specific medicine and dosage. The system verifies that this medicine and dosage has been ordered for this patient. If there is not a match, the nurse receives an audible warning signal.[7]

HealthConnect also empowers healthcare plan subscribers to take more responsibility for managing their own health care. Kaiser subscribers can access HealthConnect via a Web portal at *kp.org*. Here they are able to view most of their personal health records online, including their lab results, medication history, and treatment summaries. Patients can enter their own readings from blood pressure and glucose meters.[8] They can also securely email their healthcare providers, which cuts down on the number of office visits and the amount of time patients spend on hold waiting to speak to a doctor. Each month some 80,000 patients send over 850,000 emails to their doctors and healthcare teams through this component of the system.[9]

Kaiser discovered that use of a comprehensive EHR improves health plan subscribers' satisfaction with the healthcare delivery system. At the same time, healthcare provider efficiency is improved with no decrease in clinical quality.[10] In addition to receiving alerts regarding potential adverse drug interactions or allergic reactions, physicians also receive automatic notifications about how lab test results should affect medication orders. The availability of data online has greatly reduced the number of patient phone calls and the mailing out of test results. In addition, the number of outpatient visits has dropped an average of 8 percent in the one and one-half years following EHR implementation at each hospital. This reduces the need to hire additional staff even as enrollment in the HMO increases.[11]

HealthConnect enables physicians to benchmark their performance against colleagues on a number of fronts—efficiency, quality, safety, and service. Hospitals can also benchmark each other on measures such as adverse events and complications. "Best in class" practices can be identified and physicians and hospitals can borrow these best practices from one another to further improve the overall quality of care.[12]

Questions to Consider

1. What do you think are the greatest benefits of the HealthConnect system for Kaiser Permanente subscribers? Can you identify any potential risks or ethical issues associated with the use of this system for Kaiser healthcare plan subscribers? How would you answer these questions from the perspective of a physician or nurse?

2. This system took over seven years to implement and is estimated to have cost at least $4.2 billion. Would you say that this was a wise investment of resources for Kaiser Permanente? Why or why not?

LEARNING OBJECTIVES

As you read this chapter, consider the following questions:

1. What impact has IT had on the standard of living and worker productivity?
2. What is being done to reduce the negative influence of the digital divide?
3. What impact can IT have on improving productivity by reducing costs and/or improving quality?
4. What ethical issues are raised because some entities can afford to make significant investments in IT while others cannot and thus are blocked in their efforts to raise productivity and quality?

THE IMPACT OF IT ON THE STANDARD OF LIVING AND WORKER PRODUCTIVITY

The standard of living varies greatly among groups within a country as well as from nation to nation. The most widely used measurement of the material standard of living is gross domestic product (GDP) per capita. National GDP represents the total annual output of a nation's economy. Overall, industrialized nations tend to have a higher standard of living than developing countries.

In the United States and in most developed countries, the standard of living has been improving over time. However, its rate of change varies as a result of business cycles that affect prices, wages, employment levels, and the production of goods and services. Major disasters—such as earthquakes, hurricanes, tsunamis, and war—can negatively impact the standard of living. The worst economic downturn in U.S. history occurred during the Great Depression, when the GDP declined by about 50 percent from 1929 to 1932; by 1932, the unemployment rate had reached 25 percent.[13] By way of comparison, during the latest recession in the United States (which began in 2007), the GDP growth rate declined by 6.8 percent during the fourth quarter of 2008[14] and the U.S. employment rate hit 10.2 percent in October 2009.[15]

IT Investment and Productivity

Productivity is defined as the amount of output produced per unit of input, and it is measured in many different ways. For example, productivity in a factory might be measured by the number of labor hours it takes to produce one item, while productivity in a service sector company might be measured by the annual revenue an employee generates divided by the employee's annual salary. Most countries have continually been able to produce more goods and services over time—not through a proportional increase in input but rather by making production more efficient. These gains in productivity have led to increases in the GDP-based standard of living because the average hour of labor produced more goods and services. The Bureau of Labor Statistics tracks U.S. productivity on a quarterly basis. In the United States, labor productivity growth has averaged about 2 percent per year for the past century, meaning that living standards have doubled about every 36 years.[16]

Figure 8-1 shows the annual change in U.S. nonfarm labor productivity since 1947. The increase in productivity averaged 2.8 percent per year from 1947 to 1973 as modern management techniques and automated technology made workers far more productive. Productivity dropped off in the mid-1970s, but has been steadily rising since.[17]

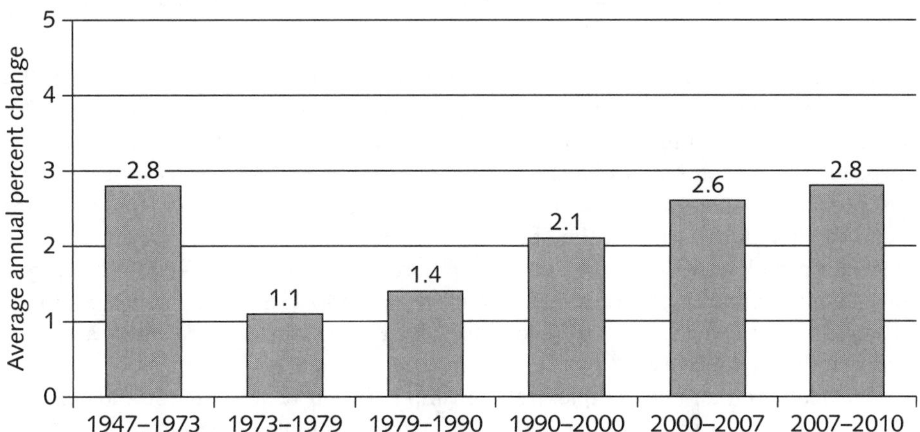

FIGURE 8-1 U.S. nonfarm labor productivity

Source Line: United States Department of Labor, Bureau of Labor Statistics, "Labor Productivity and Costs, 1947–2010," www.bls.gov/lpc/prodybar.htm.

Innovation is a key factor in productivity improvement, and IT has played an important role in enabling innovation. Progressive management teams use IT, as well as other new technology and capital investment, to implement innovations in products, processes, and services.

In the early days of IT in the 1960s, productivity improvements were easy to measure. For example, midsized companies often had a dozen or more accountants focused solely on payroll-related accounting. When businesses implemented automated payroll systems, fewer accounting employees were needed. The productivity gains from such IT investments were obvious.

Today, organizations are trying to further improve IT systems and business processes that have already gone through several rounds of improvement. Organizations are also adding new IT capabilities to help workers who already have an assortment of personal productivity applications on their desktop computers, laptops, and smartphones, such as the Blackberry®, Droid®, and iPhone®. Instead of eliminating workers, IT enhancements are saving workers small amounts of time each day. Whether these saved minutes actually result in improved worker productivity is a matter for debate. Many analysts argue that workers merely use the extra time to do some small task they didn't have time to do before, such as respond to email they would have otherwise ignored. These minor gains make it harder to quantify the benefits of today's IT investments on worker productivity.

The relationship between investment in information technology and U.S. productivity growth is more complex than you might think. Consider the following facts:

- The rate of productivity from 1990 to 2000 of 2.1 percent is only slightly higher than the long-term U.S. rate of 2 percent and not nearly as high as it was during the 26 years following World War II. So, although the recent increase in productivity is welcome, it is not statistically significant.
- Labor productivity in the United States increased despite a reduced level of investment in IT from 2000 to 2007. If there were a simple, direct relationship, the productivity rate should have decreased.[18]

One possible explanation for the previous points is that there is a lag time between the application of innovative IT solutions and the capture of significant productivity gains. IT can enhance productivity in fundamental ways by allowing firms to make radical changes in work processes, but such major changes can take years to complete because firms must make substantial complementary investments in retraining, reorganizing, changing reward systems, and the like. Furthermore, the effort to make such a conversion can divert resources from normal activities, which can actually reduce productivity—at least temporarily. For example, researchers examined data from 527 large U.S. firms from 1987 to 1994 and found that it can take five to seven years for IT investment to result in a substantial increase in productivity.[19]

Another explanation for the complex relationship between IT investment and U.S. productivity growth lies in the fact that many other factors influence worker productivity rates besides IT. Table 8-1 summarizes fundamental ways in which companies can increase productivity.

TABLE 8-1 Fundamental drivers for productivity performance

Reduce the amount of input required to produce a given output by:	Increase the value of the output produced by a given amount of input by:
Consolidating operations to better leverage economies of scale	Selling higher-value goods
Improving performance by becoming more efficient	Selling more goods to increase capacity and use of existing resources

Source Line: Course Technology/Cengage Learning.

The following list summarizes additional factors that can affect national productivity rates:

- Labor productivity growth rates differ according to where a country is in the business cycle—expansion or contraction. Times of expansion enable firms to gain full advantage of economies of scale and full production. Times of contraction present fewer investment opportunities.
- Outsourcing can skew productivity if the contracting firms have different productivity rates than the outsourcing firms.

- U.S. regulations make it easier for companies to hire and fire workers and to start and end business activities compared to many other industrialized nations. This flexibility makes it easier for markets to relocate workers to more productive firms and sectors.
- More competitive markets for goods and services can provide greater incentives for technological innovation and adoption as firms strive to keep ahead of competitors.
- In today's service-based economy, it is difficult to measure the real output of such services as accounting, customer service, and consulting.
- IT investments don't always yield tangible results, such as cost savings and reduced head count; instead, many produce intangible benefits, such as improved quality, reliability, and service.

As you can see, it is difficult to quantify how much the use of IT has contributed to worker productivity. Ultimately, however, the issue is academic. There is no way to compare organizations that don't use IT with those that do, because there is no such thing as a noncomputerized airline, financial institution, manufacturer, or retailer.

Businesspeople analyze the expected return on investment to choose which IT option to implement, but at this point, trying to measure its precise impact on worker productivity is like trying to measure the impact of telephones or electricity.

Telework

Telework (also known as telecommuting) is a work arrangement in which an employee works away from the office—at home, at a client's office, in a hotel—literally, anywhere. In telework, an employee uses various forms of electronic communication, including email, audio conferencing, video conferencing, and instant messaging. Teleworkers access the Internet via cell phone, smartphone, laptop, and similar devices to retrieve computer files; log on to software applications; and communicate with fellow employees, managers, customers, and suppliers. The goal of telework is to allow employees to be effective and productive from wherever they are. It is estimated that 84 percent of all companies allow employees to telework at least one day per week.[20] More than 34 million U.S. workers telework at least occasionally, and that number is expected to grow to 43 percent of U.S. workers—or 63 million people—by 2016.[21]

Factors that have increased the prevalence of telework include advances in technology that enable people to communicate and access the Internet from almost anywhere, the increasing number of broadband connections in homes and retail locations, high levels of traffic congestion, rising gasoline prices, and growing concern over the effects of automobile CO2 emissions. Another key factor is that "scarce, highly skilled workers have begun to demand more flexible work arrangements, especially as they choose to live farther and farther from their employers."[22]

A number of states and the federal government have passed laws to encourage telework. For example, Virginia set a goal of having 20 percent of eligible state workers teleworking by 2010. The state defines someone who works out of the office at least once per week or 32 hours per month as a teleworker. Several states—California and Virginia, among others—and the federal government have based their pandemic-preparedness plans on the widespread use of telework.[23] The Telework Improvement Act of 2010 makes all federal employees eligible to telework one day per week (unless their manager determines they are ineligible for telework.)[24] Currently, only about 6 percent of federal government workers telework.[25]

Organizations should prepare guidelines and policies to define the types of positions and workers who represent ideal telework opportunities. Clear guidelines must be set for how and when work will be given to and collected from teleworkers. If there are certain hours during which the teleworker must be available, these too must be defined. Employee work expectations and performance criteria must also be delineated.

In 2008, Western Union launched a telework program called iFlex, which allowed 15 percent of its south Florida workforce to telework up to three days per week. The firm estimates that it will save about $3.2 million over five years by reducing real estate costs and other expenses. HR director Sara Baker states: "We discovered telework is an excellent solution from a capital standpoint. We also found from an employee perspective that it not only saves the employee money, but they are more productive, more engaged, and use less sick leave."[26]

A Sensis® Consumer Report found that while the majority of teleworkers (62%) were positive about teleworking, nearly 25 percent felt teleworking had no real impact on their lives, and 13 percent of teleworkers reported a negative impact. The negative feelings were primarily because workers felt that teleworking had not really improved their productivity. In addition, many teleworkers felt that there was increased pressure to work outside the normal business hours so that they now worked longer hours, thus taking time away from their families.[27]

Some positions—such as management positions or those in which face-to-face communication with other employees or customers is required—may not be well suited for telework. In addition, some individuals are not well suited to be teleworkers. Telework opportunities need to be weighed based on the characteristics of the individual as well as the requirements of the position. Table 8-2 and Table 8-3 list some of the advantages and disadvantages of telework from the perspectives of employees and organizations, respectively.

TABLE 8-2 Advantages/disadvantages of teleworking for employees

Advantages	Disadvantages
People with disabilities who otherwise find public transportation and office accommodations a barrier to work may now be able to join the workforce.	Some employees are unable to be productive workers away from the office.
Teleworkers avoid long, stressful commutes and gain time for additional work or personal activities.	Teleworkers may suffer from isolation and may not really feel "part of the team."
Telework minimizes the need for employees to take time off to stay home to care for a sick family member.	Workers who are out of sight also tend to be out of mind. The contributions of teleworkers may not be fully recognized and credited.
Teleworkers have an opportunity to experience an improved work/family balance.	Teleworkers must guard from working too many hours per day because work is always there.
Telework reduces ad hoc work requests and disruptions from fellow workers.	The cost of the necessary equipment and communication services can be considerable if the organization does not cover these.

Source Line: Course Technology/Cengage Learning.

TABLE 8-3 Advantages/disadvantages of teleworking for organizations

Advantages	Disadvantages
As more employees telework, there is less need for office and parking space; this can lead to lower costs.	Allowing teleworkers to access organizational data and systems from remote sites creates potential security issues.
Allowing employees to telework can improve morale and reduce turnover.	Informal, spontaneous meetings become more difficult if not impossible.
Telework allows for the continuity of business operations in the event of a local or national disaster and supports national pandemic-preparedness planning.	Managers may have a harder time monitoring the quality and quantity of the work performed by teleworkers, wondering, for instance, if they really "put in a full day."
The opportunity to telework can be seen as an additional perk that can help in recruiting.	Increased planning is required by managers to accommodate and include teleworkers.
There may be an actual gain in worker productivity.	There are additional costs associated with providing equipment, services, and support for people who work away from the office.
Telework can decrease an organization's carbon footprint by reducing daily commuting.	Telework increases the potential for lost or stolen equipment.

Source Line: Course Technology/Cengage Learning.

The Digital Divide

When people talk about standard of living, they are often referring to a level of material comfort measured by the goods, services, and luxuries available to a person, group, or nation—factors beyond the GDP-based measurement of standard of living. Some of these indicators are:

- Average number of calories consumed per person per day
- Availability of clean drinking water
- Average life expectancy
- Literacy rate
- Availability of basic freedoms
- Number of people per doctor
- Infant mortality rate
- Crime rate
- Rate of home ownership
- Availability of educational opportunities

Another indicator of standard of living is the availability of technology. The **digital divide** is a term used to describe the gulf between those who do and those who don't have access to modern information and communications technology such as cell phones, personal computers, and the Internet. There are roughly 2 billion Internet users worldwide,[28] but the distribution of Internet users varies greatly from country to country, as shown in Figure 8-2.

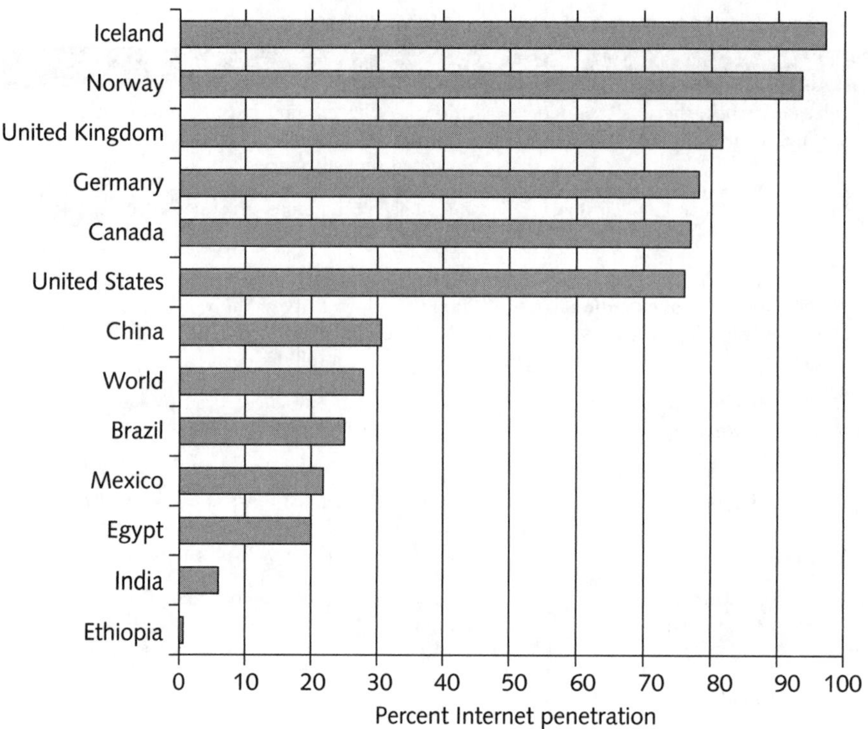

FIGURE 8-2 Degree of Internet penetration by select country

Source Line: Miniwatts Marketing Group, "Internet Usage Statistics: The Internet Big Picture," © March 31, 2011, www.internetworldstats.com/stats.htm.

The digital divide exists not only between more and less developed countries but also within countries—among age groups, economic classes, and people who live in cities versus those in rural areas.

- In the United States 94 percent of households with income over $100,000 have access to broadband Internet access, but only 36 percent of households with income under $25,000 have high-speed access.[29]
- There is a digital divide among the more than 800 million people of Europe, where 42 percent of the population is not connected to the Internet. The two extremes are Iceland, where 98 percent of the population regularly uses the Internet, and Kosovo, where only 21 percent of the population are Internet users.[30,31]
- There is no economical way to provide telecommunications service to the 700 million rural people in India who have an annual GDP of less than $500 per capita. In many of India's rural communities, one must travel more than 5 miles to reach the nearest telephone. Only 7 percent of the population regularly uses the Internet.[32,33]

- Much of Africa lacks basic facilities, such as drinking water, roads, and electricity. Easy access to modern computers and information technology is simply not realistic. As a result, only 11 percent of the population of Africa has access to the Internet.[34]

Many people believe that the digital divide must be bridged for a number of reasons. Clearly, health, crime, and other emergencies could be resolved more quickly if a person in trouble had easy access to a communications network. Access to IT and communications technology can also greatly enhance learning and provide a wealth of educational and economic opportunities as well as influence cultural, social, and political conditions. Much of the vital information people need to manage their career, retirement, health, and safety is increasingly provided by the Internet.

The E-Rate program was designed to help eliminate the digital divide within the United States. This program, as well as the availability of low-cost computers and cell phones, is discussed in the following sections.

E-Rate Program

The **Education Rate (E-Rate) program** was created through the Telecommunications Act of 1996. The full name of the program is The Schools and Libraries Program of the Universal Service Fund. E-Rate helps schools and libraries obtain broadband Internet services to advance the availability of educational and informational resources. The program provides cost discounts that range from 20 percent to 90 percent for eligible telecommunications services, depending on location (urban or rural) and economic need. The level of discount is based on the percentage of students eligible for participation in the National School Lunch Program. The Federal Communications Commission (FCC) ruled that the program would be supported with up to $2.25 billion per year from a fee charged to telephone customers.

The FCC established the private, nonprofit Universal Service Administrative Company (USAC) to administer the program.[35] E-Rate reimburses telecommunications, Internet access, and internal connections providers for discounts on eligible services provided to schools and libraries. Schools and libraries must apply for the discounts, and the USAC works with the service providers to make sure that the discounts are passed along to program participants.[36]

Recently, the E-Rate program was expanded to provide affordable, extremely fast fiber-optic connections for schools and libraries. In addition, the FCC is testing a pilot program to provide off-campus wireless Internet access for mobile learning devices such as digital textbooks.[37]

Unfortunately, the program has not gone well. Following a year-long investigation, a House subcommittee in 2005 approved a bipartisan staff report detailing abuse, fraud, and waste in E-Rate. In one infamous example, USAC disbursed $101 million between 1998 and 2001 to equip Puerto Rico's 1,540 schools with high-speed Internet access, but a review found that very few computers were actually connected to the Internet. In fact, $23 million worth of equipment was found in unopened boxes in a warehouse.[38] Eventually, the former Puerto Rican secretary of education was found guilty of fraud, sentenced to three years in prison, and fined $4 million.[39] A University of Chicago study examined the impact of the E-Rate program in California and found that the number of students in poor schools going online had indeed increased dramatically. However, the study found no evidence that the program had any effect on students' performance on any of the six subjects (math, reading, science, language, spelling, and social studies) covered in the Stanford Achievement Test. Researchers concluded that either the schools did not know how to make effective use of the Internet or that Internet use was simply not a productive way to boost test scores.[40] The E-Rate program continues today; with funding for 2011 budgeted at $2.25 billion.[41]

Low-Cost Computers

As noted above, it is estimated that nearly 2 billion people worldwide have Internet access as of June 2010.[42] Although that number is impressive, it still leaves nearly 5 billion people (72% of the world's population) unconnected. What most of those 5 billion people have in common is low income. Increasing the availability of low-cost computers can help reduce the digital divide.

One Laptop per Child (OLPC)

The nonprofit organization **One Laptop per Child (OLPC)** has a goal of providing children around the world with low-cost laptop computers to aid in their education. The first version of its laptop, the OLPC XO, was made available to third-world countries in 2007 and came with a hand crank for generating power in places where electricity is not readily available. It was distributed at a cost of around $200. The current version of the OLPC is the XO-3 (see Figure 8-3), which was designed to require just 1 watt of electricity per hour.

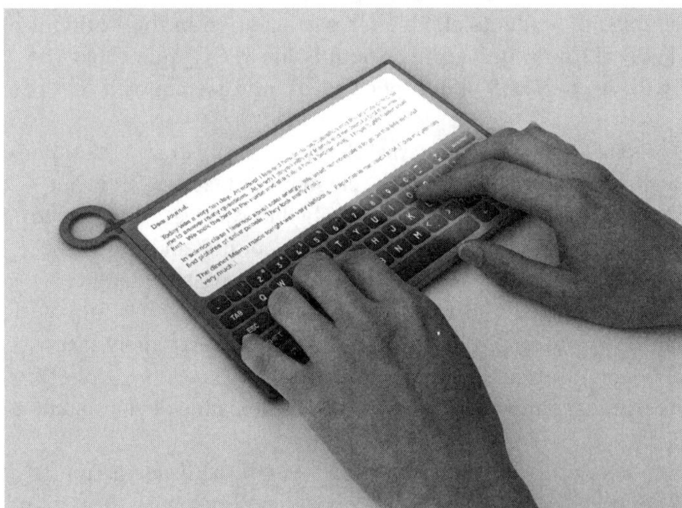

FIGURE 8-3 The OLPC XO-3 tablet computer

Credit: Image courtesy of fuseproject.

The OLPC XO-3 comes with Wi-Fi capability, integrated video, and a still camera; it also supports two-way conferencing. The system is offered to children and institutions supported by the OLPC at a price of $75; that low price is subsidized by sales of the device at higher prices in developed countries.[43] In commenting on the XO-3, Nicholas

Negroponte, chairman of OLPC, states: "We're going to push down on price, we're going to push on nonbreakable, we're going to push particularly on power because we want to hand crank these things. Our characteristics are ones that the market wouldn't do normally, but that we will bring sooner or prove that can be done."[44]

Nepal is one of the poorest countries in the world, with an average annual per capita income of about $475 (U.S.)[45] and an unemployment rate of 46 percent.[46] Much of its rural population has limited access to even the most basic of social services. The OLPC program in Nepal started in 2008 as a small pilot project in two schools. Within three years, it expanded to 32 schools with a total enrollment of 3,300 students in grades 2 to 6. The goal is to improve the quality of education and the access to instructional materials. Some 180 teachers are learning how to integrate the technology into their teaching practices and developing curriculum-based computer educational activities. The OLPC program in Nepal has shown that technology-based education can be successfully introduced into rural schools using prepared local teachers already in the school system.[47]

Classmate+

In 2006, Intel introduced a low-cost laptop called the Classmate PC. The first generation of this notebook computer cost under $400 and was designed for use in kindergarten through high school classrooms in developing countries. Intel does not manufacture the computers but provides free blueprints to manufacturers and independent dealers.[48] The computer began shipping in early 2007 to 25 countries, including Brazil, Chile, Nigeria, China, India, and Vietnam.[49] Since then, Intel and Lenovo have partnered to introduce the Classmate+ laptop targeted for sale in bulk quantities to educational institutions and agencies in third-world countries. With a cost between $300 and $400, the machine has a 10-inch screen, runs the Windows 7 operating system, comes with 1 GB of RAM, has a built-in camera, and has wireless capability.[50] Over 158,000 of the devices are set to be deployed in schools in Buenos Aires, Argentina.[51]

Eee Laptop

Asustek Computer, Inc. is a Taiwanese multinational manufacturer of computers and computer components. Its Eee Pad Transformer computer is a tablet computer that can convert into a laptop with the addition of an optional keyboard. The device is priced at $399—or about $100 less than the Apple iPad tablet computer. At this price, the computer is a competitor to the OLPC and Intel's Classmate+ for widespread deployment in developing countries. The system runs on Google's Android operating system and does not run Microsoft Windows programs.[52]

Mobile Phone the Tool to Bridge the Digital Divide?

Even though the mobile phone cannot do all that a personal computer can, many industry observers think that it will be the cell phone that will ultimately bridge the digital divide. Cell phones have the following advantages over PCs in developing countries:

- Cell phones come in a wide range of capabilities and costs, but are cheaper than personal computers. For example, one can purchase a cell phone in India for the equivalent of $40 or less, and 500 minutes of use cost about $10 per month.

- In developing countries, many more people have access to cell phones than they do PCs (see Table 8-4).

TABLE 8-4 Comparison of PCs and cell phones per 1,000 people

Country	Number of PCs per 1,000 people	Number of cell phone subscribers per 1,000 people
Afghanistan	0.4	143
Cambodia	2	181
China	122	411
India	36	203
Kenya	15	301
Pakistan	5	524
Bangladesh	N/A	224
Swaziland	27	337

Source Line: "Mobile Phone Subscribers per 1,000 People," World Sites Atlas, © 2008, www.sitesatlas.com/Thematic-Maps/Mobile-phone-subscribers-per-capita.html.

- Cell phones are more portable and convenient than the smallest laptop computer.
- Cell phones come with an extended battery life (much longer than any PC battery), which makes the cell phone more reliable in regions where access to electricity is inadequate or nonexistent.
- The infrastructure needed to connect wireless devices to the Internet is easier and less expensive to build.
- There is almost no learning curve required to master the use of a cell phone.
- Basic cell phones require no costly or burdensome applications that must be loaded and updated.
- There are essentially no technical-support challenges to overcome when using a cell phone.

When IT is available to everyone—regardless of economic status, geographic location, language, or social status—it can enhance the sharing of ideas, culture, and knowledge. How much will the benefits of IT raise the standard of living in underdeveloped countries? Could the end of the digital divide change the way people think about themselves in relation to the rest of the world? Could such enlightenment, coupled with a better standard of living, contribute to a reduction in violence, poverty, poor health, and even terrorism?

THE IMPACT OF IT ON HEALTHCARE COSTS

The rapidly rising cost of health care is one of the twenty-first century's major challenges. U.S. healthcare spending is expected to increase an average of 6.3 percent a year (a rate much higher than overall inflation) and grow from $2.6 trillion to $4.6 trillion by

2019 according to the Centers for Medicare and Medicaid Services.[53] The Health Care and Education Reconciliation Act of 2010 will likely have little effect on total costs. Although almost 93 percent of U.S. residents will have health insurance by 2019 versus 84 percent in 2010, the cost of this expanded coverage for some 32 million Americans is expected to be largely offset by new fees and taxes, plus reductions in what Medicare pays private health insurers, hospitals, and others providers.[54]

The development and use of new medical technology, such as new diagnostic procedures and treatments (see Figure 8-4), accounts for most of the increase in healthcare spending in excess of general inflation.[55, 56, 57] Although many new diagnostic procedures and treatments are at least moderately more effective than their older counterparts, they are also more costly. In addition, even if new procedures and treatments cost less (for example, magnetic resonance imaging), they may stimulate much higher rates of use because they are more effective or cause less discomfort to patients.

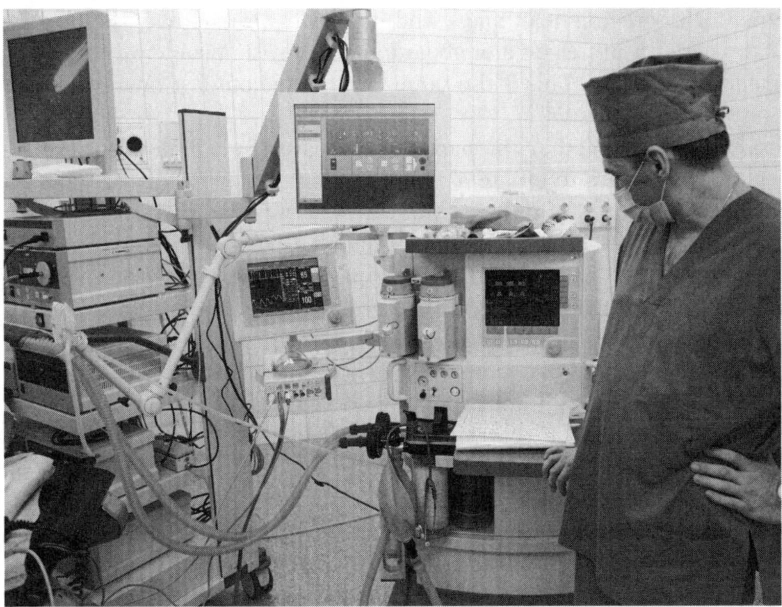

FIGURE 8-4 The development and use of new technology has increased healthcare spending

Credit: Image copyright Farferros, 2009. Used under license from Shutterstock.com.

Patients sometimes overuse medical resources that appear to be free or almost free thanks to the share of medical bills that is paid by third parties, such as insurance companies and government programs. A patient who doesn't have to pay for a medical test or procedure is probably less likely to consider its cost-to-benefit ratio. Attempts by insurance companies to rein in those costs have led to a blizzard of paperwork but have proven ineffective.

To really gain control over soaring healthcare costs, patient awareness must be raised and technology costs must be managed more carefully. In the meantime, however, the improved use of IT in the healthcare industry can lead to significant cost reductions in a number of ways.

Electronic Health Records

Although the healthcare industry depends on highly sophisticated technology for diagnostics and treatment, it has been slow to implement IT solutions to improve productivity and efficiency. As of 2005, the healthcare industry invested about $3,000 in IT for each worker, compared with about $7,000 per worker in private industry generally, and nearly $15,000 per worker in the banking industry.[58] However, the healthcare industry has now greatly increased its investment in IT—spending over $88 billion annually to implement EHRs, convert to a new coding system (known as ICD-10) for diagnosis and inpatient codes, and begin use of a new Food and Drug Administration Web portal to report deaths and injuries caused by medical devices.[59]

As noted in the opening vignette, using IT to capture and record patient data provides a significant opportunity for improving health care and increasing productivity. Before seeing a physician, many patients are given a clipboard and pen with a standard form to complete. Some people must wonder: "This is the same form I filled out last time; what did they do with the data from my last visit?"

It is nearly impossible to pull together the paper trail created by a patient's interactions with various healthcare entities to create a clear, meaningful, consolidated view of that person's health history. This lack of patient data transparency can result in diagnostic and medication errors as well as the ordering of duplicate tests, which dramatically increase healthcare costs. It can even compromise patient safety. For example, physicians in the emergency room must often treat a patient who is unconscious and incapable of providing essential medical information, such as the name of his or her primary care physician, information about recent illnesses or surgeries, medications taken, allergies, and other useful data. Without such data, the ER physician is essentially taking a gamble in treating the patient. If the United States had a comprehensive healthcare information network, such medical data could be readily available for all patients at any medical facility.

Studies over the last 10 years have estimated that at least 98,000 people die in hospitals each year due to preventable medical mistakes.[60] A 2009 Consumers Union report claimed that we have made no real progress toward reducing the number of such deaths.[61] However, as far back as 2004, healthcare experts agreed that "going digital" could eliminate many of these needless deaths.[62]

An **electronic health record (EHR)** is a computer readable record of health-related information on an individual. An EHR can include patient demographics, medical history, family history, immunization records, laboratory data, health problems, progress notes, medications, vital signs, and radiology reports. Data can be added to an EHR based on each patient encounter in any healthcare delivery setting. EHRs can incorporate data from any healthcare entity a patient uses and make the data easily accessible to other healthcare professionals. Healthcare professionals can use an EHR to generate a complete electronic record of a clinical patient encounter (see Figure 8-5).[63]

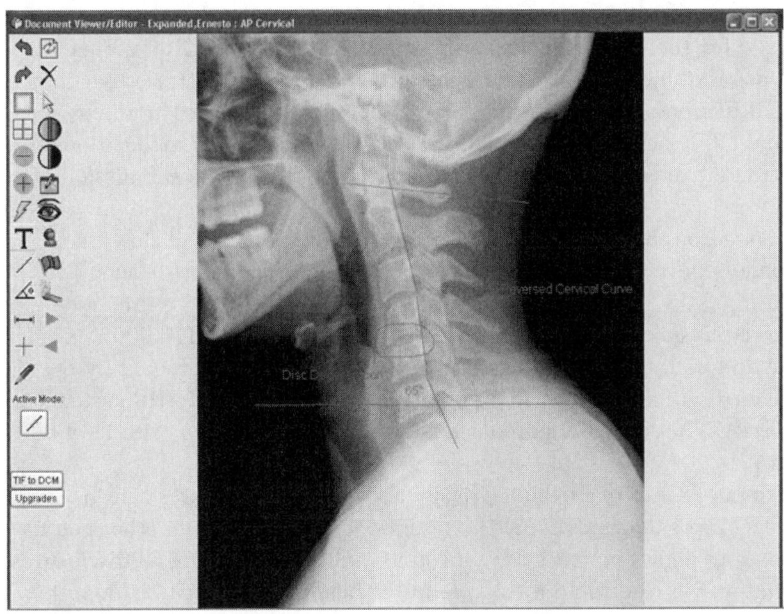

FIGURE 8-5 An EHR is a computer readable record of health-related information on an individual

Credit: Copyright Future Health 2011, www.FutureHealthSoftware.com.

The **Health Information Technology for Economic and Clinical Health Act (HITECH Act)** was passed as part of the $787 billion 2009 American Recovery and Reinvestment Act economic stimulus plan. HITECH is intended to increase the use of health information technology by: (1) requiring the government to develop standards for the nationwide electronic exchange and use of health information; (2) providing $20 billion in incentives to encourage doctors and hospitals to use EHR to electronically exchange patient healthcare data; (3) saving the government $10 billion through improvements in the quality of care, care coordination, reductions in medical errors, and duplicate care; and (4) strengthening the protection of identifiable health information. Under this act, increased Medicaid or Medicare reimbursements will be made to doctors and hospitals that demonstrate "meaningful use" of EHR technology. Meaningful use is defined as EHR technology that enables a hospital to prescribe electronically, exchange data with other providers, and generate certain "clinical quality measure" reports.[64]

The PricewaterhouseCoopers LLP Health Research Institute estimates that a 500-bed hospital could receive $6.1 million in HITECH incentives to purchase, deploy, and maintain an EHR system. On the other hand, failure to implement such a system by 2015 could cause the hospital to lose $3.2 million in funding annually, depending on the hospital's volume of Medicare, Medicaid, and charity-care patients. Hospitals are expected to increase their use of comprehensive EHR systems from 10 percent in 2009 to 55 percent in 2014.[65]

Individual physicians are eligible to receive as much as $44,000 under Medicare and $63,750 under Medicaid for the implementation and meaningful use of EHR systems. To meet the meaningful use requirement, physicians must be able to demonstrate that they are using certified EHR technology in ways that lead to significant and measurable results in achieving health and efficiency improvements, such as e-prescribing of medications and treatments, electronic exchange of health information, and electronic submission of clinical quality data.[66]

Physicians have concerns that EHR systems are difficult to use and will slow them down. In addition, some believe that the data collected by the government through EHR reporting will be used to justify a decrease in their Medicare and Medicaid reimbursements.[67] A typical three-physician practice will need to spend between $173,000 and $296,000 to purchase and maintain an EHR system.[68] A 2009 survey conducted by the Centers for Disease Control indicated that 48.3 percent of physicians used EHR systems in their office-based practices, but only 6.9 percent had systems that met the criteria of a fully functional system.[69]

HITECH also attempts to improve federal privacy and security measures safeguarding health information. It requires that individuals be notified if there is any unauthorized use of their health information, allows patients to request an audit trail showing all disclosures of their health information via electronic means, requires health providers to gain authorization from patients to use their health information for marketing and fundraising activities, and increases penalties for violations and provides greater resources for enforcement and oversight activities.[70]

There have been mixed results from studies evaluating the use of EHR systems in improving patient care and reducing costs.

In a Commonwealth Fund study of 41 Texas hospitals that treat a diverse group of patients suffering from a variety of medical conditions, researchers found that a 10 percent increase in the use of electronic notes and medical records was associated with a 15 percent reduction in the likelihood of patient death. When physicians electronically entered patient care instructions, there was a 55 percent reduction in the likelihood of death related to certain procedures.[71]

However, a second study published by the RAND Corporation in late 2010 that involved half the acute-care hospitals in the United States, found that except for basic systems used to treat congestive heart failure patients, EHRs are not improving process of care measures for many large hospitals.[72]

An earlier study by the RAND Corporation concluded that if 90 percent of doctors and hospitals adopted EHR systems, the United States would save at least $81 billion per year in healthcare costs—with $77 billion coming from a reduction in hospital stays as well as a drop in duplicate and unnecessary testing. The other $4 billion in savings would come from decreased medication errors and treatment of adverse side effects.[73] The RAND Corporation also estimated it would take 15 years and cost hospitals roughly $98 billion and physicians about $17 billion to implement EHR systems. The hospitals and physicians who must make the investment will not, however, reap the full amount of savings from this new technology. The RAND Corporation estimated that Medicare would receive about $23 billion of the savings each year, and private insurers about $31 billion a year.[74]

A much more conservative estimate by the federal government predicts a savings of $15.5 billion in healthcare spending from 2016 to 2019 ($5.1 billion per year) due to improved efficiencies and reduced reimbursements to healthcare providers who are "not meaningful" users of EHR.[75]

One cannot help but wonder where we will be in 10 years in terms of healthcare spending. Will we have made a meaningful reduction in the number of avoidable deaths? Will we have earned a worthwhile return on the investment in EHR? Are those who are advocating the adoption of EHR acting ethically or are they pushing some other agenda?

Use of Mobile and Wireless Technology in the Healthcare Industry

Although slow to invest in IT, the healthcare industry was actually a leader in adopting mobile and wireless technology, perhaps because of the frequent urgency of communications with doctors and nurses, who are almost always on the move. For example, doctors were among the first large groups to start using personal digital assistants (PDAs) on the job. Other common uses of wireless technology in the healthcare field include:

- Providing a means to access and update EHRs at patients' bedsides to ensure accurate and current patient data
- Enabling nurses to scan bar codes on patient wristbands and on medications to help them administer the right drug in the proper dosage at the correct time of day (an attached computer on a nearby cart is linked via a wireless network to a database containing physician medication orders)
- Using wireless devices to communicate with healthcare employees wherever they may be

Telemedicine

Telemedicine employs modern telecommunications and information technologies to provide medical care to people who live far away from healthcare providers. This technology reduces the need for patients to travel for treatment and allows healthcare professionals to serve more patients in a broader geographic area. There are two basic forms of telemedicine: store-and-forward and live.

Store-and-forward telemedicine involves acquiring data, sound, images, and video from a patient and then transmitting everything to a medical specialist for later evaluation. This type of monitoring does not require the presence of the patient and care provider at the same time. Yet, having access to such information can enable healthcare professionals to recognize problems and intervene with remote patients before high-risk situations become life threatening. For example, patients who have chronic diseases often don't recognize early warning signs that indicate an impending health crisis. A sudden weight gain by a patient who has suffered congestive heart failure could indicate retention of fluids, which could lead to a traumatic trip to the emergency room or even loss of life. A physician who uses telemedicine to keep tabs on such patients could make a vital difference.

Teleradiology involves the transfer of CT scans, MRIs, X-rays, and other forms of images from one location to another—such as hospital to hospital or an imaging center to a radiologist specializing in image reading (see Figure 8-6). The radiologic images must be captured or converted into digital format and then sent over a network to a display device for viewing by the image specialist. The CT scan of an auto accident victim lying in a coma at 3 a.m. in a rural hospital in Ohio can be transmitted to an image specialist in India and treatment based on an analysis of the scan can begin within 20 minutes.

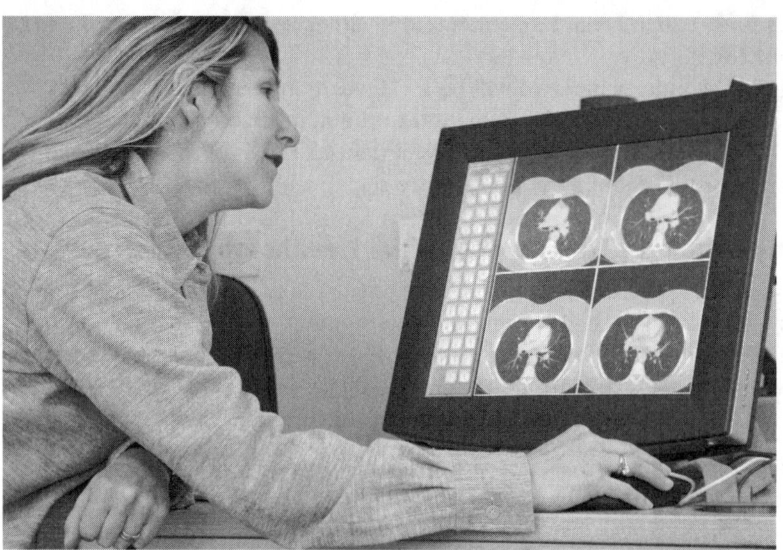

FIGURE 8-6 Image specialist at work
Credit: ©iStockphoto.com/WILLSIE.

Live telemedicine requires the presence of patients and healthcare providers at different sites at the same time and often involves a video conference link between the two sites. For example, work on an oil rig can be extremely dangerous and the nearest hospital is often hundreds of miles away. Oil companies are increasingly relying on live telemedicine to connect a nurse or EMT on an oil platform to emergency physicians at a major medical center. "Access to health care via telemedicine is an excellent application of technology and can save lives and money according to Charles R. Doarn, co-Editor-in-Chief of Telemedicine and e-Health."[76]

Mobile telemedicine is being used by Cincinnati's Children's Emergency Department with a system called Transport AV that attaches to the victim's stretcher and provides hospital-based caregivers with video, audio, and telemetry signals while the patient is in transit. "What we are trying to do is get the patient linked to a physician at the earliest time—at first touch so to speak—when decisions are the most critical" says Roger Downey, communications manager at GlobalMedia, which helped develop the system. Hospitals in Tucson, AZ and Baton Rouge, LA have similar systems; however, those rely on municipal Wi-Fi capabilities. The Transport AV system transmits over 3G and 4G mobile data networks, which results in improved range and coverage, although image quality can be inconsistent, with video streams more likely to get interrupted.[77]

Telemedicine is also being used to provide two hospitals and three primary care clinics on the Turks and Caicos Islands access to medical specialists who were previously unavailable on the islands. Use of telemedicine has improved the quality of care and reduces travel for many patients living on the islands.[78]

The use of telemedicine does raise some new ethical issues. Must the physicians providing advice to patients at a remote location be licensed to perform medicine at that location—perhaps a different state or country? Must a healthcare system be required to possess a license from a state in which it has a "virtual" facility, such as a video conferencing room? Will the various states require some form of assurance that minimum technological standards (such as the minimum resolution of network-transmitted images) are being met? What sort of system certification and verification is necessary to ensure that a critical system performs as expected in crises situations, and what are the ramifications if it does not?

In addition, recent studies have shown that there is reluctance on the part of many doctors and nurses for remote doctors to have anything more than minimal involvement with their patients. There is concern that patient involvement with remote doctors may have a negative effect on the local doctors' relationships with their patients and could adversely affect patient care.[79]

Medical Information Web Sites for Laypeople

Healthy people as well as those who suffer from illness need reliable information on a wide range of medical topics to learn more about healthcare services and to take more responsibility for their health. Clearly, laypeople cannot become as informed as trained medical practitioners, but a tremendous amount of healthcare information is available via the Web. These sites have a critical responsibility to publish current, reliable, and objective information. Table 8-5 provides just a small sample of Web sites that offer information on a variety of medical-related topics.

TABLE 8-5 Health information Web sites

URL	Site
www.americanheart.org	American Heart Association
www.cancer.org	American Cancer Society
www.cdc.gov	Centers for Disease Control and Prevention
www.diabetes.org	American Diabetes Association
www.heartburn.about.com	Information on the causes of heartburn and how to prevent it
www.heartdisease.about.com	Basic information about heart disease and cardiology
www.medicinenet.com	Source for medical information on a variety of topics, including symptoms, procedures, tests, and medications, as well as a medical dictionary
www.nia.nih.gov/Alzheimers	National Institute on Aging—Alzheimer's Disease Education and Referral Center
www.niddk.nih.gov	National Institute of Diabetes and Digestive and Kidney Diseases
www.oncolink.upenn.edu	Abramson Cancer Center of the University of Pennsylvania
www.osteo.org	National Institutes of Health—Osteoporosis and Related Bone Diseases National Resource Center
www.urologychannel.com	Information about urologic conditions, including erectile dysfunction, HIV, AIDS, kidney stones, and STDs; site contains overviews, symptoms, causes, diagnostic procedures, and treatment options
www.webmd.com	Access to medical reference material and online professional publications

Source Line: Course Technology/Cengage Learning.

The contents of a medical information Web site, such as text, graphics, and images, are for informational purposes only. These Web sites are not intended to be substitutes for professional medical advice, diagnosis, or treatment. Individuals should always seek the advice of a physician or other qualified healthcare provider with any questions regarding a medical condition. A patient should never disregard professional medical advice or delay seeking it because of something he or she reads on a medical information Web site.

In addition to publicly available information on the Web, many healthcare providers, employers, and medical insurers offer useful online tools that go beyond basic health information. These tools enable patients to go online and compare the quality, safety, and cost information on hospitals nationwide. You can also find risk indicators for specific health treatment options and nationwide average prices of drugs and treatment options. In addition, the coverage and costs for treatments by in-network and out-of-network healthcare providers can be found on many of these Web sites.

For example, an individual who needs a hip replacement can go online and find information about the surgery; other available treatment options; a list of questions to ask the physician; potential risks; nearby hospitals that perform the surgery; and quality-of-service information about the hospitals, such as the number of reported postoperative infections and other complications.

Summary

- The most widely used measurement of the material standard of living is gross domestic product (GDP) per capita.
- In the United States, as in most developed nations, the standard of living has been improving over time. However, its rate of change varies as a result of business cycles that affect prices, wages, employment levels, and the production of goods and services.
- Productivity is defined as the amount of output produced per unit of input.
- Most countries have continually been able to produce more goods and services over time—not through a proportional increase in input but by making production more efficient. These gains in productivity have led to increases in the GDP-based standard of living because the average hour of labor produced more goods and services.
- Progressive management teams use IT, other new technology, and capital investment to implement innovations in products, processes, and services.
- It can be difficult to quantify the benefits of IT investments on worker productivity because there can be a considerable lag between the application of innovative IT solutions and the capture of significant productivity gains. In addition, many factors other than IT influence worker productivity rates.
- Many organizations offer telework opportunities to their employees as a means of reducing costs; increasing productivity; reducing the organization's carbon footprint; and preparing for potential local or widespread disasters.
- Telework opportunities provide many advantages for employees, such as avoiding long, stressful commutes and providing more flexibility to balance the needs of work and family life.
- The *digital divide* is a term used to describe the gulf between those who do and those who don't have access to modern information and technology, such as computers and the Internet.
- The digital divide exists not only between more and less developed countries but also within countries—among age groups, economic classes, and people who live in cities versus those in rural areas.
- The E-Rate program helps schools and libraries helps schools and libraries obtain broadband Internet services to advance the availability of educational and informational resources.
- One Laptop per Child is a nonprofit organization whose goal is to provide children around the world with low-cost laptop computers to aid their education.
- Many people think that it will be the cell phone and not the computer that will ultimately bridge the digital divide.
- Healthcare costs are soaring out of control and expected to increase an average of 6.3 percent per year from today until 2019.
- To gain control over healthcare costs, patients will need to gain a much greater awareness of medical costs, and new technology costs will need to be managed more carefully.

- Improved use of IT in the healthcare industry can lead to significantly reduced costs in a number of ways: electronic health records (EHRs) of patient information can be generated from each patient visit in every healthcare setting, and wireless technology can be used to access and update EHRs at patients' bedsides, match bar-coded patient wristbands and medication packages to physician orders, and communicate with healthcare employees wherever they may be.
- Telemedicine employs modern telecommunications and information technologies to provide medical care to people who live or work far away from healthcare providers. This technology reduces the need for patients to travel for treatment and allows healthcare professionals to serve more patients in a broader geographic area.
- Web-based health information can help people inform themselves about medical topics.

Key Terms

digital divide
Education Rate (E-Rate) program
electronic health record (EHR)
Health Information Technology for Economic and Clinical Health Act (HITECH Act)
live telemedicine
One Laptop per Child (OLPC)
productivity
store-and-forward telemedicine
telemedicine
telework

Self-Assessment Questions

The answers to the Self-Assessment Questions can be found in Appendix E.

1. Which of the following statements about the standard of living is *not* true?
   a. It varies greatly from nation to nation.
   b. It varies little among groups within the same country.
   c. Industrialized nations generally have a higher standard of living than developing countries.
   d. It is frequently measured using the GDP per capita.
2. The amount of output produced per unit of input is called ____________.
3. The long-term rate of U.S. productivity improvement over the past century is 3 percent per year. True or False?
4. A study of 527 large U.S. firms from 1987 to 1994 found that the benefits of applying IT grow over time and that an IT investment can take:
   a. one to three years to break even.
   b. three to five years for its users to become efficient in its use.
   c. five to seven years to result in a substantial increase in productivity.
   d. over seven years to fully recover the initial investment costs.

5. ____________ is a term used to describe the gulf between those who do and those who don't have access to modern information and communications technology such as cell phones, personal computers, and the Internet.

6. ____________ is a work arrangement in which an employee works away from the office.

7. Which of the following is a valid reason for trying to reduce the digital divide?

   a. Health, crime, and other emergencies could be resolved more quickly if people in trouble had access to a communications network.

   b. Much of the vital information that people need to manage their retirement, health, and safety is increasingly provided by the Internet.

   c. Ready access to information and communications technology can provide a country with a wealth of economic opportunities and give its industries a competitive advantage.

   d. All of the above.

8. Iceland, Norway, United Kingdom, Germany, and Canada all have a higher degree of Internet penetration than the United States. True or False?

9. The new version of the OLPC XO-3 is sold at a price in developed nations that will help subsidize a version of the system offered to children and institutions supported by the OLPC at a price of:

   a. under $50

   b. around $75

   c. just over $100

   d. about $125

10. Which of the following statements about healthcare spending is *not* true?

    a. U.S. spending on health care is expected to grow to $4.6 trillion by 2019.

    b. U.S. spending on health care is expected to increase an average of 6.3 percent from now until 2019.

    c. It is estimated that almost 93 percent of U.S. residents will have health insurance by 2019.

    d. The development and use of new medical technology in the United States has clearly led to a reduction in healthcare costs.

11. The two main reasons advanced as the cause of rising healthcare costs are the use of more expensive technology and the shielding of patients from the true costs of medical care. True or False?

12. Studies over the last 10 years have estimated that at least ____________ people die in hospitals each year due to preventable mistakes.

13. A(n) ____________ is a summary of health information generated by each patient encounter in any healthcare delivery setting.

14. Under the Health Information Technology for Economic and Clinical Health Act, increased Medicaid or Medicare reimbursements will be made to doctors and hospitals that demonstrate ____________ of EHR technology.

Discussion Questions

1. Identify three factors that can affect national productivity rates.
2. Explain how increases in worker productivity can lead to an increase in the standard of living.
3. Why is it harder to quantify the benefits of today's IT investments than it was in the 1960s?
4. What factors have increased the prevalence of telework?
5. Would you accept a telework position in which you would work from home three or four days per week? Why or why not?
6. Identify and briefly discuss three good reasons why it is important to bridge the digital divide.
7. As mentioned in the text, a University of Chicago study examined the impact of the E-Rate program and found that the number of poor schools going online had increased dramatically. However, the study found no evidence that the program had any effect on students' performance on the Stanford Achievement Test. Should this be sufficient evidence to discontinue this program, which costs U.S. taxpayers roughly $2 billion per year?
8. The development and use of new medical technology, such as new diagnostic procedures and treatments, has increased spending and accounts for one-half to two-thirds of the increase in healthcare spending in excess of general inflation. Should the medical industry place more emphasis on using older medical technologies and containing medical costs? Which approach to dealing with moral issues discussed in Chapter 1 would you use to analyze this question? What decision did you come to as a result of your analysis?
9. Medical information that you obtain from Web sites must be accurate and reliable. Identify three earmarks of a credible Web site.

What Would You Do?

1. You are in your local computer store and see a "low-cost" laptop selling for just $199. There is a note on the price tag that $50 of the purchase price will be used to subsidize the cost of this computer to students in developing countries. How do you feel about paying an extra $50 for this purpose? Would you attempt to negotiate a lower price? Would you be willing to pay the additional $50?
2. You are a midlevel manager at a major metropolitan hospital and are responsible for capturing and reporting statistics regarding the cost and quality of patient care. You believe in a strict interpretation when defining various reportable incidents; as a result, your hospital's rating on a number of quality issues has declined in the six months you have held the position. Your predecessor was more lenient and was inclined to let minor incidents go unreported or to classify some serious incidents as less serious. The quarterly quality

meeting is next week, and you know that your reporting will be challenged by the chief of staff and other members of the quality review board. How should you prepare for this meeting? Should you defend your strict reporting procedures or revert to the former reporting process for the "sake of consistency in the numbers," as several people have urged?

3. You have been diagnosed with a rare bone marrow disorder that affects only 2 people out of 1 million. The disease is potentially life threatening, but your symptoms are currently only mild and do not yet present a major concern. Your physician recommends that you go to the Mayo Clinic in Rochester, Minnesota for further diagnosis and possible treatment. As you do some research on the Internet, you find a support group for those afflicted by this rare disease. You are alarmed to hear that the disease can cause a very rapid decrease in the quality of one's life, with many victims confined to a wheelchair or bed and in great discomfort for the last months of their life. When you meet with specialists at the Mayo Clinic, they provide a much more optimistic outlook and claim that medical breakthroughs in treating the disease have been made. You do not know what to believe. You wonder about reaching out to the support group to get further information.
4. As a second-year teacher at a lowly rated inner-city elementary school, you have been asked to form and lead a three-person committee to define and obtain funding for an E-Rate program for your school. Do some research on the Internet and outline a process you would follow to request funding.
5. You are an elected official in a small, third-world country's house of parliament, which is responsible for initiating revenue spending bills. Your country is very poor; unemployment is high; most families cannot afford a healthy diet; there is an insufficient amount of doctors and healthcare services; and there is an inadequate infrastructure for water, telephone, and power. Recently, senior executives from technology firms have approached you and lobbied you strongly to support increased spending on information technology infrastructure, including the placement of 1 million low-cost computers in your nation's schools. They make a strong case that the computers will increase the educational opportunities for your nation's children. They are willing to subsidize one-half of the estimated $1 billion (U.S.) required to implement this program successfully. While their idea provides hope for a better life for the children, your country has many needs. How would you proceed to evaluate this opportunity and weigh its costs against your country's other needs?
6. You are on the administrative staff of a large midwestern hospital taking its first steps to implement a comprehensive EHR system. You have been asked to survey the medical personnel to determine their level of acceptance regarding the implementation of the system. The goal is to identify any issues and to reduce any resistance. How would you propose surveying the staff? What sort of resistance would you expect to encounter? What actions might you take to reduce resistance?
7. You have volunteered to lead a group of citizens in approaching the board of directors of the nearest hospital (55 miles away) about establishing some sort of telemedicine-based monitoring of 50 or so chronically ill people in your small community. What sort of facts do you need to gather to make a sound recommendation to the board? What are some specific items that you would request?

8. You have been offered a position as a software support analyst. If you accept, you will have three weeks of on-site training, after which you will work from your home full-time, answering customer service calls. What questions would you want answered before you decide whether or not to take this position?

Cases

1. Problems in the E-Rate Program

As discussed in this chapter, the E-Rate Program is a program overseen by the Federal Communications Commission (FCC) that provides money to connect schools and libraries to the Internet. The program is funded by fees collected from telephone users.

In November 2010, the U.S. Department of Justice settled two lawsuits for a total of $16.25 million against Hewlett-Packard (HP) in connection with the awarding of E-Rate technology and service contracts in the Dallas (DISD) and Houston Independent School Districts (HISD). The lawsuits alleged that between 2002 and 2005 contractors working with HP offered bribes in order to win very profitable contracts that included some $17 million in HP equipment.[80]

According to Austin Schlick, general counsel for the FCC, the settlement agreement "ensures that HP will train its employees thoroughly on the FCC's gift and other E-Rate rules, and provides for audits of HP's E-Rate program. If HP fails to monitor its E-Rate activities closely and abide by E-Rate Program requirements, it will face substantial penalties."[81]

An HP spokesperson declared: "HP requires that all employees and partners adhere to lawful and ethical business practices. The activities at the center of this investigation occurred more than five years ago, the partner relationships have been terminated, and the employees involved are no longer with the company."[82]

DISD and HISD bought most of their HP equipment from resellers, and the actions of two of those resellers were the focus of the FCC investigation. As a result of the settlement, HP now requires that all partners who bid on E-Rate projects certify that their employees have been trained on all E-Rate rules and regulations. The FCC has required that HP establish an interactive E-Rate Web site with annual mandatory training of its partners.

In 2006, HP banned the two resellers who were under investigation—Analytical Computer Services (ACS) and Micro Systems Engineering (MSE)—from selling its equipment. The two companies allegedly provided illegal inducements in the form of the use of a private yacht, sporting tickets, and other gifts to school district employees while the companies were bidding on the DISD and HISD E-Rate program contracts.[83, 84]

In 2008, the former chief technology officer of DISD and the former chief executive officer of MSE were found guilty of bribery for their illegal conduct related to the DISD contract and sentenced to over a decade in prison. In 2009, the DISD was fined $750,000 and agreed to drop its requests for more than $150 million in federal funding.[85] The HISD was fined $850,000.[86]

As part of its ongoing antifraud efforts, in September 2010 the FCC adopted an order that "bolsters and clarifies the prohibition against E-Rate applicants soliciting or receiving gifts, and against service providers offering or providing gifts."[87]An HP solution provider declares that "five years ago it was nothing to give someone a $100 ticket to go to a game. Today people are not accepting that. They are making sure that it is a valued relationship. You've got to be very careful today. You don't want to do something stupid."[88]

Discussion Questions

1. What are the key points from this case that apply to anyone who is involved in competing for or awarding contracts for products or services?
2. In addition to mandatory training, what measures should HP put in place to ensure that there are no future cases of bribery involving the sale of its products and services—either directly or through resellers—in connection with the E-Rate program?
3. Imagine you are a salesperson who will be awarded over $1 million in commissions if your firm is awarded a major contract. What ethical actions can you take to ensure that you and your firm are viewed favorably by the key decision maker in the deal?

2. Decision Support for Healthcare Diagnosis

Diagnosis errors are a frequent and serious problem in the healthcare industry. Analysis of malpractice claims reveals that diagnosis errors far outnumber medication errors as a cause of malpractice claims by a margin of at least two to one.[89, 90] Failure to fully diagnose a patient's condition puts the patient at risk of suffering a recurrence of the problem such as incurring further damage from another accident caused by, for example, an undiagnosed brain injury. Misdiagnosis of a patient's condition can lead to costly, painful, potentially harmful, and inappropriate treatments. A delay in the diagnosis of a patient can allow an otherwise reversible condition to advance to the point that it is no longer treatable.

Over the past decade, several decision support systems to aid in healthcare diagnosis have been developed, including DiagnosisPro®, DXPlain®, First Consult©, PEPID, and Isabel©. A decision support system is an interactive computer application that aids in decision making by gathering data from a wide range of sources and presenting that data in a way that aids in decision making. Isabel, one of the more advanced healthcare decision support systems, is a Web-based system developed in the United Kingdom. Isabel uses key facts from the patient's history, physical exam, or laboratory findings to identify the most likely diagnosis based on pattern matches in the system's database. The system can interface with electronic medical records systems to obtain patient data or the data can be entered manually. Each diagnosis is linked to information in commonly used medical reference sources such as *The 5 Minute Clinical Consult*, *Oxford Textbook of Medicine*, and Medline, the U.S. National Laboratory of Medicine's online bibliographic database. Isabel can also suggest bioterrorism agents that might be responsible for a patient's symptoms, as well as identify drugs or drug combinations that might be the cause.[91] The cost of using Isabel ranges from a few thousand dollars for a family practice to as much as $400,000 for a health system.[92]

United Hospital, a large hospital in St. Paul, Minnesota, recently implemented the Isabel system to help physicians investigate and diagnose patient cases. The system will integrate directly with the hospital's electronic medical record system and physicians will be able to access Isabel from mobile devices.[93]

On another front, medical researchers at Columbia University are busy feeding data from medical textbooks and journals into IBM's Watson supercomputer to create a world-class healthcare diagnostic tool. Watson is the same supercomputer that gained recognition for beating the world's best players on the TV show *Jeopardy!* in 2011. Watson is now being programmed to understand plain language so that it can absorb data about a patient's symptoms and medical history, form a diagnosis, and suggest an appropriate course of treatment. When presented with a set of symptoms, Watson will provide several diagnoses, rank ordered in order of its confidence. Sometime within the next couple years, physicians will be able to interact with Watson by simply speaking into a microphone or entering data via a smartphone.[94]

One incentive hospitals have to adopt such systems is concern that a failure to adopt new technology could subject the hospital to liability in cases where it could be shown that adoption of the technology would not have been overly costly and could have prevented patient injury.[95]

Discussion Questions

1. What concerns might a physician have about using a decision support system such as Isabel or Watson to make a medical diagnosis? How might those concerns be alleviated?
2. Is it possible that in a decade this type of technology could be easily accessible by laypeople who could then perform self-diagnosis, thus helping to reduce the cost of medical care?
3. Does the use of decision support systems to support healthcare decisions seem like an effective way to reduce healthcare costs? Why or why not?

3. Does IT Investment Pay Off?

Tween Brands sells fashion merchandise and accessories for girls aged 7–14 through its 900 Justice stores located in the United States, Puerto Rico, Europe, and the Middle East. It also offers its fashions through its Web site. Recent annual revenue for the company was just under $1 billion.[96] In 2007, the company implemented a "put-to-light" system that improved the productivity of its warehouse workers by 25 percent. Workers scan identifiers on products received at the warehouse and then follow flashing light signals at stocking locations that indicate where to put the scanned product. In many instances, the incoming product can be unloaded from one truck and loaded directly onto another outbound truck headed for a store. Such cross-docking opportunities eliminate the need to store product in inventory and greatly streamline operations. Just prior to installing the "put-to-light" system, Tween Brands upgraded its warehouse management system, which tracks inventory levels for each item.[97]

MIT economist Erik Brynjolfsson has reported a strong correlation between IT capital per worker and a company's productivity. He says there is a growing consensus that IT is the most important factor in increasing productivity.[98] Considering the impact that technology has had on the workplace, this correlation seems obvious.

Or is it? What about the cost of training Tween Brands workers to use the new inventory and warehouse management systems? What about the costs incurred when the hardware breaks down? What if there is a bug in the software? What if a new system makes the current one obsolete in two or three years?

Paul Strassmann, a former CIO of both Xerox and NASA, disagrees with Brynjolfsson and argues that IT has not improved labor productivity at all. Back in the 1990s, Strassmann developed a method to measure increased productivity based on microeconomics rather than GDP and other large-scale measurements. Strassmann argues that productivity should be taken as the ratio of the cost of goods to transaction costs, which includes administrative and other costs not directly related to the production or delivery of a good or service. Strassmann and others at Alinean LLC have found that this ratio has remained constant for the past 10 years.[99]

In response, macroeconomists argue that Strassmann's calculations do not account for increases in customer value, such as timely delivery or the creation of innovative products and services. In addition, they say technology has allowed companies to meet regulatory standards and reporting requirements established by government agencies.

Strassmann and other proponents of the microeconomic approach don't deny this oversight and don't argue against IT investment. No company can remain competitive and fail to innovate. Strassmann and his colleagues are simply hoping to help companies avoid ill-conceived IT investments by doing away with short "build and junk" cycles, in which new information systems are scrapped before ever breaking even. They want to develop microeconomic parameters that CIOs can take to their boards of directors as proof that they are cutting costs.[100] Thus, while economists still debate the impact of IT investment, the discussions are at least producing tools that help businesses improve their IT decision making and perhaps even increase their productivity.

Discussion Questions

1. Apart from the annual rate of output per worker, what are other ways of measuring labor productivity?
2. What factors determine whether a new information system will increase or decrease labor productivity?
3. Why is it so difficult to determine whether IT has increased labor productivity?

End Notes

[1] Philip Maddocks, "Struggling to Find a New Meaning to its Existence, the Dow Hires a Self-Help Guru," Wicked Local Natick, March 6, 2009, www.wickedlocal.com/natick/news/lifestyle/columnists/x949368656/MADDOCKS-Struggling-to-find-a-new-meaning-to-its-existence-the-Dow-hires-a-self-help-guru#axzz1PV3DIfN7.

[2] Howard J. Anderson, "Kaiser's Long and Winding Road," *Health Data Management*, August 1, 2009, www.healthdatamanagement.com/issues/2009_69/-38718-1.html.

[3] Neil Versel, "As EHR Installation Nears Completion, Kaiser Recommends 'Big Bang,'" FierceEMR, July 30, 2009, www.fierceemr.com/story/ehr-installation-nears-completion-kaiser-recommends-big-bang/2009-07-30.

[4] Howard J. Anderson, "Kaiser's Long and Winding Road," *Health Data Management*, August 1, 2009, www.healthdatamanagement.com/issues/2009_69/-38718-1.html.

[5] Kaiser Permanente, "Kaiser Permanente HealthConnect® Electronic Health Record," xnet.kp.org/newscenter/aboutkp/healthconnect/index.html (accessed May 31, 2011).

[6] Howard J. Anderson, "Kaiser's Long and Winding Road," *Health Data Management*, August 1, 2009, www.healthdatamanagement.com/issues/2009_69/-38718-1.html.

[7] Kaiser Permanente, "Kaiser Permanente HealthConnect® Electronic Health Record," xnet.kp.org/newscenter/aboutkp/healthconnect/index.html (accessed May 31, 2011).

8 Howard J. Anderson, "Kaiser's Long and Winding Road," *Health Data Management*, August 1, 2009, www.healthdatamanagement.com/issues/2009_69/-38718-1.html.

9 Kaiser Permanente "Kaiser Permanente HealthConnect® Electronic Health Record," xnet.kp.org/newscenter/aboutkp/healthconnect/index.html (accessed May 31, 2011).

10 Howard J. Anderson, "Kaiser's Long and Winding Road," *Health Data Management*, August 1, 2009, www.healthdatamanagement.com/issues/2009_69/-38718-1.html.

11 Howard J. Anderson, "Kaiser's Long and Winding Road," *Health Data Management*, August 1, 2009, www.healthdatamanagement.com/issues/2009_69/-38718-1.html.

12 Jane Sarasohn-Kahn, "The Story of Kaiser Permanente's EHR," *Health Populi*, September 15, 2010, http://healthpopuli.com/2010/09/15/the-story-of-kaiser-permanentes-ehr.

13 Kimberly Amadeo, "The Great Depression of 1929," *About.com*, http://useconomy.about.com/od/grossdomesticproduct/p/1929_Depression.htm.

14 Kimberly Amadeo, "GDP 2008 Statistics," *US Economy*, April 30, 2011, http://useconomy.about.com/od/GDP-by-Year/a/2008-GDP-statistics.htm.

15 Peter S. Goodman, "Unemployment Rate Hits 10.2%, Highest in 26 Years," *New York Times*, November 6, 2009.

16 Stephen D. Oliner and William L. Wascher, "Is a Productivity Revolution Underway in the United States?," *Challenge*, November–December 1995, www.questia.com/googleScholar.qst;jsessionid=K2sGFsv9Dtjpghh25XQYNGcZcWNVv4hGWLJJ1sLX0wJtGGX2RZZq!1481560549!1622387428?docId=5000361656.

17 United States Department of Labor, Bureau of Labor Statistics, Labor Productivity and Costs, Productivity Change in the Nonfarm Business Sector, 1947–2010, www.bls.gov/lpc/prodybar.htm (accessed May 6, 2011).

18 United States Department of Labor, Bureau of Labor Statistics, Labor Productivity and Costs, Productivity Change in the Nonfarm Business Sector, 1947–2010, www.bls.gov/lpc/prodybar.htm (accessed May 6, 2011).

19 Erik Brynjolfsson and Lorin M. Hitt, "Computing Productivity: Firm-Level Evidence," November 2002, http://opim.wharton.upenn.edu/~lhitt/cpg.pdf.

20 Laureen Miles Brunelli, "Tops for Telecommuting: Fortune's '100 Best Companies to Work For,'" *About.com*, January 7, 2010, http://workathomemoms.about.com/b/2010/01/27/tops-for-telecommuting-fortunes-100-best-companies-to-work-for.htm.

21 Cindy Krischer Goodman, "More Companies Opt to Have Their Workers Telecommute," *Star Telegram*, June 2, 2010, www.star-telegram.com/2010/06/01/2231873/more-companies-opt-to-have-their.html.

22 Lisa Shaw, *Telecommute! Go to Work Without Leaving Home* (New Jersey: Wiley, 1996).

23 Matt Williams, "Swine Flu: Agencies Scramble to Update Telecommuting Policies," *Government Technology*, April 30, 2009, www.govtech.com/gt/652450.

24 Jim Garrettson, "Telework Improvements Act of 2010 Passes, More Government Workers to Telecommute," *GovConExecutive*, July 20, 2010, www.govconexecutive.com/2010/07/telework-improvements-act-of-2010-passes-more-government-workers-to-telecommute.

[25] Ed O'Keefe, "Federal Eye: Report—Less Than 6% of Federal Workers Telecommute," *Washington Post*, February 18, 2011, http://voices.washingtonpost.com/federal-eye/2011/02/report_less_than_6_percent_of.html.

[26] Cindy Krischer Goodman, "More Companies Opt to Have Their Workers Telecommute," *Star Telegram*, June 2, 2010, www.star-telegram.com/2010/06/01/2231873/more-companies-opt-to-have-their.html.

[27] "The Impact of Teleworking—The Views of Teleworkers," Australian Government Department of Communications, Information Technology and the Arts, February 5, 2008, www.archive.dcita.gov.au/2007/11/australian_telework_advisory_committee/sensis_insights_report_teleworking/the_impact_of_teleworking__the_views_of_businesses.

[28] "Global Digital Divide Persists But Is Narrowing," *EuroMonitor International*, February 11, 2011, http://blog.euromonitor.com/2011/02/global-digital-divide-persists-but-is-narrowing-1.html.

[29] "Socio-Economic Factors Continue to Impact Digital Divide in the US," *eGov Monitor*, www.egovmonitor.com/node/39411.

[30] Mark Bradshaw, "Europe's Digital Divide," *Krakow Post*, April 28, 2008, www.krakowpost.com/article/1069.

[31] Miniwatts Marketing Group, "Internet Usage Statistics: The Internet Big Picture," www.internetworldstats.com/stats.htm (accessed May 8, 2011).

[32] Mark Bradshaw, "Europe's Digital Divide," *Krakow Post*, April 28, 2008, www.krakowpost.com/article/1069.

[33] Miniwatts Marketing Group, "Internet Usage Statistics: The Internet Big Picture," www.internetworldstats.com/stats.htm (accessed May 8, 2011).

[34] Miniwatts Marketing Group, "Internet Usage Statistics: The Internet Big Picture," www.internetworldstats.com/stats.htm (accessed May 8, 2011).

[35] U.S. Department of Education, "E-Rate Program—Discounted Telecommunications Services," www.ed.gov/about/offices/list/oii/nonpublic/erate.html (accessed May 9, 2011).

[36] USAC Universal Service Administrative Company, www.universalservice.org/sl (accessed May 9, 2011).

[37] W. David Gardner, "FCC Upgrades E-Rate Program," *InformationWeek*, September 24, 2010, www.informationweek.com/news/hardware/desktop/227500668.

[38] Andrew T. LeFevre, "Report Finds Fraud, Waste, and Abuse in Federal E-Rate Program," Heartland Institute, January 1, 2006, www.heartland.org/publications/school%20reform/article/18217/Report_Finds_Fraud_Waste_and_Abuse_in_Federal_ERate_Program.html.

[39] "E-Rate Fraud: Connectivity at What Cost?", Education Reporter, March 6, 2006, www.eagleforum.org/educate/2006/mar06/e-rate.html.

[40] Antone Gonsalves, "Study: Internet Has No Impact on Student Performance," *InformationWeek*, November 21, 2005.

[41] "Your Guide to Securing 2011 e-Rate Dollars," *eSchool News*, October 18, 2010, www.eschoolnews.com/2010/10/18/your-guide-to-securing-2011-e-rate-dollars.

[42] "Internet World Statistics as of June 2010," www.internetworldstats.com/stats.htm (accessed May 26, 2011).

[43] Donald Bell, "Marvel Backs Ambitious $100 OLPC Tablet," *CNET*, May 27, 2010, http://news.cnet.com/8301-17938_105-20006211-1.html.

[44] Nick Barber, "OLPC XO-3 Tablet Delayed," *PCWorld*, November 3, 2010, www.pcworld.com/article/209631/olpc_xo3_tablet_delayed.html.

[45] "Nepal," Foundation Nepal, www.foundation-nepal.org/content/nepal (accessed May 26, 2011).

[46] "Nepal Unemployment," Index Mundi, www.indexmundi.com/nepal/unemployment_rate.html (accessed May 26, 2011).

[47] OLPC, "Open Learning Exchange Nepal Enters Into Its Fourth Year," May 23, 2011, www.olpcnews.com/countries/nepal/open_learning_exchange_nepal_e.html.

[48] Michael Kanellos, "Intel's Bridge for the Digital Divide, *CNET*, June 15, 2006, http://news.cnet.com/Intels-bridge-for-the-digital-divide/2100-1005_3-6084250.html.

[49] Intel Corporation, "Product Brief: The Classmate PC Powered by Intel," www.intel.com/intel/worldahead/pdf/CMPCbrochure.pdf.

[50] Chris Nuttall, "Lenovo to Sell Intel Classmate PC," *FT Tech Hub* (blog), *Financial Times*, March 10, 2011, http://blogs.ft.com/fttechhub/2011/03/intel-classmate-pc-to-be-sold-by-lenovo.

[51] Joanna Stern, "Intel and Lenovo Release the Classmate+ PC for the Kids," *engadget*, March 10, 2011, www.engadget.com/2011/03/10/intel-and-lenovo-release-the-classmate-for-the-kids/.

[52] "Eee Pad Tablet Transforms Into Laptop," *Fox News*, May 19, 2011, www.foxnews.com/scitech/2011/05/19/eee-pad-tablet-transforms-laptop.

[53] Janet Adamy, "Health Outlays Still Seen Rising," *Wall Street Journal*, September 8, 2010, http://online.wsj.com/article/SB10001424052748704362404575480161749608830.html.

[54] Noam N. Levey, "U.S. Healthcare Costs Projected to Continue to Climb," *Los Angeles Times*, September 9, 2010, http://articles.latimes.com/2010/sep/09/nation/la-na-health-costs-20100909.

[55] Len M. Nichols, "Can Defined Contribution Health Insurance Reduce Cost Growth?," Employee Benefit Research Institute Issue Brief, no. 246, June 2002, http://papers.ssrn.com/sol3/papers.cfm?abstract_id=318824.

[56] "How Changes in Medical Technology Affect Health Care Costs," The Henry J. Kaiser Foundation, March 2007, www.kff.org/insurance/snapshot/chcm030807oth.cfm.

[57] Jason Fodeman, M.D. and Robert A. Book, PhD, "Bending the Curve: What Really Drives Healthcare Spending," The Heritage Foundation, February 17, 2010, www.heritage.org/Research/Reports/2010/02/Bending-the-Curve-What-Really-Drives-Health-Care-Spending.

[58] Steve Lohr, "Health Industry Under Pressure to Computerize," *New York Times*, February 19, 2005.

[59] Rebecca Adams, "Washington Health Policy Week in Review Health IT Investments Will Top Industry Concerns in 2011," The Commonwealth Fund, December 20, 2010,

www.commonwealthfund.org/Content/Newsletters/Washington-Health-Policy-in-Review/2010/Dec/December-27-2010/Health-IT-Investments-Will-Top-Industry-Concerns-in-2011.aspx.

60 Knapp & Roberts, "Press Release: Preventable Medical Mistakes: A National Nightmare," *24-7 Press Release*, January 6, 2010, www.24-7pressrelease.com/press-release/preventable-medical-mistakes-a-national-nightmare-131199.php.

61 Amanda Frayer, "To Err is Human, To Delay Is Deadly," Consumers Union Safe Patient Project, May 1, 2009, www.safepatientproject.org/2009/05/to_err_is_human_to_delay_is_de.html.

62 Reuters, "US Pushes Digital Medical Records," *Boston Globe*, July 22, 2004.

63 Reuters, "US Pushes Digital Medical Records," *Boston Globe*, July 22, 2004.

64 "HITECH Act: The Health Information Technology for Economic and Clinical Health (HITECH) Act, ARRA Components," January 6, 2009," HITECH Survival Guide, www.hipaasurvivalguide.com/hitech-act-text.php.

65 "EHR Adoption Rate in US Hospitals to Rise Through 2014?," *InfoGrok Healthcare*, April 16, 2010, www.infogrok.com/index.php/prediction-healthcare/ehr-adoption-rate-in-us-hospitals-to-rise-through-2012.html.

66 "CMS EHR Meaningful Use Overview," U.S. Department of Health and Human Services, Centers for Medicare and Medicaid Services, www.cms.gov/EHRIncentivePrograms/30_Meaningful_Use.asp#TopOfPage (accessed June 19, 2011).

67 Lucas Mearian, "Only 10% of Doctors Using Complete e-Health Record Systems, Survey Finds," *Computerworld*, January 18, 2011, www.computerworld.com/s/article/9205238/Only_10_of_doctors_using_complete_e_health_record_systems_surveys_find.

68 "Rock and a Hard Place: An Analysis of the $36 Billion Impact from Health IT Spending," PricewaterhouseCoopers' Health Research Institute, www.pwc.com/us/en/healthcare/publications/rock-and-a-hard-place.jhtml.

69 Lucas Mearian, "Only 10% of Doctors Using Complete e-Health Record Systems, Survey Finds," *Computerworld*, January 18, 2011, www.computerworld.com/s/article/9205238/Only_10_of_doctors_using_complete_e_health_record_systems_surveys_find.

70 "HITECH Act: The Health Information Technology for Economic and Clinical Health (HITECH) Act, ARRA Components, January 6, 2009," HITECH Survival Guide, www.hipaasurvivalguide.com/hitech-act-text.php.

71 "Physician Use of HIT in Hospitals Linked to Fewer Deaths and Complications, Lower Costs," *e! Science News*, January 27, 2009, http://esciencenews.com/articles/2009/01/27/physician.use.hit.hospitals.linked.fewer.deaths.and.complications.lower.costs.

72 Cheryl Clark, "EHR Effectiveness for Hospital Care Questioned," *Health Leaders Media*, December 29, 2010, www.healthleadersmedia.com/content/TEC-260743/EHR-Effectiveness-for-Hospital-Care-Questioned.

73 Lauran Neergaard, "Study: Doctors Resist Use of Electronic Medical Records," *SignOnSanDiego.com*, September 13, 2005, www.signonsandiego.com/news/health/20050913-2252-computerizedmedicine.html.

[74] Lauran Neergaard, "Study: Doctors Resist Use of Electronic Medical Records," *SignOnSanDiego.com*, September 13, 2005, www.signonsandiego.com/news/health/20050913-2252-computerizedmedicine.html.

[75] John Commins, "Hospital CIOs: EHR Carrot Too Small, Stick Too Big," *HealthLeaders Media*, April 16, 2009, www.healthleadersmedia.com/content/231613/topic/WS_HLM2_TEC/Hospital-CIOs-EHR-Carrot-Too-Small-Stick-Too-Big.html.

[76] "Telemedicine for Maintaining Health of Oil Rig Workers," *Science Daily*, August 9, 2010, www.sciencedaily.com/releases/2010/08/100809171529.htm.

[77] Neil Versel, "Mobile Telemedicine System Beams Live Video from Ambulance to Cincinnati Children's," Fierce Mobil Healthcare, August 17, 2010, www.fiercemobilehealthcare.com/story/mobile-telemedicine-system-beams-live-video-ambulance-cincinnati-childrens/2010-08-17.

[78] "Press Release: AMD Global Telemedicine Inc. Chosen by InterHealth Canada as the Telemedicine Provider for Turks and Caicos Hospital Projects," *PRWeb*, May 13, 2010, www.prweb.com/releases/2010/05/prweb3992104.htm.

[79] Pauline W. Chen, M.D., "Are Doctors Ready for Virtual Visits," *New York Times*, January 7, 2010, www.nytimes.com/2010/01/07/health/07chen.html.

[80] Robert Wilonsky, "DOJ, FCC Settle with HP (for $16.25 Mil!) Over E-Rate Fraud in Dallas, Houston ISDs," *Crime* (blog), *Dallas Observer*, November 10, 2010, http://blogs.dallasobserver.com/unfairpark/2010/11/doj_fcc_settle_with_hp_for_162.php.

[81] Chloe Albanesius, "HP to Pay $16.25M to Settle E-Rate Fraud Charges," *PC Magazine*, November 10, 2010, www.pcmag.com/article2/0,2817,2372488,00.asp.

[82] Chloe Albanesius, "HP to Pay $16.25M to Settle E-Rate Fraud Charges," *PC Magazine*, November 10, 2010, www.pcmag.com/article2/0,2817,2372488,00.asp.

[83] Lynn Walsh, "Gift-Giving Culture Flourished at HISD; Vendors Lavished Cash, Dinners and Tickets on Employees," July 20, 2010, www.texaswatchdog.org/taxonomy/term/143.

[84] Lynn Walsh, "Hewlett-Packard To Pay $16M+ As More Problems Come Out in HISD, DISD E-Rate Program," *Texas Watchdog*, November 11, 2010, www.texaswatchdog.org/2010/11/erate/1289514507.column.

[85] Grant Gross, "Dallas School District Settles E-Rate Fraud Case," *CIO*, June 26, 2009, www.cio.com/article/496107/Dallas_School_District_Settles_E_Rate_Fraud_Case.

[86] "Whistleblower Realtors Dan Cain, Pamela Tingley, Dave Richardson & Dave Gillis Win HP E-Rate Program Settlement," *Outlook Series*, November 11, 2010, www.outlookseries.com/A0999/Financial/3801_Whistleblower_Relators_Dan_Cain_Pamela_Tingley_Dave_Richardson_Dave_Gillis_Win_HP_E-Rate_Program_Settlement.htm.

[87] Steven Burke, "HP to Require Training for Partners in Wake of $16M E-Rate Settlement," *CRN*, November 11, 2010, www.crn.com/news/channel-programs/228200787/hp-to-require-training-for-partners-in-wake-of-16m-e-rate-settlement.htm;jsessionid=l9wooL8lxDxX3P1Z8wh66Q.ecappj03.

88 Steven Burke, "HP to Require Training for Partners in Wake of $16M E-Rate Settlement," *CRN*, November 11, 2010, www.crn.com/news/channel-programs/228200787/hp-to-require-training-for-partners-in-wake-of-16m-e-rate-settlement.htm;jsessionid=I9wooL8IxDxX3P1Z8wh66Q**.ecappj03.

89 L. Sato, "Evidence-Based Patient Safety and Risk Management Technology," *Journal of Quality Improvement* 2001; 27:435.

90 R. Phillips, L. Bartholomew, S. Dovey, et al., "Learning from Malpractices Claims About Negligent, Adverse Events in Primary Care in the United States," *Quality and Safety in Health Care* 2004; 13: 121–126.

91 Gary Kantor, M.D., "Guest Software Review: Isabel Diagnosis Software," *HISTalk* (blog), January 31, 2006, http://histalk.blog-city.com/guest_software_review_isabel_diagnosis_software_by_gary_kant.htm.

92 "Watson Computer Prepped for Medicine," *Newsmaxhealth.com*, May 23, 2011, www.newsmaxhealth.com/health_stories/Watson_Jeopardy_Medicine/2011/05/23/390166.html.

93 Isabel Healthcare, "Press Release: United Hospital Adopts Isabel Healthcare for Its Diagnosis Decision Support," *PRWeb*, March 22, 2011, www.prweb.com/releases/2011/3/prweb8211181.htm.

94 "Watson Computer Prepped for Medicine," *Newsmaxhealth.com*, May 23, 2011, www.newsmaxhealth.com/health_stories/Watson_Jeopardy_Medicine/2011/05/23/390166.html.

95 George J. Annas, "The Patient's Right to Safety—Improving the Quality of Care Through Litigation Against Hospitals," *New England Journal of Medicine*, May 11, 2006; 354(19): 2063–6.

96 Tween Brands, Inc. "Tween Brands 2008 Annual Report," www.annualreports.com/Company/4984 (accessed June 19, 2011).

97 "Web-Enabled Fulfillment System Increases Productivity 25% at Tween Brands," *Internet Retailer*, January 8, 2005, www.internetretailer.com/2007/01/08/web-enabled-fulfillment-system-increases-productivity-25-at-twe.

98 Erik Brynjolfsson, "The IT Productivity Gap," *InformationWeek's Optimize 21* (July 2003).

99 Paul A. Strassman, *Information Productivity: Assessing the Management Costs of U.S. Industrial Corporations* (New Canaan, CT: Information Economics Press, 1999).

100 Paul A. Strassman, *Information Productivity: Assessing the Management Costs of U.S. Industrial Corporations* (New Canaan, CT: Information Economics Press, 1999).

CHAPTER 9

SOCIAL NETWORKING

QUOTE

The difference between PR [public relations] and social media is that PR is about positioning, and social media is about becoming, being, and improving.
—Chris Brogan, author of *Trust Agents: Using the Web to Build Influence, Improve Reputation, and Earn Trust*

VIGNETTE

Facebook Raises Privacy Issues

Since its founding in 2004, Facebook has changed dramatically—expanding from a tiny start-up firm to an Internet powerhouse, with an estimated market value of over $50 billion and earnings in excess of $2 billion in 2011.[1]

Facebook has also changed the way it manages its users' privacy—going from being a private Web site where you could communicate with just the people you chose to becoming a forum where much of your information is made public. An excerpt from the earliest Facebook privacy policy reads: *No personal information that you submit to Facebook will be available to any user of the Web site who does not belong to at least one of the groups specified by you in your privacy settings.*[2] Contrast this with an excerpt from Facebook's privacy policy in 2011: *When you connect with an application or website it will have access to General Information about you. The term General Information includes you and your friends' names, profile pictures, gender, user IDs, connections, and any content shared using the Everyone privacy setting....*

The default privacy setting for certain types of information you post on Facebook is set to "everyone."[3] This change in its approach to privacy is causing Facebook to undergo increasing scrutiny of its policies and actions in terms of safeguarding the information of its more than 750 million active users.[4]

Most of the Facebook applications, or *apps*, that enable users to play games and share common interests are created by independent software developers. It was recently uncovered that many of these apps (e.g., FarmVille, Texas HoldEm, and FrontierVille) were transmitting users' unique Facebook user IDs to at least two dozen marketing and database firms, where they were used to build profiles on users' online activities.[5]

Because a Facebook user ID is a public part of all Facebook profiles, knowledge of user IDs enables a company to determine the users' names even if they set all their Facebook information to be private; companies can simply perform a search using the ID to find the person's name. The user IDs may also reveal the users' age, address, and occupation, and can allow a company to access photos for those users who did not specify the most restrictive privacy setting. At least one of the companies linked the Facebook data to its own database of Internet user's data, which it sells to other marketers.

Interestingly, Facebook prohibits app developers from sharing data about users to outside marketing and database companies. However, it has been difficult for Facebook to enforce its rules for the more than half million apps that run on its Web site. Facebook itself was caught transmitting user ID numbers under certain conditions when users clicked on an ad; Facebook discontinued the practice after it was reported in the press.[6]

In December 2010, Facebook implemented a new Tag Suggestions feature for photos. When users add photos to their Facebook pages, the Tag Suggestions feature uses facial recognition software to suggest the names of people in the photos based on photos in which they have already been tagged. It is estimated that more than 100 million people tag photos every day on Facebook.[7] Although users could always tag photos of their Facebook "Friends," the Tag Suggestions feature process is now semiautomated, with Facebook providing suggestions regarding which of your friends is in a photo.[8]

Initially the feature was made available to just users in the United States, perhaps because the privacy laws in many other countries are much stricter. However, within five months, Facebook began to rollout the Tag Suggestions feature worldwide. European Union data protection regulators are now studying the new feature for possible privacy rule violations.[9]

Facebook also changed its users' privacy settings to make the facial recognition feature a default; users must opt out of having their names suggested for photos by changing their privacy settings to disable the feature. However, Facebook does not give its users the option to not be tagged in any photos; a user's Facebook Friends can still tag a user in a photo manually, even if the user has disabled the Tag Suggestions feature. Users who do not want their name associated with a photo must manually "untag" themselves in each photo. This approach of automatically enrolling users into new features without their knowledge or consent has become standard practice for the firm as a means of ensuring that users experience the full effects of Facebook.[10]

Google had developed similar facial recognition technology for smartphones for its Google Goggles application, but did not release the facial recognition portion of that app. Google chairman Eric Schmidt said: "As far as I know, it's the only technology that Google built and after looking at it, we decided to stop. People could use this stuff in a very, very bad way as well as in a good way."[11]

Questions to Consider

1. Do you agree with Facebook's philosophy of automatically enrolling users in new features without their knowledge or consent? Why or why not?
2. What concerns might a Facebook user have with the new Tag Suggestions feature?

LEARNING OBJECTIVES

As you read this chapter, consider the following questions:

1. What are social networks, how do people use them, and what are some of their practical business uses?
2. What are some of the key ethical issues associated with the use of social networking Web sites?
3. What is a virtual life community, and what are some of the ethical issues associated with such a community?

WHAT IS A SOCIAL NETWORKING WEB SITE?

A **social networking Web site** is a site that creates an online community of Internet users that enables members to break down barriers created by time, distance, and cultural differences. Social networking Web sites allow people to interact with others online by sharing opinions, insights, information, interests, and experiences. Members of an online social network may use the site to interact with friends, family members, and colleagues—people they already know—but they may also make use of the site to develop new personal and professional relationships.

With over 2 billion Internet users, there is an endless range of interests represented online, and a correspondingly wide range of social networking Web sites catering to those interests.[12] There are thousands of social networking Web sites worldwide; Table 9-1 lists some of the most popular ones, based on the number of unique visitors per month.

TABLE 9-1 Popular social networking Web sites

Social networking Web site	Description	Estimated unique monthly visitors
Facebook	Social networking site for keeping up with friends, uploading photos, sharing links and videos, and meeting new people online	700 million
Twitter	A real-time information service for friends, family members, and coworkers looking to stay connected through the exchange of messages that are a maximum of 140 characters	200 million
LinkedIn	Business-oriented social networking site used for professional networking; users create a network made up of people they know and trust in business.	100 million
MySpace	General social networking Web site used by teenagers and adults worldwide; allows members to communicate with friends via personal profiles, blogs, and groups, as well as to post photos, music, and videos to their personal pages	80.5 million

(Continued)

Social networking Web site	Description	Estimated unique monthly visitors
Ning	Platform that enables users to create their own social network following a simple process to name the network, choose a color scheme, and allow for unique profile questions; serves as a portal to access tens of thousands of user-created social networks	60 million
Tagged	Social network with a focus on helping members meet new people; suggests new friends based on shared interests; allows members to browse people, share tags and virtual gifts, and play games	25 million
Google Plus[13]	Social network operated by Google that integrates social services such as Google Profiles and Google Buzz, and introduces new services such as Circles (enables users to organize contacts into groups for sharing), Hangouts (URLs used to facilitate group video chat), Sparks (enables users to identify topics in which they are interested), and Huddles (allows instant messaging within Circles)	25 million

Source Line: "Top 15 Most Popular Social Networking Websites/July 2011," eBiz/MBA, www.ebizmba.com/articles/social-networking-websites. "Google Plus Reaches 25 Million Users, Activity Declines," Search Engine Journal, © August 3, 2011, www.searchenginejournal.com/google-plus-reaches-25-million-users-activity-declines/31500.

357

The market research company Nielsen estimates that 22 percent of all time spent online is spent on social networking and blog sites, with the average visitor spending almost six hours per month on such sites. Brazil has the highest percentage of Internet consumers visiting social networking and blog sites—at 86 percent. Australian Internet users spend the most time on social networking and blog sites, with an average of 7 hours and 19 minutes per month.[14] Facebook is the most popular social network destination worldwide; the average Facebook user logs in 19 times per month and spends nearly 6 hours on the site each month. Worldwide, there are over 314.5 million social network users.[15] Although social networking is most popular among younger people, its popularity is increasing most rapidly with those aged 50 and older, whose use jumped from 22 percent in May 2009 to 42 percent in May 2010. Increasingly, these people use social networking to communicate and share photos with grandchildren and to reconnect with people from their past.[16]

BUSINESS APPLICATIONS OF ONLINE SOCIAL NETWORKING

Although social networking Web sites are primarily used for nonbusiness purposes, a number of forward-thinking organizations are employing this technology to advertise, assess job candidates, and sell products. There is an increase in the number of business-oriented social networking sites designed to encourage and support relationships with consumers, clients, potential employees, suppliers, and business partners around the world, as shown in Table 9-2.

TABLE 9-2 Popular business-oriented social networks

Social networking Web site	Description
BT Tradespace	Site that brings businesses and customers together to buy, sell, share, and do business with people they can get to know and trust
Huddle	Networking site that can be used to manage projects, share files, and collaborate with people both inside and outside a company in a secure manner
LinkedIn	Site focused on helping professionals build and maintain a list of contacts; frequently used by job seekers and recruiters alike to link professionals with job opportunities
Ryze	Professional networking site targeted specifically at entrepreneurs, enabling them to build up a personal network and find new jobs; companies can also use Ryze to create a business community.
Viadeo	Site that identifies useful contacts through members' extended networks to enable the sharing of knowledge and to identify and recruit talent
Xing	Social network platform that helps professionals build and maintain a list of contacts; frequently used by job seekers and recruiters alike to link professionals with job opportunities

Source Line: Course Technology/Cengage Learning.

Social Network Advertising

Social network advertising involves the use of social networks to communicate and promote the benefits of products and services. Social network advertising has become big business in the United States. It is estimated that U.S. marketers will spend just over $3 billion to advertise on social networking sites in 2011. Worldwide, social network ad spending is estimated to be just under $6 billion, and ad spending on Facebook alone is estimated at $4 billion worldwide for 2011.[17]

Organizations are increasingly looking to new forms of advertising to reach their target markets. Two significant advantages of social networking advertising over more traditional advertising media (e.g., radio, TV, and newspapers) are as follows: (1) advertisers can create an opportunity to generate a conversation with viewers of the ad, and (2) ads can be targeted to reach people with the desired demographic characteristics.

There are several social network advertising strategies, and organizations may employ one or more of the strategies detailed in the following sections.

Direct Advertising

Direct advertising involves placing banner ads on a social networking Web site. An ad can either be displayed to every visitor to the Web site, or, by using the information in user profiles, an ad can be directed toward those members who would likely find the product most appealing. Thus, an ad for a new magazine on mountain biking could be directed to individuals on a social networking Web site who are male, who are 18 to 35 years old, and who express an interest in mountain biking. Others on the social networking Web site would not see the ad.

Advertising Using an Individual's Network of Friends

Companies can use social networking Web sites to advertise to an individual's network of contacts. When you sign on to your favorite social networking Web site, you might see a message saying, "Jared [your friend] just went to see *Hangover III*—awesome, he says!" This can be an extremely persuasive message, as people frequently make decisions to do something or purchase something based on input from their close group of friends. This might be a spontaneous message sent by Jared, or Jared might be getting paid by an online promotion firm to send messages about certain products. There are many ethical issues with this approach, as some people consider this to be exploiting an individual's personal relationships for the financial benefit of a company.

Indirect Advertising Through Groups

Innovative companies are also making use of a new marketing technique by creating groups on social networking Web sites that interested users can join by becoming "fans." These groups can quickly grow in terms of numbers of fans to become a marketing tool for a company looking to market contests, promote new products, or simply increase brand awareness. Often, though, the fans gained in this manner do not remain loyal and are simply interested in earning discounts or special promotions.

In a survey by digital marketing company Razorfish focusing on online consumer behavior, 40 percent of the 1,000 respondents indicated they had "friended" brands on Facebook, and some 25 percent said they followed brands on Twitter. Their main reason for doing so was to access exclusive deals or offers.[18] However, another survey found that 40 percent of consumers who visit Facebook at least once a month and are a "fan" of at least one company or brand don't believe that marketers are welcome in social networks.[19] In order to overcome this barrier, marketers must be extremely careful in how they present themselves. One effective approach for a company can be to use its expertise to meet a consumer need. For example, Betty Crocker discovered that many people needed help baking a cake, so the company capitalized on this consumer need and created cake-making videos that it posted on the company's Web site—thus using the brand's knowledge and product without aggressively pushing the brand.

In its ongoing fight for market share in the beverage industry, Coca-Cola has implemented a number of social networking initiatives to promote its brands, including the following:

- Coke has its own corporate blog called *Coca-Cola Conversations* that covers its brand history and provides information about Coca-Cola collectibles.
- Coke started a competition for the residents of the Second Life virtual world, challenging them to design a vending machine that dispenses the essence of Coke.
- The company placed a video on YouTube called "Mean Joe Greene—The Making of the Commercial," documenting the making of one of Coke's most famous TV commercials.

- Two fans of Coke, neither of whom had any official connection to the company, launched a Coca-Cola Facebook page in 2008. Within a few weeks, the page had attracted over 750,000 fans. As the number of fans grew into the millions, the page's creators agreed to turn over administration of the page to Coca-Cola.[20] The site is monitored by software filters for offensive words and phrases, and live moderators check its pages for anything truly offensive. Other than that, Coca-Cola managers generally let Facebook fans say what they want on the site.[21] The result has been nothing short of amazing. The Facebook fan page quickly grew to over 3 million members worldwide.[22]
- In late 2010, Coca-Cola launched a nationwide campaign called Coke Secret Formula using SCVNGR (a social location-based gaming platform for mobile phones); the campaign encouraged young teens to look for hidden shopping experiences at shopping malls in the form of check-ins and photos. As participants completed challenges, they earned points toward American Express gift cards and Coke-branded merchandise.[23]

Company-Owned Social Networking Web Site

A variation on the above approach is for a company to form its own social networking Web site. Dell created its own social networking Web site, IdeaStorm, in February 2007 as a means for its millions of customers in more than 100 countries to talk about what new products, services, or improvements they would like to see Dell develop. As of July 2011, the IdeaStorm community has suggested 15,838 ideas and posted 92,461 comments; Dell has implemented 446 customer-submitted ideas.[24]

Viral Marketing

Viral marketing encourages individuals to pass along a marketing message to others, thus creating the potential for exponential growth in the message's exposure and influence as one person tells two people, each of those two people tell two or three more people, and so on. The goal of a viral marketing campaign is to create a buzz about a product or idea that spreads wide and fast. A successful viral marketing campaign requires little effort on the part of the advertiser; however, the success of such campaigns can be very difficult to predict.

Hotmail created what is recognized by many as the most successful viral marketing campaign ever when it first launched its service in 1996. Every email sent by a Hotmail user contained a short message at the end of the email that promoted Hotmail's free email service. As a result, almost 12 million new users signed up for Hotmail over a period of 18 months.[25]

The Use of Social Networks in the Hiring Process

A 2011 survey found that 89 percent of respondents are already using some form of social media in their recruiting or plan to in the next year.[26] In addition, employers can and do look at the social networking profiles of job candidates when making hiring decisions (see Figure 9-1). Some human resource executives feel that they can use social networking Web sites to "learn a little about the candidate's cultural fit and professionalism."[27] According to a survey by CareerBuilder.com, 22 percent of hiring managers use social networking Web sites as a source of information about candidates, and an additional 9 percent are planning to do so.

Of those managers who use social networking Web sites to screen candidates, 34 percent have found information that made them drop a candidate from consideration. Companies may reject candidates who post information about their drinking or drug use habits or those who post provocative or inappropriate photos. Candidates are also sometimes rejected due to postings containing discriminatory remarks relating to race, gender, and religion or because of postings that reveal confidential information from previous employers.[28]

FIGURE 9-1 Employers often use social networking Web sites as a source of information about job candidates

Credit: Image copyright Drazen Vukelic, 2009. Used under license from Shutterstock.com.

Members of social networking Web sites frequently provide sex, age, marital status, sexual orientation, religion, and political affiliation data in their profile. Users who upload personal photos may reveal a disability or their race or ethnicity; therefore, without even thinking about it, an individual may have revealed data about personal characteristics that are protected by civil rights legislation. Employers can legally reject a job applicant based on the contents of the individual's social networking profile only if the company is not violating federal or state discrimination laws. For example, an employer cannot legally screen applicants based on race or ethnicity. Or suppose that by checking a social networking Web site, a hiring manager finds out that a job candidate is pregnant and makes a decision not to hire that person based on that information. That employer would be at risk of a job employment discrimination lawsuit because refusing to hire on the basis of pregnancy is prohibited by the Pregnancy Discrimination Act, which amended Title VII of the Civil Rights Act of 1964.

Another survey done by CollegeGrad.com revealed that 47 percent of college graduates who use social networking Web sites change the contents of their pages as a result of their job search.[29] Many jobseekers delete their Facebook or MySpace account altogether because they know employers check such sites. More graduates are beginning to realize that pictures and words posted online, once intended for friends only, are reaching a much larger audience and could have an impact on their job search.[30]

The Use of Social Media to Improve Customer Service

In the past, companies relied heavily on their market research and customer service organizations to provide them with insights into what customers think about their products and services. Many consumer goods companies put toll-free 800 phone numbers on their products so that consumers could call in and speak with trained customer service reps to share their comments and complaints.

Increasingly, consumers are using social networks to share their experiences, both good and bad, with others. They also seek help and advice on how to use products more effectively and how to deal with special situations encountered when using a product. Unless organizations monitor social networks, their customers are left to resolve their issues and questions on their own, often in ways that are not ideal. The end result can be dissatisfaction with the product and loss of customers and future sales.

Even though one survey shows that 70 percent of marketers have very little understanding of how their brands are faring on social media,[31] some progressive companies are focusing more resources on monitoring issues and assisting customers via social networks. One of the major challenges with these efforts is in filtering the few nuggets of actionable data from the volumes of chatter and converting these key findings into useful business actions. Here are a few examples of companies whose efforts have been effective in this area:

- Jet Blue has a customer service group that monitors Twitter on a 24×7 basis and responds immediately to tweets about operational problems. For example, social media monitors will call gate agents to suggest more information be announced if a customer tweets about a lack of information regarding a flight delay.[32]
- When browsing online media, most readers rely solely on an article headline when deciding whether or not to read on. Knowing this, *The Huffington Post* initially shows two headlines for the same story, and after five minutes of monitoring readers' reactions, they discard the less popular one.[33]
- In 2009, GM formed a new customer service center including five U.S. call centers that communicate with over 25,000 dealers and customers each day. The team also searches social media Web sites such as Facebook and Twitter looking for negative comments about the firm and its products and offering solutions to problems when possible.[34]
- Every day, Dell uses social media monitoring tools to track some 25,000 social media conversations that mention the brand, and the company participates in as many of these conversations as possible. Dell's goal is to listen to consumers, act on what it learns to introduce new products, improve its marketing, and provide better customer support. All this helps to retain customers and increase future sales.[35]

Social Shopping Web Sites

A **social shopping Web site** brings shoppers and sellers together in a social networking environment in which members can share information and make recommendations while shopping online. Thus, these sites combine two highly popular online activities—shopping and social networking. Social shopping Web site members can typically build their own pages to collect information and photos about items in which they are interested. On many social shopping Web sites, users can offer opinions on other members' purchases or potential purchases. The social shopping Web site Stuffpit has implemented a reward system for members, in which they are paid a commission each time another shopper acts on their recommendation to purchase a specific item.[36]

There are numerous social shopping Web sites, a few of which are summarized in Table 9-3.

TABLE 9-3 Sample of social shopping Web sites

Social shopping site	Description
Buzzillions	Product review Web site with over 15 million reviews across a wide range of products, with product rankings based on feedback from customers
Crowdstorm	Price comparison shopping resource that aggregates product information from various online buyers guides, reviews, and blog postings
JustBoughtIT!	Facebook and Twitter app for capturing product recommendations from the online community; users can post a photo or screenshot online, share their purchases, and comment on what others are buying.
Kaboodle	Social shopping site where members can discover, recommend and share new products, provide advice, share feedback, get discounts, and locate bargains
MyDeco	Site with a focus on interior design and home décor; users can mock up virtual rooms using their favorite products
OSOYOU	UK-based social shopping site for women with an interest in fashion and beauty products

Source Line: Course Technology/Cengage Learning.

Social shopping Web sites generate revenue through retailer advertising. Some also earn money by sharing with retailers data about their members' likes and dislikes.

Social shopping Web sites can be a great way for small businesses to boost their sales. Amenity Home—a tiny start-up with just three products, four employees, and no advertising budget—became a retailer on ThisNext.com, a social shopping Web site whose goal is to link shoppers with hard-to-find products. Shoppers at ThisNext.com found the Amenity Home products, copied photos of the products to their own blog pages, and brought the tiny firm some much-needed recognition. Amenity Home products started getting more and more hits on ThisNext.com, and the company has continued to grow and add more products to its online offerings.[37]

Retailers can purchase member data and comments from some social shopping Web sites to find out what consumers like and don't like and what they are looking for in items sold by the retailer. This can help the retailer design product improvements and come up with ideas for new product lines.

SOCIAL NETWORKING ETHICAL ISSUES

When you have a community of tens of millions of users, not everyone is going to be a good "neighbor" and abide by the rules of the community. Many will stretch or exceed the bounds of generally accepted behavior. Some common ethical issues that arise for members of social networking Web sites are cyberbullying, cyberstalking, encounters with sexual predators, and the uploading of inappropriate material.

Cyberbullying

Cyberbullying is the harassment, torment, humiliation, or threatening of one minor by another minor or group of minors via the Internet or cell phone. A 2010 survey of 4,441 students between the ages of 10 and 18 in 37 different schools showed that 20.8 percent had experienced cyberbullying in their lifetime. Adolescent girls are more likely than boys to have experienced cyberbullying (25.1% of girls versus 16.6% of boys); see Figure 9-2. Boys are more likely to post harmful pictures or videos while girls are more likely to spread rumors about others.[38]

FIGURE 9-2 Cyberbullying is more common among teenage females

Credit: Image copyright Ana Blazic, 2009. Used under license from Shutterstock.com.

Cyberbullying has sometimes become so intense that some children have committed suicide as a result:

- Ryan Halligan, a 13-year-old boy in Vermont, committed suicide after bullying from his schoolmates and cyberbullying online.[39]
- Megan Meier, a 13-year-old girl from Missouri, also committed suicide after she was harassed online by a fictitious boy created by neighbors with whom she had had a falling out.[40]

- Jeffrey Johnston, a boy from Florida, was bullied for three years both in school and online before finally committing suicide at age 15.[41]
- Phoebe Prince, a 15-year old Massachusetts girl, committed suicide after three months of incessant cyberbullying by high school classmates via text messages and Facebook.[42]

Even with all these incidents of cyberbullying, only seven states have enacted cyberbullying laws as of July 2011. Most laws directed against cyberbullying focus on the school system as the most effective force and call on the school districts to develop policies regarding cyberbullying detection and punishment.[43]

There are numerous forms of cyberbullying, such as the following:

- Sending mean-spirited or threatening messages to the victim
- Sending thousands of text messages to the victim's cell phone and running up a huge cell phone bill
- Impersonating the victim and sending inappropriate messages to others
- Stealing the victim's password and modifying his or her profile to include racist, homophobic, sexual, or other inappropriate data that offends others or attracts the attention of undesirable people
- Posting mean, personal, or false information about the victim in the cyberbully's blog or on a social networking page
- Creating a Web site or social networking profile whose purpose is to humiliate or threaten the victim
- Taking inappropriate photos of the victim and either posting them online or sending them to others via cell phone
- Setting up an Internet poll to elicit responses to embarrassing questions, such as "Who's the biggest geek in Miss Adams's homeroom?" and "Who is the biggest loser in the senior class?"
- Sending inappropriate messages while playing interactive games that enable participants to communicate with one another

Because cyberbullying can take many forms, it can be difficult to identify and stop. Ideally, minors would inform their parents if they became a victim of cyberbullying. Unfortunately, this often does not happen. When school authorities do get involved in an effort to discipline students for cyberbullying, they are sometimes sued for violating the student's right to free speech, especially if the activity occurred off school premises. As a result, some schools have modified their discipline policy to reserve the right to punish a student for actions taken off school premises if they adversely affect the safety and well-being of a student while in school.

All children should be educated about the potential serious impacts of cyberbullying, how to identify cyberbullying, and why it is important for them to refrain from cyberbullying. Children should be encouraged not to retaliate to mean-spirited messages, as doing so may cause the harassment to increase. Children need to understand that they can become inadvertent cyberbullies if they fail to think through the consequences of their actions. They should also be counseled against posting any data that is too personal, such as phone numbers, their home address, their school, or any other information that could allow a stranger to locate the child.

Cyberstalking

Cyberstalking is threatening behavior or unwanted advances directed at an adult using the Internet or other forms of online and electronic communications; it is the adult version of cyberbullying. Online stalking can be a serious problem for victims, terrifying them and causing mental anguish. It is not unusual for cyberstalking to escalate into abusive or excessive phone calls, threatening or obscene mail, trespassing, vandalism, physical stalking, and even physical assault. Over three dozen states have passed laws prohibiting cyberstalking.[44] Internet safety groups such as Working to Halt Online Abuse and Cyber Angels have reported increasing numbers of cyberstalking reports and requests for help from victims. In addition, many researchers feel that it is likely that the true extent of cyberstalking has been underestimated because the number of people online is increasing each year, and many cases still go unreported.[45] Although current federal statutes address some forms of cyberstalking, there are large gaps in current federal and state law (see Table 9-4).

TABLE 9-4 Federal laws addressing cyberstalking

Federal law	Scope	Gap
18 U.S.C. 2425	Protects children against online stalking by making it a federal crime to communicate with any person with the intent to solicit or entice a child into unlawful sexual activity	Does not address harassing phone calls to minors absent a showing of intent to entice or solicit the child for illicit sexual purposes
18 U.S.C. 2261A	Makes it a federal crime for any person to travel across state lines with the intent to injure or harass another person	The requirement that the stalker physically travel across state lines makes the law largely inapplicable to cyberstalking cases.
47 U.S.C. 223	Makes it a federal crime to use a telephone or telecommunications device to annoy, abuse, harass, or threaten any person at the called number	Applies only to direct communications between the perpetrator and the victim; thus, it would not address a cyberstalking situation in which a person harasses or terrorizes another person by posting messages on a bulletin board or in a chat room encouraging others to harass or annoy another person.
18 U.S.C. 875(c)	Makes it a federal crime to transmit any communication in interstate or foreign commerce containing a threat to injure another person	Applies only to communications of actual threats; thus, it would not apply in a situation where a cyberstalker engaged in a pattern of conduct intended to harass or annoy another (absent some threat). It is also not clear if it would apply to situations in which a person harasses or terrorizes another by posting messages on a bulletin board or in a chat room encouraging others to harass or annoy another person.

Source Line: "1999 Report on Cyberstalking: A New Challenge for Law Enforcement and Industry," U.S. Department of Justice, www.justice.gov/criminal/cybercrime/cyberstalking.htm.

In 2010, a Maine man was arrested for posting provocative solicitations on social networking sites that were supposedly from his ex-girlfriend, inviting men to her home or workplace to engage in sexual activities. The postings lasted for over four years and continued even after the ex-girlfriend changed her name and moved out of state. The woman had to take steps to counter what the man was posting. She left notes on her door saying the Internet invitations were fake and devised special knocks for her friends so she knew when it was safe to open the door. Tracking down who was responsible for the posts was difficult because the perpetrator used various public wireless networks, rather than posting the solicitations from his home. The man faces significant jail time for violating a previous protection-from-abuse order and because some of the alleged activity crossed state lines.[46]

The National Center for Victims of Crime offers a detailed set of recommended actions for combating cyberstalking, including the following:

- When the offender is known, victims should send the stalker a written notice that their contact is unwanted and that all further contact must cease.
- Evidence of all contacts should be saved.
- Victims of cyberstalking should inform their ISP provider as well as the stalker's ISP, if possible.
- Victims should consider speaking to law enforcement officers.
- Above all else, victims of cyberstalking should never agree to meet with the stalker to "talk things out."[47]

Encounters with Sexual Predators

Some social networking Web sites have been criticized for not doing enough to protect minors from encounters with sexual predators. MySpace spent two years purging potential problem members from its site, including 90,000 registered sex offenders banned from its site in early 2009. (It is estimated that there are over 700,000 sex offenders in the United States).[48] "Almost 100,000 convicted sex offenders mixing with children on MySpace ... is absolutely appalling and totally unacceptable," stated Connecticut's then–Attorney General Richard Blumenthal, who pushed social networking Web sites to adopt stronger safety measures.[49]

A surprising 2009 report released by the Internet Safety Technical Task Force—which includes 49 state attorneys general—concluded that "social network sites do not appear to have increased the overall risk for solicitation" or exposure to sexual predators. However, not all members of the task force supported this conclusion, including Richard Blumenthal, who stated, "Children are solicited every day online. Some fall prey, and the results are tragic. That harsh reality defies the statistical academic research underlying the report."[50] Meanwhile, Ernie Allen, CEO of the National Center for Missing and Exploited Children, argues that the explosive use of social networks, mobile devices, and online games has meant that there are "more opportunities for potential offenders to engage with children."[51]

Uploading of Inappropriate Material

Most social networking Web sites have policies against uploading videos depicting violence or obscenity. Facebook, MySpace, and most other social networking Web sites have terms of use agreements, a privacy policy, or a content code of conduct that summarizes key legal aspects regarding use of the Web site. Typically, the terms state that the Web site has the right to delete material and terminate user accounts that violate the site's policies. The policies set specific limits on content that is sexually explicit, defamatory, hateful, or violent, or that promotes illegal activity.

Policies do not stop all members of the community from attempting to post inappropriate material, and most Web sites do not have sufficient resources to review all material submitted for posting. For example, more than 48 hours of video are uploaded to YouTube every minute (this is the equivalent of 240,000 full-length movies each week).[52] Quite often, it is only after other members of a social networking Web site complain about objectionable material that such material is taken down. This can be days or even weeks. Ideally, reviewers would also look at the text content submitted to a networking site—not just photos and videos. A posting to a teenage-oriented Web site may advocate underage drinking, sex, and drug use but may not use photos or videos to do so.

Individuals who appear in photos or videos doing inappropriate or illegal things may find themselves in trouble with authorities if those photos and videos end up on the Internet:

- U.S. Representative Anthony Weiner was forced to resign in 2011 after admitting he sent sexually explicit photos of himself to several women using social networks including Facebook and Twitter.[53]
- Seven Pennsylvania students filmed themselves beating and attempting to hang another 13-year old student. (The term "wolfpacking" is used to describe such acts of violence by gangs of bullies.) The students were arrested for alleged kidnapping and assault.[54]
- A brutal beating of a one-year-old pit bull was captured on a cell phone camera and posted on YouTube. Police were able to use the video to quickly track down the person responsible for the beating. The individual was arrested and charged with animal cruelty. The dog was taken into custody by the Society for the Prevention of Cruelty to Animals (SPCA).[55]
- A third-year political science student at the University of California, Los Angeles withdrew from the school after she posted her video rant against Asian students on YouTube. The student was complaining about Asian students in the school library using their cell phone to reach family members after the tsunami in Japan.[56]

ONLINE VIRTUAL WORLDS

An **online virtual world** is a shared multimedia, computer-generated environment in which users, represented by avatars, can act, communicate, create, retain ownership of what they create, and exchange assets, including currency, with each other. An **avatar** (see Figure 9-3) is a character in the form or a human, animal, or mythical creature.

FIGURE 9-3 An avatar is a representation of a virtual world visitor
Credit: Image copyright Ralf Juergen Kraft, 2009. Used under license from Shutterstock.com.

Virtual worlds are usually thought of as alternative worlds where visitors go to entertain themselves and interact with others. CityVille, Entropia Universe, FarmVille, and Second Life are all examples of online virtual worlds.

A **massively multiplayer online game (MMOG)** is a multiplayer video game capable of supporting hundreds and even thousands of concurrent players. The games are accessible via the Internet, with players using personal computers; game consoles such as Xbox 360, Wii, and Play Station 3; and even smartphones. **Massive multiplayer online role playing games (MMORPG)** is a subcategory of MMOG that provides a huge online world in which players take on the role of a character and control that character's action. Characters can interact with one another to compete in online games and challenges that unfold according to the online world's rules and storyline. Happy Farm and World of Warcraft are two examples of popular MMOGs.

Avatars in many virtual worlds can shop, hold jobs, run for political office, develop relationships with other avatars, take a test drive in a virtual world car, and even engage in criminal activities. Avatars may promote events and hold them in the virtual world (e.g., garage sales, concerts, conferences). Avatars can even start up new businesses and create or purchase new entities, such as houses, furnishings for their houses, clothing, jewelry, and other products. Avatars use the virtual world's currency to purchase goods and services in the virtual world. The value of objects in a virtual world is usually related to their usefulness and the difficulty of obtaining them. The ownership of such items is recognized by other avatars in the virtual world—for example, this is John's house; others may not occupy it without his permission.

Avatars can earn virtual world money by performing tasks in the virtual world, or their owners can purchase virtual world money for them using real world cash. In some virtual worlds, avatars can convert their virtual world money back into real dollars at whatever the going exchange rate is by using their credit card at online currency exchanges. Virtual world items may also be sold to other virtual world players for real world money. For example, John Jacobs, whose avatar is the character Neverdie in the Entropia Universe MMOG, sold his Club Neverdie on the virtual asteroid orbiting Planet Calypso for $635,000.[57] It is estimated that the U.S. virtual goods market was $1.6 billion in 2010, and the total worldwide market may have been as much as $10 billion.[58]

Crime in Virtual Worlds

It seems the freedom and anonymity afforded avatars in a virtual world encourages some individuals to unleash their darker side. Thus, virtual worlds raise many interesting questions regarding what is a criminal act and whether law enforcement—real or virtual—should get involved in acts that occur in virtual worlds.

Some virtual activities are clear violations of real world law and need to be reported to law enforcement authorities—for example, avatars trafficking in actual drugs or stolen credit cards. Other virtual activities, such as online muggings and sex crimes, can cause real life anguish for the human owners of the avatars involved but generally do not rise to the level of a real life crime. Although most virtual worlds have rules against offensive behavior in public, such as using racial slurs or performing overtly sexual actions, consenting adults can travel to private areas and engage in all sorts of socially unacceptable behavior. Bad deeds done online are often mediated by the game administrators, who can take action according to the rules of the game and with consequences internal to the game.

Some virtual world activities fall into a vast gray area. For example, in the real world, gambling games within casinos are inspected and regulated by state gaming commissions to ensure that the games are "fair." However, such regulations do not exist in the virtual world, and the potential for unfair games stacked in favor of the operator is high.

Educational and Business Uses of Virtual Worlds

Virtual online worlds are also being used for education and business purposes. The New Media Consortium (NMC) is an international consortium of hundreds of colleges, universities, museums, and research centers exploring the use of new media and technologies to improve teaching, learning, and creative expression. Members of NMC can conduct classes and meetings from within a growing number of virtual learning worlds. They can also build custom virtual learning worlds, simulations, and learning games. The virtual reality experience provides participants with a real sense of being there when attending a virtual class or conference. Experienced designers can develop virtual classes that immerse and engage students in the same way that today's video games grab and keep the attention of players.[59]

Second Life Work Microsites enable businesses and government agencies to use Second Life for virtual meetings, events, training, and simulations to stimulate innovation while minimizing the cost and environmental impact of travel. Second Life Education Microsites are designed for educators to offer virtual education solutions that stimulate engaged, collaborative learning to augment their traditional curriculum.[60]

Germany's TUEV NORD Group is an international provider of safety solutions—including certification and testing—with 8,500 employees in over 70 countries. The firm has been using Second Life since 2007 to recruit, conduct meetings, and hold game-based education.[61]

Chevron is a U.S. multinational company engaged in the gas, oil, and geothermal industries. The firm is using a virtual world model of its Salt Lake refinery for training new operators, some of whom have never even been in a refinery unit. Trainees guide their avatars through a 3D virtual model of the refinery to learn the basics of safe operation and how to deal with typical operational issues. Practicing in the virtual environment enables trainees to be exposed to many more operations scenarios in much less time than waiting for similar situations to arise in the real world.[62]

Summary

- A social networking Web site creates an online community of Internet users that enables members to break down barriers created by time, distance, and cultural differences; such a site allows people to interact with others online by sharing opinions, insights, information, interests, and experiences.
- Nielsen estimates that 22 percent of all time spent online is spent on social networking and blog sites, with the average visitor spending almost six hours per month on such sites.
- There is an increase in the number of business-oriented social networking sites designed to encourage and support relationships with consumers, clients, potential employees, suppliers, and business partners around the world.
- Social network advertising enables advertisers to generate a conversation with viewers of their ads and to target ads to reach people with the desired demographic characteristics.
- There are several social network advertising strategies, including direct advertising, advertising using an individual's network of friends, indirect advertising through social networking groups, advertising via company-owned social networking Web sites, and viral marketing.
- Employers often look at the social networking Web site profiles of job candidates when making hiring decisions.
- Employers can legally reject a job applicant based on the contents of the individual's social networking profile as long as the company is not violating federal or state discrimination laws.
- Students who use social networking Web sites should review and make appropriate changes to their profiles before starting a job search.
- Many organizations monitor social media networks as a means of improving customer service, retaining customers, and increasing sales.
- A social shopping Web site brings shoppers and sellers together in a social networking environment in which members share information and make recommendations while shopping online.
- Cyberbullying is the harassment, torment, humiliation, or threatening of one minor by another minor or group of minors via the Internet or cell phone. Approximately 20.8 percent of 10- to 18-year-olds have experienced cyberbullying sometime in their life.
- Cyberstalking is threatening behavior or unwanted advances directed at an adult using the Internet or other forms of online and electronic communications; it is the adult version of cyberbullying.
- Although current federal statutes address some forms of cyberstalking, there are still large gaps in current federal and state law.
- There are about 700,000 registered sex offenders in the United States; 90,000 of them were onetime members of MySpace.
- Many social networking Web sites have policies against uploading violent or obscene material; however, these policies are difficult to enforce.
- An online virtual world is a shared multimedia, computer-generated environment in which users, represented by avatars, can act, communicate, create, retain ownership of what they create, and exchange assets, including currency.

- Virtual worlds raise many interesting questions regarding what is a criminal act and whether law enforcement, real or virtual, should get involved in acts that occur in virtual worlds.
- Virtual online worlds are increasingly being used for education and business purposes.

Key Terms

avatar
cyberbullying
cyberstalking
massively multiplayer online game (MMOG)
massive multiplayer online role playing games (MMORPG)
online virtual world
social network advertising
social networking Web site
social shopping Web site
viral marketing

Self-Assessment Questions

The answers to the Self-Assessment Questions can be found in Appendix E.

1. Nielsen estimates that 22 percent of all time spent online is spent on social networking and blog sites, with the average visitor spending almost six hours per month on such sites. True or False?
2. The country with the highest percentage of Internet consumers visiting a social network, at 86 percent, is ____________.
   a. Australia
   b. Brazil
   c. Canada
   d. the United States
3. ____________ is a popular business-oriented Web site used for professional networking, with over 100 million unique visitors each month.
4. Ad spending on Facebook alone is estimated at $4 billion worldwide for 2011. True or False?
5. ____________ encourages individuals to pass along a marketing message to others, thus creating the potential for exponential growth in the message's exposure and influence as one person tells two people, each of those two people tell two or three more people, and so on.
   a. Direct advertising
   b. Viral marketing
   c. Indirect advertising through groups
   d. A company-owned social networking Web site
6. Employers can legally reject a job applicant based on the content of the individual's social networking Web site as long as the company is not violating discrimination laws. True or False?

7. There are around 700,000 registered sex offenders in the United States, and 90,000 of them were found on and subsequently banned from the social networking Web site ____________.
8. Cyberbullying is more common among 15- and 16-year-old males than any other group of social networking users. True or False?
9. Current federal statutes thoroughly address all aspects of cyberstalking, with no gaps in federal and state laws. True or False?
10. Which of the following measures is employed by social networking Web sites to avoid the posting of objectionable material?
    a. The terms of use agreement for most social networking Web sites states that the Web site reserves the right to delete material or terminate user accounts that violate the site's policies.
    b. The Web sites employ people to review material submitted.
    c. Other users sometimes report objectionable material.
    d. All of the above
11. To date, no practical business applications of online virtual worlds have been implemented. True or False?
12. Social shopping Web sites generate money primarily through advertising and by selling ____________.

Discussion Questions

1. MIT professor Sherry Turkle has written a book, *Alone Together*, which is highly critical of social networking. She argues that the manner in which some people frenetically communicate online using Facebook, Twitter, and instant messaging is a form of modern madness. Turkle thinks that under the illusion of enabling improved communications, technology is actually isolating us from true human interactions. Others disagree and argue that Facebook, Twitter, and instant messaging have led to more communications, not less. What do you think?
2. Keep track of the time that you spend on social networking Web sites for one week. Do you think that this is time well spent? Why or why not?
3. Do you think that college instructor–student friendships on social networking Web sites are appropriate? Why or why not?
4. Develop an idea for a social media marketing campaign for one of your favorite consumer products. Document how you would turn your message viral.
5. Discuss the following idea: The information posted on social networking Web sites about news events occurring in foreign countries is an excellent source of up-to-the-minute news.
6. Identify two significant advantages that social network advertising has over other forms of advertising.
7. What advice would you give a friend who is the victim of cyberstalking?

8. Studies have shown that people using their phones while driving are four times more likely to cause an accident than nonphone users. Yet 75 percent of adults with cell phones have talked on their phone while driving and 47 percent admit to sending text messages while driving.[63] Do you think that more needs to be done to discourage the use of cell phones when driving? If so, what additional actions could be taken? If not, why not?
9. What measures would you use to gauge the success of a social networking promotion designed to get people to try a new consumer product?
10. Can role playing illegal and violent fantasies in a virtual world affect individuals and society in the real world? What are the social and ethical implications of such role playing? Should limits be placed on what players can do in virtual worlds?
11. Do a search of the Web and develop a list of six companies that have created their own social networking Web sites.
12. What type of online information about a job candidate should employment managers consider when screening candidates for an interview? Give three examples of information that might be found that should automatically disqualify a candidate from a job offer. Give three examples of online information that should increase a candidate's chances of a job offer.
13. Review your user profile on your most frequently used social networking Web site. Do you think you need to make any changes to this profile? If so, what changes?
14. Check out the privacy policy of three social shopping Web sites to see if they say anything about selling user data to retailers. Write a couple of sentences summarizing your findings.

What Would You Do?

1. You are extremely disappointed to receive a rejection letter from the ABCXYZ Company. The job interviewer for the firm had told you that you "aced" the job interview and that you should expect a job offer after the formality of a background check was completed. What would you do?
2. You are shocked to receive a sexually provocative, but not entirely revealing, photo on your cell phone from one of your neighbors in the apartment building that you moved into last weekend. You really are not interested in the individual and are concerned that the person might become more aggressive. What would you do?
3. You are a new hire at a firm that manufactures and markets athletic equipment. You and several other new hires have been asked by your employer to begin posting positive messages about your firm's latest product—a $225 running shoe with state-of-the-art features—on Facebook, Stuffpit, and Twitter. While you are interested in the product, you cannot afford to buy it and have never tried it. Your manager says not to worry, the Marketing Department will provide you with prewritten statements for you to post. What would you do?
4. You have just received a second invitation to join John's friend list. You met John three weeks ago at a group study session prior to last semester's Calculus II finals. He came

across as very quiet and sort of strange. You did nothing to encourage his attention, but now you keep running into him at the oddest places and strangest times—at the self-service car wash, the 24-hour gym at 1:00 a.m., and the bakery at 7:00 a.m. He always flashes you a smile but has nothing to say. You think you've caught him taking snapshots of you a couple of times with his cell phone. He is starting to creep you out. What would you do?

5. A coworker who is a recruiter told you that she is going to drop a job candidate because she feels that he is totally irresponsible. She found out through research on Facebook that the candidate married and divorced his high school sweetheart before graduating from college and once had his car repossessed. What would you say to your coworker about this?
6. A friend of yours has asked you to help him and a group of three or four others shoot a video and upload it to YouTube. The subject of the video is "Happenings at Work," and it will include several vignettes about funny incidents at work. What would you do?
7. You are a new player in Second Life and are surprised when another avatar asks if you want to buy some drugs. You are not sure if the person is merely role playing or is serious. What would you do?

Cases

1. Procter & Gamble Turns to Social Networking

Spending big on advertising has always been a key strategy for consumer products giant Procter & Gamble (P&G). With recent annual global ad spending in the neighborhood of $9 billion, P&G is one of the world's top advertisers.[64] P&G was an early sponsor and producer of daytime radio and TV dramas—for which the term *soap opera* was coined. Its *Guiding Light* program began airing on radio in 1937 and moved to TV in 1952.[65] The P&G produced and sponsored *As the World Turns* was the leading soap opera for decades, winning many daytime Emmy awards. However, over the years, women—the target audience for such programs—made a huge shift in their TV viewing habits as they moved into the workplace, became more interested in talk and reality shows, and, more recently, began spending more of their leisure time online. The daytime soap opera died when P&G finally pulled the plug on *As the World Turns* in 2010 after 53 years.[66]

In 2006, P&G began working with Facebook to promote its brands using standard banner ads and promoting Facebook groups seeking fans for the company and/or its products. However, this approach was not very successful. P&G's biggest success was a Crest Whitestrips promotion that invited college students to become "fans" on its Facebook page. The company offered thousands of free movie screenings and sponsored concerts but still only attracted 14,000 fans to the product's Facebook page. When the promotions ended, the fans left too. The problem, according to Web guru Seth Goldstein, is: "Advertisers distract users; users ignore advertisers; advertisers distract better; users ignore better."[67]

P&G's next experiment with social networking was in 2007, when it collaborated with Yahoo! and the ZiZi Group to create Capessa.com, an online social network targeted at women. Capessa enabled women to post their stories and discuss topics such as parenting, managing their careers, getting in shape, and dealing with illnesses. One of the goals of this experiment was to identify ways in which social networking could be used to gain a better understanding of women's likes, dislikes, interests, and needs. Ultimately, no direct connection between Capessa and an increase in sales could be made.[68]

Some people are concerned about the blurring of lines between editorial and advertising content. They also are skeptical as to whether the stories posted on social networks are from real people or from paid actors or authors.[69]

Recognizing that its current approach was not working, P&G invited Google, Facebook, Twitter, and other social media experts to work with it to explore how online and digital media could more effectively support its marketing program. This spawned several new approaches. P&G has expanded its efforts to sell some top brands (Pampers, Olay, and Pantene) by offering shopping through a Facebook app. Consumers click a Shop Now tab on the page to complete their orders and then check out through Amazon.com.[70] And its "Smell Like a Man, Man" commercials began appearing on YouTube. These commercials starred the beefy ex-football player Isaiah Mustafa wearing only a towel. The commercials were a big hit, drawing over 140 million views and helping its Old Spice brand sales to expand at a double-digit growth rate.[71, 72]

As a result of these successes, P&G set a goal that each of its brands develop a meaningful presence on Facebook. In addition, it is creating smartphone applications for its consumers; one free app available at Apple's App Store and Google's Android Market provides women with guidance on which P&G beauty products to use and how to use them to get a desired look.[73]

Marketing experts say many of the large companies were slow to adopt the use of social networks; however, they also agree that consumer goods companies Coca-Cola, Pepsi, P&G, Unilever, and Johnson & Johnson are now coming on strong.[74] The decision to drop soap operas and move advertising dollars to social media was not a popular one with some senior P&G managers (who had spent much of their careers supporting advertising on soap operas) nor with many soap opera fans (who were very attached to these programs), but the decision was a clear indication that P&G recognizes that "the times they are a-changing."

Discussion Questions

1. Should the success of a social networking marketing campaign be measured simply by an increase in units sold? Why or why not?
2. What key arguments might have been used to convince P&G marketing executives to drop their long-running use of soap operas and replace them with social network advertising?
3. Develop a list of five key criteria that P&G might use to assess both the appropriateness and effectiveness of its YouTube commercials.

2. Social Networks for Chronic Health Condition Sufferers

For people living with chronic health conditions (one that is persistent and long lasting, such as arthritis, asthma, high blood pressure, cancer, or HIV/AIDS), social networking can play an important role. A 2010 national telephone survey of 3,001 adults found that 23 percent of Internet users living with a chronic health condition have gone online to find others with similar health conditions and to share experiences and seek information.[75] Indeed, there are a number of patient networking sites that enable users to connect directly to one another based on medical condition, including Alliance Health Networks, CureTogether, DiabeticConnect, Disaboom,

FacetoFace Health, HealthCentral, Inspire, and PatientsLikeMe. Some chronic conditions sufferers have even taken the initiative to start their own social networking sites focused on their condition by using social network creation services, such as Ning and Wetpaint.

Some patients use social networking to discover no-nonsense tips about coping with their disease or disability that physicians and their family cannot provide simply because they have not lived with it (for example, which restaurants and movie theaters have the best wheelchair access). Others network to become better informed about their condition and long-term prognosis, learn about alternative or experimental drugs and treatments, or discuss medical costs and insurance. Certainly doctors, nurses, and other health professional should continue to be the primary source of health information, but social networks can be a useful source of information as well.

Many patients network as a means to deal with the anxiety, depression, and stress that frequently accompany chronic conditions. For example, Sean Fogerty, a 50-year-old with multiple sclerosis who is recovering from brain cancer, spends over an hour most nights chatting with other patients online. Sean says social networking has "literally saved my life, just to be able to connect with other people."[76]

Social networking sites targeted at those suffering from chronic health conditions do raise some potential ethical concerns. Obviously people must be cautious about sharing too much private health information and personal data online, especially among a group of often anonymous users. There is also a risk that members of the group may unwittingly share erroneous or out-of-date information on which others may act. In addition, if the general tone of the network site is overly pessimistic about the chronic condition, participants can experience an even deeper depression over their situation. Thus although such social networking sites have the potential to help sufferers with chronic health conditions, they must be used with care.

Discussion Questions

1. Imagine that your grandmother has suffered from asthma all her life and within the past year was diagnosed with type II diabetes. She has been quite depressed as she must now be extremely careful with her diet, administer insulin shots to herself before each meal, and take two types of oral medication. She hates sticking herself to take her sugar level two or three times a day and has found it difficult to watch her diet and keep her blood sugar level within normal ranges. Would you suggest she join a social network for people with diabetes? Why or why not?
2. Do research and try to find social networks that are designed for people who suffer from alcohol, drug, or gambling addiction. Are there additional potential ethical issues for social networking sites targeted at those suffering from an addiction? Write a brief paragraph or two summarizing your findings.
3. What issues might arise trying to discuss with your primary care physician an experimental treatment or drug you discovered on a social network? How might you be able to broach this topic without upsetting your physician?

3. Cyberbullying Results in Death

Megan Meier was a seventh grader who had been diagnosed with depression and attention deficit disorder; she had been under the care of a psychiatrist since the third grade. As a result of her depression, she had suffered from some suicidal thoughts. However, Megan was delighted when 16-year-old Josh Evans began exchanging messages with her shortly after she joined MySpace. They became online friends even though they never met face-to-face or spoke to one another. For about six weeks things went well, and Megan's family noticed that the online friendship seemed to have lifted her spirits.

But one day the tone of Josh's messages changed. He sent a message to Megan saying, "I don't know if I want to be friends with you anymore because I've heard that you are not very nice to your friends." Josh followed up with similar messages, and he shared some of Megan's messages to him with others. Quickly, other people began posting abusive messages and online bulletins about Megan.[77]

Megan told her mother that she had suddenly become the target of an increasing number of mean online messages. When Megan's mother saw some of the replies that Megan had posted, they got into an argument over the vulgar language Megan had used in some of her responses. Her mother was also upset to find that Megan had not logged off the computer when she had been told to do so earlier that afternoon. Megan ran upstairs and shut herself in her room. Twenty minutes later her mother found her dead; she had committed suicide by hanging.

The last message Megan received from Josh read, "You are a bad person and everybody hates you. Have a shitty rest of your life. The world would be a better place without you." Megan responded, "You're the kind of boy a girl would kill herself over."[78]

Six weeks after Megan's death, her parents were informed by a neighbor that Lori Drew, a 49-year-old mother who lived just four doors down from the Meier family, had created a fake "Josh Evans" MySpace account. The neighbor said that "Lori laughed about it" and said she intended to use it to "mess with Megan." Megan and Lori's 13-year-old daughter, Sarah, had been friends but had had a falling out because Megan allegedly had spread gossip about Sarah.[79]

The local police and the FBI asked the Meier family not to say anything publicly while the Drew family was being investigated. In December 2007, some 13 months after the suicide, Jack Banas, the county's prosecuting attorney, held a press conference. He stated that Ashley Grills, an 18-year-old employee in Lori Drew's advertising business, admitted that she wrote most of the messages to Megan, including the final message. After Megan's death, Grills was hospitalized and underwent psychiatric treatment for her involvement in the cyberbullying incident. Banas went on to say that Sarah Drew had moved and was now attending a new school in a different city. Lori Drew would not reveal where her daughter was living out of fear of what might happen to her. Neighbors avoided the Drews, and business advertisers in Lori Drew's coupon book business were shunned. After reviewing the case, the county prosecutor decided not to file any criminal charges related to the hoax because he determined that none of Lori Drew's actions met the criminal threshold.[80]

In January 2008, a federal grand jury began issuing subpoenas to MySpace and other witnesses as federal prosecutors considered whether to bring federal charges against Lori Drew. The grand jury in Los Angeles was given jurisdiction because MySpace has its headquarters in Beverly Hills.[81] The U.S. attorney granted immunity to Ashley Grills in exchange for her testimony against Lori Drew.[82] Eventually, the grand jury indicted Lori Drew on one count of conspiracy and three counts of unauthorized access for accessing protected computers—a violation of the Computer Fraud and Abuse Act. The case was to go forward not as a homicide but as a computer fraud prosecution.

The Computer Fraud and Abuse Act is intended to prosecute those who hack into what are known as protected computers. It applies to cases with a compelling federal interest, in which computers of the federal government or certain financial institutions are involved, the crime itself is interstate in nature, or the computers are used in interstate and foreign commerce. Section (b) of the act punishes not just anyone who commits or attempts to commit an offense under the Computer Fraud and Abuse Act but also those who conspire to do so.

The basis for the charges against Lori Drew was her alleged violation of the MySpace terms of service agreement by creating a fake profile for the nonexistent Josh Evans. Prosecutors argued that this was the legal equivalent of hacking into a protected computer. Drew's attorneys filed motions to dismiss the case on the basis that violating the MySpace terms of service did not constitute illegal access of a protected computer. Judge George Wu ruled against the motion and ordered the case to go to court.[83]

The trial began in November 2008. The prosecution's case against Drew was dealt a setback when Ashley Grills admitted that it was her idea to create the MySpace account and that it was she who clicked to agree to the MySpace terms of service. According to Grills, neither she nor Lori nor Sarah Drew looked at the terms of service. In addition, Grills had previously admitted that she sent most of the messages from Josh Evans. However, several witnesses said that Drew admitted that she had created the MySpace account with Grills and Sarah, and that she had sent some of the messages herself. Grills did state that Drew had encouraged her to create the fake profile and told her not to worry because "people do it all the time."[84]

Drew's attorney acknowledged that the case was indeed sad but that "it doesn't amount to a violation of a computer statute. The reality is that there's lots of blame to go around in this case."[85]

After just one day of deliberation, the jury was deadlocked on the conspiracy charge. They acquitted Drew on the three felony counts of accessing a computer without authorization to inflict emotional harm. The jury did convict Drew of three lesser charges of accessing computers without authorization. Probation authorities recommended probation and a $5,000 fine, while the prosecution asked for a sentence of three years and a $300,000 fine.[86]

Drew's attorney asked Judge George Wu to grant a motion for a directed acquittal based on the defense's view that the prosecution failed to prove that Drew knew that the MySpace terms of service existed, and that she knew what they said and intentionally violated them.[87] Wu postponed his ruling until July 2, 2009, at which time he dismissed the case. Wu expressed concern that if Drew's conviction was allowed to stand, it would create a precedent that violating a Web site's terms of service agreement is the legal equivalent of computer hacking and could result in criminal prosecution for what would have been in the past a misdemeanor for civil breach of contract.[88]

Discussion Questions

1. Do you believe that knowingly violating the terms of a Web site's service agreement should be punishable as a serious crime, with potential penalties of substantial fines and jail time? Why or why not?
2. Imagine that you are the defense counsel for Lori Drew. Present your strongest argument for why your defendant should not be convicted for violation of the Computer Fraud and

Abuse Act. Now imagine that you are the prosecutor. Present your strongest argument for why Drew should be convicted. Which argument do you believe is stronger?

3. Do you think that Lori Drew was responsible for Megan's death? Do you think that justice was served in this case? Should new laws be created to address similar future cases?

End Notes

[1] Jessica Hall, "Facebook's Growth Exceed Expectations: Report," *Reuters*, May 1, 2011, www.reuters.com/article/2011/05/02/us-facebook-wsj-idUSTRE74106A20110502.

[2] Facebook, "Facebook Privacy Policy," www.facebook.com/group.php?gid=118478201508582 (accessed July 13, 2001).

[3] Facebook, "Facebook Privacy Policy," www.facebook.com/group.php?gid=118478201508582 (accessed July 13, 2001).

[4] Facebook, "Statistics," www.facebook.com/press/info.php?statistics (accessed July 13, 2011).

[5] Emily Steel and Geoffrey A. Fowler, "Facebook in Privacy Breach," *Wall Street Journal*, October 18, 2010, http://online.wsj.com/article/SB10001424052702304772804575558484075236968.html.

[6] Emily Steel and Geoffrey A. Fowler, "Facebook in Privacy Breach," *Wall Street Journal*, October 18, 2010, http://online.wsj.com/article/SB10001424052702304772804575558484075236968.html.

[7] Daniel Ionescu, "Facebook Adds Facial Recognition to Make Photo Tagging Easier," *PCWorld*, December 16, 2010, www.pcworld.com/article/213894/facebook_adds_facial_recognition_to_make_photo_tagging_easier.html.

[8] Ed Oswald, "Facebook Facial Recognition: Security Firm Issues Alert," *PCWorld*, June 8, 2011, www.pcworld.com/article/229689/facebook_facial_recognition_security_firm_issues_alert.html.

[9] Ed Oswald, "Facebook Facial Recognition: Security Firm Issues Alert," *PCWorld*, June 8, 2011, www.pcworld.com/article/229689/facebook_facial_recognition_security_firm_issues_alert.html.

[10] Nick Bilton, "Facebook Changes Privacy Settings to Enable Facial Recognition," *Bits, New York Times* (blog), June 7, 2011, http://bits.blogs.nytimes.com/2011/06/07/facebook-changes-privacy-settings-to-enable-facial-recognition.

[11] Geoffrey A. Fowler and Christopher Lawton, *Wall Street Journal*, June 8, 2011, http://online.wsj.com/article/SB10001424052702304778304576373730948200592.html.

[12] Miniwatts Marketing Group, "Internet World Stats, Usage and Population Statistics," www.internetworldstats.com/stats.htm, March 31, 2011.

[13] "Google Plus Reaches 25 Million Users, Activity Declines," *Search Engine Journal*, August 3, 2011, www.searchenginejournal.com/google-plus-reaches-25-million-users-activity-declines/31500.

14 "Social Networks/Blogs Now Account for One in Every Four and a Half Minutes Online," *NielsenWire* (blog), June 15, 2010, http://blog.nielsen.com/nielsenwire/global/social-media-accounts-for-22-percent-of-time-online.

15 Jennifer Van Grove, "Social Networking Usage Surges Globally [STATS]," *Mashable*, March 16, 2010, http://mashable.com/2010/03/19/global-social-media-usage.

16 Alex Mindlin, "Social Networks Exhibit Senior Appeal," *New York Times*, September 13, 2010, www.nytimes.com/2010/09/13/technology/13drill.html.

17 "Facebook Drives US Social Network Ad Spending Past $3 Billion in 2011," *eMarketer*, January 18, 2011, www.emarketer.com/Articles/Print.aspx?1008180.

18 Robert MacManus, "40% of People 'Friend' Brands on Facebook," *ReadWriteWeb*, November 10, 2009, www.readwriteweb.com/archives/survey_brands_making_big_impact_on_facebook_twitter.php.

19 Morgan Stewart, "Why 70% of Facebook 'Fans' Don't Want Marketing, and What You Can Do About It," *MarketingProfs*, November 10, 2009, www.marketingprofs.com/articles/2009/3149/why-70-of-facebook-fans-dont-want-marketing-and-what-you-can-do-about-it.

20 Joe Guy Collier, "Coke Fans' Facebook Page Draws Millions of Users," *Atlanta Journal-Constitution*, March 30, 2009, www.ajc.com/business/content/business/coke/stories/2009/03/30/coke_facebook_page.html.

21 Theresa Howard, "Seeking Teens, Marketers Take Risks by Emulating MySpace," *USA Today*, May 22, 2006, www.usatoday.com/tech/news/2006-05-01-myspace-marketers_x.htm.

22 "How Fortune 1000 Companies are Harnessing the Power of Social Media," White Paper, 2009, www.dna13.com/company/downloads/white-papers/how-f1000-leverage-social-media.

23 Jennifer Van Grove, "Coke Targets Teens with Black Friday SCVNGR Promotion," *Mashable*, November 18, 2010, http://mashable.com/2010/11/19/scvngr-coke-rewards.

24 Dell, Inc., "IdeaStorm," www.ideastorm.com (accessed July 14, 2011).

25 Dr. Ralph F. Wilson, "The Six Simple Principles of Viral Marketing," *Web Marketing Today*, February 1, 2005, www.wilsonweb.com/wmt5/viral-principles.htm.

26 John Zappe, "New Survey Finds That 89% Are Using Social Media in Recruiting," *TLNT*, July 12, 2011, www.tlnt.com/2011/07/12/new-survey-finds-that-89-are-using-or-will-use-social-media-in-recruiting.

27 Mike Hargis, "Social Networking Sites Dos and Don'ts," *CNN*, November 5, 2008, www.cnn.com/2008/LIVING/worklife/11/05/cb.social.networking/index.html.

28 CareerBuilder.com, "One-in-Five Employers Use Social Network Sites to Research Job Candidates, CareerBuilder.com Survey Finds," www.niacc.edu/careercenter/pdfs/One-in-Five_Employers_Use_Social_Networking_Sites_to_Research_Job_Candidates.pdf (accessed June 15, 2011).

29 Mallory Terrence, "Employers View Social Networking Sites Before Hiring Employees," *Loquitur*, April 17, 2008, http://theloquitur.com/?p=6968.

30 Mallory Terrence, "Employers View Social Networking Sites Before Hiring Employees," *Loquitur*, April 17, 2008, http://theloquitur.com/?p=6968.

31 "Why Should Marketers Monitor Social Media," *Position*^{2} (blog), January 28, 2011, http://blogs.position2.com/integrating-social-media-monitoring-with-business-functions-marketing.

32 "Companies Expand Their Customer Support to Include Social Media Monitoring," *Position*^{2} (blog), January 13, 2011, http://blogs.position2.com/integrating-social-media-monitoring-with-business-functions-customer-service.

33 Nestor Bailly, "Huffington Post Uses A/B Testing on Its Headlines," *editorsweblog.org* (blog), October 15, 2009, www.editorsweblog.org/multimedia/2009/10/huffington_post_uses_ab_testing_on_its_h.php.

34 Drew Johnson, "GM Keeping Customers Happy Through Social Media Monitoring," *Left Lane*, March 26, 2010, www.leftlanenews.com/gm-keeping-customers-happy-through-social-media-monitoring.html.

35 Peter Cervieri, "Dell Social Media – Linking Conversations to Sales," *ScribeMedia.org*, April 18, 2011, www.scribemedia.org/2011/04/18/dell-social-media-linking-conversations-to-sales.

36 Stuffpit, "Earn Money Recommending Products," www.stuffpit.com/stuff/earn (accessed July 2, 2011).

37 Bob Tedeschi, "Like Shopping? Social Networking? Try Social Shopping," *New York Times*, September 11, 2006, www.nytimes.com/2006/09/11/technology/11ecom.html.

38 Cyberbullying Research Center, http://cyberbullying.us/research.php (accessed June 11, 2011).

39 Tracy Ness, "Chatham Learns About Cyberbullying," *nj.com*, November 16, 2009, www.nj.com/independentpress/index.ssf/2009/11/chatham_learns_about_cyber_bul.html.

40 "How Lori Drew Became America's Most Reviled Mother," *The Age*, December 1, 2007, www.theage.com.au/articles/2007/11/30/1196394672124.html.

41 Valli Finney, "Facebook Prank Ends with Two Estero Teens Arrested," *naplesnews.com*, January 13, 2011, www.naplesnews.com/news/2011/jan/13/facebook-prank-naked-teens-girls-estero.

42 Russell Goldman, "Teens Indicted After Allegedly Taunting Girl Who Hanged Herself," *ABC News*, http://abcnews.go.com/Technology/TheLaw/teens-charged-bullying-mass-girl-kill/story?id=10231357.

43 Sameer Hinduja, Ph.D. and Justin W. Patchin, Ph.D., "State Cyberbullying Laws," Cyberbullying Research Center, June 2011, www.cyberbullying.us/Bullying_and_Cyberbullying_Laws.pdf.

44 "State Cyberstalking, Cyberharassment, and Cyberbullying Laws," National Conference of State Legislatures, January 26, 2011, www.ncsl.org/IssuesResearch/TelecommunicationsInformationTechnology/CyberstalkingLaws/tabid/13495/Default.aspx

45 Claire M. Renzetti and Jeffrey L. Edleson, eds., *Encyclopedia of Interpersonal Violence* (Thousand Oaks, CA: Sage Publications, 2008), 164.

46 "David Hench, "Investigation Yields Arrest in Years Long Cyberstalking Case," *Morning Sentinel*, July 3, 2010, www.onlinesentinel.com/news/investigation-yields-arrest-in-years long-cyberstalking-case_2010-07-02.html.

47 The National Center for Victims of Crime, "Cyberstalking," www.ncvc.org/ncvc/main.aspx?dbName=DocumentViewer&DocumentID=32458 (accessed July 13, 2011).

48 Jenna Wortham, "MySpace Turns Over 90,000 Names of Registered Sex Offenders," *New York Times*, February 3, 2009, www.nytimes.com/2009/02/04/technology/internet/04myspace.html.

49 Jenna Wortham, "MySpace Turns Over 90,000 Names of Registered Sex Offenders," *New York Times*, February 3, 2009, www.nytimes.com/2009/02/04/technology/internet/04myspace.html.

50 "Are Internet Predator Worries Overblown," *PBS*, February 20, 2009, www.pbs.org/now/shows/508/internet-predators.html.

51 Byron Acohido, "Sex Predators Target Children Using Social Media," *USA Today*, March 1, 2011, www.usatoday.com/tech/news/2011-02-28-online-pedophiles_N.htm.

52 YouTube, "Statistics," www.youtube.com/t/press_statistics (accessed June 26, 2011).

53 Catalina Camia, "Rep. Anthony Weiner Resigns After Online Sex Scandal," *USA Today*, June 16, 2011, http://content.usatoday.com/communities/onpolitics/post/2011/06/anthony-weiner-sex-scandal-resignation-/1.

54 Sheryl Young, "YouTube Video of Bullying Incident Gets Students Arrested," *Associated Content*, February 4, 2011, www.associatedcontent.com/article/7719236/youtube_video_of_bullying_incident.html?cat=15&fb_xd_fragment#?=&cb=f2f3a1379e44338&relation=parent.parent&transport=fragment&type=resize&height=20&ackData[id]=1&width=90.

55 Gene Warner, "Beating of Dog on YouTube Aids Rescue," *Buffalo News.com*, June 20, 2011, www.buffalonews.com/city/police-courts/police-blotter/article461258.ece.

56 Ian Lovett, "UCLA Student's Video Rant Against Asians Fuels Firestorm," *New York Times*, March 15, 2011, www.nytimes.com/2011/03/16/us/16ucla.html.

57 Oliver Chiang, "Meet the Man Who Just Made a Half Million from the Sale of Virtual Property," *SelectStart, Forbes* (blog), November 13, 2010, http://blogs.forbes.com/oliverchiang/2010/11/13/meet-the-man-who-just-made-a-cool-half-million-from-the-sale-of-virtual-property.

58 Oliver J. Chiang, "The World's Most Expensive Island–Online," *Forbes*, February 17, 2010, www.forbes.com/2010/02/17/farmville-facebook-zynga-technology-business-intelligence-virtual-goods.html.

59 NMC, "NMC Benefits and Services," www.nmc.org/services (accessed June 26, 2011).

60 Amanda Linden, "Goodbye SecondLifeGrid.net and Hello SL Microsites," SecondLife (blog), May 13, 2010, http://community.secondlife.com/t5/Working-Inworld-General/bg-p/2007.

61 Catherine Linden, "Case Study: Making the Real World Safer—TUEV NORD Group in Second Life," SecondLife (blog), April 16, 2010, http://community.secondlife.com/t5/Working-Inworld-General/bg-p/2007.

[62] Chevron, "Working in a Virtual World," www.chevron.com/next/digitalrefinery (accessed June 26, 2011).

[63] Jim Motavalli, "Teenagers Not Tops In Texting," *Wheels; The Blog, New York Times*, June 27, 2010, http://query.nytimes.com/gst/fullpage.html?res=9B01E0DF1138F934A1575 5C0A9669D8B63&ref=pewinternetandamericanlifeproject.

[64] David Holthaus, "P&G Ad Spending Climbs," http://news.cincinnati.com/article/20110620/ BIZ01/306200037/P-G-ad-spending-climbs, June 20, 2011.

[65] "Guiding Light," *She Knows Soaps* (blog), http://soaps.sheknows.com/guidinglight (accessed June 15, 2011).

[66] Dan Swell, "Procter & Gamble Moves from Soap Operas to Tweets," *ABC News*, December 9, 2010, http://abcnews.go.com/Entertainment/wireStory?id=12354844.

[67] Randall Stross, "Advertisers Face Hurdles on Social Networking Sites," *New York Times*, December 13, 2008, www.nytimes.com/2008/12/14/business/media/14digi.html.

[68] Alyce Lomax, "Capessa Set to Discover What Women Want," *The Motley Fool*, January 9, 2007, www.fool.com/investing/value/2007/01/09/capessa-focuses-on-women.aspx.

[69] Alyce Lomax, "Capessa Set to Discover What Women Want," *The Motley Fool*, January 9, 2007, www.fool.com/investing/value/2007/01/09/capessa-focuses-on-women.aspx.

[70] Katie Deatsch, "Procter & Gamble Sells on Facebook with Help from Amazon," *Internet Retailer*, October 1, 2010, www.internetretailer.com/2010/10/01/procter-gamble-sells-facebook-help-amazon.

[71] Michael W. Jones, "What's the Revenue Model for Social Networking Sites?," *Tech. Blorge*, December 14, 2008, http://tech.blorge.com/Structure:%20/2008/12/14/whats-the-revenue-model-for-social-networking-sites.

[72] Dan Swell, "Procter & Gamble Moves from Soap Operas to Tweets," *ABC News*, December 9, 2010, http://abcnews.go.com/Entertainment/wireStory?id=12354844.

[73] Lauren Johnson, "Procter & Gamble Targets On-The-Go Women with Beauty App," *Mobile Marketer*, June 16, 2011, www.mobilemarketer.com/cms/news/advertising/10211.html.

[74] Dan Swell, "Procter & Gamble Moves from Soap Operas to Tweets," *ABC News*, December 9, 2010, http://abcnews.go.com/Entertainment/wireStory?id=12354844.

[75] Susannah Fox, "Peer-to-Peer Healthcare," Pew Research Center's Internet & American Life Project, February 11, 2011, pewinternet.org/Reports/2011/P2PHealthcare.aspx, http://pewinternet.org/Reports/2011/P2PHealthcare.aspx.

[76] Claire Cain Miller, "Social Networks a Lifeline for the Chronically Ill," *New York Times*, March 24, 2010, www.nytimes.com/2010/03/25/technology/25disable.html.

[77] Stephen Pokin, "UPDATE: No Charges to Be Filed over Meier Suicide," Suburban Journals, December 3, 2007, www.stltoday.com/suburban-journals/article_fd48db3e-b0ad-5332-b5a5-4ac231bc378c.html.

[78] Gordon Tokumatsu and Jonathan Lloyd "MySpace Case: 'You're the Kind of Boy a Girl would Kill Herself Over'," *NBC Los Angeles*, January 26, 2009, www.nbclosangeles.com/ news/local/Woman-Testifies-About-Final-Message-Sent-to-Teen.html.

[79] "How Lori Drew Became America's Most Reviled Mother," *The Age*, December 1, 2007, www.theage.com.au/articles/2007/11/30/1196394672124.html.

[80] Stephen Pokin, "UPDATE: No Charges to Be Filed over Meier Suicide," *Suburban Journals*, December 3, 2007, www.stltoday.com/suburban-journals/article_fd48db3e-b0ad-5332-b5a5-4ac231bc378c.html.

[81] Scott Glover and P. J. Huffstutter, "L.A. Grand Jury Issues Subpoenas in Web Suicide Case," *Los Angeles Times*, January 9, 2008, www.latimes.com/news/local/la-me-myspace 9jan09,0,5809715.story.

[82] Jonann Brady, "Exclusive: Teen Talks About Her Role in Web Hoax That Led to Suicide," *Good Morning America*, April 1, 2008, http://abcnews.go.com/GMA/story?id= 4560582&page=1.

[83] Kim Zetter, "Judge Postpones Lori Drew Sentencing, Weighs Dismissal," *Wired*, May 18, 2009, www.wired.com/threatlevel/2009/05/drew_sentenced.

[84] Kim Zetter, "Judge Postpones Lori Drew Sentencing, Weighs Dismissal," *Wired*, May 18, 2009, www.wired.com/threatlevel/2009/05/drew_sentenced.

[85] Kim Zetter, "Prosecution: Lori Drew Schemed to Humiliate Teen Girl," *Wired*, November 25, 2008, www.wired.com/threatlevel/2008/11/defense-lori-dr.

[86] Kim Zetter, "Judge Postpones Lori Drew Sentencing, Weighs Dismissal," *Wired*, May 18, 2009, www.wired.com/threatlevel/2008/11/defense-lori-dr.

[87] Kim Zetter, "Can Lori Drew Verdict Survive the 9th Circuit Court?," *Wired*, December 1, 2008, www.wired.com/threatlevel/2008/12/can-lori-drew-v.

[88] Alexandra Zavis, "Judge Tentatively Dismisses Case in MySpace Hoax That Led to Teenage Girl's Suicide," *Los Angeles Times*, July 2, 2009, http://latimesblogs.latimes.com/lanow/2009/07/myspace-sentencing.html.

CHAPTER **10**

ETHICS OF IT ORGANIZATIONS

QUOTE

There is a perception that ethical practices and profitability are mutually exclusive. In fact, they are dependent on one another. Responsible corporate behavior builds trust. It is an investment like any other sound business practice—and it pays substantial, measurable dividends to our shareholders.

—Stanley S. Litow, IBM Vice President of Corporate Citizenship and Corporate Affairs

VIGNETTE

IBM Committed to Green Computing

IBM has publicly committed itself to making its products environmentally friendly, energy efficient, reusable, recyclable, and safely disposable. It has had an environmental program since 1990 "during periods when the environment was not always as popular a subject as it is today; during profound changes in global economy, our industry, and our business model; and during periods of differing financial results," according to Wayne S. Balta, IBM's vice president for Corporate Environmental Affairs and Product Safety.[1]

It took over five years and extensive coordination with many suppliers, but in 2010, IBM became the first computer manufacturer to eliminate the use of perfluorooctane sulfonate and perfluorooctanoic acid compounds from its chip manufacturing processes. The two compounds are known to be toxic to both humans and animals.[2]

In 2009, as part of IBM's ongoing commitment to environmental sustainability, the company spent $14.3 million in capital and $102.3 million in operating expense to build, maintain, and upgrade plants and research laboratories to improve environmental performance and manage environmental programs. The costs were offset by an estimated savings of $152.4 million, including savings from energy, material, and water conservation; packaging and process improvements; and miscellaneous cost avoidance.[3] IBM estimates that over the long term its various environmental programs have captured savings and reduced costs at a rate of $1.60 for every $1.00 spent.[4]

Since 2008, the firm has conducted over 3,100 energy conservation projects at 350 IBM facilities in 49 countries that reduced total electric energy consumption by 523,000 megawatt hours and reduced energy costs by $50 million. The company has a goal of conserving a total of 1.1 million megawatts hours of energy consumption by 2013.[5,6]

In addition, IBM has developed tools that create 3D images to identify hot and cold spots in its data centers and other facilities; virtualization technology to ensure that its servers are operating at top energy efficiency; and analytics software to manage energy consumption across its data centers.[7] Based on this experience, IBM is working in partnership with Johnson Controls to implement the IBM Intelligent Building Management solution, which it sells to customers.[8]

By the end of 2009, IBM manufacturing operations had achieved an average annual water savings of 3.1 percent over the past five years.[9] And in 2010, IBM recycled 79 percent of the nonhazardous waste that it generated.[10]

IBM is one of 1,500 organizations worldwide that follows the guidelines of the Global Reporting Initiative (GRI) to communicate, measure, and track its environmental performance goals and results such as discussed above. The guidelines define basic principles and performance indicators that provide IBM with a standard benchmark for comparing its performance over time and for measuring results relative to various laws, norms, standards and voluntary initiatives. Use of the guidelines helps ensure consistency and transparency of reporting.[11]

Questions to Consider

1. Identify four key areas in which organizations might set environmental and/or natural conservation goals.
2. A considerable number of rules and processes have been put in place to ensure that an organization's financial reports are accurate and complete. Should similar efforts be made to ensure the accuracy and completeness of an organization's reporting of its results in the areas of corporate responsibility, environmental and energy conservation, and product safety? Why or why not?

LEARNING OBJECTIVES

As you read this chapter, consider the following questions:

1. What are contingent workers, and how are they employed in the information technology industry?
2. What key ethical issues are associated with the use of contingent workers, including H-1B visa holders and offshore outsourcing companies?
3. What is whistle-blowing, and what ethical issues are associated with it?
4. What is an effective whistle-blowing process?
5. What measures are members of the electronics manufacturing industry taking to ensure the ethical behavior of the many participants in their long and complex supply chains?
6. What is green computing, and what are organizations doing to support this initiative?

KEY ETHICAL ISSUES FOR ORGANIZATIONS

This chapter will touch on the following ethical topics that are pertinent to organizations in the IT industry, as well as to organizations that make use of IT:

- The use of nontraditional workers, including temporary workers, contractors, consulting firms, H-1B visa workers, and outsourced offshore workers, gives an organization more flexibility in meeting its staffing needs, often at a lower cost than if the organization used traditional workers. The use of nontraditional workers also raises ethical issues for organizations. When should such nontraditional workers be employed, and how does such employment affect an organization's ability to grow and develop its own employees? How does the use of nontraditional workers impact the wages of the organization's own employees?

- Whistle-blowing, as discussed in Chapter 2, is an effort to attract public attention to a negligent, illegal, unethical, abusive, or dangerous act by a company or some other organization. It is an important ethical issue for individuals and organizations. How does one safely and effectively report misconduct, and how should managers handle a whistle-blowing incident?
- **Green computing** is a term applied to a variety of efforts directed toward the efficient design, manufacture, operation, and disposal of IT-related products, including personal computers, laptops, servers, printers, and printer supplies. Computer manufacturers and end users are faced with many questions about when and how to transition to green computing, and at what cost.
- The electronics and information and communications technology (ICT) industry recognizes the need for a code to address ethical issues in the areas of worker safety and fairness, environmental responsibility, and business efficiency. What has been done so far, and what still needs to be done?

Let's begin with a discussion of the use of nontraditional workers and the ethical issues raised by this practice.

The Need for Nontraditional Workers

According to the Computing Research Association, the number of undergraduate degrees awarded in computer science and computer engineering at doctoral-granting computer science departments decreased dramatically from around 21,000 in 2004 to less than 10,000 in 2009. In 2010, however, the number increased slightly, to just over 10,000.[12] This 50 percent decrease in degrees awarded occurred in spite of the federal government's forecast of an increased need for workers in computer science-related fields.

The Bureau of Labor Statistics has forecasted that network systems and data communications analysts will be the second-fastest-growing occupation in the economy between 2008 and 2018. In addition, a large number of new jobs are expected to be created during that time period for computer software engineers, as shown in Table 10-1.

TABLE 10-1 IT-related jobs forecasted to have strong growth

Occupation	Percent growth 2008–2018	Number of new jobs (in thousands)	Median wages for all workers in this field (in 2008)
Network systems and data communications analysts	53%	155.8	$71,000
Computer software engineers	34%	175.1	$85,430

Source Line: United States Department of Labor, Bureau of Labor Statistics, "Overview of the 2008–2018 Projections, Occupational Outlook Handbook, 2010–2011 Edition," bls.gov/oco/oco2003.htm.

As a result of the decline in undergraduate degrees being awarded in computer science and engineering fields, IT firms and organizations that use IT products and services are concerned about a shortfall in the number of U.S. workers to fill these positions.

Consequently, computer science (with an average starting salary of $63,017), computer engineering ($60,112), systems engineering ($57,497), and business systems networking/ telecommunications ($56,808) were all rated in the top-ten paid majors for 2010–2011 bachelor's degree graduates.[13]

Facing a likely long-term shortage of trained and experienced workers, employers are increasingly turning to nontraditional sources to find IT workers with skills that meet their needs; these sources include contingent workers, H-1B workers, and outsourced offshore workers. As employers consider these options, they must confront ethical decisions about whether to recruit new and more skilled workers from these sources or to develop their own staff to meet the needs of their business.

CONTINGENT WORKERS

The Bureau of Labor Statistics defines **contingent work** as a job situation in which an individual does not have an explicit or implicit contract for long-term employment. The contingent workforce includes independent contractors, temporary workers hired through employment agencies, on-call or day laborers, and on-site workers whose services are provided through contract firms.

A firm is likely to use contingent IT workers if it experiences pronounced fluctuations in its technical staffing needs. Workers are often hired on a contingent basis as consultants on an organizational restructuring project, as technical experts on a product development team, and as supplemental staff for many other short-term projects, such as the design and installation of new information systems.

Typically, these workers join a team of full-time employees and other contingent workers for the life of the project and then move on to their next assignment. Whether they work, when they work, and how much they work depends on the company's need for them. They have neither an explicit nor an implicit contract for continuing employment.

Organizations can obtain contingent workers through temporary staffing firms or employee leasing organizations. Temporary staffing firms recruit, train, and test job seekers in a wide range of job categories and skill levels, and then assign them to clients as needed. Temporary employees are often used to fill in during staff vacations and illnesses, handle seasonal workloads, and help staff special projects. However, they are not considered official employees of the company, so they are not eligible for company benefits such as vacation, sick pay, and medical insurance. Because temporary workers do not receive additional compensation through company benefits, they are often paid a higher hourly wage than full-time employees doing equivalent work. Temporary working arrangements sometimes appeal to people who want maximum flexibility in their work schedule as well as a variety of work experiences. Other workers take temporary work assignments only because they are unable to find more permanent work.

In **employee leasing**, a business (called the subscribing firm) transfers all or part of its workforce to another firm (called the leasing firm), which handles all human-resource-related activities and costs, such as payroll, training, and the administration of employee benefits. The subscribing firm leases these workers, but they remain employees of the leasing firm. Employee leasing firms operate with minimal administrative, sales, and marketing staff to keep down overall costs, and they pass the savings on to their clients. Employee leasing is a type of **coemployment relationship**, in which two employers have actual or potential legal rights and duties with respect to the same employee or group of employees. Employee leasing firms are subject to special regulations regarding workers' compensation and unemployment insurance. Because the workers are technically employees of the leasing firm, they may be eligible for some company benefits through the firm.

Organizations can also obtain temporary IT employees by hiring a consulting firm. Consulting organizations maintain a staff of employees with a wide range of skills and experience, up to and including world-renowned industry experts; thus, they can often provide the exact skills and expertise that an organization requires for a particular project. Consulting firms work with their clients on engagements for which there are typically well-defined expected results or deliverables that must be produced (e.g., an IT strategic plan, implementation of an Enterprise Resource Planning [ERP] system, or selection of a hardware vendor). The contract with a consulting firm typically specifies the length of the engagement and the rate of pay for each of the consultants, who are directed on the engagement by a senior manager or director from the consulting firm. Table 10-2 shows the world's largest IT consulting firms (in alphabetical order).

TABLE 10-2 Large IT consulting firms

Firm	Headquarters
Accenture	Dublin, Ireland
Deloitte Touche Tohmatsu	New York, New York
Electronic Data Systems	Plano, Texas
Ernst & Young	New York, New York
HP Enterprise Business	Palo Alto, CA
IBM Global Business Services	Armonk, New York
Infosys	Bangalore, India
KPMG	Amstelveen, Netherlands
Tata Consultancy Services	Mumbai, India
Wipro Technologies	Bangalore, India

Source Line: Course Technology/Cengage Learning.

Advantages of Using Contingent Workers

When a firm employs a contingent worker, it does not usually have to provide benefits such as insurance, paid time off, and contributions to a retirement plan. A company can easily adjust the number of contingent workers it uses to meet its business needs, and can release contingent workers when they are no longer needed. An organization cannot usually do the same with full-time employees without creating a great deal of ill will and negatively impacting employee morale. Moreover, because many contingent workers are already specialists in a particular task, a firm does not customarily incur training costs for contingent workers. Therefore, the use of contingent workers can enable a firm to meet its staffing needs more efficiently, lower its labor costs, and respond more quickly to changing market conditions.

Disadvantages of Using Contingent Workers

One downside to using contingent workers is that they may not feel a strong connection to the company for which they are working. This can result in a low commitment to the company and its projects, along with a high turnover rate. Although contingent workers may already have the necessary technical training for a temporary job, many contingent workers gain additional skills and knowledge while working for a particular company; those assets are lost to the company when a contingent worker departs at a project's completion.

Deciding When to Use Contingent Workers

When an organization decides to use contingent workers for a project, it should recognize the trade-off it is making between completing a single project quickly and cheaply versus developing people within its own organization. If the project requires unique skills that are probably not necessary for future projects, there may be little reason to invest the additional time and costs required to develop those skills in full-time employees. Or, if a particular project requires only temporary help that will not be needed for future projects, the use of contingent workers is a good approach. In such a situation, using contingent workers avoids the need to hire new employees and then fire them when staffing needs decrease.

Organizations should carefully consider whether or not to use contingent workers when those workers are likely to learn corporate processes and strategies that are key to the company's success. It is next to impossible to prevent contingent workers from passing on such information to subsequent employers. This can be damaging if the worker's next employer is a major competitor.

Although using contingent workers is often the most flexible and cost-effective way to get a job done, their use can raise ethical and legal issues about the relationships among the staffing firm, its employees, and its customers—including the potential liability of a staffing firm's customers for withholding payroll taxes, payment of employee retirement benefits and health insurance premiums, and administration of workers' compensation to the staffing firm's employees. Depending on how closely workers are supervised and how the job is structured, contingent workers may be viewed as permanent employees by the Internal Revenue Service, the Labor Department, or a state's workers' compensation and unemployment agencies.

For example, in 2001, Microsoft agreed to pay a $97 million settlement to some 10,000 "permatemps"—temporary workers who were employed for an extended length of time as software testers, graphic designers, editors, technical writers, receptionists, and office support staffers. Some had worked at Microsoft for several years. The *Vizcaino v. Microsoft* class action was filed in 1992 by eight former workers who claimed that they—and thousands more permatemps—had been illegally shut out of a stock purchase plan that allowed employees to buy Microsoft stock at a 15 percent discount. Microsoft shares had skyrocketed in value throughout the 1990s. The sharp appreciation in the stock price meant that had they been eligible, some temporary workers in the lawsuit could have earned more money from stock gains than they received in salary while at Microsoft.[14]

The *Vizcaino v. Microsoft* lawsuit dramatically illustrated the cost of misclassifying employees and violating laws that cover compensation, taxes, unemployment insurance, and overtime. The key lesson of this case is that even if workers sign an agreement indicating that they are contractors and not employees, the deciding factor is not the agreement but the degree of control the company exercises over the employees. The following questions can help determine whether a worker is an employee:

- Does the worker have the right to control the manner and means of accomplishing the desired result?
- How much work experience does the person have?
- Does the worker provide his own tools and equipment?
- Is the worker engaged in a distinct occupation or an independently established business?
- Is the method of payment by the hour or by the job?
- What degree of skill is required to complete the job?
- Does the worker hire employees to help?

The Microsoft ruling means that employers must exercise care in their treatment of contingent workers. If a company wants to hire contingent workers through an agency, then the agency must hire and fire the workers, promote and discipline them, do performance reviews, decide wages, and tell them what to do on a daily basis.

Read the manager's checklist in Table 10-3 for questions that pertain to the use of contingent workers. The preferred answer to each question is *yes*.

TABLE 10-3 Manager's checklist for the use of contingent employees

Question	Yes	No
Have you reviewed the definition of an employee in your company's policies and pension plan documents to ensure it is not so broad that it encompasses contingent workers, thus entitling them to benefits?		
Are you careful not to use contingent workers on an extended basis? Do you make sure the assignments are finite, with break periods in between?		
Do you use contracts that specifically designate workers as contingent workers?		
Are you aware that the actual circumstances of the working relationship determine whether a worker is considered an employee in various contexts, and that a company's definition of a contingent worker may not be accepted as accurate by a government agency or court?		
Do you avoid telling contingent workers where, when, and how to do their jobs and instead work through the contingent worker's manager to communicate job requirements?		
Do you request that contingent workers use their own equipment and resources, such as computers and email accounts?		
Do you avoid training your contingent workers?		
When leasing employees from an agency, do you let the agency do its job? Do you avoid asking to see résumés and getting involved with compensation, performance feedback, counseling, or day-to-day supervision?		
If you lease employees, do you use a leasing firm that offers its own benefits plan, deducts payroll taxes, and provides required insurance?		

Source Line: Course Technology/Cengage Learning.

H-1B WORKERS

An **H-1B visa** is a temporary work visa granted by the U.S. Citizenship and Immigration Services (USCIS) for people who work in specialty occupations—jobs that require at least a four-year bachelor's degree in a specific field, or equivalent experience. Many companies turn to H-1B workers to meet critical business needs or to obtain essential technical skills and knowledge that cannot be readily found in the United States. H-1B workers may also be used when there are temporary shortages of needed skills. Employers often need H-1B professionals to provide special expertise in overseas markets or on projects that enable U.S. businesses to compete globally. A key requirement for using H-1B workers is that employers must pay H-1B workers the prevailing wage for the work being performed.

A person can work for a U.S. employer as an H-1B employee for a maximum continuous period of six years. After a worker's H-1B visa expires, the foreign worker must remain outside the United States for one year before another H-1B petition will be approved. Table 10-4 shows the employers who received approval for the most H-1B visas for the decade 2001 to 2010.

TABLE 10-4 Top H-1B visa employers in 2001–2010

Company	Total H-1B visas granted (2001–2010)
Satyam Computer Services	24,775
Microsoft	24,044
Enterprise Business Solutions	12,789
IBM	12,693
Patni Computer Systems	9,971
Oracle	8,518
Ernst & Young	7,322
Intel	7,034
Infosys Technologies	7,018

Source Line: GaramChai.com, "Top H-1B Employers 2009–2010," www.garamchai.com/TopH1b.htm. GaramChai.com, "Top H-1B Employers for 2001–2009," www.garamchai.com/TopH1b2009.htm.

The top five countries of birth for H-1B workers in 2009 were India with 48% of all approved H-1B petitions, China (10%), Canada (4%), the Philippines (4%), and Korea (4%).[15]

Each year the U.S. Congress sets an annual cap on the number of H-1B visas to be granted—although the number of visas issued often varies greatly from this cap. Since 2004, the cap has been set at 65,000, with an additional 20,000 visas available for foreign graduates of U.S. universities with advanced degrees.[16] The cap only applies to certain IT professionals, such as programmers and engineers at private technology companies. A large number of foreign workers are exempt from the cap, including scientists hired to teach at American universities, work in government research labs, or work for nonprofit organizations.

The administration of President George W. Bush tried unsuccessfully to get Congress to increase the H-1B visa cap. So, in 2008, the administration extended the amount of time that a foreign student with a science, engineering, math, or technology degree could work without a visa—from 12 to 29 months. Some 20,000 students have applied for this Optional Practical Training extension in the first two years of the program. Many of these students complete their studies in the United States, gain two and one-half years of work experience, and then go back to their home country and obtain great jobs. Some people have proposed an alternative strategy of extending H-1B visas to such students—enabling them to be absorbed into the U.S. economy instead of sending them home and losing their brainpower and skills.[17]

When considering the use of H-1B visa workers, companies should take into account that even highly skilled and experienced H-1B workers may require help with their English skills. Communication in many business settings is fast paced and full of idiomatic expressions; workers who are not fluent in English may find it difficult and uncomfortable to participate. As a result, some H-1B workers might become isolated. Even worse, H-1B workers who are not comfortable with English may gradually stop trying to acclimate and may create their own cliques, which can hurt a project team's morale and lead to division. Managers and coworkers should make it a priority to assist H-1B workers looking to improve their English skills and to develop beneficial working relationships based on a mutual respect for any cultural differences that may exist. H-1B workers must feel at ease and be able to interact easily and feel like true members of their team.

As concern increases about employment in the IT sector, displaced workers and other critics challenge whether the United States needs to continue importing thousands of H-1B workers every year. Many business managers, however, say such criticisms conceal the real issue, which is the struggle to find qualified people, wherever they are, for increasingly challenging work. Heads of U.S. companies continue to complain that they have trouble finding enough qualified IT employees and have urged the USCIS to loosen the reins on visas for qualified workers. They warn that reducing the number of visas will encourage them to move work to foreign countries, where they can find the workforce they need. Some human resource managers and educators are concerned that the continued use of H-1B workers may be a symptom of a larger, more fundamental problem—that the United States is not developing a sufficient number of IT employees with the right skills to meet its corporate needs.

Many IT workers in the United States have expressed concern that the use of H-1B workers has a negative impact on their wages. Researchers from New York University and the Wharton School of the University of Pennsylvania examined tens of thousands of résumés as well as demographic and wage data on 156,000 IT workers employed at 7,500 publicly held U.S. firms. Their conclusion was that "H-1B admissions at current levels are associated with a 5% to 6% drop in wages for computer programmers, systems analysts, and software engineers." The researchers also concluded that "there is substantial evidence that H-1B admissions appear to directly improve levels of innovation and entrepreneurship, which in the long term should create new jobs and raise demand for technology workers in other areas."[18]

H-1B Application Process

Most companies make ethical hiring decisions—based on how well an applicant fulfills the job qualifications. Such companies consider the need to obtain an H-1B visa *after* deciding to hire the best available candidate. To receive an H-1B visa, the person must have a job offer from an employer who is also willing to offer sponsorship.

Once a decision has been made to hire a worker who will require an H-1B visa, an employer must begin the application process. There are two application stages: the Labor Condition Application (LCA) and the H-1B visa application. The company files an LCA with the Department of Labor (DOL), stating the job title, the geographic area in which the worker is needed, as well as the salary to be paid. The DOL's Wage and Hour Division reviews the LCA to ensure that the foreign worker's wages will not undercut those of an American worker. After the LCA is certified, the employer may then apply to the USCIS for the H-1B visa, identifying who will fill the position and stating the person's skills and qualifications for the job. A candidate cannot be hired until the USCIS has processed the application, which can take several days or several months.[19]

A company whose contingent of H1-B workers makes up more than 15 percent of its workforce faces hurdles before it can hire any more. To do so, the company must prove that it first tried to find U.S. workers—for example, the company can show copies of employment ads it placed online or in newspapers. A company must also confirm that it is not hiring an H-1B worker after having laid off a similar U.S. worker. Employers must attest to such protections by affirmatively filing with the DOL and maintaining a public file. Failure to comply with DOL regulations can result in an audit and fines in excess of $1,000 per violation, payment of back wages to the employee, and ineligibility to participate in immigration programs.[20]

In order to cut down on the amount of time it can take to hire an H-1B worker, a provision was added to the American Competitiveness in the Twenty-First Century Act of 2000 that allows current H-1B holders to start working for employers as soon as their petitions are filed. Therefore, a company looking to hire a critical person who already has H-1B status can do so in a matter of weeks.

Using H-1B Workers Instead of U.S. Workers

In order to compete in the global economy, U.S. firms must be able to attract the best and brightest workers from all over the world. Most H-1B workers are brought to the United States to fill a legitimate gap that cannot be filled from the existing pool of workers. However, there are some managers who reason that as long as skilled foreign workers can be found to fill critical positions, why invest thousands of dollars and months of training to develop their current U.S. workers? Although such logic may appear sound for short-term hiring decisions, it does nothing to develop the strong core of permanent IT workers that the United States will need in the future. Heavy reliance on the use of H-1B workers can lessen the incentive for U.S. companies to educate and develop their own workforces.

Potential Exploitation of H-1B Workers

Even though companies applying for H-1B visas must offer a wage that is not 5 percent less than the average salary for the occupation, some companies use H-1B visas as a way to lower salaries. Because wages in the IT field vary greatly, unethical companies can get around the average salary requirement. Determining an appropriate wage is an imprecise science at best. For example, an H-1B worker may be classified as an entry-level IT employee and yet fill a position of an experienced worker who would make $10,000 to

$30,000 more per year. Unethical companies can also find other ways to get around the salary protections included in the H-1B program, as shown in the charges filed against the Vision Systems Group. In 2009, the company was charged with H-1B visa fraud for allegedly stating that certain H-1B workers in its New Jersey office were actually working in Iowa, where the company had another office and where the prevailing wage is much less. The president and another officer of the firm were sentenced to three years' probation, and the company was forced to pay $236,250 to the U.S. Citizenship and Immigration Services.[21]

Until Congress approved the Visa Reform Act of 2004, there were few investigations into H-1B salary abuses. The act increased the H-1B application fee by $2,000, of which $500 was earmarked for antifraud efforts; the act also defined a modified wage-rate system, allowing for greater variances in pay to visa holders. Investigations are typically triggered by complaints from H-1B holders, but the government can conduct random audits or launch an investigation based on information from third-party sources.

Companies using H-1B workers, as well as the workers themselves, must also consider what will happen at the end of the six-year H-1B visa term. The stopgap nature of the visa program can be challenging for both sponsoring companies and applicants. If a worker is not granted a green card, the firm can lose a worker without having developed a permanent employee. Many of these foreign workers, finding that they are suddenly unemployed, are forced to uproot their families and return home.

B-1 VISA CONTROVERSY

A "visitor" visa is a nonimmigrant visa for people who wish to enter the United States temporarily for business (B-1), for pleasure or medical treatment (B-2), or for both purposes (B-1/B-2). Business travelers who travel for short periods of time to consult with business associates; attend a convention or conference; negotiate a contract; or install, service, or repair machinery must obtain a B-1 visa.

According to the U.S. Department of State, some companies may be using the B-1 visa to get around the cap and wage restrictions of the H-1B program. For instance, Infosys Technologies (a major Indian software firm with operations in the United States) makes extensive use of the H-1B program and is now the subject of a federal grand jury investigation relating to its use of B-1 visas. The grand jury probe was touched off by a lawsuit filed by one the company's U.S. employees, who alleges he was harassed for refusing to assist in a plan to use Infosys employees holding B-1 visas for work he argues requires an H-1B visa.[22] The B-1 visa is faster, easier, and cheaper to obtain and has no systematic oversight or audits from federal agencies. There is a lot of gray area in the use of the B-1 visa that the Infosys case may help to define.

OUTSOURCING

Outsourcing is another approach to meeting staffing needs. **Outsourcing** is a long-term business arrangement in which a company contracts for services with an outside organization that has expertise in providing a specific function. A company may contract with an organization to provide services such as operating a data center, supporting a telecommunications network, or staffing a computer help desk.

Coemployment legal problems with outsourcing are minimal, because the company that contracts for the services does not generally supervise or control the contractor's employees.

The primary rationale for outsourcing is to lower costs, but companies also use it to obtain strategic flexibility and to keep their staff focused on the company's core competencies.

In the 1970s, IT executives started the trend toward outsourcing as they began to supplement their IT staff with contractors and consultants. This trend eventually led to companies outsourcing entire IT business units to such organizations as Accenture, Electronic Data Systems, and IBM, which could take over the operation of a company's data center as well as perform other IT functions.

Offshore Outsourcing

Offshore outsourcing is a form of outsourcing in which the services are provided by an organization whose employees are in a foreign country. Any work done at a relatively high cost in the United States may become a candidate for offshore outsourcing—not just IT work. However, IT professionals in particular can do much of their work anywhere—on a company's premises or thousands of miles away in a foreign country. In addition, companies can reap large financial benefits by reducing labor costs through offshore outsourcing. As a result, and because a large supply of experienced IT professionals is readily available in certain foreign countries, the use of offshore outsourcing in the IT field is increasing. A 2010 survey indicated that 93 percent of multinational companies had undertaken some sort of IT outsourcing project.[23] American Express, Aetna, Compaq, General Electric, IBM, Microsoft, Motorola, Shell, Sprint, and 3M are examples of big companies that employ offshore outsourcing for functions such as help-desk support, network management, and information systems development.

As more businesses move their key processes offshore, U.S. IT service providers are forced to lower prices. Many U.S. software firms set up development centers in low-cost foreign countries where they have access to a large pool of well-trained candidates. Intuit—maker of the Quicken tax preparation software—currently has facilities in Canada, Great Britain, and India. Accenture, IBM, and Microsoft all maintain large development centers in India. Cognizant Technology Solutions is headquartered in Teaneck, New Jersey, but operates primarily from technology centers in India.

Because of the high salaries earned by application developers in the United States and the ease with which customers and suppliers can communicate, it is now quite common to use offshore outsourcing for major programming projects. According to the Gartner Group, some of the top sources of contract programming include Argentina, Australia, Belarus, Brazil, Bulgaria, Canada, China, India, Ireland, Israel, Malaysia, Malta, Mexico, Nepal, the Philippines, Poland, Russia, Serbia, Singapore, Sri Lanka, and Vietnam.[24] India, with its rich talent pool (a high percentage of whom speak English) and low labor costs, is considered one of the best sources of programming skills outside Europe and North America.

Organizations must consider many factors when deciding where to locate outsourcing activities. For example, political unrest in Egypt reduced the attractiveness of that country as a source of IT outsourcing, particularly after the government there temporarily blocked all Internet and cell phone service.[25] Global management consulting firm A.T. Kearney publishes an annual Global Services Location Index™ that ranks the 50 most attractive offshoring destinations based on 39 measures across three primary categories: financial attractiveness, people and skills availability, and overall business environment. Table 10-5 lists the top ten countries from the index's 2011 rankings.

TABLE 10-5 Most attractive offshoring destinations (Based on A.T. Kearney rating methodology)

Country
1. India
2. China
3. Malaysia
4. Egypt
5. Indonesia
6. Mexico
7. Thailand
8. Vietnam
9. Philippines
10. Chile

Source Line: A.T. Kearney, Inc., "A.T. Kearney's Global Services Location Index™," © 2011, www.atkearney.com/index.php/Publications/at-kearneys-global-services-location-index-volume-xiii-number-2-2010.html.

In 2011, Nokia, the Finnish mobile device manufacturer, and Accenture, the global consulting company headquartered in Ireland, agreed to a major outsourcing deal in which Accenture will provide Nokia with software development and support services for the once popular Symbian mobile operating system and computing platform. Some 2,800 Nokia employees in China, Denmark, Finland, India, the United Kingdom, and the United States will transfer to Accenture as part of the deal.[26] Symbian has fallen out of favor with phone handset manufacturers, and Nokia has decided to transition to a Windows Phone platform for its line of smartphones. As part of the outsourcing agreement, Accenture will become the preferred supplier to Nokia to aid its transition from Symbian to the Windows Phone platform.[27]

Table 10-6 lists the top IT outsourcing firms according to the International Association of Outsourcing Professionals.

TABLE 10-6 Top-rated IT outsourcing firms according to the International Association of Outsourcing Professionals

Firm	Headquarters location
Accenture	Dublin, Ireland
Infosys Technologies	Bangalore, India
CSC	Falls Church, Virginia
Wipro Technologies	Bangalore, India
Capgemini S.A.	Paris, France
PCCW Solutions	Hong Kong

(Continued)

Firm	Headquarters location
CGI Group	Montreal, Quebec, Canada
HCL Technologies	New Delhi, India
ITC Infotech	Bangalore, India

Source Line: International Association of Outsourcing Professionals, "The 2011 Global Outsourcing 100," © 2011, www.iaop.org/Content/19/165/1793/Default.aspx.

Pros and Cons of Offshore Outsourcing

Wages that an American worker might consider low represent an excellent salary in many other parts of the world, and some companies feel they would be foolish not to exploit such an opportunity. Why pay a U.S. IT worker a six-figure salary, they reason, when they can use offshore outsourcing to hire three India-based workers for the same cost? However, this attitude might represent a short-term point of view—offshore demand is driving up salaries in India by roughly 15 percent per year. Because of this, Indian offshore suppliers have begun to charge more for their services. The cost advantage for offshore outsourcing to India used to be 6:1 or more—you could hire six Indian IT workers for the cost of one U.S. IT worker. The cost advantage is shrinking, and once it reaches about 1.5:1, the cost savings will no longer be much of an incentive for U.S. offshore outsourcing to India.

Another benefit of offshore outsourcing is its potential to dramatically speed up software development efforts. For example, the state of New Mexico contracted the development of a tax system to Syntel, one of the first U.S. firms to successfully launch a global delivery model that enables workers to work on a project around the clock. With technical teams working from networked facilities in different time zones, Syntel executes a virtual "24-hour workday" that saves its customers money, speeds projects to completion, and provides continuous support for key software applications.

While offshore outsourcing can save a company in terms of labor costs, it will also result in other expenses. In determining how much money and time a company will save with offshore outsourcing, the firm must take into account the additional time that will be required to select an offshore vendor as well as the additional costs that will be incurred for travel and communications. In addition, organizations often find it takes years of ongoing effort and a large up-front investment to develop a good working relationship with an offshore outsourcing firm. Finding a reputable vendor can be especially difficult for a small or midsized firm that lacks experience in identifying and vetting contractors.

Many of the ethical issues that arise when considering whether to use H-1B and contingent workers also apply to offshore outsourcing. For example, managers must consider the trade-offs between using offshore outsourcing firms and devoting money and time to retain and develop their own staff. Often, companies that begin offshoring also lay off portions of their own staff as part of that move. For example, Dex One Corporation, whose products include yellow pages print directories and an online ad network, outsourced much of its IT work to HCL Technologies to speed up development of new digital offerings while simultaneously cutting operational costs. As a result of this deal, about 30 percent of the Dex One IT staff will be eliminated.[28] Offshore outsourcing tends to upset domestic staff when a company begins to lay off employees in favor of low-wage workers outside the United States. The remaining members of a department may become bitter and nonproductive, and morale may be affected.

Cultural and language differences can cause misunderstandings among project members in different countries. For example, in some cultures, shaking one's head up and down simply means "Yes, I understand what you are saying." It does not necessarily mean "Yes, I agree with what you are saying." And the difficulty of communicating directly with people over long distances can make offshore outsourcing perilous, especially when key team members speak English as their second language.

The compromising of customer data is yet another potential outsourcing issue. For example, Atlanta's Grady Memorial Hospital discovered that 45 patient records—including doctor's notes, diagnoses, and medical conditions—were accessible on an unsecured, publicly available Web site for a few weeks due to an error by outsourcing firm in India. The hospital had outsourced the job of transcribing patient records to a Georgia firm, which outsourced it to a Nevada contractor, which in turn outsourced the job to the company in India.[29] Clearly organizations that outsource must take precautions to protect private data, regardless of where it is stored or processed.

Another downside to offshore outsourcing is that a company loses the knowledge and experience gained by outsourced workers when those workers are reassigned after a project's completion. Finally, offshore outsourcing does not advance the development of permanent IT workers in the United States, which increases its dependency on foreign workers to build the IT infrastructure of the future. Many of the jobs that go overseas are entry-level positions that help develop employees for future, more responsible positions.

Strategies for Successful Offshore Outsourcing

Successful projects require day-to-day interaction between software development and business teams, so it is essential for the hiring company to take a hands-on approach to project management. Companies cannot afford to outsource responsibility and accountability.

To improve the chances that an offshore outsourcing project will succeed, a company must carefully evaluate whether an outsourcing firm can provide the following:

- Employees with the required expertise in the technologies involved in the project
- A project manager who speaks the employer company's native language
- A pool of staff large enough to meet the needs of the project
- A state-of-the-art telecommunications setup
- High-quality on-site managers and supervisors

To ensure that company data is protected in an outsourcing arrangement, companies can use the Statement on Auditing Standards (SAS) No. 70, Service Organizations, an internationally recognized standard developed by the American Institute of Certified Public Accountants (AICPA). A successful SAS No. 70 audit report demonstrates that an outsourcing firm has effective internal controls in accordance with the Sarbanes–Oxley Act of 2002.

The following list provides several tips for companies that are considering offshore outsourcing:

- Set clear, firm business specifications for the work to be done.
- Assess the probability of political upheavals or factors that might interfere with information flow, and ensure the risks are acceptable.
- Assess the basic stability and economic soundness of the outsourcing vendor and what might occur if the vendor encounters a severe financial downturn.
- Establish reliable satellite or broadband communications between your site and the outsourcer's location.
- Implement a formal version-control process, coordinated through a quality assurance person.
- Develop and use a dictionary of terms to encourage a common understanding of technical jargon.
- Require vendors to have project managers at the client site to overcome cultural barriers and facilitate communication with offshore programmers.
- Require a network manager at the vendor site to coordinate the logistics of using several communications providers around the world.
- Obtain advance agreement on the structure and content of documentation to ensure that manuals explain how the system was built, as well as how to maintain it.
- Carefully review a current copy of the outsourcing firm's SAS No. 70 audit report to ascertain its level of control over information technology and related processes.

WHISTLE-BLOWING

Like the subject of contingent workers, whistle-blowing is a significant topic in any discussion of ethics in IT. Both issues raise ethical questions and have social and economic implications. How these issues are addressed can have a long-lasting impact not only on the people and employers involved, but also on the entire IT industry.

As noted previously, whistle-blowing is an effort to attract public attention to a negligent, illegal, unethical, abusive, or dangerous act by a company or some other organization. In some cases, whistle-blowers are employees who act as informants on their company, revealing information to enrich themselves or to gain revenge for a perceived wrong. In most cases, however, whistle-blowers act ethically in an attempt to correct what they think is a major wrongdoing, often at great personal risk.

A whistle-blower usually has personal knowledge of what is happening inside the offending organization because of his or her role as an employee of the organization. Sometimes the whistle-blower is not an employee but a person with special knowledge gained from a position as an auditor or business partner.

In going public with the information they have, whistle-blowers often risk their own careers and sometimes even affect the lives of their friends and family. In extreme situations, whistle-blowers must choose between protecting society and remaining silent.

Protection for Whistle-Blowers

Whistle-blower protection laws allow employees to alert the proper authorities to employer actions that are unethical, illegal, or unsafe, or that violate specific public policies. Unfortunately, no comprehensive federal law protects all whistle-blowers from retaliatory acts. Instead, numerous laws protect a certain class of specific whistle-blowing acts in various industries. To make things even more complicated, each law has different filing provisions, administrative and judicial remedies, and statutes of limitations (which set time limits for legal action). Thus, the first step in reviewing a whistle-blower's claim of retaliation is for an experienced attorney to analyze the various laws and determine if and how the employee is protected. Once that is known, the attorney can determine what procedures to follow in filing a claim.

From the whistle-blower's perspective, a short statute of limitations is a major weakness of many whistle-blower protection laws. Failure to comply with the statute of limitations is a favorite defense of firms accused of wrongdoing in whistle-blower cases.

The **False Claims Act**, also known as the Lincoln Law, was enacted during the U.S. Civil War to combat fraud by companies that sold supplies to the Union Army. War profiteers sometimes shipped boxes of sawdust instead of guns, for instance, and some swindled the Union Army into purchasing the same cavalry horses several times. When it was enacted, the act's goal was to entice whistle-blowers to come forward by offering them a share of the money recovered.

The **qui tam** ("who sues on behalf of the king as well as for himself") provision of the False Claims Act allows a private citizen to file a suit in the name of the U.S. government, charging fraud by government contractors and other entities who receive or use government funds. In qui tam actions, the government has the right to intervene and join the legal proceedings. If the government declines, the private plaintiff may proceed alone. Some states have passed similar laws concerning fraud in state government contracts.[30]

Qui tam actions can be based on a variety of charges, including mischarging for services, product and service substitution, false certification of entitlement for benefits, and false negotiation to justify an inflated contract. Mischarging is the most common charge in qui tam cases.[31] For example, an IT contractor might overcharge hundreds of hours of programming time as part of a government contract, or a physician might charge the government for medical services that a nurse actually performed.

Violators of the False Claims Act are liable for three times the dollar amount for which the government was defrauded. They can also be fined civil penalties of $5,000 to $10,000 for each instance of a false claim. A qui tam plaintiff can receive between 15 and 30 percent of the total recovery from the defendant, depending on how helpful the person was to the success of the case.[32]

In a lawsuit initially brought by a whistle-blower in 2007 and settled in 2011, Verizon agreed to pay $93.5 million to settle allegations that for years it overcharged the federal government on voice and data communications contracts. The whistle-blower alleged Verizon had billed the government for "tax-like" surcharges to which the government was not subject.[33] In this case, the government refused to pay the whistle-blower the statutory minimum amount of 15 percent ($14 million) because it disputed "the extent to which the relater 'substantially contributed' to the $93.5 million settlement between Verizon and the United States." Instead the whistle-blower was awarded just $4 million, an amount he is challenging in court.[34]

The False Claims Act provides strong whistle-blower protection. Any person who is discharged, demoted, harassed, or otherwise discriminated against because of lawful acts of whistle-blowing is entitled to all relief necessary "to make the employee whole." Such relief may include job reinstatement; double back pay; and compensation for any special damages, including litigation costs and reasonable attorney's fees.[35]

The provisions of the False Claims Act are complicated, so it is unwise to pursue a claim without legal counsel. However, because the potential for significant financial recovery is good, attorneys are generally willing to assist.

Whistle-Blowing Protection for Private-Sector Workers

Under state law, an employee could traditionally be terminated for any reason, or no reason, in the absence of an employment contract. However, many states have created laws that prevent workers from being fired because of an employee's participation in "protected" activities. One such activity is the filing of a qui tam lawsuit under the provisions of the False Claims Act. States that recognize the public benefit of such cases offer protection to whistle-blowers; for example, whistle-blowers may be able to file claims against their employers for retaliatory termination and may be entitled to a jury trial. If successful, they may receive punitive damage awards.

Dealing with a Whistle-Blowing Situation

Each potential whistle-blowing case involves different circumstances, issues, and personalities. Two people working together in the same company may have different values and concerns that cause them to react in different ways to a particular situation—and both reactions might be ethical. It is impossible to outline a definitive step-by-step procedure of how to behave in a whistle-blowing situation. This section provides a general sequence of events, and highlights key issues that a potential whistle-blower should consider.

Assess the Seriousness of the Situation

Before considering whistle-blowing, a person should have specific knowledge that his or her company or a coworker is acting unethically and that the action represents a *serious* threat to the public interest. The employee should carefully and informally seek trusted resources outside the company and ask for their assessment. Do they also see the situation as serious? Their point of view may help the employee see the situation from a different perspective and alleviate concerns. On the other hand, the outside resources may reinforce the employee's initial suspicions, forcing a series of difficult ethical decisions.

Begin Documentation

An employee who identifies an illegal or unethical practice should begin to compile adequate documentation to establish wrongdoing. The documentation should record all events and facts as well as the employee's insights about the situation. This record helps construct a chronology of events if legal testimony is required in the future. An employee should identify and copy all supporting memos, correspondence, manuals, and other documents *before* taking the next step. Otherwise, records may disappear and become inaccessible. The employee should maintain documentation and keep it up to date throughout the process.

Attempt to Address the Situation Internally

An employee should next attempt to address the problem internally by providing a written summary to the appropriate managers. Ideally, the employee can expose the problem and deal with it from inside the organization. The focus should be on disclosing the facts and how the situation affects others. The employee's goal should be to fix the problem, not to place blame. Given the potential negative impact of whistle-blowing on the employee's future, this step should not be dismissed or taken lightly.

Fortunately, many problems are solved at this point, and further, more drastic actions by the employee are unnecessary. The appropriate managers get involved and resolve the issue that initiated the whistle-blower's action.

On the other hand, managers who are engaged in unethical or illegal behavior might not welcome an employee's questions or concerns. In such cases, the whistle-blower can expect to be strongly discouraged from taking further action. Employee demotion or termination on false or exaggerated claims can occur. Attempts at discrediting the employee can also be expected. As an extreme example, Dr. Jeffrey Wigand, former vice president of research and development at Brown & Williamson, disclosed wrongdoings involving the use of cancer-causing ingredients in the tobacco industry. As a result, he received several anonymous deaths threats; however, none of the threats could be traced back to their source.[36]

Consider Escalating the Situation Within the Company

The employee's initial attempt to deal with a situation internally may be unsuccessful. At this point, the employee may rationalize that he or she has done all that is required by raising the issue. Others may feel so strongly about the situation that they are compelled to take further action. Thus, a determined and conscientious employee may feel forced to choose between escalating the problem and going over the manager's head, or going outside the organization to deal with the problem. The employee may feel obligated to sound the alarm on the company because there appears to be no chance to solve the problem internally.

Going over an immediate manager's head can put one's career in jeopardy. Supervisors may retaliate against a challenge to their management, although some organizations may have an effective corporate ethics officer who can be trusted to give the employee a fair and objective hearing. Alternatively, a senior manager with a reputation for fairness and some responsibility for the area of concern might step in. However, in many work environments, the challenger is likely to be fired, demoted, or reassigned to a less desirable position or job location. Such actions send a loud signal throughout an organization that loyalty is highly valued and that challengers will be dealt with harshly. Whether reprisal is ethical depends in large part on the legitimacy of the employee's issue. If the employee is truly overreacting to a minor issue, then the employee may deserve some sort of reprimand for exercising poor judgment.

If senior managers refuse to deal with a legitimate problem, the employee can decide to drop the matter or go outside the organization to try to remedy the situation. Even if a senior manager agrees with the employee's position and overrules the employee's immediate supervisor, the employee may want to request a transfer to avoid working for the same person.

Assess the Implications of Becoming a Whistle-Blower

If whistle-blowers feel they have made a strong attempt to resolve the problem internally without results, they must stop and fully assess whether they are prepared to go forward and blow the whistle on the company. Depending on the situation, an employee may incur significant legal fees in order to air or bring charges against an agency or company that may have access to an array of legal resources as well as a lot more money than the individual employee. An employee who chooses to proceed might be accused of having a grievance with the employer or of trying to profit from the accusations. The employee may be fired and may lose the confidence of coworkers, friends, and even family members.

A potential whistle-blower must attempt to answer many ethical questions before making a decision on how to proceed:

- Given the potentially high price, do I really want to proceed?
- Have I exhausted all means of dealing with the problem? Is whistle-blowing all that is left?
- Am I violating an obligation to be loyal to my employer and work for its best interests?
- Will the public exposure of corruption and mismanagement in the organization really correct the underlying cause of these problems and protect others from harm?

From the moment an employee becomes known as a whistle-blower, a public battle may ensue. Whistle-blowers can expect attacks on their personal integrity and character as well as negative publicity in the media. Friends and family members will hear these accusations, and ideally, they should be notified beforehand and consulted for advice before the whistle-blower goes public. This notification helps prevent friends and family members from being surprised at future actions by the whistle-blower or the employer.

The whistle-blower should also consider consulting support groups, elected officials, and professional organizations. For example, the National Whistleblowers Center provides referrals for legal counseling and education about the rights of whistle-blowers.

Use Experienced Resources to Develop an Action Plan

A whistle-blower should consult with competent legal counsel who has experience in whistle-blowing cases. He or she will determine which statutes and laws apply, depending on the agency, employer, state involved, and nature of the case. Counsel should also know the statute of limitations for reporting the offense, as well as the whistle-blower's protection under the law. Before blowing the whistle publicly, the employee should get an honest assessment of the soundness of his or her legal position and an estimate of the costs of a lawsuit.

Execute the Action Plan

A whistle-blower who chooses to pursue a matter legally should do so based on the research and guidance of legal counsel. If the whistle-blower wants to remain unknown, the safest course of action is to leak information anonymously to the press. The problem with this approach, however, is that anonymous claims are often not taken seriously. In most cases, working directly with appropriate regulatory agencies and legal authorities is more likely to get results, including the imposition of fines, the halting of operations, or other actions that draw the offending organization's immediate attention.

Live with the Consequences

Whistle-blowers must be on guard against retaliation, such as being discredited by coworkers, threatened, or set up; for example, management may attempt to have the whistle-blower transferred, demoted, or fired for breaking some minor rule, such as arriving late to work or leaving early. To justify their actions, management may argue that such behavior has been ongoing. The whistle-blower might need a good strategy and a good attorney to counteract such actions and take recourse under the law.

A massive computer-data breach at TJX (the parent company of T.J. Maxx, Marshalls, and other stores) affecting 94 million Visa and MasterCard accounts occurred in June 2005.[37] A college student who was an hourly worker at TJX noticed many computer-related security problems at the firm prior to the data breach. He reported these verbally to TJX managers and also posted information about the breaches on an online security forum, *http://sla.ckers.org*. In the forum, he revealed serious security weaknesses in sufficient detail that the information could be of use to hackers. The employee spoke to store managers and the district loss prevention manager before the data breach occurred, but nothing was done. Eventually the worker was fired over the public disclosures and violation of his nondisclosure agreement.[38] This is a perfect example of how *not* to be a whistle-blower.

GREEN COMPUTING

Many computer manufacturers today are talking about building a "green PC," by which they usually mean one that uses less electricity to run than the standard computer; thus, its carbon footprint on the planet is smaller. However, to manufacturer a truly green PC, a hardware company must also reduce the amount of hazardous materials and dramatically increase the amount of reusable or recyclable materials used in its manufacturing and packaging processes. The manufacturers must also help consumers dispose of their products in an environmentally safe manner at the end of their useful life.

Electronic devices such as personal computers and cell phones contain hundreds or even thousands of components. The components, in turn, are composed of many different materials, including some that are known to be potentially harmful to humans and the environment, such as beryllium, cadmium, lead, mercury, brominated flame retardants (BFRs), selenium, and polyvinyl chloride.[39]

Electronics manufacturing employees and suppliers at all steps along the supply chain and manufacturing process are at risk of unhealthy exposure to these raw materials. Users of these products can also be exposed to these materials when using poorly designed or improperly manufactured devices. Care must also be taken when recycling or destroying these devices to avoid contaminating the environment. The United States has no federal law prohibiting the export of toxic waste, so many used electronic devices intended for recycling are sold to companies in developing countries that try to repair the components or extract valuable metals from them, using methods that release carcinogens and other toxins into the air and the water supply.[40]

Electronic Product Environmental Assessment Tool (EPEAT) is a system that enables purchasers to evaluate, compare, and select electronic products based on a total of 51 environmental criteria. Over 45 participating manufacturers currently register some 3,200 unique products across 41 countries. Products are ranked in EPEAT according to three tiers of environmental performance: Bronze (meets all 23 required criteria), Silver (meets all 23 of the required criteria plus at least 50 percent of the optional criteria), and Gold (meets all 23 required criteria plus at least 75 percent of the optional criteria).[41] Individual purchasers of home computers as well as corporate purchasers of thousands of computers can use the EPEAT Web site (*www.epeat.net*) to screen manufacturers and computer models based on certain environmental attributes.

The European Union's Restriction of Hazardous Substances Directive, which took effect in 2006, restricts the use of many hazardous materials in computer manufacturing. The directive also requires that manufacturers use at least 65 percent reusable or recyclable components, implement a plan to manage products at the end of their life cycle in an environmentally safe manner, and reduce or eliminate toxic material in their packaging. The state of California has passed a similar law, called the Electronic Waste Recycling Act. Because of these two acts, manufacturers had a strong motivation to remove brominated flame retardants from their PC casings. By the start of 2010, the Apple iPad was free of arsenic, mercury, PVC (polyvinyl chloride), and BFRs. In addition, according to Apple, the iPad's aluminum and glass enclosure is "highly recyclable."[42]

It is estimated that each year over 40 million computers become obsolete.[43] How should users safely dispose of their obsolete computers? Over half the states have established statewide programs for some recycling of obsolete computers. These statutes either impose a fee for each unit sold at retail or require manufacturers to reclaim the equipment at disposal.[44]

Some electronics manufacturers have developed programs to assist their customers in disposing of old equipment. For example, Dell offers a free worldwide recycling program for consumers. It also provides no-charge recycling of any brand of used computer or printer with the purchase of a new Dell computer or printer. This equipment is recycled in an environmentally responsible manner, using Dell's stringent and global recycling guidelines.[45] HP and other manufacturers offer similar programs.

The environmental activist organization Greenpeace issues ratings of the top manufacturers of personal computers, mobile phones, TVs, and game consoles according to the manufacturers' policies on toxic chemicals, recycling, and climate change. Table 10-7 shows the companies with the top ten Greenpeace ratings in October 2010. With 10 being a perfect score, it is clear that these manufacturers have a long way to go in meeting the very high "green" standards of Greenpeace.

TABLE 10-7 Greenpeace ratings of the top ten electronics manufacturers

Organization	October 2010 rating
Nokia	7.5
Sony Ericsson	6.9
Phillips	5.5
HP	5.5
Samsung	5.3
Motorola	5.1
Panasonic	5.1
Sony	5.1
Apple	4.9
Dell	4.9

Source Line: Greenpeace, "Guide to Greener Electronics," © October 2010, www.greenpeace.org/international/campaigns/toxics/electronics/how-the-companies-line-up#.

ICT INDUSTRY CODE OF CONDUCT

The **Electronic Industry Citizenship Coalition (EICC)** was established to promote a common code of conduct for the electronics and ICT industry.[46] The EICC focuses on the areas of worker safety and fairness, environmental responsibility, and business efficiency. ICT organizations, electronic manufacturers, software firms, and manufacturing service providers may voluntarily join the coalition.

The EICC has established a code of conduct that defines performance, compliance, auditing, and reporting guidelines across five areas of social responsibility: labor, health and safety, environment, management system, and ethics. Adopting organizations apply the code across their entire worldwide supply chain and require their first-tier suppliers to acknowledge and implement it.[47] As of July 2009, the code has been formally adopted by over 53 EICC member organizations, including Adobe, Cisco, Dell, HP, IBM, Intel, Lenovo, Microsoft, Oracle, Philips, Samsung, Sony and Xerox. The following are the five areas of social responsibility and guiding principles covered by the code:[48]

1. **Labor**: "Participants are committed to uphold the human rights of workers, and to treat them with dignity and respect as understood by the international community."
2. **Health and Safety**: "Participants recognize that in addition to minimizing the incidence of work-related injury and illness, a safe and healthy work environment enhances the quality of products and services, consistency of production and worker retention and morale. Participants also recognize that ongoing worker input and education is essential to identifying and solving health and safety issues in the workplace."

3. **Environmental**: "Participants recognize that environmental responsibility is integral to producing world class products. In manufacturing operations, adverse effects on the community, environment, and natural resources are to be minimized while safeguarding the health and safety of the public."
4. **Management System**: "Participants shall adopt or establish a management system whose scope is related to the content of this Code. The management system shall be designed to ensure (a) compliance with applicable laws, regulations and customer requirements related to the participant's operations and products; (b) conformance with this Code; and (c) identification and mitigation of operational risks related to this Code. It should also facilitate continual improvement."
5. **Ethics**: "To meet social responsibilities and to achieve success in the marketplace, participants and their agents are to uphold the highest standards of ethics including: business integrity; no improper advantage; disclosure of information; intellectual property; fair business, advertising, and competition; and protection of identity."

Prior to the adoption of the EICC Code of Conduct, many electronic manufacturing companies developed their own codes of conduct and used them to audit their suppliers. Thus, suppliers could be subjected to multiple, independent audits based on different criteria. The adoption of a single, global code of conduct by members of the EICC enables those companies to provide leadership in the area of corporate social responsibility. It also exerts pressure on suppliers to meet a common set of social principles.

The EICC has developed an audit program for member organizations in which audits are conducted by certified, third-party audit firms. EICC members use the audits to measure supplier compliance with the EICC Code of Conduct and to identify areas for improvement.

Summary

- IT firms and organizations that use IT products and services are concerned about a shortfall in the number of U.S. workers to fill these positions. As a result, they are turning to nontraditional sources to find IT workers with skills that meet their needs.
- Contingent work is a job situation in which an individual does not have an explicit or implicit contract for long-term employment. The contingent workforce includes independent contractors, temporary workers hired through employment agencies, on-call or day laborers, and on-site workers whose services are provided through contract firms.
- An H-1B is a temporary work visa granted by the U.S. Citizenship and Immigration Services (USCIS) for people who work in specialty occupations—jobs that require at least a four-year bachelor's degree in a specific field, or equivalent experience.
- Employers hire H-1B workers to meet critical business needs or to obtain essential technical skills or knowledge that cannot be readily found in the United States. H-1B workers may also be used when there are temporary shortages of needed skills.
- Some people contend that employers exploit contingent workers, especially H-1B foreign workers, to obtain skilled labor at less than competitive salaries. Others believe that the use of H-1B workers is required to keep the United States competitive.
- Employers must make ethical decisions about whether to recruit new and more skilled workers from these sources or to spend the time and money to develop their current staff to meet the needs of their business.
- Outsourcing is a long-term business arrangement in which a company contracts for services with an outside organization that has expertise in providing a specific function. Offshore outsourcing is a form of outsourcing in which the services are provided by an organization whose employees are in a foreign country.
- Outsourcing and offshore outsourcing are used to meet staffing needs while potentially reducing costs and speeding up project schedules.
- Many of the same ethical issues that arise when considering whether to hire H-1B and contingent workers apply to outsourcing and offshore outsourcing.
- Whistle-blowing is an effort to attract public attention to a negligent, illegal, unethical, abusive, or dangerous act by a company or some other organization.
- A potential whistle-blower must consider many ethical implications, including whether the high price of whistle-blowing is worth it; whether all other means of dealing with the problem have been exhausted; whether whistle-blowing violates the obligation of loyalty that the employee owes to his or her employer; and whether public exposure of the problem will actually correct its underlying cause and protect others from harm.
- An effective whistle-blowing process includes the following steps: (1) assess the seriousness of the situation, (2) begin documentation, (3) attempt to address the situation internally, (4) consider escalating the situation within the company, (5) assess the implications of becoming a whistle-blower, (6) use experienced resources to develop an action plan, (7) execute the action plan, and (8) live with the consequences.
- Computer companies looking to manufacture green computers are challenged to produce computers that use less electricity, include fewer hazardous materials that may harm

people or pollute the environment, and contain a high percentage of reusable or recyclable material. These companies should also provide programs to help consumers dispose of their products in an environmentally safe manner at the end of their useful life.

- EPEAT (Electronic Product Environmental Assessment Tool) is a system that enables purchasers to evaluate, compare, and select electronic products based on 51 environmental criteria.
- The European Union passed the Restriction of Hazardous Substances Directive to restrict the use of many hazardous materials in computer manufacturing, require manufacturers to use at least 65 percent reusable or recyclable components, implement a plan to manage products at the end of their life cycle in an environmentally safe manner, and reduce or eliminate toxic material in their packaging.
- The Electronic Industry Citizenship Coalition (EICC) has established a code of conduct that defines performance, compliance, auditing, and reporting guidelines across five areas of social responsibility: labor, health and safety, environment, management system, and ethics.
- A number of electronics manufacturers have applied this code across their entire worldwide supply chain and also require their first-tier suppliers to acknowledge and implement the code.

Key Terms

coemployment relationship
contingent work
employee leasing
Electronic Industry Citizenship Coalition (EICC)
Electronic Product Environmental Assessment Tool (EPEAT)
False Claims Act
green computing
H-1B visa
offshore outsourcing
outsourcing
qui tam

Self-Assessment Questions

The answers to the Self-Assessment Questions can be found in Appendix E.

1. Between 2004 and 2009, there was a 50 percent decrease in undergraduate degrees awarded in computer science and computer engineering at doctoral-granting computer science departments. True or False?
2. Which of the following statements is true about future job prospects in the IT industry?
    a. The Bureau of Labor Statistics has forecasted that very few new jobs are expected to be created for network systems, data communications analysts, and computer software engineers.
    b. Computer science, computer engineering, systems engineering, and business systems networking/telecommunications all rated in the top-ten paid majors for 2010–2011 bachelor's degree graduates.

c. Employers should have no problem recruiting IT workers with the skills that meet their needs.

d. All of the above

3. Which of the following is *not* an advantage for organizations that employ contingent workers?

   a. The firm does not have to offer employee benefits to contingent workers.

   b. Training costs are kept to a minimum.

   c. It provides a way to meet fluctuating staffing needs.

   d. The contingent worker's experience may be useful to the next firm that hires him or her.

4. Depending on how closely workers are supervised and how the job is structured, contingent workers can be viewed as permanent employees by the IRS, the Labor Department, or a state's workers' compensation and unemployment agencies. True or False?

5. A temporary working visa granted by the U.S. Citizenship and Immigration Services for people who work in specialty occupations—jobs that require at least a four-year bachelor's degree in a specific field, or equivalent experience—is called a (an) ____________ visa.

6. It appears that some companies may be using the ____________ visa to get around the cap and wage restrictions of the H-1B program.

7. Because of the high cost of U.S.-based application developers and because a large number of IT professionals is readily available in certain foreign countries, it is now quite common to use offshore outsourcing for major software programming projects. True or False?

8. According to A.T. Kearney, the top three countries that are the most attractive offshoring destinations are:

   a. India, Egypt, and Philippines

   b. China, Mexico, and Thailand

   c. Vietnam, Philippines, and Chile

   d. India, China, and Malaysia

9. The cost advantage for Indian workers over U.S. workers is shrinking. True or False?

10. Which of the following statements about whistle-blowing is true?

    a. From the moment an employee becomes known as a whistle-blower, a public battle may ensue, with negative publicity attacks on the individual's personal integrity.

    b. Whistle-blowing is an effective approach to take in dealing with all work-related matters, from the serious to mundane.

    c. Violators of the False Claim Act are liable for four times the dollar amount that the government is defrauded.

    d. A whistle-blower must be an employee of the company that is the source of the problem.

11. Which of the following are desirable characteristics of a "green computer"?

    a. It runs on less electricity than the typical computer.

    b. It contains a high percentage of reusable or recyclable materials.

c. Its manufacturer has a program to help consumers dispose of it at the end of its life.

d. All of the above

12. Personal computers and cell phones contain hundreds if not thousands of components, which, in turn, are composed of many materials. Many of these materials are potentially harmful to humans and the environment. True or False?
13. Apple, Dell, and HP have earned the highest possible marks from Greenpeace for their excellent corporate policies in regard to toxic chemicals, recycling, and climate change. True or False?
14. Products are ranked in EPEAT according to three tiers of environmental performance, with ____________ being the highest.

Discussion Questions

1. Visit the EPEAT Web site (*www.epeat.net*) and use the tool to select your next laptop computer. How would you make trade-offs between an expensive machine with a Gold rating and a less expensive machine with the same features and performance but only a Bronze rating?
2. During a time in which unemployment in the United States exceeds 9 percent, some people feel that it is unethical to hire H-1B workers to work in the United States. Prepare a brief summary of reasons why hiring H-1B workers might be considered unethical. Make a list of reasons why hiring H-1B workers should be considered ethical. What set of reasons is stronger? Why?
3. Which steps of the whistle-blowing process are most important? Why?
4. Apple, Dell, and HP all received low ratings from Greenpeace for their efforts in green computing. Choose one of these companies, visit its Web site, and do research to find out what, if anything, the company is doing to improve its green computing results.
5. What factors must one consider in deciding whether to employ offshore outsourcing on a project?
6. You work for an electronics manufacturer that does not belong to the EICC. Present a strong argument for your firm to join. Then present a strong argument for why it makes sense for your firm not to be a member.
7. While labor savings associated with offshore outsourcing may look attractive, what cost increases and other problems can one expect with such projects?
8. Why do companies that make use of a lot of contingent workers fear getting involved in a coemployment situation? What steps should they take to avoid this situation?
9. Your company has decided to offshore-outsource a $50 million project to an experienced, reputable firm in India. This is the first offshore outsourcing project of significant size that your company has run. What steps should your company take to minimize the potential for problems?
10. Briefly describe a situation that could occur at your employer or your school that would rise to the level of a potential whistle-blower situation. What steps would you take and to whom would you speak to call this matter to the attention of appropriate members of management?

What Would You Do?

1. A coworker complains to you that he is sick of seeing the company pollute the waters of a nearby stream by dumping runoff water into it from the manufacturing process. He plans to send an anonymous email to the EPA to inform the agency of the situation. What would you do?
2. As a relatively new hire within a large multinational firm, you are extremely pleased with the many challenging assignments that have come your way. Now another new hire with whom you have become friends is seeking your input on an important decision that she must make within the next week. She has been challenged to cut costs in her department by outsourcing a large portion of the department's work to an offshore resource firm that has an excellent reputation. Your friend would remain with the firm to oversee the outsourcing work. What advice would you offer your friend?
3. Your firm has just added six H-1B workers to your 50-person department. You have been asked to help get one of the workers "on board." Your manager wants you to introduce him to other team members, provide him with some basic company background and information, and explain to him how work gets done within your organization. Your manager has also asked you to help your new coworker become familiar with the community, including residential areas, shopping centers, restaurants, and recreational activities. Your goal would be to help the new worker be productive and comfortable with his new surroundings as soon as possible. How would you feel about taking on this responsibility? How would you help the new employee?
4. Dr. Jeffrey Wigand is a whistle-blower who was fired from his position of vice president of research and development at Brown & Williamson Tobacco Corporation in 1993. He was interviewed for a segment of the CBS show *60 Minutes* in August 1995, but the network made a highly controversial decision not to air the interview as initially scheduled. The segment was pulled because CBS management was worried about the possibility of a multibillion-dollar lawsuit for tortuous interference; that is, interfering with Wigand's confidentiality agreement with Brown & Williamson. The interview finally aired on February 4, 1996, after the *Wall Street Journal* published a confidential November 1995 deposition that Wigand gave in a Mississippi case against the tobacco industry, which repeated many of the charges he made to CBS. In the interview, Wigand said that Brown & Williamson had scrapped plans to make a safer cigarette and continued to use a flavoring in pipe tobacco that was known to cause cancer in laboratory animals. Wigand also charged that tobacco industry executives testified untruthfully before Congress about tobacco product safety. Wigand suffered greatly for his actions; he lost his job, his home, his family, and his friends.

   Visit Wigand's Web site at *www.jeffreywigand.com* and answer the following questions. (You may also want to watch *The Insider*, a 1999 movie based on Wigand's experience.)

   - What motivated Wigand to take an executive position at a tobacco company and then five years later to denounce the industry's efforts to minimize the health and safety issues of tobacco use?
   - What whistle-blower actions did Dr. Wigand take?
   - If you were in Dr. Wigand's position, what would you have done?

5. You are in the last stages of evaluating laptop vendors for a major hardware upgrade and standardization project for your firm. You will be purchasing a total of 1,200 new laptops to deploy to the worldwide sales force. One vendor's product carries a Bronze EPEAT rating; the other vendor's product would cost an additional $96,000 but carries a Gold EPEAT rating. The two products are very evenly matched on other key factors, such as performance, features, reliability, and support costs. How would you decide between the two vendors' products?
6. You are the manager of a large IT project that will involve two dozen offshore outsourced workers who will do program development and testing. Identify several potential issues that could arise on this project due to the outsourcing arrangement. What specific steps would you take to improve the likelihood of success of the project?
7. Microsoft relies heavily on temporary workers. To minimize potential legal issues, Microsoft sought to ensure that the temporary workers were not mistaken about their place within the company. Therefore, temporary agencies provided these workers with handbooks that laid out the ground rules in explicit detail. Temporary workers were barred from using company-owned athletic fields—for insurance reasons, the handbooks explained. At some Microsoft facilities, temporary workers were told they could not drive their cars to work because it would create parking problems for regular workers. Instead, they were told to take the bus. Temporary workers were also told not to buy goods at the company store or participate in company social clubs such as chess, tai chi, or rock climbing, which were open to regular employees. They were not permitted to attend parties given for regular employees, a private screening of the latest *Star Wars* film, or company meetings at the Kingdome stadium in Seattle. In addition, their email addresses were to contain an *a-* to indicate their nonpermanent status in the company.

   Imagine that you are a senior manager in the Human Resources Department at Microsoft and that you have been asked to respond to temporary workers' complaints about working conditions. What sort of complaints would you expect to hear? How would you handle this?
8. Catalytic Software—a U.S.-based IT outsourcing firm with offices in both Redmond, Washington, and Hyderabad, India—wants to tap India's large supply of engineers as contract software developers for IT projects. However, instead of just outsourcing projects to local Indian software development companies, as is the common practice of U.S. companies, Catalytic has developed a self-contained company community near Hyderabad. Spread over 500 acres, the community of New Oroville is a self-sustaining residential and office community designed to house about 4,000 software developers and their families, as well as 300 support personnel who supply sanitation, police, and fire services.

   The goal of this high-tech city is to knock down barriers that large-scale technology businesses encounter in India. By building a company community, Catalytic is trying to ensure that it has enough qualified employees to staff round-the-clock shifts. The company expects this facility to attract and keep top professionals from all over the world. Building a company town also solved the problem of transportation, which can be challenging for such a large workforce. Because of the terrible state of the local roads, the commute from Hyderabad—25 kilometers from New Oroville—would take almost an hour.

   Catalytic provides private homes with private gardens, all within a short walk of work, school, recreation, shopping, and public facilities. Each house includes cable television, telephones, and a fiber-optic data pipeline that connects to the Internet so that employees can work efficiently even at home. (Employees are awarded bonuses for working overtime.) New Oroville was designed to include four indoor recreational complexes, six large retail complexes, and ample green space, including five parks for outdoor exercise and recreation.

You have just completed a job interview with Catalytic Software for a position as project manager and have been offered a 25 percent raise to join the company. Your position will be based in Redmond and will involve managing U.S.-based projects for customers. The position requires that you spend the first year with Catalytic in the New Oroville facility to learn its methods, culture, and people. (You can take your entire family or accept a pair of three-week company-paid trips back to Redmond.)

Why do you think the temporary assignment in New Oroville is a requirement? What else would you need to know in considering this position? Would you accept it? Why or why not?

Cases

1. E-Verify

U.S. law requires that companies operating in the United States only employ individuals who are legally permitted to work in the United States. To this end, all U.S. employers must complete and retain a Form I-9: Employment Eligibility Verification for each individual (both citizens and noncitizens) they hire in the United States. An employer must examine the employment eligibility and identity documents presented by a new employee to determine whether the documents reasonably appear to be genuine and relate to the individual; the employer must then record the document information on the Form I-9 for that employee. The list of acceptable documents includes documents that (1) establish both identity and employment authorization (U.S. passport, permanent resident card, etc.), (2) establish identity only (driver's license or photo ID card issued by a local, state, or federal government), or (3) establish authorization to work only (social security card, certificate of birth, etc.).[49]

E-Verify is an Internet-based system run by the U.S. Department of Homeland Security that allows employers to electronically verify the employment authorization of their newly hired employees.[50] The program compares information from an employee's Form I-9 to data in U.S. government records. If the data matches, the employee is eligible to work in the United States. If there is a mismatch, E-Verify alerts the employer, and the employee has eight work days to contact either the Social Security Administration or the Department of Homeland Security (depending on the source of the data mismatch) to start resolving the problem. During this eight-day period, and during the time it takes for the data mismatch to be further researched by the government agencies, the employee cannot be terminated.

In a 2011 ruling involving Arizona state law, the U.S. Supreme Court affirmed that states may constitutionally mandate the use of E-Verify for all employers within a given state. The states are greatly divided on this issue—some states have passed laws requiring all employers to use E-Verify to determine the eligibility of new hires; some require just public employers and government contractors to use E-Verify; and some require just private employers with more than a specified number of workers to use E-Verify. In some states, the decision to use E-Verify is being made at the county and city level. For example, in California, San Diego County is one of many city and county governments exploring the use of E-Verify in the hiring process. In early

2011, the city of Escondido was the first city in the county to implement E-Verify for all new city employees and contractors.[51]

It is estimated that as of June 2011, about 271,500 employers nationwide use E-Verify, with some 1,300 new employers signing up each week.[52,53] U.S. Representative Lamar Smith of Texas (Chairman of the House Judicial Committee) has proposed legislation that would make E-Verify mandatory for all employers in the United States.[54]

According to a 2010 analysis of E-Verify conducted by an outside consulting firm on behalf of the U.S. Citizenship and Immigration Services, 4.1 percent of initial responses from the E-Verify system were inconsistent with the worker's actual employment eligibility status. Of those erroneous responses, 0.7 percent related to workers who were initially determined to be not authorized to work but who were in fact authorized, and 3.3 percent related to workers who were determined to be employment authorized but who were not actually eligible to do so. Of the 6.2 percent of workers who truly were not authorized to work in the United States, the E-Verify system failed to discover more than half. (Many of these unauthorized workers beat the system by submitting I-9 documents of a person who is authorized to work.[55])

The system's less than 100 percent accuracy rate and failure to identify unauthorized workers has raised many political concerns and emotional responses. In order to reduce errors due to identity theft, the U.S. Citizenship and Immigration Services is considering implementing a photo tool that will match the photo submitted with the I-9 documents with photos in government records and with the actual employee. Additional government databases may be integrated into the E-Verify matching process to improve accuracy.[56]

In addition to accuracy issues, there is concern that mandatory use of E-Verify will harm authorized workers and lead to discrimination. Opponents also fear it will create additional work for human resource departments in terms of updating their personnel records and initiating and following up on requests to various government agencies.

Meanwhile, proponents argue that the accuracy of the E-Verify system will improve over time and as further enhancements are made. They believe that is fair to ask employers to do a quick check of each employee to ensure that they are hiring authorized workers. With the high level of unemployment, supporters of E-Verify believe that steps should be taken to ensure that jobs go to authorized workers.

Discussion Questions

1. Do you support the implementation of enhancements such as photo matching and access to additional government databases to improve the accuracy of the E-Verify system? Why or why not?
2. If you were the owner of a small business, would you use the E-Verify system to screen prospective workers? Why or why not?
3. Would you favor mandatory use of the E-Verify system at large corporations and government agencies? Why or why not?

2. Problems with Suppliers

Many computer hardware manufacturers rely on foreign companies to provide raw materials; build computer parts; and assemble hard drives, monitors, keyboards, and other components. While there are many advantages to dealing with foreign suppliers, hardware manufacturers may find certain aspects of their business (such as quality and cost control, shipping, and communication) more complicated when dealing with a supplier in another country.

In addition to these fairly common business problems, hardware manufacturers are sometimes faced with serious ethical issues relating to their foreign suppliers. Two such issues that have recently surfaced involve (1) suppliers who run their factory in a manner that is unsafe or unfair to their workers and (2) raw materials suppliers who funnel money to groups engaged in armed conflict, including some that commit crimes and human rights abuses.

In February 2009, alarming information came to light about the Meitai Plastics and Electronics factory in Dongguan City, in China's Guangdong province. This factory, in fact, represents an extreme example of a supplier who runs its factory in an unsafe and unfair manner. Meitai Plastics employs 2,000 workers, mostly young women, who make computer equipment and peripherals—such as printer cases and keyboards—for Dell, IBM, Lenovo, Microsoft, and Hewlett-Packard products.[57] Based on research conducted between June 2008 and January 2009, the National Labor Committee (a human rights organization based in the United States) published a report in February 2009 highly critical of the work environment at the factory.[58] According to the report, young workers sit on hard wooden stools for 12 hours a day, working on an assembly line that never stops. Workers are prohibited from talking, listening to music, raising their heads from their work, or putting their hands in their pockets. Employees are fined for being even one minute late, not trimming their fingernails, or stepping on the grass of the factory grounds. A worker who needs to use the restroom must wait until there is a group break.

The average workweek consists of 74 hours, with a take-home pay of $57.19—well below the amount necessary to meet subsistence-level needs in China. If a worker takes a Sunday off, she is docked one-and-a-half-day's wages. Workers are housed 10 to 12 per dorm room. The dorms have no air conditioning, and temperatures in the rooms can reach the high 90s in the summer. Workers must walk down several floors to get hot water in a small bucket to use for personal hygiene.[59]

Manufacturers who use rare raw materials face another ethical issue related to the use of foreign suppliers: how to ensure that their suppliers do not funnel money to groups that engage in armed conflict or commit crimes and human rights abuses. Manufacturers of computers, digital cameras, cell phones, and other electronics frequently purchase rare minerals such as gold, tin, tantalum, and tungsten for use in their products. Unfortunately, some of these purchases are helping to finance the deadliest conflict in the world today—the war in the Democratic Republic of Congo. The war began in 1998 and has dragged on long after a peace agreement was signed in 2003. During the war and its aftermath, over 5 million people have died—mostly from disease and starvation—making it the deadliest conflict since World War II.[60]

In Congo, many mines are controlled by groups that engage in armed conflict and inflict human rights abuses on local populations. The Enough Project's "Raise Hope for Congo" campaign is trying to get large electronics firms to trace and audit their supply chains to ensure that their suppliers do not source minerals from mines in Congo that are controlled by armed groups. This is often easier said than done because of the long, complex supply chain and often disreputable middlemen involved in the minerals trade. As manufacturers struggle with these issues, some are trying to use their influence to demand that their suppliers stop sourcing from mines that continue to fund violence in Congo and elsewhere.[61]

Discussion Questions

1. What responsibility does an organization have to ensure that its suppliers and business partners behave ethically? To whom is this responsibility owed?
2. How can an organization monitor the business practices of its suppliers and business partners to determine if they are behaving in an ethical manner?
3. Is it good business practice to refuse to do business with a supplier who provides good quality materials at a low cost but who behaves in an unethical manner? How can senior management justify its decision to do business instead with a supplier who provides lower quality or higher priced materials but behaves in an ethical manner?

3. The Census Bureau's Outsourcing Debacle

In 2008, just one year before the Census Bureau was scheduled to begin street-canvassing operations for the 2010 decennial census, news broke that the bureau's mobile initiative had flopped. It was revealed that the Field Data Collection Automation (FDCA) system, which was supposed to save taxpayers $1 billion, would now raise the total cost of conducting the census from $11.5 to $14.5 billion.[62,63] Many reporters quickly described the failure as just one more case of federal IT mismanagement. Yet these reports overlooked the bureau's historic role in introducing cutting-edge IT and database development.

The U.S. Constitution requires that a census of the country's population be taken every 10 years. This data is used not only to decide the distribution of congressional seats within the United States but also to determine how to allocate federal funds. Census figures are also used by state and local governments as a basis for deciding where to build roads, schools, job-training centers, and so forth. The bureau has a history of promoting new technology to accommodate both a growing population and the proliferation of uses for census data. For example, after spending seven years collecting data for the 1880 census, the bureau became the first organization to use Herman Hollerith's automatic tabulating machine, which performed so successfully that updated versions were used in subsequent censuses as well. (Hollerith's firm, the Tabulating Machine Company, later merged with three other companies to form IBM.) UNIVAC, the first commercial computer, was originally designed for use in the 1950 census.[64]

In the 1980s, the Census Bureau created the Topologically Integrated Geographic Encoding and Referencing (TIGER) database, which provides automated access to and retrieval of relevant geographic information about the United States and its territories.[65] TIGER maps roads as well as state, county, and city boundaries; railroads; and every body of water in the country. For each entity, TIGER lists attributes, such as name, alternative name, longitude, and latitude. For roads, TIGER defines address ranges and associates zip codes with street addresses.[66] This information allows the bureau to mail surveys to each citizen, thus cutting down on the number of census workers, called "enumerators," who have to go into the field to locate housing units and conduct interviews. But TIGER's impact extended way beyond the bureau; it jump-started the geographic information system (GIS) industry, facilitating the development of products such as MapQuest, Google Maps, and GPS navigation for automobiles.

After the 1990 census, the bureau created the Master Address File (MAF) database, composed of data collected during the census, as well as from the U.S. Postal Service and local, state, and tribal governments.[67] MAF contained a more complete listing of housing units than

TIGER did, so the two database systems were used together. Following the 2000 census, the two databases were merged onto an Oracle database platform.

At the same time, the Census Bureau was also embracing outsourcing to meet its mounting IT needs. "By the 1990s, we recognized that we needed to move away from Census home-grown technologies. The capabilities of a robust, nascent IT industry had by then exceeded our internal abilities," reported Census Bureau director Charles Kincannon at a 2007 Congressional hearing.[68]

The bureau relies on technology to support its wide range of operations—enhancing legacy systems and acquiring new IT systems when necessary. For each decennial census, the bureau must first identify where to count by collecting all known addresses of all people living within the United States. While the MAF/TIGER system provides much of this data, street canvassing must be carried out to update the information in the database. After street canvassing is complete and all addresses are identified, the next step is to collect information from all households. The Census Bureau mails a survey to each household. If the survey is not returned, the bureau must send out an enumerator to the address to interview the resident. The data assembled from the surveys and the reports of the enumerators is integrated by the Decennial Response Integration System (DRIS). The final step of tabulating and summarizing the results is carried out through the Data Access and Dissemination System (DADS II).[69]

Following the 2000 census, the bureau elected to develop a mobile system that would allow enumerators working in the field to collect and transmit data back to the offices. In 2006, the Census Bureau outsourced the work and awarded a $600 million contract to Harris Corporation to create the Field Data Collection Automation (FDCA) system to automate the collection of field data.[70] Harris Corporation is a communications and IT company with recent annual revenue of $6 billion and over 16,000 employees.[71] The company had already helped the bureau integrate MAF and TIGER, and through the new contract, Harris was supposed to provide the IT system and the mobile hardware—handheld computers—used by enumerators. The bureau planned to use this system not only to collect and transmit data during the canvassing and the interviews, but also to manage the field operations.

On May 1, 2008, the Census Bureau conducted a test of the new FDCA system. The results revealed that the mobile handheld computers were slow, sometimes froze up, and did not transmit data consistently.[72] In addition, the test showed problems with the program designed to manage operations in the field. The bureau announced that it would need to push back the development schedule and allocate considerably more funds to complete the project. In August 2009, it announced that enumerators would not be using the mobile devices during the interview process and would instead rely exclusively on paper-based operations. The Census Bureau agreed to a new contract with Harris that would cover just the handheld-driven update of addresses; even with this greatly reduced scope, the cost was $200 million more than the original budget for the entire system.[73]

The question that both reporters and members of the federal government asked was: What went wrong with this outsourcing project? The U.S. Government Accountability Office (GAO) had been carefully monitoring preparations for the 2010 census and identified the following project management shortcomings:

- Failure to identify key project deliverables and milestones
- Failure to gain stakeholder buy-in on the project plan, including key project parameters such as estimated costs and schedule

- Failure to validate key project requirements
- Failure to assign responsibility for risks and to prepare mitigation plans
- Failure to define key metrics for contract tracking and executive oversight[74]

In a March 2009 report, the Commerce Department's inspector general stated that a root cause of the project failure was "the failure of senior Census Bureau managers in place at the time to anticipate the complex IT requirements involved in automating the census." Indeed, the Census Bureau kept changing the requirements, with each change adding to the cost and further delaying the project. In addition, the Census Bureau set up a cost-plus contract with Harris Corporation instead of a fixed-cost contract. With a cost-plus contract, Harris could increase the cost each time the Census Bureau changed its mind on what it wanted.[75]

While some 140,000 enumerators had to use pencil and paper to collect data from the interviews, the bureau used the FDCA system to conduct street canvassing to identify and correct the locations of housing units. So the millions spent on this project will not be entirely for naught. And perhaps the bureau will learn from these mistakes and develop procedures to manage future outsourced IT projects more effectively.[76]

Discussion Questions

1. After many years of conducting successful IT projects, why did the Census Bureau decide to outsource creation of the FDCA system to automate the collection of field data?
2. Go to the Harris Corporation Web site (*www.harris.com*) to gain an understanding of the broad range of projects that Harris is working on for the U.S. government. Do you think it is appropriate that the government continues to spend so much money with this firm based on the Census Bureau experience? Why or why not?
3. Make a list of three key principles for the successful outsourcing of IT projects, based on the Census Bureau's experience.

End Notes

1 GreenBiz Staff, "IBM Racks Up Nearly $27M in Energy Savings," *GreenBiz*, June 20, 2010, www.greenbiz.com/news/2010/06/30/ibm-racks-up-nearly-27m-energy-savings.

2 GreenerComputing Staff, "IBM Achieves First Full Phase-Out of Toxic Compounds," *GreenerComputing*, March 2, 2010, www.greenbiz.com/news/2010/03/02/ibm-achieves-industry-first-phase-out-toxic-compounds.

3 IBM, "2010 Corporate Responsibility Report: A Commitment to Environmental Leadership," www.ibm.com/ibm/responsibility/report/2010/bin/downloads/IBM_CorpResp_2010_Environment.pdf (accessed August 6, 2011).

4 IBM, "Press Release: IBM Reduces $50 Million in Electricity Expenses, Boosts Conservation Efforts," June 23, 2011, www-03.ibm.com/press/us/en/pressrelease/34879.wss.

5 IBM, "One of the Biggest Line Items on Your Balance Sheet Could be Your Building," www.ibm.com/smarterplanet/us/en/green_buildings/overview/index.html (accessed July 9, 2011).

[6] IBM, "Press Release: IBM Reduces $50 Million in Electricity Expenses, Boosts Conservation Efforts," *IBM Press Release*, June 23, 2011, www-03.ibm.com/press/us/en/pressrelease/34879.wss.

[7] IBM, "Press Release: IBM Reduces $50 Million in Electricity Expenses, Boosts Conservation Efforts," June 23, 2011, www-03.ibm.com/press/us/en/ pressrelease/34879.wss.

[8] IBM, "One of the Biggest Line Items on Your Balance Sheet Could be Your Building," www.ibm.com/smarterplanet/us/en/green_buildings/overview/index.html (accessed July 9, 2011).

[9] GreenBiz Staff, "IBM Racks Up Nearly $27M in Energy Savings," *GreenBiz*, June 20, 2010, www.greenbiz.com/news/2010/06/30/ibm-racks-up-nearly-27m-energy-savings.

[10] IBM, "Press Release: IBM Reduces $50 Million in Electricity Expenses, Boosts Conservation Efforts," June 23, 2011, www-03.ibm.com/press/us/en/ pressrelease/34879.wss.

[11] Global Reporting Initiative, "What is GRI," www.globalreporting.org/AboutGRI/WhatIsGRI (accessed August 15, 2011).

[12] Stuart Zweben, "2009–2010 Taulbee Survey: Undergraduate CS Degree Production Rises; Doctoral Production Steady," *Computing Research News*, May 2011, www.cra.org/uploads/documents/resources/crndocs/issues/0511.pdf.

[13] National Association of Colleges and Employers, "Top-Paid Majors for the Class of 2011," www.naceweb.org/Press/Releases/Top-Paid_Majors_for_the_Class_of_2011.aspx.

[14] Bill Virgin, "Microsoft Settles 'Permatemp' Suits," *Seattle-Post Intelligencer*, December 13, 2000.

[15] Jeanne Batalova, "H-1B Temporary Skilled Worker Program," Migration Information Source, October 2010, www.migrationinformation.org/USfocus/display.cfm?ID=801.

[16] The HMA Law Firm, "The H-1B Cap: Putting a Quota on Intelligence," www.hmalegal.com/the-h-1b-quota-putting-a-cap-on-intelligence.html (accessed July 7, 2011).

[17] Patrick Thibodeau, "H-1B at 20: How the 'Tech Worker Visa' Is Remaking IT in America," *Computerworld*, November 17, 2010, www.computerworld.com/s/article/9196738/H_1B_at_20_How_the_tech_worker_visa_is_remaking_IT_in_America.

[18] Patrick Thibodeau, "Hiring H-1B Visa Workers Trims U.S. Tech Workers' Wages," *PCWorld*, April 19, 2009, www.pcworld.com/article/163383/hiring_h1b_visa_workers_trims_us_tech_workers_wages.html.

[19] United States Department of Labor, Employment Law Guide, "Work Authorization for non-U.S. Citizens: Workers in Professional and Specialty Occupations (H-1B and H-1B1 Visas)," www.dol.gov/compliance/guide/h1b.htm#who (accessed July 5, 2011).

[20] United States Department of Labor, Employment Law Guide, "Work Authorization for non-U.S. Citizens: Workers in Professional and Specialty Occupations (H-1B and H-1B1 Visas)," www.dol.gov/compliance/guide/h1b.htm#who.

[21] Patrick Thibodeau, "Troubled H-1B Fraud Case Ends Quietly," *Computerworld*, May 16, 2011, www.computerworld.com/s/article/9216694/Troubled_H_1B_fraud_case_ends_quietly.

[22] Patrick Thibodeau, "Infosys Faces Grand Jury as Visa Probe Broadens," *Computerworld*, May 25, 2011, www.computerworld.com/s/article/9217064/Infosys_faces_grand_jury_as_visa_probe_broadens.

[23] "Most Multinational Companies Use IT Outsourcing: Study," *IT World Canada*, February 25, 2010, www.itworldcanada.com/news/most-multinational-companies-use-it-outsourcing-study/140078.

[24] "Where Is Your Software From?," *TechByter Worldwide* (blog), January 11, 2009, www.techbyter.com/2009/20090111.html.

[25] Stephanie Overby, "Egypt Unrest Threatens Status as Rising Outsourcing Star," *CIO*, January 28, 2011, www.cio.com/article/659463/Egypt_Unrest_Threatens_Status_as_Rising_Outsourcing_Star.

[26] Peter Sayer, "Nokia to Cut 7,000 Jobs Through Outsourcing, Layoffs," *Computerworld*, April 27, 2011, www.computerworld.com/s/article/9216203/Nokia_to_cut_7_000_jobs_through_outsourcing_layoffs.

[27] Don Reisinger, "Nokia Passes Symbian Torch to Accenture," *CNET* June 22, 2011, http://news.cnet.com/8301-13506_3-20073211-17/nokia-passes-symbian-torch-to-accenture.

[28] Patrick Thibodeau, "N.C. Outsourcing Deal Leads to IT Layoffs," *Network World*, June 30, 2011, www.networkworld.com/news/2011/060311-nc-outsourcing-deal-leads-to.html?source=nww_rss.

[29] Craig Schneider, "Human Error to Blame for Grady Data Breach," *Atlanta Journal-Constitution*, September 23, 2008, www.ajc.com/search/content/metro/stories/2008/09/23/grady.html.

[30] "False Claims," Cornell University Law School, www.law.cornell.edu/uscode/31/usc_sec_31_00003729----000-.html (accessed July 8, 2011).

[31] "False Claims," Cornell University Law School, www.law.cornell.edu/uscode/31/usc_sec_31_00003729----000-.html (accessed July 8, 2011).

[32] "False Claims," Cornell University Law School, www.law.cornell.edu/uscode/31/usc_sec_31_00003729----000-.html (accessed July 8, 2011).

[33] Associated Press, "Verizon Pays $93.5M to Settle Whistleblower Suit," *ABC News*, April 5, 2011, www.cbsnews.com/stories/2011/04/05/ap/business/main20050941.shtml.

[34] Mike Scarcella, "Whistleblower in Verizon Case Demands Bigger Cut of $93.5M Settlement," *The BLT: The Blog of Legal Times, LegalTimes*, June 22, 2011, http://legaltimes.typepad.com/blt/2011/06/whistleblower-in-verizon-case-demands-bigger-cut-of-935m-settlement.html.

[35] "False Claims," Cornell University Law School, www.law.cornell.edu/uscode/31/usc_sec_31_00003729----000-.html (accessed July 8, 2011).

[36] Federal Accountability Initiative for Reform, "The Whistleblower's Ordeal," http://fairwhistleblower.ca/wbers/wb_ordeal.html.

[37] Jaikumar Vijayan, "Scope of TJX Data Breach Doubles: 94 Million Cards Now Said to Be Affected," *Computerworld*, October 24, 2007.

[38] Steve Ragan, "TJX Fires Whistleblower—Was It Justified Action or Something Else?," *Tech Herald*, May 26, 2008, www.thetechherald.com/article.php/200821/1070/TJX-fires-whistleblower-%E2%80%93-was-it-justified-action-or-something-else.

[39] Brad Wells, "What Truly Makes a Computer 'Green'?," *OnEarth* (blog), September 8, 2008, www.onearth.org/node/658.

[40] Elizabeth Royte, "E-Gad! Americans Discard More Than 100 Million Computers, Cellphones and Other Electronic Devices Each Year. As 'E-Waste' Piles Up, So Does Concern about This Growing Threat to the Environment," *Smithsonian*, August 1, 2005, www.smithsonianmag.com/arts-culture/e-gad.html.

[41] EPEAT, "New Branding for EPEAT Green Electronics Rating System, June 23 2011, www.epeat.net/2011/06/news/new-branding-for-epeat-green-electronics-rating-system.

[42] Wendy Koch, "Is Apple's 'Recyclable iPad Really Green? Do You Care?," *USA Today*, February 1, 2010, http://content.usatoday.com/communities/greenhouse/post/2010/01/is-apples-recyclable-chemical-free-ipad-really-green-/1.

[43] Senate Fiscal Agency, "Electronic Waste Recycling," Michigan Legislature, November 21, 2008, www.legislature.mi.gov/documents/2007-2008/billanalysis/Senate/pdf/2007-SFA-0897-B.pdf (accessed August 2, 2011).

[44] "State Legislation: States Are Passing E-Waste Legislation," Electronics TakeBack Coalition, March 20, 2008, www.electronicstakeback.com/legislation/state_legislation.htm (accessed August 2, 2011).

[45] "Dell's Worldwide Technology Recycling Options," http://content.dell.com/us/en/corp/d/corp-comm/GlobalRecycling.aspx (accessed July 11, 2011).

[46] Electronic Industry Citizenship Coalition, "Membership," www.eicc.info/MEMBERSHIP.htm

[47] Electronic Industry Citizenship Coalition, "Electronic Industry Code of Conduct Version 3.0 (2009), www.eicc.info/PDF/EICC%20Code%20of%20Conduct%20English.pdf (accessed August 2, 2011).

[48] Electronic Industry Citizenship Coalition, "Electronic Industry Code of Conduct Version 3.0 (2009), www.eicc.info/PDF/EICC%20Code%20of%20Conduct%20English.pdf (accessed August 2, 2011).

[49] Form I-9, Employment Eligibility Verification, Department of Homeland Security, U.S. Citizenship and Immigration Services, www.uscis.gov/files/form/i-9.pdf (accessed July 6, 2011).

[50] U.S. Department of Homeland Security, U.S. Citizenship and Immigration Services, "E-Verify: Questions and Answers," www.uscis.gov/portal/site/uscis/menuitem.eb1d4c2a3e5b9ac89243c6a7543f6d1a/?vgnextoid=51e6fb41c8596210VgnVCM100000b92ca60aRCRD&vgnextchannel=51e6fb41c8596210VgnVCM100000b92ca60aRCRD#General (accessed August 6, 2011).

[51] Christopher Cadelago, "County Moves Toward E-Verify System, *SignOn San Diego*, June 28, 2011, www.signonsandiego.com/news/2011/jun/28/county-may-tap-database-check-worker-legal-status.

[52] Christopher Cadelago, "County Moves Toward E-Verify System, *SignOn San Diego*, June 28, 2011, www.signonsandiego.com/news/2011/jun/28/county-may-tap-database-check-worker-legal-status.

[53] Representative Lamar Smith, "E-Verify Will Help American Jobs Go To Legal Workers," *The Hill's Congress Blog, The Hill*, June 21, 2011, http://thehill.com/blogs/congress-blog/politics/167509-e-verify-will-help-american-jobs-go-to-legal-workers.

[54] Representative Lamar Smith, "E-Verify Will Help American Jobs Go To Legal Workers," *The Hill's Congress Blog, The Hill*, June 21, 2011, http://thehill.com/blogs/congress-blog/politics/167509-e-verify-will-help-american-jobs-go-to-legal-workers.

[55] Capital Immigration Law Group "Reports Highlights E-Verify Accuracy Problems," February 25, 2010, www.cilawgroup.com/news/2010/02/25/report-highlights-e-verify-accuracy-problems.

[56] Capital Immigration Law Group "Reports Highlights E-Verify Accuracy Problems," February 25, 2010, www.cilawgroup.com/news/2010/02/25/report-highlights-e-verify-accuracy-problems.

[57] Jason Gooljar, "Chinese Factory That Supplies IBM, Microsoft, Dell, Lenovo and Hewlett-Packard to Be Investigated," *Dissent is Patriotic* (blog), February 15, 2009, www.jasongooljar.com/?tag=meitai-plastic-and-electronics.

[58] Tom Espiner, "Tech Coalition Launches Sweatshop Probe," *CNET*, February 14, 2009, http://news.cnet.com/8301-1001_3-10164325-92.html.

[59] National Labor Committee, "High Tech Misery in China: The Dehumanization of Young Workers Producing Our Computer Keyboards," February 2009, www.nlcnet.org/article.php?id=613.

[60] Joe Bavier, "Congo War-Driven Crisis Kills 45,000 A Month—Study," *Reuters*, January 22, 2008, www.reuters.com/article/2008/01/22/idUSL22802012._CH_.2400.

[61] The Enough Project, "Electronics Companies Respond to the Enough Project," www.raisehopeforcongo.org/responses (accessed July 4, 2011).

[62] Harris Corporation, "Harris Corporation Demonstrates Functionality of Handheld Mobile Computing Device for 2010 Decennial Census," January 3, 2007, www.harris.com/view_pressrelease.asp?act=lookup&pr_id=2031.

[63] Jean Thilmany, "Behind the Census Bureau's Mobile SNAFU," *CIO Insight*, May 20, 2008, www.cioinsight.com/c/a/Case-Studies/Census-Mobile-SNAFU.

[64] Charles Louis Kincannon, "Prepared Statement of Charles Louis Kincannon, Director US Census Bureau: Hearing to Examine Issues Relating to the Census Bureau's Risk Management of Key 2010 Information Technology Acquisitions, Before the Subcommittee on Information Policy, Census, and National Archives, U.S. House of Representatives," December 11, 2007, www.ogc.doc.gov/ogc/legreg/testimon/110f/Kincannon121107.pdf.

[65] U.S. Census Bureau, "TIGER® Overview," August 31, 2005, www.census.gov/geo/www/tiger/overview.html.

66 Jacob S. Siegel, *Applied Demography* (San Diego, CA. Academic Press, 2002), 201–3, http://books.google.com/books?id=a5Ax1oRbkDMC&pg=PA201&lpg=PA201&dq=TIGER+Census+Bureau+1983+DIME&source=bl&ots=8fCjHu7ILF&sig=pECg3Pz6WJc7QUGw09ztwrYV1GA&hl=en&ei=1-4jSvnYJMyntgezhcSvBg&sa=X&oi=book_result&ct=result&resnum=4#PPA202,M1.

67 Shawana P. Johnson and J. Edward Kunz, "Private Sector Makes Census Bureau's TIGER Roar," *GPS World*, May 1, 2005, www.gpsworld.com/gis/local-government/private-sector-makes-census-bureau039s-tiger-roar-5360.

68 Charles Louis Kincannon, "Prepared Statement of Charles Louis Kincannon, Director, US Census Bureau: Hearing to Examine Issues Relating to the Census Bureau's Risk Management of Key 2010 Information Technology Acquisitions, Before the Subcommittee on Information Policy, Census, and National Archives, U.S. House of Representatives," December 11, 2007, http://informationpolicy.oversight.house.gov/documents/20080318161001.pdf.

69 GAO-09-262, "Information Technology: Census Bureau Testing of 2010 Decennial Systems Can Be Strengthened," Government Accountability Office, March 2009, www.gao.gov/htext/d09262.html.

70 Jean Thilmany, "Behind the Census Bureau's Mobile SNAFU," *CIO Insight*, May 20, 2008, www.cioinsight.com/c/a/Case-Studies/Census-Mobile-SNAFU.

71 Harris Corporation, "News and Information," www.harris.com/whats_new.asp?act=search&srchyr=-1#srchrslts (accessed August 7, 2011).

72 GAO-09-262, "Information Technology: Census Bureau Testing of 2010 Decennial Systems Can Be Strengthened," Government Accountability Office, March 5, 2009, www.gao.gov/htext/d09262.html.

73 Brian Friel, "The Right Stuff," *MyTwoCensus.com*, May 20, 2009, www.mytwocensus.com/2009/05/20/investigative-series-spotlight-on-harris-corp-part-1.

74 Michael Krigsman, "Billion-Dollar IT Failure at Census Bureau," *IT Project Failures, ZDNet* (blog), March 20, 2008, www.zdnet.com/blog/projectfailures/billion-dollar-it-failure-at-census-bureau/660?tag=mantle_skin;content.

75 U.S.. Department of Commerce, Office of Inspector General, "U.S. Census Bureau: Census 2010: Revised Field Data Collection Automation Contract Incorporated OIG Recommendations, But Concerns Remain Over Fee Awarded During Negotiations," www.oig.doc.gov/OIGPublications/CAR-18702.pdf (accessed August, 2011).

76 Sybase, "Press Release: Sybase and Harris Corporation Announce the Success of the 2010 Census Address Canvassing Operation", July 29, 2009, www.sybase.com/detail?id=1065049.

A BRIEF INTRODUCTION TO MORALITY

By Clancy Martin, Assistant Professor of Philosophy, University of Missouri—Kansas City

INTRODUCTION

This appendix offers a quick survey of various attempts by Western civilization to make sense of the ethical question "What is the good?" As you will recall from Chapter 1, *ethics* is the discipline dealing with what is good and bad and with moral duty and obligation. How should we live our lives? How should we act? Which goals are worth pursuing and which are not? What do we owe to ourselves and to others? These are all ethical questions.

The answers to these questions are provided in what we call *moralities* or *moral codes*. The Judeo-Christian morality, for example, attempts to tell us how we should live our lives, the difference between right and wrong, how we ought to act toward others, and so on. If you ask a question like "Is it wrong to lie?," the Judeo-Christian morality has a ready answer: "Yes, it is wrong to lie; it is right to tell the truth." Speaking loosely, we could also say that, according to Judeo-Christian morality, it is *immoral* to lie and *moral* to tell the truth.

Moralities, or moral codes, differ by time and place. According to some people—8th-century BC Greeks, for example—it is not always wrong to lie, and it is not always right to tell the truth. So we are confronted with the *ethical* problem of choosing between different *moralities*. Some moralities may be better than others. It may even be true—as many thinkers have argued—that only *one* system of morality is ultimately acceptable. Thinking about ethics means thinking about the strengths and weaknesses of moralities, understanding why we might endorse one morality and reject another, and searching for better systems of morality or even "the best" morality. Especially in our own day, when globalization and accelerating advances in communication have created a cultural blending (and cultural conflicts) like never before, our ability to understand different moralities is crucial.

This appendix introduces you to the way various Western philosophers have answered the ethical question "What is the good?" Because the Western tradition is complicated enough, we have not addressed Eastern moralities and the ethical thinking of many fascinating Eastern philosophers. One of the interesting things about studying ethics is the enormous variety of moralities that humans have created and the many similarities between competing moralities. Unlike the rest of your textbook, this appendix is not specifically focused on the ethical problems created by technology. But as you read through the various moralities in the appendix, ask yourself how you would deal with the moral dilemmas you have studied and confronted in your own life.

THE KNOTTY QUESTION OF GOODNESS

Achilles kills Hector outside the gates of Troy. He binds Hector's corpse by the ankles, ties the ankles to the back of his chariot, and drags the body around the city walls. The treatment of the fallen Trojan hero by his victorious Greek enemy is so outrageous that not only Trojans, but most of Achilles' Greek allies and even the Gods, are shocked. But what is wrong with Achilles' action?

To an ancient Greek of the time, the answer would not have been obvious. When the poet **Homer (8th century BC)** tells this story in his epic *The Iliad*, his purpose is to illustrate a failure in the morality of his own day. Among Greeks of Homer's day, the prevailing moral code was: "Help to friends and harm to enemies." That code may sound naïve or ridiculously simplistic today. But for the collection of small and largely independent city-states that was ancient Greece, it was a moral code that had worked reasonably well for centuries. Yet Homer saw that different times were on the way. When the Greeks banded together, as they did to combat the Trojans, the old morality looked barbaric. There was nothing heroic about the lone Achilles dragging his vanquished enemy behind him. On the contrary, he seemed like a savage.

When a society is passing from an old moral code to a new one, or when two different cultures clash in their moral codes, the extraordinarily difficult question of which moral code is correct inevitably appears. *Ethics*, the systematic study of moral codes, is the attempt to answer that question. Almost every philosopher and most thinking people will agree that some moral codes are better than others; many philosophers and others will argue that a particular moral code is the best.

Perhaps the most famous philosopher of all time, **Socrates (470–399 BC)**, argued that there was only one true moral code, and it was simple: "No person should ever willingly do evil." Socrates thought that no harm could come to a person who always sought the good, because what truly counted in life was the caretaking of one's self or soul. But Socrates also acknowledged that identifying the good was rarely easy, and his method of constantly interrogating his friends and fellow citizens—what came to be called Socratic questioning, or the Socratic dialectic—tried to improve everyone's thinking about what one ought and ought not do.

Socrates never wrote down any of his philosophy. But his student **Plato (427–347 BC)** made Socrates the hero of almost all of his many philosophical dialogues. Plato was the first "professional" philosopher in the West: he established a school of philosophy called the Academy (where we get the word *academic*), published a great number of books both for general readers and his own students, and formed arguments on virtually every subject in philosophy (not only morality). In fact, Plato possessed such breadth that the 20th-century philosopher Lord Alfred North Whitehead wrote that "all subsequent philosophy is only a footnote to Plato."

In many of his dialogues Plato raises the question: "What is the good?" Like Homer (who was one of Plato's favorite writers), Plato lived in a time when great political, social, and cultural changes were occurring. Athens had lost the first major war in its history, trade was accelerating across the Mediterranean, and people were traveling deeper into Asia and Africa and discovering new cultures, religions, and values. Many candidates for "the good" were being offered by different thinkers: some thought that "pleasure" was the highest good, others argued that "peace" (both personal and social) and what contributed to it was the best, others argued for "flourishing" and material wealth and power, while still others endorsed "honor and fame." But Plato responded that, while all of these things might be examples of goodness, they were not good itself. What is it that makes them good? What is the nature of the property "goodness" that they all share? And because we recognize that most "goods" may also mislead us into badness—the good of pleasure is an obvious example—how shall we sort the good from the bad?

Plato's idea is that we cannot reliably say what is good and what is not until we know what goodness is. Once we have identified goodness itself, we can discriminate among particular goods and particular activities that are designed to seek the good. We will judge what is "good" and "better" by comparing it with what is "best": the truly and wholly good. And the truly and wholly good ought always and everywhere to be good. Could we say that something was truly, wholly good if it was good only in some countries and not others, during some times and not others? So, if we can identify goodness as such, Plato said, we can solve every problem posed by the clash between good and bad; that is, we can solve every problem of morality.

One way to think about Plato's insight is to see the moral importance of *standards*. We have standards for good hamburgers, for good businesses, and for good hammers, so why not have standards for good people and good actions? A standard is one way of providing a *justification* for an evaluation. Suppose Rebecca insists, "It is always wrong to kill an innocent human being." And Thomas replies, "But why?" Rebecca may justify her evaluation by appealing to a standard of rightness and wrongness. Of course, identifying that standard may prove more difficult than appealing to it, and the history of ethics, again, may be seen as the struggle to provide such a standard. The philosophers you will read about in the following sections attempted to answer Plato's knotty questions in their own ways.

RELATIVISM: WHY "COMMON SENSE" WON'T WORK

What about simply using common sense to find the good? Some 20th-century philosophers argued for what they called moral "intuitions": a kind of "consult your conscience" approach to morality. This view is initially compelling for most people; it holds that the standard for goodness demanded by Plato is accessible to all of us if we simply think through our moral decisions carefully enough. (Socrates may have been arguing for the same view.) There is a "voice" in our heads that tells us what is morally right and wrong, and if you honestly and thoroughly interrogate yourself about what you ought to do, that "voice" will praise the right action and warn you against the wrong one. Someone who says "Do the right thing!" is invoking this common-sense notion. We all know what the right thing is, a moral intuitionist argues, if we use our common sense and are tough on ourselves. The difficulty is that we don't always want to use common sense or ask ourselves tough questions. Therefore, the problem of right and wrong is not so much that of moral knowledge as it is weakness of will. We *know* what we ought to do, but it is hard to make ourselves *do* it.

A crippling difficulty with this view is called the problem of relativism. *Cultural relativism* is the simple observation that different cultures employ different norms (or standards). Implicit in this view is that it is morally legitimate for different cultures to create and embrace different norms. So, for example, among the Greeks of Homer's day, lying was considered to be a virtue. Odysseus was praised specifically for his ability to lie well. In 18th-century Germany, on the other hand, lying was widely considered as morally reprehensible as theft. Some philosophers even argued that lying was just as morally foul as murder. For the relativist, lying is neither right nor wrong; rather, it can be right at a certain time and place and wrong in another. Another example is bribery. Although people in many nations condemn bribery, it is perfectly acceptable in other countries, particularly in Latin America. The relativist would say: "Bribery itself is not right or wrong. Rather, some people at some times and in some places say it is wrong, and other people say it is right, depending on the circumstances. Bribery is therefore wrong for some people, right for others."

You have probably encountered this relativism with something as simple as email. The conventions that govern email etiquette vary dramatically from user to user, group to group, and culture to culture. The emoticon-laced email you send to a friend would be wholly inappropriate if sent to a professor. The kind of language you use in an email to a college admissions officer is not what you would use to email your parents or an email pal in India. A practical platitude that embodies this idea is: "When in Rome, do as Romans." What is *appropriate* and what counts as a "good" email (as opposed to a "bad" or offensive email) depends on the conventions within its cultural context. Even emails have *norms*.

Moral relativists argue that all norms and values are relative to the cultures in which they are created and expressed. For the moral relativist, it makes no sense to say that there are any transcultural or transhistorical values, and that any attempt to construct them would still be informed by the particular cultural values of a person or group. All you can talk about are the values "on the ground": the values that particular cultures embrace. And common sense may be one of the best tools for discovering those values. Common sense may be the psychological embodiment of the complex structure of rules, standards, and values that are the substance of every robust culture.

But moral relativists run into trouble, because there are some moral claims they cannot consistently make. Moral relativists can say "slavery is wrong in my society" or "slavery is wrong in the 20th century," but they cannot say that slavery is always wrong. Furthermore, because they cannot appeal to transcultural standards for morality, they cannot speak of *moral progress*. Moral values (like all other values) change over time for the relativist, but they do not improve or degenerate. Yet, most of us would agree that the growing worldwide prohibition against slavery and torture, for example, is not merely a change, it is moral progress. And if we believe in moral progress, we cannot be relativists.

Egoism vs. Altruism

Throughout this book we have seen that ethics deals with the question of how we should treat one another. But some thinkers would say we have already misconstrued the question when we ask "How should we treat others?" For an *egoist*, the salient moral question is "How do I best benefit myself?," and the answer to Plato's question "What is the good?" is simply "The good is whatever is pleasing to *me*."

Egoism is usually divided into two types. *Psychological egoism* is the thesis that people always act from selfish motives, whether they should or not. *Ethical egoism* is the more controversial thesis that, whether people always act from selfish motives, they should if they want to be moral.

There is a superficial plausibility to psychological egoism, because it might appear that most of us make many of our choices for self-interested reasons. You probably decided that you wanted to go to college rather than immediately finding a job. You might respond: "No, I went to college because my parents wanted me to!" But the psychological egoist would reply: "That simply means that, for you, pleasing your parents is more important than other things that would have kept you out of college."

However, some of the problems with psychological egoism already are glaringly apparent. First, though we may make many decisions based on our own interests, it is far from obvious that *all* of our decisions are motivated by self-interest. We make many decisions, including decidedly uncomfortable ones, because we are thinking of the interests of others. It is silly to suppose that our own interests must always and implicitly conflict with those of others, as a psychological egoist believes. Why did you go to college? Because you wanted to, and your parents, teachers, and friends wanted you to. Everyone's interests happily coincided, and it is oversimplifying your complex choice to say, as a psychological egoist would, "I did it because *I* wanted to."

While considering ethical egoism, we should also look at its opposite: *altruism*. The altruist argues that the morally correct action always best serves the interest of others. Wouldn't the world be a better place, the altruist asks, if we worried about ourselves less and tried to help other people?

No one will deny that everyone benefits from altruism, but problems arise if we try to adopt altruism as a moral code. Practically speaking, it is sometimes difficult to know what best serves the interest of another, beyond helping people with the basic necessities of life. For example, a devout Southern Baptist might sincerely believe that his neighbors are condemned to hell unless they accept his religious views, and might feel an altruistic urge to convert them, despite their hesitation. Another more famous example involves a boat full of altruists lost at sea. They can only survive if one of them volunteers to be eaten, but if the only moral action is to serve the interests of others, how can any of the adrift altruists be truly moral when one of them has to die to save the rest?

Problems like these help to motivate advocates of ethical egoism. We do not reliably know the interests of others, the ethical egoist says, but we certainly know our own. And, unlike altruists, whose satisfaction is in helping others, ethical egoists try to create a happy and moral world by seeking good for themselves. The hacker who thinks she can morally break the rules because she has the smarts to do so is both a psychological egoist ("you would break the rules too, if you could") and an ethical egoist ("everyone who can break the rules to help themselves should do so"). Given the choice between self-interest and altruism, the ethical egoist takes the former.

Of course, the only choice is not between ethical egoism and altruism. Most moral codes and most people recognize the importance of both self-interest and the interest of others. The more telling objection to ethical egoism is that it does not respect our deepest intuitions about moral goodness. If an ethical egoist can serve his own interest by performing some horrific act against another human being, and be guaranteed that the act will not interfere with his self-interest, he is morally permitted to perform that act. In fact, if he finds that he can *only* serve his interest by performing the horrific act and getting away with it, he is morally *required* to do so. An employer who could benefit from spying on her employee's email would be morally required to do it if it served her long-term interest. But for most of us, such examples are sufficient to defeat ethical egoism. Moral codes are plausible only if they accommodate basic intuitions about our sense of right and wrong, and ethical egoism fails on that ground.

DEONTOLOGY, OR THE ETHICS OF LOGICAL CONSISTENCY AND DUTY

Most people find they cannot accept relativism as a moral code because of their moral intuitions that some things are *always* wrong (like slavery or the torture of innocents). For this reason, they must also abandon a "common sense" approach to morality, which relies on embedded knowledge of cultural norms. The problems with egoism and altruism are even more glaring. But don't despair—there are lots more moral theories to consider. The rest of this appendix reviews several modern attempts to articulate a consistent morality.

Immanuel Kant (1724–1804) is generally considered the most important philosopher since Aristotle. Kant's moral theory is an attempt to refine and provide a sound philosophical foundation for the strict Judeo-Christian morality of his own day. Most people, when they begin thinking about ethics in a philosophical way, find that they are some brand of Kantian. Kant's theory is called *deontology*, from the Greek word *deon*, meaning *duty*. For Kant, to do what is morally right is to do one's duty.

Understanding what one's duty requires is the difficult part, of course. Kant begins with the idea that the only thing in the world that is wholly good, without any qualification, is good will. Most good things may be turned to evil or undesirable ends, or are mixed with bad qualities. Human beings do not seem wholly good: they are a mix of good and bad. Money is a good that most of us seek, while "love of money is the root of all evil." But the will to do good—the desire or intention—must be wholly good. If we think through what we mean by "moral goodness," Kant argues, we realize that the notion of moral goodness is just another name for this will to goodness. Kant recognizes that, as the old saying goes, "the road to Hell is paved with good intentions"; he is not saying that good will must always have good consequences. (In general, Kant is suspicious of the moral worth of consequences.) But the intention to do good, before it gets tangled up in the difficulties of the world, must itself be purely good.

Morality, therefore, comes from our ability to intend that certain things happen: that is, from our ability to choose. The good choice will come from a good will. But how do we sort the good choice from the bad? Kant, following the ancient Greek philosopher Aristotle, believed that the property that makes human beings unique, and that propels us into the moral sphere, is the faculty of *reason*. Kant saw human beings as constantly torn between their passions, drives, and desires (what he called "inclinations") and the rational ability to make good choices on the basis of good and defensible reasons. For Kant, with his dim view of human nature, what we *want* to do is very rarely what we *ought* to do. But we can recognize what we ought to do by the application of reason.

Kant's derivation of the *categorical imperative*, which he argued is the fundamental principle of all morality, is notoriously complex. But the key idea is simple: reason demands consistency and rejects contradiction. Accordingly, Kant argued that the moral principle we should follow must preserve consistency in all cases and prevent any possibility of contradiction. This moral principle might be expressed as: "Act only on that maxim such that the maxim of your action can be willed to be a universal law." (Although Kant offered several different formulations of the categorical imperative, this is the most famous and most basic formulation.) Kant's prose is dense and confusing, and the categorical imperative is no exception. What does Kant mean?

Kant observed that we make choices according to rules. We tell the truth even when it is inconvenient or embarrassing because we have a rule in our heads that tells us to do so. This is an example of what Kant calls a "subjective principle of action" or a *maxim*. Other examples of maxims are "don't steal" and "keep your promises." Our heads are full of rules that we use to guide our choices. When we worry about *moral* choices, Kant tells us in the categorical imperative that we should act only on choices that "can be willed to be a universal law." That is, before acting on a maxim that informs a moral choice, one must ask: "Could this rule (this maxim) be applied to everyone, everywhere, for all time?" Kant argues that, by *universalizing* a maxim, one can see whether it generates a contradiction. If it generates a contradiction, it cannot be rational, and so it is not a legitimate expression of a good will. If it does not generate a contradiction, it looks morally permissible. When we follow the categorical imperative, Kant thinks, we are doing our (moral) duty.

Take a couple of examples. Suppose you decide to borrow money without intending to pay it back. Your maxim might be: "If I need to borrow money I should do so, even though I know I will never pay it back." Now universalize this maxim according to the categorical imperative. Suppose everyone, everywhere, always borrowed money without the intention of paying it back? Obviously no one would lend money and the very possibility of borrowing would be eliminated. It is rationally contradictory to choose to borrow money without intending to pay it back.

Or, suppose you are caught cheating and try to lie your way out of it. Your maxim is: "When caught cheating, I should lie to get out of trouble." But suppose everyone, everywhere, always lied to get out of trouble when caught cheating? To lie you must hide the truth, and in this situation, were it universalized, it would be impossible to hide the truth. Lies depend on being exceptions to the rule of truthful communication; if lies are no longer the exception but the rule, there is no more truthful communication, and a lie becomes impossible. Again, this is a rational contradiction, and we see that the lie is immoral.

Suppose, however, that you try a maxim like "Thou shalt not kill." What if everyone, everywhere, always avoided killing others? No contradiction is generated. There may be many impractical consequences of universal not-killing, but there are no logical problems with it. If you try a maxim of "Thou shalt kill," on the other hand, you see how quickly it falls apart.

It is not difficult to generate objections to this theory. If one makes maxims specific enough, it is easy to justify apparently immoral actions while following the rule of universal maxims. For example, one can easily universalize a maxim like "a woman with no money whose children are dying of pneumonia should steal penicillin if necessary to save her children's lives," yet Kant would maintain that theft is always wrong and irrational.

Kant also maintains that it is always irrational and wrong to lie, even in the attempt to save an innocent life. But to most of us that sounds absurd. Should a mother never lie, even if it means saving the life of her child? Should the Danes who lied to the Nazis about whether they were protecting Jews have told the truth? Surely not.

Perhaps the most controversial aspect of Kant's moral theory is his distinction between moral duty and happiness. Kant argues that choosing freely on the basis of what we rationally see is right—following the categorical imperative, acting from duty—is the only way we can choose *morally*. But suppose we are acting a certain way solely because it makes us happy, even though those actions happen to agree with what would otherwise be our duty. For Kant, actions motivated by inclination (with the result of happiness) are not motivated by duty, and so we should not consider them *moral* actions. For example, a suicidal person who does not shoot herself because she recognizes that it would be irrational (and thus contrary to her duty) is acting morally. However, another person who fleetingly considers shooting himself but then declines because he loves his life is not acting morally; he is merely inclining toward his happiness.

But if moral duty and happiness are opposed, it seems that only miserable people can be moral. Wouldn't it be nicer if we could have both moral worth in our actions and happy lives? This leads us to *utilitarianism*, the theory of morality that responds specifically to deontology by insisting that morality and happiness are not opposites, but the very same thing.

HAPPY CONSEQUENCES, OR UTILITARIANISM

Hedonism is the notion, first advocated by the Greek philosopher Epicurus (342–270 BC), that pleasure is the greatest good for human beings. (Epicurus is the source of the word *epicurean*.) To be moral is to live the life that produces the most pleasure and avoids pain. But we should not suppose that Epicurus was arguing for a life of debauchery. Drinking too much wine, for example, though fun while it lasts, produces more pain than pleasure in the end, so Epicurus sorted pleasures into categories:

- Natural and necessary, like sleeping and moderate eating
- Natural but unnecessary, like drinking wine or playing chess
- Unnatural and unnecessary, which hurt one's body (for example, smoking cigarettes)
- Unnatural but necessary (but there are no such pleasures)

Epicurus said that we should cultivate natural and necessary pleasures, enjoy natural but unnecessary pleasures in moderation, and avoid all other sorts. The true hedonist does not seek what is immediately pleasurable, but looks for pleasures that will guarantee a long, healthy life full of them. For this reason, *friendship* is Epicurus' favorite example of a pleasure that everyone should cultivate; friendship was consistently considered one of the highest human goods among ancient Greeks.

Jeremy Bentham (1748–1832) adopted Epicurus' basic principles when he developed the theory that later became known as *utilitarianism*. In response to Plato's question "What is the good?," Bentham argued that it is easy to see what humans consider good because they are always seeking it: pleasure. But Bentham was not an egoist, and he argued that the highest good would result from a maximum of pleasure for all people concerned in any moral decision. Decisions that promote *utility* are those that create the most pleasure (the words *utility* and *pleasure* were virtually interchangeable to Bentham, though later utilitarians would ascribe many different meanings to *utility*). Whenever making a decision, the person who desires a moral result should weigh all possible outcomes, and choose the action that produces the most pleasure for everyone concerned. Bentham called this weighing of outcomes a "utilitarian calculus."

Bentham's new moral theory enjoyed enormous popularity, but brought inevitable objections. Some philosophers argued that such a theory made people look no better than swine (because they were just pursuing pleasure). Others objected that people would surely frame their moral decisions to enable them to do whatever they pleased. **John Stuart Mill (1806–1873)** responded to these objections and gave us the form of utilitarianism that, in its fundamentals, is the same moral theory that so many philosophers and economists still endorse today.

Mill argued that the good that human beings seek is not so much pleasure as happiness, and that the basic principle of utilitarianism was what he called the "Greatest Happiness Principle": that action is good which creates the greatest happiness, and the least unhappiness, for the greatest number. He also insisted that people who used this principle must adopt a disinterested view when deciding what would create the greatest happiness. He called this the perspective of "the perfectly disinterested benevolent spectator."

When making a moral decision, then, people will consider the various outcomes and make the choice that produces the most happiness for themselves and everyone else. This is not the same as asking which choice will produce the most pleasure. Accepting a job selling computer software for $55,000 a year might produce more short-term pleasure than going to graduate school, but it might not produce the most happiness. You might be broke and hungry in graduate school, but still very happy because you are progressing toward a goal and finding intellectual stimulation along the way.

The utilitarian must also ask: does this decision produce the most happiness for everyone else, and am I evaluating their happiness fairly and reasonably? Suppose that the recent graduate is again deliberating whether to go to graduate school. Her mother and her father, both attorneys, very much want her to go into the law. But she is fed up with school and will be miserable sitting in a classroom all day. She is sick of eating Ramen noodles and having roommates, and would like to drink a nice bottle of wine once in a while and buy a new car. It is true that her parents' happiness is relevant to the decision, but she must try to weigh the happiness of everyone involved. How unhappy will her parents be if she takes a few years off? How unhappy will she be back in a lecture hall? Utilitarians admit that finding the good is not always easy, but they insist that they offer a practical method for finding the good that anyone can use to solve a moral dilemma.

Utilitarianism is a kind of *consequentialism*, because we evaluate the morality of actions on the basis of their probable outcomes or consequences. For this reason utilitarianism is also what we call a *teleological* theory. Coming from the Greek word *telos*, meaning *purpose* or *end*, teleology refers to the notion that some things and processes are best understood by considering their goals. For utilitarians the goal of life is happiness, and thus they argue that the good (moral) life for humans is the happy life.

Utilitarianism is probably the most popular moral theory of the last hundred years. It is widely used by economists, because one easy way of measuring utility is by assigning dollar signs to outcomes. Today's most famous advocate of animal rights, Peter Singer, is also a well-known utilitarian. Many different versions of utilitarianism have been advanced. In *rule-utilitarianism*, we first rationally determine the general rules that will produce good outcomes, and then follow those rules. In *preference-utilitarianism*, we solve the difficult problem of what will create the most happiness for others by simply asking every person involved for their preference.

But there are many strong objections to utilitarianism. One was raised by the German philosopher **Friedrich Nietzsche (1844–1900)** in his masterpiece *Thus Spoke Zarathustra*. At the end of the book, Zarathustra asks himself if his efforts to find the good for human beings and for himself have increased his personal happiness. He responds to himself: "Happiness? Why should I strive for happiness? I strive for my work!" Nietzsche's point is that many profound and praiseworthy human goals are unquestionably moral, and yet they cannot be said to contribute to the happiness of the person who has those goals, and perhaps not even to the happiness of the greater number. It is true that Van Gogh's paintings, though they destroyed him, created a greater happiness for the rest of us. But that did not count for him as a reason to paint them—he had no idea of his own legacy. For a utilitarian, such self-sacrifice is not only confused but immoral. And yet if our moral theory has difficulty accounting for the value of Van Gogh sacrificing his happiness and everything he loved to his art, we might be in trouble.

Perhaps the most telling objection to utilitarianism is that it could be used to morally sanction a "tyranny of the majority." Suppose you could solve all of the suffering of the world and create universal happiness by flipping a switch on a black box. But, in order to power the box, you had to place one person inside it, who would suffer unspeakably painful torture. None of us would be willing to flip that switch, and yet for a utilitarian such an action would not only be permissible, it would be morally demanded.

A related objection comes from the British philosopher Bernard Williams. Suppose you are an explorer in the Amazon basin and you stumble on a tribe that is about to slaughter 20 captured warriors from another tribe. You interrupt the gruesome execution, and the tribal chief offers to release 19 prisoners in your honor, on the condition that you accept the ceremonial role of choosing one victim and killing him yourself. A utilitarian would be morally required to accept, but most of us would be morally appalled at the idea of killing a complete stranger who presented no threat to us.

Utilitarians have responded to such objections by introducing the notion of certain irreducible human *rights* into utilitarianism. The discussion now turns to these rights and their origins in *social contract theory*.

Promises and Contracts

Although there are good reasons for being suspicious of egoism of any stamp, **Thomas Hobbes (1588–1679)** was a psychological egoist, which was essential to his moral and political theory. Hobbes argued that there are two fundamental facts about human beings: (1) we are all selfish, and (2) we can only survive by banding together. You may have heard Hobbes' famous dictum that human life outside of a society—that is, in his imagined "state of nature"—is "solitary, poor, nasty, brutish, and short." We form groups for self-interested reasons because we need one another to survive and prosper. But the fact that we band together as selfish beings inevitably results in tension between people. Because resources are always scarce, there is competition, and competition creates conflict. Accordingly, if we are to survive as a group, we need rules that everyone promises to follow. These rules, which may be simple at first but become enormously complex, are an exchange of protections for freedoms. "I promise not to punch you in the nose as long as you promise not to punch me in the nose" is precisely such an exchange. You trade the freedom to throw your fists wherever you please for the protection of not being punched yourself. These rules of mutual agreement are, of course, called *laws*, and they guarantee our protections or *rights*. The system of laws and rights that make up the society is called the *social contract*.

Social contract theory builds on the Greek notion that good people are most likely encouraged by a good society. Few social contract theorists would argue that morality can be reduced to societal laws. But most would insist that it is extremely difficult to be a good person unless you are in a good society with good laws. Hobbes argued that the habit of exchanging liberties for protections would extend itself into all dimensions of a good citizen's behavior. The law is an expression of the reciprocity expressed in the Golden Rule—"do unto others as you would have them do unto you"—and so through repeated obedience to the law we would develop the habit, Hobbes thought, of treating others as we would like to be treated.

The most famous American social contract theorist was **John Rawls (1921–2002)**. Rawls argued that "justice is fairness," and for him the morally praiseworthy society distributes its goods in a way that helps the least advantaged of its members. Rawls asked us to imagine what rules we would propose for a society if, when we thought about the rules, we imagined that we had no idea what our own role in that society would be. What rules would we want for our society if we did not know whether we would be poor or rich, African-American or Native Indian, man or woman, or a teacher, plumber, or famous actor? Rawls imagined that this thought experiment—which he called standing behind "the veil of ignorance"—would guarantee fairness in the formulation of the social contract. Existing social rules and laws that did not pass this test—that no rational person would endorse if standing behind the veil of ignorance—were obviously unfair and should be changed or discarded.

Strictly speaking, social contract theory is not a moral code. But because so many of our moral decisions are made in the context of laws and rights, we should understand that the foundation of those laws and rights is a system of promises that have been made, either implicitly or explicitly, by every citizen who freely chooses to live in and benefit from a commonwealth.

A RETURN TO THE GREEKS: THE GOOD LIFE OF VIRTUE

In the 20th century, many philosophers grew increasingly suspicious of the possibility of founding a workable moral system upon rules or principles. The problem with moral rules or principles is that they self-consciously ignore the particulars of the situations in which people actually make moral decisions. For the dominant moralities of the 20th century, deontology and utilitarianism, what is moral for one person is moral for another, regardless of the many differences that undoubtedly exist between their lives, personalities, and stations. This serious weakness in prevailing moral systems caused philosophers to turn once more to the ancient Greeks for help.

Aristotle (384–307 BC) argued that it does not make sense to speak of good actions unless one recognizes that good actions are performed by good people. But good people deliberate over their actions in particular situations, each of which may differ importantly from other situations in which a person has to make a moral choice. But what is a good person?

Aristotle would have responded to this question with his famous "function argument," which posits that the goodness of anything is expressed in its proper function. A good hammer is good because it pounds nails well. A good ship is good if it sails securely across the sea. A bad ship, on the other hand, will take on water and drift aimlessly across the waves. Moreover, we can recognize the function of a thing by identifying what makes it different from other things. The difference between a door and a curtain lies fundamentally in the way they do their jobs. Human function, the particular ability that makes mankind different from all other species, is the ability to reason. The good life is the life of the mind: to be a good person is to actively think.

But to pursue the life of the mind, we need many things. We need health; we need the protection and services of a good society; we need friends for conversation. We need leisure time and enough money to satisfy our physical needs (but not so much as to distract or worry us); we need education, books, art, music, culture, and pleasant distractions to relax the mind.

This does sound like the good *life*. But how does the thinking person *act*? Presumably, Aristotle's happy citizen will encounter moral conflicts and dilemmas like the rest of us. How do we resolve these dilemmas? What guides our choices?

Aristotle did not believe that human beings confront each choice as though it were the first they ever made. Rather, he thought, we develop habits that guide our choices. There are good habits and bad habits. Good habits contribute to our flourishing and are called *virtues* (Aristotle's word, *arête*, may also be translated as *excellence*). Bad habits diminish our happiness and are called *vices*. And happily, for Aristotle, the thinking person will see that there is a practical method for sorting between virtues and vices built into the nature of human beings. Aristotle insisted that human beings are animals, like any other warm-blooded creature on the earth; just as a tiger can act in ways that cause it to flourish or fail, so human beings have a natural guide to their betterment. This has come to be called Aristotle's "golden mean": the notion that our good lies between the extremes of the deficiency of an activity and its excess. Healthy virtue lies in moderation.

An example will help. Suppose you are sitting in the classroom with your professor and fellow students when a wild buffalo storms into the room. The buffalo is enraged and ready to gore all comers. What do you do? An excessive action would be to attack the buffalo with your bare hands: this would, for Aristotle, show the vice of rashness. A deficient action would be to cower behind your desk and shriek for help: this would show the vice of cowardice. But a moderate action would be to make a loud noise to frighten the buffalo, or perhaps to distract it so that others could make for the door, or to do whatever might reasonably reduce the danger to others and yourself. This moderate course of action exemplifies the virtue of courage. Notice, however, that the courageous course of action would change if an enraged tomcat came spitting into the room. Then the moderate and virtuous choice might be to trap the feline with a handy trash basket.

Aristotle's list of virtues includes courage, temperance, justice, liberality or generosity, magnificence (living well), pride, high-mindedness, aspiration, gentleness, truthfulness, friendliness, modesty, righteous indignation, and wittiness. But one could write many such lists, depending on one's own society and way of life. Aristotle would doubtless argue that at least some of these virtues are virtuous for any human being in any place or time, but a strength of his theory is that others' virtues depend on the when, where, and how of differing human practices and communities. One appeal of virtue ethics is that it insists on the context of our moral deliberations.

But is human goodness fully expressed by moderation? Or by being a good citizen? And what about people who lack Aristotle's material requirements of health, friends, and a little property? Aristotle is committed to the idea that such people cannot live fully moral lives, but can that be right? As powerful as it is, one weakness of Aristotle's virtue ethics is that it seems to overemphasize the importance of "fitting" into one's society. The rebel, the outcast, or the romantic chasing an iconoclastic ideal has no place. And Aristotle's theory may sanction some gross moral injustices—such as slavery—if they contribute to the flourishing of society as a whole. Aristotle himself would have had no problem with this: his theory was explicitly designed for the aristocratic way of life. But today we would insist that the good life, if it is to be truly *good* for any of us, must at least in principle be available to every member of our society.

Feminism and the Ethics of Care

Psychologist and philosopher **Carol Gilligan** discovered that moral concepts develop differently in young children. Boys tend to emphasize reasons, rules, and justifications; girls tend to emphasize relationships, the good of the group, and mutual nurturing. From these empirical studies Gilligan developed what came to be called the "ethics of care": the idea that morality might be better grounded on the kind of mutual nurturing and love that takes place in close friendships and family groups. The ethical ideal, according to Gilligan, is a good mother.

Gilligan's ethics of care is compelling because it seems to reflect how many of us make our daily moral decisions. Consider the moral decisions you face in a typical day: telling the truth or lying to a parent or sibling, skipping a party to take care of a heartsick friend or going to see that cute guy, keeping a promise to another student to copy your notes or saying "oops, I forgot." We often confront the moral difficulties of being a good son, sister, friend, or colleague. Generally speaking, we do not settle these moral issues on the basis of impersonal moral principles—we wonder whether it would even be appropriate to do so, given that we are personally involved in these decisions. Should you treat your best friend in precisely the same way you treat a stranger on the street? Some moralists would say, "Of course!" Yet, many of us would consider such behavior odd or psychologically impossible.

The feminist attack on traditional ethics does not accuse one Western morality or another, but indicts its whole history. Western morality has insisted on rationality at the expense of emotions, on impartiality at the expense of relationships, on punishment at the expense of forgiveness, and on "universal principles" at the expense of real, concrete moral problems. In a phrase, morality has been male at the expense of the female. Thus, the feminist argues, a radical rethinking of the entire history of morality is necessary.

As a negative attack on traditional morality, it is hard to disagree with feminism. Our moral tradition does have a suspiciously masculine cast; it is not surprising that virtually every philosopher mentioned in this appendix was a man. But feminism has struggled to develop a positive ethics of its own. Many consider Gilligan's ethics of care to be the best attempt so far, and it works well in family contexts. But when we try to extend the ethics of care into larger spheres, we run into trouble. Gilligan insists on the moral urgency of partiality (as a mother is partial to her children, and even among children). But you would object if you were a defendant in a lawsuit and saw the plaintiff enter, wave genially to the judge, and say, "Hi Mom!" The point, of course, is that in many situations we insist on *impartiality*, and for good reasons. And we all agree that people we have never met may still exercise moral demands upon us. We believe that a man rotting in prison on the other side of the world ought not be tortured, and maybe that we should do something about it if he is (if only by donating money to Amnesty International). Everyone deserves protection from torture for reasons that apply equally to all of us.

PLURALISM

When the German philosopher Nietzsche famously proclaimed that "God is dead," he was not proposing that the nature of the universe had changed. Rather, he was proposing that a change had taken place in the way we view ourselves in the universe. He meant that the Judeo-Christian tradition that has informed all of our values in the West can no longer do the job for us that it used to do. Part of that tradition, Nietzsche thought, was the unfortunate Platonic idea that there is an answer to the question "What is the good?" There is no more one "good" than there is one "God" or one "truth": there are, Nietzsche insisted, many goods, like there are many truths. Nietzsche argued the moral position that we now call *pluralism*.

Pluralism is the idea that there are many goods and many sources of value. Pluralism is explicitly opposed to Plato's insistence that all good things and actions must share some quality that makes them all good. But does this make the pluralist a relativist? No, because the pluralist argues for the moral significance of two ideas that the relativist rejects: (1) that some aspects of human nature are transcultural and transhistorical, and (2) that some methods of inquiry reveal transcultural and transhistorical human values.

When we look at human history, we see goods that repeatedly contribute to human flourishing and evils that interfere with it. War is almost always viewed as an evil in history that has consistently interfered with human flourishing; health, on the other hand, is almost always viewed as a good (with the exception of aberrant religious practices like asceticism). "Avoid war and seek health" is not a moral code—although it might go further than we think—but it does provide an example of what a pluralist is looking for. The pluralist wants concrete goods and practices that actually enrich human life. For the pluralist, the choice between Plato's absolutism and moral relativism is a false dichotomy. Just because there is no absolute "good" does not mean that all goods or values are relative to the time, place, and culture in which we find them. Some things and practices are usually bad for humans, others are usually good, and the discovery and encouragement of the good things and practices is the game the smart ethicist plays.

For this reason, pluralists emphasize the importance of investigating and questioning. Is our present culture enhancing or diminishing us as human beings? Is the American attitude toward sexuality, say, improving the human condition or interfering with it? (And before we can answer that question, what *is* the American attitude toward sexuality? Or are there many attitudes?) The ethical contribution to the history of philosophy made by the fascinating 20th-century movement called *existentialism* is its insistence on this kind of vigorous, ruthlessly honest interrogation of oneself and one's culture. The danger of hypocrisy and self-deception, or what the leading existentialist **Jean-Paul Sartre (1912–1984)** called *bad faith*, is rampant in every culture: challenging our values is uncomfortable. It is much easier for us, like the subjects of the nude ruler in H. C. Andersen's fable *The Emperor's New Clothes*, to collectively pretend that something is good (even if we know there is really nothing there at all). Thus, the project of becoming a good person becomes not just a matter of following the rules, doing one's duty, seeking happiness, becoming virtuous, or caring for others. It is also the lifelong project of discovering *if, when, and why* the apparently good things we seek are what we ought to pursue.

SUMMARY

After reading this appendix, a reasonable student might ask: "But which of these moralities is the *right* one?" Admittedly, philosophers are better at posing problems than solving them. But the lesson was not in demonstrating that one or another morality is the one a person ought to follow. Rather, this appendix has attempted to show you how different people have struggled with the enormously difficult questions of ethics. Many people think they simply know the difference between right and wrong, or unreflectively accept the definitions of right and wrong offered by their parents, churches, communities, or societies. This appendix tried to show that there is nothing simple about ethics. To understand ethics means to think, to challenge, to question, and to reflect. Accordingly, being a good person might mean attempting your own struggle with, and attempting to find your own answer to, what we called Plato's knotty question of goodness.

SOFTWARE ENGINEERING CODE OF ETHICS AND PROFESSIONAL PRACTICE

(Version 5.2) as recommended by the IEEE-CS/ACM Joint Task Force on Software Engineering Ethics and Professional Practices and Jointly approved by the ACM and the IEEE-CS as the standard for teaching and practicing software engineering.

SHORT VERSION

Preamble

The short version of the code summarizes aspirations at a high level of the abstraction; the clauses that are included in the full version give examples and details of how these aspirations change the way we act as software engineering professionals. Without the aspirations, the details can become legalistic and tedious; without the details, the aspirations can become high sounding but empty; together, the aspirations and the details form a cohesive code.

Software engineers shall commit themselves to making the analysis, specification, design, development, testing and maintenance of software a beneficial and respected profession. In accordance with their commitment to the health, safety and welfare of the public, software engineers shall adhere to the following Eight Principles:

1. PUBLIC - Software engineers shall act consistently with the public interest.
2. CLIENT AND EMPLOYER - Software engineers shall act in a manner that is in the best interests of their client and employer consistent with the public interest.
3. PRODUCT - Software engineers shall ensure that their products and related modifications meet the highest professional standards possible.
4. JUDGMENT - Software engineers shall maintain integrity and independence in their professional judgment.
5. MANAGEMENT - Software engineering managers and leaders shall subscribe to and promote an ethical approach to the management of software development and maintenance.
6. PROFESSION - Software engineers shall advance the integrity and reputation of the profession consistent with the public interest.

7. COLLEAGUES - Software engineers shall be fair to and supportive of their colleagues.
8. SELF - Software engineers shall participate in lifelong learning regarding the practice of their profession and shall promote an ethical approach to the practice of the profession.

FULL VERSION

Preamble

Computers have a central and growing role in commerce, industry, government, medicine, education, entertainment and society at large. Software engineers are those who contribute by direct participation or by teaching, to the analysis, specification, design, development, certification, maintenance and testing of software systems. Because of their roles in developing software systems, software engineers have significant opportunities to do good or cause harm, to enable others to do good or cause harm, or to influence others to do good or cause harm. To ensure, as much as possible, that their efforts will be used for good, software engineers must commit themselves to making software engineering a beneficial and respected profession. In accordance with that commitment, software engineers shall adhere to the following Code of Ethics and Professional Practice.

The Code contains eight Principles related to the behavior of and decisions made by professional software engineers, including practitioners, educators, managers, supervisors and policy makers, as well as trainees and students of the profession. The Principles identify the ethically responsible relationships in which individuals, groups, and organizations participate and the primary obligations within these relationships. The Clauses of each Principle are illustrations of some of the obligations included in these relationships. These obligations are founded in the software engineer's humanity, in special care owed to people affected by the work of software engineers, and in the unique elements of the practice of software engineering. The Code prescribes these as obligations of anyone claiming to be or aspiring to be a software engineer.

It is not intended that the individual parts of the Code be used in isolation to justify errors of omission or commission. The list of Principles and Clauses is not exhaustive. The Clauses should not be read as separating the acceptable from the unacceptable in professional conduct in all practical situations. The Code is not a simple ethical algorithm that generates ethical decisions. In some situations, standards may be in tension with each other or with standards from other sources. These situations require the software engineer to use ethical judgment to act in a manner which is most consistent with the spirit of the Code of Ethics and Professional Practice, given the circumstances.

Ethical tensions can best be addressed by thoughtful consideration of fundamental principles, rather than blind reliance on detailed regulations. These Principles should influence software engineers to consider broadly who is affected by their work; to examine if they and their colleagues are treating other human beings with due respect; to consider how the public, if reasonably well informed, would view their decisions; to analyze how the least empowered will be affected by their decisions; and to consider whether their acts would be judged worthy of the ideal professional working as a software engineer. In all these judgments concern for the health, safety and welfare of the public is primary; that is, the "Public Interest" is central to this Code.

The dynamic and demanding context of software engineering requires a code that is adaptable and relevant to new situations as they occur. However, even in this generality, the Code provides support for software engineers and managers of software engineers who need to take positive action in a specific case by documenting the ethical stance of the profession. The Code provides an ethical foundation to which individuals within teams and the team as a whole can appeal. The Code helps to define those actions that are ethically improper to request of a software engineer or teams of software engineers.

The Code is not simply for adjudicating the nature of questionable acts; it also has an important educational function. As this Code expresses the consensus of the profession on ethical issues, it is a means to educate both the public and aspiring professionals about the ethical obligations of all software engineers.

PRINCIPLES

Principle 1 Public

Software engineers shall act consistently with the public interest. In particular, software engineers shall, as appropriate:

1.01. Accept full responsibility for their own work.

1.02. Moderate the interests of the software engineer, the employer, the client and the users with the public good.

1.03. Approve software only if they have a well-founded belief that it is safe, meets specifications, passes appropriate tests, and does not diminish quality of life, diminish privacy or harm the environment. The ultimate effect of the work should be to the public good.

1.04. Disclose to appropriate persons or authorities any actual or potential danger to the user, the public, or the environment, that they reasonably believe to be associated with software or related documents.

1.05. Cooperate in efforts to address matters of grave public concern caused by software, its installation, maintenance, support or documentation.

1.06. Be fair and avoid deception in all statements, particularly public ones, concerning software or related documents, methods and tools.

1.07. Consider issues of physical disabilities, allocation of resources, economic disadvantage and other factors that can diminish access to the benefits of software.

1.08. Be encouraged to volunteer professional skills to good causes and to contribute to public education concerning the discipline.

Principle 2 Client and Employer

Software engineers shall act in a manner that is in the best interests of their client and employer, consistent with the public interest. In particular, software engineers shall, as appropriate:

2.01. Provide service in their areas of competence, being honest and forthright about any limitations of their experience and education.

2.02. Not knowingly use software that is obtained or retained either illegally or unethically.

2.03. Use the property of a client or employer only in ways properly authorized, and with the client's or employer's knowledge and consent.

2.04. Ensure that any document upon which they rely has been approved, when required, by someone authorized to approve it.

2.05. Keep private any confidential information gained in their professional work, where such confidentiality is consistent with the public interest and consistent with the law.

2.06. Identify, document, collect evidence and report to the client or the employer promptly if, in their opinion, a project is likely to fail, to prove too expensive, to violate intellectual property law, or otherwise to be problematic.

2.07. Identify, document, and report significant issues of social concern, of which they are aware, in software or related documents, to the employer or the client.

2.08. Accept no outside work detrimental to the work they perform for their primary employer.

2.09. Promote no interest adverse to their employer or client, unless a higher ethical concern is being compromised; in that case, inform the employer or another appropriate authority of the ethical concern.

Principle 3 Product

Software engineers shall ensure that their products and related modifications meet the highest professional standards possible. In particular, software engineers shall, as appropriate:

3.01. Strive for high quality, acceptable cost, and a reasonable schedule, ensuring significant tradeoffs are clear to and accepted by the employer and the client, and are available for consideration by the user and the public.

3.02. Ensure proper and achievable goals and objectives for any project on which they work or propose.

3.03. Identify, define and address ethical, economic, cultural, legal and environmental issues related to work projects.

3.04. Ensure that they are qualified for any project on which they work or propose to work, by an appropriate combination of education, training, and experience.

3.05. Ensure that an appropriate method is used for any project on which they work or propose to work.

3.06. Work to follow professional standards, when available, that are most appropriate for the task at hand, departing from these only when ethically or technically justified.

3.07. Strive to fully understand the specifications for software on which they work.

3.08. Ensure that specifications for software on which they work have been well documented, satisfy the users' requirements and have the appropriate approvals.

3.09. Ensure realistic quantitative estimates of cost, scheduling, personnel, quality and outcomes on any project on which they work or propose to work and provide an uncertainty assessment of these estimates.

3.10. Ensure adequate testing, debugging, and review of software and related documents on which they work.

3.11. Ensure adequate documentation, including significant problems discovered and solutions adopted, for any project on which they work.

3.12. Work to develop software and related documents that respect the privacy of those who will be affected by that software.

3.13. Be careful to use only accurate data derived by ethical and lawful means, and use it only in ways properly authorized.

3.14. Maintain the integrity of data, being sensitive to outdated or flawed occurrences.

3.15. Treat all forms of software maintenance with the same professionalism as new development.

Principle 4 Judgment

Software engineers shall maintain integrity and independence in their professional judgment. In particular, software engineers shall, as appropriate:

4.01. Temper all technical judgments by the need to support and maintain human values.

4.02. Only endorse documents either prepared under their supervision or within their areas of competence and with which they are in agreement.

4.03. Maintain professional objectivity with respect to any software or related documents they are asked to evaluate.

4.04. Not engage in deceptive financial practices such as bribery, double billing, or other improper financial practices.

4.05. Disclose to all concerned parties those conflicts of interest that cannot reasonably be avoided or escaped.

4.06. Refuse to participate, as members or advisors, in a private, governmental or professional body concerned with software related issues, in which they, their employers or their clients have undisclosed potential conflicts of interest.

Principle 5 Management

Software engineering managers and leaders shall subscribe to and promote an ethical approach to the management of software development and maintenance. In particular, those managing or leading software engineers shall, as appropriate:

5.01. Ensure good management for any project on which they work, including effective procedures for promotion of quality and reduction of risk.

5.02. Ensure that software engineers are informed of standards before being held to them.

5.03. Ensure that software engineers know the employer's policies and procedures for protecting passwords, files and information that is confidential to the employer or confidential to others.

5.04. Assign work only after taking into account appropriate contributions of education and experience tempered with a desire to further that education and experience.

5.05. Ensure realistic quantitative estimates of cost, scheduling, personnel, quality and outcomes on any project on which they work or propose to work, and provide an uncertainty assessment of these estimates.

5.06. Attract potential software engineers only by full and accurate description of the conditions of employment.

5.07. Offer fair and just remuneration.

5.08. Not unjustly prevent someone from taking a position for which that person is suitably qualified.

5.09. Ensure that there is a fair agreement concerning ownership of any software, processes, research, writing, or other intellectual property to which a software engineer has contributed.

5.10. Provide for due process in hearing charges of violation of an employer's policy or of this Code.

5.11. Not ask a software engineer to do anything inconsistent with this Code.

5.12. Not punish anyone for expressing ethical concerns about a project.

Principle 6 Profession

Software engineers shall advance the integrity and reputation of the profession consistent with the public interest. In particular, software engineers shall, as appropriate:

6.01. Help develop an organizational environment favorable to acting ethically.

6.02. Promote public knowledge of software engineering.

6.03. Extend software engineering knowledge by appropriate participation in professional organizations, meetings and publications.

6.04. Support, as members of a profession, other software engineers striving to follow this Code.

6.05. Not promote their own interest at the expense of the profession, client or employer.

6.06. Obey all laws governing their work, unless, in exceptional circumstances, such compliance is inconsistent with the public interest.

6.07. Be accurate in stating the characteristics of software on which they work, avoiding not only false claims but also claims that might reasonably be supposed to be speculative, vacuous, deceptive, misleading, or doubtful.

6.08. Take responsibility for detecting, correcting, and reporting errors in software and associated documents on which they work.

6.09. Ensure that clients, employers, and supervisors know of the software engineer's commitment to this Code of ethics, and the subsequent ramifications of such commitment.

6.10. Avoid associations with businesses and organizations which are in conflict with this code.

6.11. Recognize that violations of this Code are inconsistent with being a professional software engineer.

6.12. Express concerns to the people involved when significant violations of this Code are detected unless this is impossible, counter-productive, or dangerous.

6.13. Report significant violations of this Code to appropriate authorities when it is clear that consultation with people involved in these significant violations is impossible, counter-productive or dangerous.

Principle 7 Colleagues

Software engineers shall be fair to and supportive of their colleagues. In particular, software engineers shall, as appropriate:

7.01. Encourage colleagues to adhere to this Code.

7.02. Assist colleagues in professional development.

7.03. Credit fully the work of others and refrain from taking undue credit.

7.04. Review the work of others in an objective, candid, and properly-documented way.

7.05. Give a fair hearing to the opinions, concerns, or complaints of a colleague.

7.06. Assist colleagues in being fully aware of current standard work practices including policies and procedures for protecting passwords, files and other confidential information, and security measures in general.

7.07. Not unfairly intervene in the career of any colleague; however, concern for the employer, the client or public interest may compel software engineers, in good faith, to question the competence of a colleague.

7.08. In situations outside of their own areas of competence, call upon the opinions of other professionals who have competence in that area.

Principle 8 Self

Software engineers shall participate in lifelong learning regarding the practice of their profession and shall promote an ethical approach to the practice of the profession. In particular, software engineers shall continually endeavor to:

8.01. Further their knowledge of developments in the analysis, specification, design, development, maintenance and testing of software and related documents, together with the management of the development process.

8.02. Improve their ability to create safe, reliable, and useful quality software at reasonable cost and within a reasonable time.

8.03. Improve their ability to produce accurate, informative, and well-written documentation.

8.04. Improve their understanding of the software and related documents on which they work and of the environment in which they will be used.

8.05. Improve their knowledge of relevant standards and the law governing the software and related documents on which they work.

8.06. Improve their knowledge of this Code, its interpretation, and its application to their work.

8.07. Not give unfair treatment to anyone because of any irrelevant prejudices.

8.08. Not influence others to undertake any action that involves a breach of this Code.

8.09. Recognize that personal violations of this Code are inconsistent with being a professional software engineer.

This Code was developed by the IEEE-CS/ACM joint task force on Software Engineering Ethics and Professional Practices (SEEPP):

Executive Committee: Donald Gotterbarn (Chair), Keith Miller and Simon Rogerson;

Members: Steve Barber, Peter Barnes, Ilene Burnstein, Michael Davis, Amr El-Kadi, N. Ben Fairweather, Milton Fulghum, N. Jayaram, Tom Jewett, Mark Kanko, Ernie Kallman, Duncan Langford, Joyce Currie Little, Ed Mechler, Manuel J. Norman, Douglas Phillips, Peter Ron Prinzivalli, Patrick Sullivan, John Weckert, Vivian Weil, S. Weisband and Laurie Honour Werth.

APPENDIX **C**

ASSOCIATION OF INFORMATION TECHNOLOGY PROFESSIONALS (AITP) CODE OF ETHICS AND STANDARD OF CONDUCT

CODE OF ETHICS

I acknowledge:

That I have an obligation to management, therefore, I shall promote the understanding of information processing methods and procedures to management using every resource at my command.

That I have an obligation to my fellow members, therefore, I shall uphold the high ideals of AITP as outlined in the Association Bylaws. Further, I shall cooperate with my fellow members and shall treat them with honesty and respect at all times.

That I have an obligation to society and will participate to the best of my ability in the dissemination of knowledge pertaining to the general development and understanding of information processing. Further, I shall not use knowledge of a confidential nature to further my personal interest, nor shall I violate the privacy and confidentiality of information entrusted to me or to which I may gain access.

That I have an obligation to my College or University, therefore, I shall uphold its ethical and moral principles.

That I have an obligation to my employer whose trust I hold, therefore, I shall endeavor to discharge this obligation to the best of my ability, to guard my employer's interests, and to advise him or her wisely and honestly.

That I have an obligation to my country, therefore, in my personal, business, and social contacts, I shall uphold my nation and shall honor the chosen way of life of my fellow citizens.

I accept these obligations as a personal responsibility and as a member of this Association. I shall actively discharge these obligations and I dedicate myself to that end.

STANDARD OF CONDUCT

These standards expand on the Code of Ethics by providing specific statements of behavior in support of each element of the Code. They are not objectives to be strived for, they are rules that no true professional will violate. It is first of all expected that an information processing professional will abide by the appropriate laws of their country and community. The following standards address tenets that apply to the profession.

In recognition of my obligation to management I shall:

- Keep my personal knowledge up-to-date and insure that proper expertise is available when needed.
- Share my knowledge with others and present factual and objective information to management to the best of my ability.
- Accept full responsibility for work that I perform.
- Not misuse the authority entrusted to me.
- Not misrepresent or withhold information concerning the capabilities of equipment, software or systems.
- Not take advantage of the lack of knowledge or inexperience on the part of others.

In recognition of my obligation to my fellow members and the profession I shall:

- Be honest in all my professional relationships.
- Take appropriate action in regard to any illegal or unethical practices that come to my attention. However, I will bring charges against any person only when I have reasonable basis for believing in the truth of the allegations and without any regard to personal interest.
- Endeavor to share my special knowledge.
- Cooperate with others in achieving understanding and in identifying problems.
- Not use or take credit for the work of others without specific acknowledgement and authorization.
- Not take advantage of the lack of knowledge or inexperience on the part of others for personal gain.

In recognition of my obligation to society I shall:

- Protect the privacy and confidentiality of all information entrusted to me.
- Use my skill and knowledge to inform the public in all areas of my expertise.
- To the best of my ability, insure that the products of my work are used in a socially responsible way.
- Support, respect, and abide by the appropriate local, state, provincial, and federal laws.
- Never misrepresent or withhold information that is germane to a problem or situation of public concern nor will I allow any such known information to remain unchallenged.
- Not use knowledge of a confidential or personal nature in any unauthorized manner or to achieve personal gain.

In recognition of my obligation to my employer I shall:

- Make every effort to ensure that I have the most current knowledge and that the proper expertise is available when needed.
- Avoid conflict of interest and insure that my employer is aware of any potential conflicts.
- Present a fair, honest, and objective viewpoint.
- Protect the proper interests of my employer at all times.
- Protect the privacy and confidentiality of all information entrusted to me.
- Not misrepresent or withhold information that is germane to the situation.
- Not attempt to use the resources of my employer for personal gain or for any purpose without proper approval.
- Not exploit the weakness of a computer system for personal gain or personal satisfaction.

Permission information
Association of Information Technology Professionals
401 North Michigan Avenue, Suite 2400
Chicago, IL 60611-4267
Phone: 800.224.9371 or 312.245.1070
Fax: 312.673.6659
Email: aitp_hq@aitp.org

SYSADM, AUDIT, NETWORK, SECURITY (SANS) IT CODE OF ETHICS

I will strive to know myself and be honest about my capability.

- I will strive for technical excellence in the IT profession by maintaining and enhancing my own knowledge and skills. I acknowledge that there are many free resources available on the Internet and affordable books and that the lack of my employer's training budget is not an excuse nor limits my ability to stay current in IT.
- When possible I will demonstrate my performance capability with my skills via projects, leadership, and/or accredited educational programs and will encourage others to do so as well.
- I will not hesitate to seek assistance or guidance when faced with a task beyond my abilities or experience. I will embrace other professionals' advice and learn from their experiences and mistakes. I will treat this as an opportunity to learn new techniques and approaches. When the situation arises that my assistance is called upon, I will respond willingly to share my knowledge with others.
- I will strive to convey any knowledge (specialist or otherwise) that I have gained to others so everyone gains the benefit of each other's knowledge.
- I will teach the willing and empower others with Industry Best Practices (IBP). I will offer my knowledge to show others how to become security professionals in their own right. I will strive to be perceived as and be an honest and trustworthy employee.
- I will not advance private interests at the expense of end users, colleagues, or my employer.
- I will not abuse my power. I will use my technical knowledge, user rights, and permissions only to fulfill my responsibilities to my employer.
- I will avoid and be alert to any circumstances or actions that might lead to conflicts of interest or the perception of conflicts of interest. If such circumstance occurs, I will notify my employer or business partners.

- I will not steal property, time or resources.
- I will reject bribery or kickbacks and will report such illegal activity.
- I will report on the illegal activities of myself and others without respect to the punishments involved. I will not tolerate those who lie, steal, or cheat as a means of success in IT.

I will conduct my business in a manner that assures the IT profession is considered one of integrity and professionalism.

- I will not injure others, their property, reputation, or employment by false or malicious action.
- I will not use availability and access to information for personal gains through corporate espionage.
- I distinguish between advocacy and engineering. I will not present analysis and opinion as fact.
- I will adhere to Industry Best Practices (IBP) for system design, rollout, hardening and testing.
- I am obligated to report all system vulnerabilities that might result in significant damage.
- I respect intellectual property and will be careful to give credit for other's work. I will never steal or misuse copyrighted, patented material, trade secrets or any other intangible asset.
- I will accurately document my setup procedures and any modifications I have done to equipment. This will ensure that others will be informed of procedures and changes I've made.

I respect privacy and confidentiality.

- I respect the privacy of my co-workers' information. I will not peruse or examine their information including data, files, records, or network traffic except as defined by the appointed roles, the organization's acceptable use policy, as approved by Human Resources, and without the permission of the end user.
- I will obtain permission before probing systems on a network for vulnerabilities.
- I respect the right to confidentiality with my employers, clients, and users except as dictated by applicable law. I respect human dignity.
- I treasure and will defend equality, justice and respect for others.
- I will not participate in any form of discrimination, whether due to race, color, national origin, ancestry, sex, sexual orientation, gender/sexual identity or expression, marital status, creed, religion, age, disability, veteran's status, or political ideology.

Contact info for permission:
301-654-7267

APPENDIX **E**

ANSWERS TO SELF-ASSESSMENT QUESTIONS

Chapter 1 answers: 1. morality; 2. Ethics; 3. Vices; 4. Code of principles; 5. Morals; 6. code of ethics; 7. respondeat superior or "let the master answer"; 8. reputation; 9. vision and leadership; 10. board of directors; 11. Section 406 of the Sarbanes-Oxley Act; 12. renew investor's trust in the content and preparation of disclosure documents by public companies; 13. code of ethics; 14. social audit; 15. formal ethics training; 16. problem definition; 17. fairness approach; 18. brainstorming

Chapter 2 answers: 1. d.; 2. IT staff; 3. stop the unauthorized copying of software produced by its members; 4. True; 5. Fraud; 6. Compliance; 7. d.; 8. Internal audit; 9. b.; 10. True; 11. Negligence; 12. code of ethics

Chapter 3 answers: 1. a.; 2. United States; 3. exploit; 4. Cloud computing; 5.True.; 6. Zero-day attack; 7. CAPTCHA; 8. False; 9. Smishing; 10. True; 11. botnet; 12. trustworthy computing; 13. risk assessment; 14. security policy; 15. False; 16. intrusion prevention system

Chapter 4 answers: 1. Bill of Rights (particularly the Fourth Amendment); 2. discovery; 3. True; 4. b.; 5. opt-out; 6. d.; 7. American Recovery and Reinvestment Act of 2009; 8. Katz; 9. False; 10. a.; 11. USA PATRIOT Act; 12. National Security Letter; 13. True; 14. d.; 15. Personalization software; 16. True

Chapter 5 answers: 1. c.; 2. True; 3. Miller vs. California; 4. b.; 5. Children's Internet Protection Act; 6. True; 7. a.; 8. d.; 9. False; 10. John Doe lawsuit; 11. anonymous Internet speakers; 12. True; 13. b.; 14. CAN-SPAM

Chapter 6 answers: 1. d.; 2. False; 3. fair use; 4. False; 5. Digital Millennium Copyright Act; 6. trade mark; 7. patent infringements; 8. c.; 9. True; 10. Trade-Related Aspects of Intellectual Property Rights (TRIPS) Agreement; 11. False; 12. reverse engineering; 13. Prior art; 14. True; 15. True

Chapter 7 answers: 1. False; 2. Software quality; 3. d.; 4. process control system; 5. product liability; 6. software development methodology; 7. False; 8. software quality assurance; 9. d.; 10. c.; 11. True; 12. b.; 13. Failure Mode and Effects Analysis (FMEA); 14. ISO 9000 standards; 15. negligence

Chapter 8 answers: 1. b.; 2. productivity; 3. False; 4. c.; 5. Digital divide; 6. Telework; 7. d.; 8. True; 9. b.; 10. d.; 11. True; 12. 98,000; 13. electronic health record (EHR); 14. meaningful use.

Chapter 9 answers: 1. True.; 2. b.; 3. LinkedIn; 4. True; 5. b.; 6. True; 7. MySpace; 8. False; 9. False; 10. d.; 11. False; 12. data about their members' likes and dislikes.

Chapter 10 answers: 1. True; 2. b; 3. d.; 4. True; 5. H-1B; 6. B-1 7. True; 8. d.; 9. True; 10. a.; 11. d.; 12. True; 13. False; 14. gold

GLOSSARY

affiliated Web sites A group of Web sites served by a single advertising network.

Agreement on Trade-Related Aspects of Intellectual Property Rights (TRIPS Agreement) An agreement of the World Trade Organization that requires member governments to ensure that intellectual property rights can be enforced under their laws and that penalties for infringement are tough enough to deter further violations.

American Recovery and Reinvestment Act of 2009 A wide-ranging act that authorized $787 billion in spending and tax cuts over a ten-year period and included strong privacy provisions for electronic health records, such as banning the sale of health information, promoting the use of audit trails and encryption, and providing rights of access for patients.

anonymous expression The expression of opinions by people who do not reveal their identity.

anonymous remailer A company that provides a service in which an originating IP number (computer address) is stripped from an email message before the message is sent on to its destination.

antivirus software Software that regularly scans a computer's memory and disk drives for viruses.

audit committee A subgroup of the board of directors that provides assistance to the board in fulfilling its responsibilities with respect to the oversight of the quality and integrity of the organization's accounting and reporting practices and controls including: financial statements and reports; the organization's compliance with legal and regulatory requirements; the qualifications, independence, and performance of the company's independent auditor; and the performance of the Company's internal audit function.

avatar A virtual world visitor's representation of him- or herself—usually in the form of a human but sometimes in some other form, such as an animal or mythical creature.

beacon A small piece of software that runs on a Web page and is able to track what a viewer is doing on the page, such as what is being typed or where the mouse is moving.

black-box testing A form of dynamic testing that involves viewing the software unit as a device that has expected input and output behaviors but whose internal workings are unknown (a black box). If the unit demonstrates the expected behaviors for all the input data in the test suite, it passes the test.

body of knowledge An agreed-upon set of skills and abilities that all licensed professionals in a particular type of profession must possess.

botnet A large group of computers controlled from one or more remote locations by hackers, without the knowledge or consent of their owners.

breach of contract The failure of one party to meet the terms of a contract.

breach of the duty of care The failure to act as a reasonable person would act.

breach of warranty The failure of a product to meet the terms of its warranty.

bribery The act of providing money, property, or favors to someone in business or government to obtain a business advantage.

business information system A set of interrelated components—including hardware, software, databases, networks, people, and procedures—that collects data, processes it, and disseminates the output.

Business Software Alliance (BSA) Trade group that represents the world's largest software and hardware manufacturers; its mission is to stop the unauthorized copying of software produced by its members.

Capability Maturity Model Integration (CMMI) A process improvement approach developed by the Software Engineering Institute at Carnegie Mellon that defines the essential elements of effective processes.

CAPTCHA (Completely Automated Public Turing Test to Tell Computers and Humans Apart) Software that generates and grades tests that humans can pass but all but the most sophisticated computer programs cannot.

certification A recognition that a professional possesses a particular set of skills, knowledge, or abilities—in the opinion of the certifying organization.

chief privacy officer (CPO) A senior manager within an organization whose role is to both ensure that the organization does not violate government regulations and reassure customers that their privacy will be protected.

Child Online Protection Act (COPA) A law that states "whoever knowingly and with knowledge of the character of the material, in interstate or foreign commerce by means of the World Wide Web, makes any communication for commercial purposes that is available to any minor and that includes any material that is harmful to minors shall be fined not more than $50,000, imprisoned not more than 6 months, or both." This law was eventually found to be unconstitutional.

Children's Internet Protection Act (CIPA) An act that requires federally financed schools and libraries to use some form of technological protection (such as an Internet filter) to block computer access to obscene material, pornography, and anything else considered harmful to minors.

Children's Online Privacy Protection Act (COPPA) A 1998 law that requires Web sites that cater to children to offer comprehensive privacy policies, notify parents or guardians about their data-collection practices, and receive parental consent before collecting any personal information from children under 13 years of age.

click-stream data Information gathered by monitoring a consumer's online activity through the use of electronic cookies.

cloud computing An environment in which software and data storage are services provided via the Internet (the cloud); the services are run on another organization's computer hardware and are accessed by a Web browser.

CMMI *See* Capability Maturity Model Integration (CMMI).

CMMI-Development (CMMI-DEV) An application of CMMI that is frequently used to assess and improve software development practices.

code of ethics A statement that highlights an organization's key ethical issues and identifies the overarching values and principles that are important to the organization and its decision making.

coemployment relationship An employment situation in which two employers have actual or potential legal rights and duties with respect to the same employee or group of employees.

collusion Cooperation between two or more people, often an employee and a company outsider, to commit fraud.

commoditization The transformation of goods or services into commodities that offer nothing to differentiate themselves from those offered by competitors. Commoditized goods and services are sold strictly on the basis of price.

common good approach An approach to ethical decision making based on a vision of society as a community whose members work together to achieve a common set of values and goals.

Communications Act of 1934 The law that established the Federal Communications Commission and gave it responsibility for regulating all non-federal-government use of radio and television broadcasting and all interstate telecommunications—including wire, satellite, and cable—as well as all international communications that originate or terminate in the United States.

Communications Assistance for Law Enforcement Act (CALEA) A 1994 law that amended both the Wiretap Act and ECPA; it requires the telecommunications industry to build tools into its products that federal investigators could use—after obtaining a court order—to eavesdrop on conversations and intercept electronic communications.

Communications Decency Act (CDA) A part of the 1996 Telecommunications Act directed at protecting children from online pornography; it was eventually ruled unconstitutional.

competitive intelligence Legally obtained information gathered using sources available to the public; used to help a company gain an advantage over its rivals.

compliance To be in accordance with established policies, guidelines, specifications, or legislation.

computer forensics A discipline that combines elements of law and computer science to identify, collect, examine, and preserve data from computer systems, networks, and storage devices in a manner that preserves the integrity of the data gathered so it is admissible as evidence in a court of law.

conflict of interest A conflict between a person's (or firm's) self-interest and the interests of a client.

contingent work A job situation in which an individual does not have an explicit or implicit contract for long-term employment.

contributory negligence A defense in a negligence case in which the defendant argues that the plaintiffs' own actions contributed to their injuries.

Controlling the Assault of Non-Solicited Pornography and Marketing (CAN-SPAM) Act A 2004 law that specifies requirements that commercial emailers must follow when sending out messages that advertise or promote a commercial product or service.

cookie An electronic text file that a Web site downloads to visitors' hard drives so it can identify them on subsequent visits.

copyright The exclusive right to distribute, display, perform, or reproduce an original work in copies or to prepare derivative works based on the work; granted to creators of original works of authorship.

copyright infringement A violation of the rights secured by the owner of a copyright; occurs when someone copies a substantial and material part of another's copyrighted work without permission.

corporate compliance officer *See* corporate ethics officer.

corporate ethics officer A senior-level manager who provides an organization with vision and leadership in the area of business conduct.

cracker Someone who breaks into other people's networks and systems to cause harm.

cyberbullying The harassment, torment, humiliation, or threatening of one minor by another minor or group of minors via the Internet or cell phone.

cybercriminal An individual, motivated by the potential for monetary gain, who hacks into computers to steal, often by transferring money from one account to another to another.

cybersquatter A person or company that registers domain names for famous trademarks or company names to which they have no connection, with the hope that the trademark's owner will buy the domain name for a large sum of money.

cyberstalking Threatening behavior or unwanted advances directed at an adult using the Internet or other forms of online and electronic communications; the adult version of cyberbullying.

cyberterrorist An individual who launches computer-based attacks against other computers or networks in an attempt to intimidate or coerce a government in order to advance certain political or social objectives.

data breach The unintended release of sensitive data or the access of sensitive data by unauthorized individuals

decision support system (DSS) A type of business information system used to improve decision making in a variety of industries.

defamation Making either an oral or a written statement of alleged fact that is false and harms another person.

deliverables The products of a software development process, such as statements of requirements, flowcharts, and user documentation.

digital divide The gulf between those who do and those who do not have access to modern information and communications technology such as cell phones, personal computers, and the Internet.

Digital Millennium Copyright Act (DMCA) An act that implements two WIPO treaties in the United States. It also makes it illegal to circumvent a technical protection or develop and provide tools that allow others to access a technologically protected work. It also limits the liability of online service providers for copyright infringement by their subscribers or customers.

distributed denial-of-service attack (DDoS) An attack in which a malicious hacker takes over computers via the Internet and causes them to flood a target site with demands for data and other small tasks.

duty of care The obligation to protect people against any unreasonable harm or risk.

dynamic testing An approach to software QA testing in which the code for a completed unit of software is tested by entering test data and comparing the actual results to the expected results.

Education Rate (E-Rate) program A program created through the Telecommunications Act of 1996; its primary goal is to help schools and libraries obtain access to state-of-the-art services and technologies at discounted rates.

Electronic Communications Privacy Act of 1986 (ECPA) A law focusing on three main issues: (1) the protection of communications while in transfer from sender to receiver; (2) the protection of communications held in electronic storage; and (3) the prohibition of devices to record dialing, routing, addressing, and signaling information without a search warrant.

electronic discovery (e-discovery) The collection, preparation, review, and production of electronically stored information for use in criminal and civil legal actions and proceedings.

electronic health record (EHR) A computer readable record of health-related information on an individual; can include patient demographics, medical history, family history, immunization records, laboratory data, health problems, progress notes, medications, vital signs, and radiology reports. Data can be added to an EHR based on each patient encounter in any healthcare delivery setting.

Electronic Industry Citizenship Coalition (EICC) An industry organization established to promote a common code of conduct for the electronics and information and communications technology (ICT) industry.

Electronic Product Environmental Assessment Tool (EPEAT) A system that enables purchasers to evaluate, compare, and select electronic products based on a total of 51 environmental criteria.

electronically stored information (ESI) Any form of digital information including emails, drawings, graphs, Web pages, photographs, word-processing files, sound recordings, and databases stored on any form of magnetic storage device including hard drives, CDs, and flash drives.

email spam The abuse of email systems to send unsolicited email to large numbers of people.

employee leasing A business arrangement in which an organization (called the subscribing firm) transfers all or part of its workforce to another firm (called the leasing firm), which handles all human resource-related activities and costs, such as payroll, training, and the administration of employee benefits. The subscribing firm leases these workers to an organization, but they remain employees of the leasing firm.

ethics A set of beliefs about right and wrong behavior within a society.

European Union Data Protection Directive A directive passed by the European Union in 1998 that requires any company doing business within the borders of 15 western European nations to implement a set of privacy directives on the fair and appropriate use of information; it also bars the export of data to countries that do not have comparable data privacy protection standards.

exploit An attack on an information system that takes advantage of a particular system vulnerability.

failure mode and effects analysis (FMEA) A technique used to develop ISO 9000-compliant quality systems by both evaluating reliability and determining the effects of system and equipment failures.

Fair and Accurate Credit Transactions Act An amendment to the Fair Credit Reporting Act that allows consumers to request and obtain a free credit report once each year from each of the three primary consumer credit reporting companies (Equifax, Experian, and TransUnion).

Fair Credit Reporting Act A law passed in 1970 that regulates the operations of credit-reporting bureaus, including how they collect, store, and use credit information.

Fair Information Practices A set of eight principles created by the Organisation for Economic Co-operation and Development that provides guidelines for the ethical treatment of consumer data.

fair use doctrine A legal doctrine that allows portions of copyrighted materials to be used without permission under certain circumstances. Title 17, section 107, of the U.S. Code established the following four factors that courts should consider when deciding whether a particular use of copyrighted property is fair and can be allowed without penalty: (1) the purpose and character of the use (such as commercial use or nonprofit, educational purposes); (2) the nature of the copyrighted work; (3) the portion of the copyrighted work used in relation to the work as a whole; and (4) the effect of the use on the value of the copyrighted work.

fairness approach An approach to ethical decision making that focuses on how fairly actions and policies distribute benefits and burdens among people affected by the decision.

Family Educational Rights and Privacy Act (FERPA) A federal law that assigns certain rights to parents regarding their children's educational records. These rights transfer to the student once the student attains the age of 18 or attends a school beyond the high school level.

False Claims Act A law enacted during the U.S. Civil War to combat fraud by companies that sold supplies to the Union Army; also known as the Lincoln Law. *See also* qui tam.

firewall A hardware or software device that serves as a barrier between an organization's network and the Internet; a firewall also limits access to the company's network based on the organization's Internet usage policy.

Foreign Corrupt Practices Act (FCPA) A federal law that makes it a crime to bribe a foreign official, a foreign political party official, or a candidate for foreign political office.

foreign intelligence Information relating to the capabilities, intentions, or activities of foreign governments, agents of foreign governments, or foreign organizations.

Foreign Intelligence Surveillance Act (FISA) An act passed in 1978 that describes procedures for the electronic surveillance and collection of foreign intelligence information in communications between foreign powers and agents of foreign powers.

fraud The crime of obtaining goods, services, or property through deception or trickery.

Freedom of Information Act (FOIA) A law passed in 1966 and amended in 1974 that grants citizens the right to access certain information and records of the federal government upon request.

globalization The process of interaction and integration among the people, companies, and governments of different nations.

government license A government-issued permission to engage in an activity or to operate a business; it is generally administered at the state level and often requires that the recipient pass a test of some kind.

Gramm-Leach-Bliley Act (GLBA) A 1999 bank deregulation law, also known as the Financial Services Modernization Act, which granted banks the right to offer investment, commercial banking, and insurance services through a single entity.

green computing Efforts directed toward the efficient design, manufacture, operation, and disposal of IT-related products, including personal computers, laptops, servers, printers, and printer supplies.

H-1B visa A temporary work visa granted by the U.S. Citizenship and Immigration Services (USCIS) for people who work in specialty occupations—jobs that require a four-year bachelor's degree in a specific field, or equivalent experience.

hacker Someone who tests the limitations of information systems out of intellectual curiosity—to see if he or she can gain access.

hacktivism Hacking to achieve a political or social goal.

Health Information Technology for Economic and Clinical Health Act (HITECH Act) Part of the $787 billion 2009 American Recovery and Reinvestment Act economic stimulus plan. HITECH is intended to increase the use of health information technology by: (1) requiring the government to develop standards for the nationwide electronic exchange and use of health information; (2) providing $20 billion in incentives to encourage doctors and hospitals to use EHRs to electronically exchange patient healthcare data; (3) saving the government $10 billion through improvements in the quality of care and care coordination and through reductions in medical errors and duplicate care; and (4) strengthening the protection of identifiable health information.

Health Insurance Portability and Accountability Act of 1996 (HIPAA) A law designed to improve the portability and continuity of health insurance coverage; to reduce fraud, waste, and abuse in health insurance and healthcare delivery; and to simplify the administration of health insurance.

identity theft The act of stealing key pieces of personal information to impersonate a person.

industrial espionage The use of illegal means to obtain business information not available to the general public.

industrial spy Someone who uses illegal means to obtain trade secrets from competitors.

information privacy The combination of communications privacy (the ability to communicate with others without those communications being monitored by other persons or organizations) and data privacy (the ability to limit access to one's personal data by other individuals and organizations in order to exercise a substantial degree of control over that data and its use).

integration testing A form of software testing in which individual software units are combined into an integrated subsystem that undergoes rigorous testing to ensure that the linkages among the various subsystems work successfully.

integrity Adherence to a personal code of principles.

intellectual property Works of the mind—such as art, books, films, formulas, inventions, music, and processes—that are distinct, and owned or created by a single person or group. Intellectual property is protected through copyright, patent, trade secret, and trademark laws.

intentional misrepresentation Fraud that occurs when a seller or lessor either misrepresents the quality of a product or conceals a defect in it.

Internet censorship The control or suppression of the publishing or accessing of information on the Internet.

Internet filter Software that can be used to block access to certain Web sites that contain material deemed inappropriate or offensive.

intrusion detection system (IDS) Software and/or hardware that monitors system and network resources and activities, and notifies network security personnel when it identifies possible intrusions from outside the organization or misuse from within the organization.

intrusion prevention system (IPS) A network security device that prevents an attack by blocking viruses, malformed packets, and other threats from getting into the protected network.

ISO 9001 family of standards A set of standards that serves as a guide to quality products, services, and management.

IT user A person for whom a hardware or software product is designed.

John Doe lawsuit A lawsuit in which the identity of the defendant is temporarily unknown, typically because the defendant is communicating anonymously or using a pseudonym.

lamer A technically inept hacker. *See also* script kiddie.

law A system of rules that tells us what we can and cannot do. Laws are enforced by a set of institutions.

libel A written defamatory statement.

live telemedicine A form of telemedicine in which patients and healthcare providers are present at different sites at the same time; often involves a videoconference link between the two sites.

logic bomb A type of Trojan horse that executes when it is triggered by a specific event.

massively multiplayer online game (MMOG) A multiplayer video game capable of supporting hundreds and even thousands of concurrent players.

massive multiplayer online role playing games (MMORPG) A multiplayer online game that provides a huge online world in which players take on the role of a character and control that character's action; players can interact with one another to compete in online games and challenges that unfold according to the online world's rules and storyline.

material breach of contract The failure of one party to perform certain express or implied obligations, which impairs or destroys the essence of the contract.

Miller v. California The 1973 Supreme Court case that established a test to determine if material is obscene and therefore not protected by the First Amendment.

misrepresentation The misstatement or incomplete statement of a material fact.

modularization The act of breaking down a production or business process into smaller components.

moral code A set of rules that establishes the boundaries of generally accepted behavior within a society.

morality Social conventions about right and wrong that are widely shared throughout a society.

morals One's personal beliefs about right and wrong.

negligence The failure to do what a reasonable person would do, or doing something that a reasonable person would not do.

negligent insider A poorly trained and inadequately managed employee who means well but who has the potential to cause much damage.

noncompete agreement Terms of an employment contract that prohibit an employee from working for any competitors for a specified period of time, often one to two years.

nondisclosure clause Terms of an employment contract that prohibit an employee from revealing secrets.

N-version programming A form of redundancy in which two computer systems execute a series of program instructions simultaneously.

offshore outsourcing A form of outsourcing in which the services are provided by an organization whose employees are in a foreign country. *See also* outsourcing.

One Laptop per Child (OLPC) A nonprofit organization whose goal is to provide children around the world with low-cost laptop computers to aid in their education.

online virtual world A shared multi-media computer-generated environment in which users, represented by avatars, can act, communicate, create, retain ownership of what they create, and exchange assets, including currency, with each other.

open source code Any program whose source code is made available for use or modification, as users or other developers see fit.

opt in To agree (either implicitly or by default) to allow an organization to collect and share one's personal data with other institutions.

opt out To refuse to give an organization the right to collect and share one's personal data with unaffiliated parties.

outsourcing A long-term business arrangement in which a company contracts for services with an outside organization that has expertise in providing a specific function.

patent A grant of a property right issued by the U.S. Patent and Trademark Office to an inventor; permits its owner to exclude the public from making, using, or selling a protected invention, and allows for legal action against violators.

patent farming An unethical strategy of influencing a standards organization to make use of a patented item without revealing the existence of a patent; later, the patent holder might demand royalties from all implementers of the standard.

patent infringement A violation of the rights secured by the owner of a patent; occurs when someone makes unauthorized use of another's patent.

patent troll A firm that acquires patents for the purpose of licensing the patents to others rather than manufacturing anything itself.

pen register A device that records electronic impulses to identify the numbers dialed for outgoing calls.

personalization software Software used by online marketers to optimize the number, frequency, and mixture of their ad placements as well as to evaluate how visitors react to new ads.

phishing The act of fraudulently using email to try to get the recipient to reveal personal data.

plagiarism The act of stealing someone's ideas or words and passing them off as one's own.

prior art The existing body of knowledge that is available to a person of ordinary skill in the art.

Privacy Act of 1974 A law decreeing that no agency of the U.S. government can conceal the existence of any personal data record-keeping system; under this law, any agency that maintains such a system must publicly describe both the kinds of information in it and the manner in which the information will be used.

problem statement A clear, concise description of the issue that needs to be addressed in a decision-making process.

product liability The liability of manufacturers, sellers, lessors, and others for injuries caused by defective products.

productivity The amount of output produced per unit of input.

profession A calling that requires specialized knowledge and often long and intensive academic preparation.

professional code of ethics A statement of the principles and core values that are essential to the work of a particular occupational group.

professional malpractice Breach of the duty of care by a professional.

project safety engineer An individual on a safety-critical system project who has explicit responsibility for the system's safety.

quality assurance (QA) Methods within the software development cycle designed to guarantee reliable operation of a product.

quality management Business practices that focus on defining, measuring, and refining the quality of the development process and the products developed during its various stages.

qui tam A provision of the False Claims Act that allows a private citizen to file a suit in the name of the U.S. government, charging fraud by government contractors and other entities who receive or use government funds. *See also* False Claim Act.

reasonable assurance A concept in computer security that recognizes that managers must use their judgment to ensure that the cost of control does not exceed the system's benefits or the risks involved.

reasonable person standard A legal standard that defines how an objective, careful, and conscientious person would have acted in the same circumstances.

reasonable professional standard A legal standard that defendants who have particular expertise or competence are measured against.

redundancy The use of multiple interchangeable components designed to perform a single function—in order to cope with failures and errors.

reliability The probability of a component or system performing without failure over its product life.

résumé inflation Falsely claiming competence in a skill, usually because that skill is in high demand.

reverse engineering The process of taking something apart in order to understand it, build a copy of it, or improve it.

Right to Financial Privacy Act of 1978 An act that protects the financial records of financial institution customers from unauthorized scrutiny by the federal government.

risk The probability of an undesirable event occurring times the magnitude of the event's consequences if it does happen.

risk assessment The process of assessing security-related risks from both internal and external threats to an organization's computers and networks.

rootkit A set of programs that enables its user to gain administrator level access to a computer without the end user's consent or knowledge.

safety-critical system A system whose failure may cause injury or death.

Sarbanes-Oxley Act A bill whose goal was to renew investors' trust in corporate executives and their firms' financial reports; the act led to significant reforms in the content and preparation of disclosure documents by public companies.

script kiddie A technically inept hacker. *See also* lamer.

security audit A process that evaluates whether an organization has a well-considered security policy in place and if it is being followed.

security policy A written statement that defines an organization's security requirements, as well as the controls and sanctions needed to meet those requirements.

sexting Sending sexual messages, nude or seminude photos, or sexually explicit videos over a cell phone.

slander An oral defamatory statement.

smart card A form of debit or credit card that contains a memory chip that is updated with encrypted data every time the card is used.

smishing A variation of phishing in which victims receive a legitimate-looking SMS text message on their phone telling them to call a specific phone number or to log on to a Web site.

social audit A process whereby an organization reviews how well it is meeting its ethical and social responsibility goals, and communicates its new goals for the upcoming year.

social network advertising Advertising using social networks to communicate and promote the benefits of products and services.

social networking Web site A Web site that creates an online community of Internet users that enables members to break down time, distance, and cultural barriers and interact with others by sharing opinions, insights, information, interests, and experiences.

social shopping Web site A Web site that brings shoppers and sellers together in a social networking environment in which members can share information and make recommendations while shopping online.

software defect Any error that, if not removed, could cause a software system to fail to meet its users' needs.

software development methodology A standard, proven work process that enables systems analysts, programmers, project managers, and others to make controlled and orderly progress in developing high-quality software.

software piracy The act of illegally making copies of software or enabling others to access software to which they are not entitled.

software quality The degree to which a software product meets the needs of its users.

spear-phishing A variation of phishing in which the phisher sends fraudulent emails to a certain organization's employees. The phony emails are designed to look like they came from high-level executives within the organization.

spyware Keystroke-logging software downloaded to users' computers without their knowledge or consent.

stakeholder Someone who stands to gain or lose depending on how a situation is resolved.

standard A definition that has been approved by a recognized standards organization or accepted as a de facto standard by a particular industry.

static testing The use of special software programs called static analyzers to look for suspicious patterns in programs that might indicate a defect.

store-and-forward telemedicine A form of telemedicine in which data, sound, images, and video are acquired from a patient and then transmitted to a medical specialist for later evaluation.

strict liability A type of product liability in which a defendant is held responsible for injuring another person, regardless of negligence or intent.

submarine patent A patented process or invention that is hidden within a standard and which is not made public until after the standard is broadly adopted.

sunset provision A provision that terminates or repeals a law or portions of it after a specific date, unless further legislative action is taken to extend the law.

system testing A form of software testing in which various subsystems are combined to test the entire system as a complete entity.

telemedicine The use of modern telecommunications and information technologies to provide medical care to people who live far away from healthcare providers.

telework A work arrangement in which an employee works away from the office—at home, at a client's office, in a hotel—literally anywhere; also known as telecommuting.

Title III of the Omnibus Crime Control and Safe Streets Act A component of a 1968 law (amended in 1986) that regulates the interception of wire and oral communications; also known as the Wiretap Act.

trade secret Information, generally unknown to the public, that a company has taken strong measures to keep confidential. It represents something of economic value that has required effort or cost to develop and that has some degree of uniqueness or novelty.

trademark A logo, package design, phrase, sound, or word that enables a consumer to differentiate one company's products from another's.

transparency Any attempt to reveal and clarify any information or processes that were previously hidden or unclear.

trap and trace A device that records electronic impulses to identify the originating number for incoming calls.

Trojan horse A program in which malicious code is hidden inside a seemingly harmless program.

trustworthy computing A method of computing that delivers secure, private, and reliable computing experiences based on sound business practices.

USA PATRIOT Act A law passed in 2001 that gave sweeping new powers to domestic law enforcement and to intelligence agencies, including increasing the ability of law enforcement agencies to search telephone, email, medical, financial, and other records, and easing restrictions on foreign intelligence gathering in the United States.

user acceptance testing Independent testing performed by trained end users to ensure that a system operates as expected.

utilitarian approach An approach to ethical decision making that states that you should choose the action or policy that has the best overall consequences for all people who are directly or indirectly affected.

vice A moral habit that inclines people to do what is generally unacceptable to society.

viral marketing An approach to advertising that encourages individuals to pass along a marketing message to others, thus creating the potential for exponential growth in the message's exposure and influence.

virtualization software A software program that emulates computer hardware by enabling multiple operating systems to run on one computer host.

virtual private network (VPN) A technology that uses the Internet to relay communications, maintaining privacy through security procedures and tunneling protocols, which encrypt data at the sending end and decrypt it at the receiving end.

virtue A moral habit that inclines people to do what is generally acceptable to society.

virtue ethics approach An approach to ethical decision making that focuses on how you should behave and think about relationships if you are concerned with your daily life in a community.

virus A piece of programming code, usually disguised as something else, that causes a computer to behave in an unexpected and usually undesirable manner.

virus signature A specific sequence of bytes that indicates to antivirus software that a specific virus is present.

vishing A variation of phishing in which victims receive a voicemail telling them to call a specific phone number or log on to access a specific Web site.

warranty An assurance to buyers or lessees that a product meets certain standards of quality.

whistle-blowing An effort to attract public attention to a negligent, illegal, unethical, abusive, or dangerous act by a company or some other organization.

white-box testing A form of dynamic testing that treats the software unit as a device that has expected input and output behaviors and whose internal workings are known. White-box testing involves testing all possible logic paths through the software unit, with thorough knowledge of its logic.

worm A harmful program that resides in the active memory of a computer and duplicates itself.

zero-day attack An attack that takes place before the security community or software developer knows about the vulnerability or has been able to repair it.

zombie A computer that is part of a botnet and that is controlled by a hacker without the knowledge or consent of its owner.

INDEX

A

B

C

D

E

F

G

H

I

J

K

L

M

N

O

P

Q

R

S

T

U

V

W

X

Y

Z